Fourth International Symposium on Pre-Harvest Sprouting in Cereals

International Organizing Committee

President

Dr. J. E. Kruger
Grain Research Laboratory
Canadian Grain Commission
1404-303 Main Street
Winnipeg, Manitoba
R3C 3G8 Canada

Secretary

Dr. D. J. Mares
The University of Sydney
Plant Breeding Institute
PO Box 219
Narrabri, NSW, 2390
Australia

Members

Mr. N. F. Derera
7 Lister Street
Winston Hills
NSW. 2153
Australia

Dr. M. D. Gale
Plant Breeding Institute
Maris Lane, Trumpington
Cambridge CB2 2LQ
England

Dr. K. Ringlund
Agricultural University
of Norway
Boks 41, 1432 AAS-NLH
Norway

Dr. F. Weilenmann
Swiss Federal Research
Station for Agronomy
Zurich - Reckenholz
Switzerland

Local Organization

Dr. D. J. Mares
The University of Sydney
Plant Breeding Institute
PO Box 219
Narrabri, NSW, 2390
Australia

About the Book and Editor

A major constraint on the production of high-quality grain around the world, pre-harvest sprouting in cereal crops causes substantial economic losses to producers and disrupts the processing, distribution, marketing, and storage of grain products. The solution to this problem must include a better understanding of the phenomenon, the development of varieties that can tolerate adverse weather conditions at maturity, and the adaptation of processing technology to permit the use of moderately weather-damaged grain for human food.

Since the early 1970s, progress toward these objectives has been fostered by the activities of an International Organizing Committee, including a series of international symposia. This volume, the result of the most recent symposium in the series, contains review articles and reports of current research conducted since 1982 on all aspects of pre-harvest sprouting in cereals. Bringing together physiologists, biochemists, molecular biologists, breeders, and agricultural technologists, the collection provides a comprehensive summary of current perspectives and highlights prescriptions for future research.

Daryl J. Mares is senior cereal biochemist at the I. A. Watson Wheat Research Centre at the University of Sydney, Narrabri, Australia.

Fourth International Symposium on Pre-Harvest Sprouting in Cereals

edited by Daryl J. Mares

CRC Press
Taylor & Francis Group
Boca Raton London New York

CRC Press is an imprint of the
Taylor & Francis Group, an **informa** business

First published 1987 by Westview Press, Inc.

Published 2018 by CRC Press
Taylor & Francis Group
6000 Broken Sound Parkway NW, Suite 300
Boca Raton, FL 33487-2742

CRC Press is an imprint of the Taylor & Francis Group, an informa business

Copyright © 1987 Taylor & Francis Group LLC

No claim to original U.S. Government works

Visit the Taylor & Francis Web site at
http://www.taylorandfrancis.com

and the CRC Press Web site at
http://www.crcpress.com

Library of Congress Cataloging-in-Publication Data
International Symposium on Pre-Harvest Sprouting in
    Cereals (4th : 1986 : Port Macquarie, N.S.W.)
    Fourth International Symposium on Pre-Harvest
Sprouting in Cereals.
    1. Grain--Preharvest sprouting--Congresses.
2. Grain--Development--Congresses.   3. Grain--
Breeding--Congresses.   I. Mares, Daryl.   II. Title.
SB188.2.I584   1987   633.1'0431   86-13157

ISBN 13: 978-0-367-00861-1 (hbk)
ISBN 13: 978-0-367-15848-4 (pbk)

Contents

SECTION III.   COMMERCIAL UTILIZATION OF SPROUTED
                    GRAIN

SECTION IV.   GRAIN DEVELOPMENT

SECTION VII. ENZYMES IN GERMINATING GRAINS

SECTION VIII. ASSAY METHODS AND OBJECTIVE TESTING

Preface

This volume represents the proceedings of the Fourth International Symposium on Pre-Harvest Sprouting in Cereals, held at Port Macquarie, New South Wales, Australia, in February 1986. The purpose of this meeting was to review the current status of all aspects of pre-harvest sprouting in cereals and to discuss new research and ideas related to a better understanding of and a solution to this problem. Each year pre-harvest sprouting causes substantial economic losses in the wheat, barley and rye industries and also affects production of other cereals such as triticale, oats, rice and maize. The need for continued research and international co-operation has been stressed by all sections of the cereal industry and highlighted by multi-million dollar losses in Australia in 1983/84 and 1985/86 and in both Canada and the United Kingdom in 1985.

This symposium carried on the tradition established in previous meetings of providing an informal atmosphere conducive to the maximum exchange of information and bringing together scientists and technologists from different fields. The success and growth of this fraternity is evident from the increased number of participants and participating nations. The published proceedings resulting from these meetings have been widely acknowledged as important sources of information on pre-harvest sprouting. Proceedings of previous symposia were published as follows:

I   Cereal Research Communications, Vol. 4 (2), 1976.
II Cereal Research Communications, Vol. 8 (1), 1980.
III Third International Symposium on Pre-Harvest Sprouting in Cereals. Kruger, J. E. and LaBerge, D. E. (eds.). Westview Press, Boulder, Co., USA, 1983.

For this meeting pre-harvest sprouting was divided into
eight sections:

(i)   Overviews of sprouting in cereals,
(ii)   Breeding and selection for sprouting tolerance,
(iii)   Commercial utilization of sprouted grain,
(iv)   Grain development,
(v)   Physiology, biochemistry and molecular biology of
        germination,
(vi)   Grain dormancy,
(vii)   Enzymes in germinating grains, and
(viii) Assay methods and objective testing.

A concerted effort was made to develop a well-balanced program including reviews and research contributions from
acknowledged experts in each section together with submitted
research papers. By and large this attempt has been
extremely successful, and I am indebted to the many international scientists who responded to the invitation to come
to Australia.

At the closing banquet, recognition in the form of a
plaque was given to Mr. W. George Freeman, wheat farmer and
industry administrator, for his efforts over many years to
obtain funding and support for pre-harvest sprouting
research.

I am grateful to many people for making the symposium a
success. Local volunteers led by N. Derera and R. Henry
organized the chairperson and projectionist rosters whilst
M. Gale, R. King and N. Derera made some useful comments
during the formulation of the program. Mr. George Freeman
and Professor B. D. H. Latter played an important part in
the highly successful fund-raising effort which enabled many
of the overseas visitors to attend the symposium. Mrs. E.
Chalmers assisted in the preparation of abstracts, distribution of conference information and the daily running of
the meeting. All participants are to be congratulated for
their whole-hearted involvement in the symposium both during
and after the scheduled program sessions.

Production of the proceedings would not have been possible without the dedicated and highly professional involvement of my wife Cheryl. Retyping of abstracts and contributed
papers, preparation of artwork and production of the final
camera-ready manuscript occupied a large portion of her
year.

The symposium was sponsored by a number of Australian
cereal industry organizations. On behalf of the international committee and the participants, I gratefully acknowledge the generous support of the following:

Wheat Research Committee of NSW,
Australian Wheat Board,
Prime Wheat Association Ltd.,
Barley Industry Research Council,
Barley Research Committee of NSW,
Wheat and Barley Research Committees of Queensland,
State Wheat Board, Queensland,
Grain Handling Authority of NSW, and
Perten Instruments (Australia) Pty, Ltd.

Daryl J. Mares

Section I

Overviews of Sprouting in Cereals

Pre-Harvest Sprouting in Wheat — The Australian Experience

G. J. McMaster, Australian Wheat Board,
GPO, Box 4562, Melbourne, 3001,
Australia.

INTRODUCTION

Perhaps it is significant that the Fourth International Symposium on Pre-Harvest Sprouting in Cereals is being held at a time when major wheat exporting countries have experienced adverse weather conditions at harvest. Such was the case with crops in Australia, Canada and Britain last season. Because the cereal industries of the world are such major food commodities, any factor affecting the value of these vital resources of the international market should be a major topic of research.

THE EXTENT AND NATURE OF PRE-HARVEST SPROUTING IN AUSTRALIA

In Australia white-grained wheats of spring habit are grown over the mild winter months and harvested during the summer months of November/December. Some regions of Australia's wheatbelt, namely northern New South Wales and Queensland, have a predominantly summer rainfall and thus suffer frequently from weather damage. Significant levels of weather damage may also occur in other states with unseasonal rainfall at harvest. The long term averages of weather damage in wheat over an eleven year period (1972/73 to 1983/84) are shown in TABLE 1.

If general rain at harvest is followed by cool, humid conditions then weather damage can be quite widespread. Such was the case during the 1983/84 season when 4.3 million tonnes of wheat out of a record production of 21 million tonnes was received as weather damaged. A breakdown of receivals of weather damage wheat in the 1983/84 season is given in TABLE 2.

3

TABLE 1. Receivals of weather damaged wheat in Australia

State	Tonnes per annum	Percent of total receivals
Queensland	159,690	16
New South Wales	470,658	10
Victoria	159,978	7
South Australia	28,304	2
Western Australia	129,435	3
Australia	948,065	7

TABLE 2. Receivals of general purpose weather damaged wheat (1983/84 season)

State	Receivals (tonnes)	Total receivals (tonnes)	Percent of total receival
New South Wales	2,571,000	8,037,000	32
Victoria	931,000	4,466,000	21
Queensland	202,000	1,866,000	11
Western Australia	427,000	3,984,000	11
South Australia	115,000	2,715,000	4
Australia	4,246,000	21,068,000	20

CLASSIFICATION SYSTEMS FOR SPROUTED WHEAT

Historically the Australian wheat classification system has been very conservative in the allowance of sprouted wheat into milling grades. The major class of Australian wheat, Australian Standard White (ASW) has a nil tolerance to visibly sprouted grain. For many years, the only two milling wheat grades into which sprouted wheat could be delivered were the Australian Hard No. 2 and the General Purpose (Weather Damaged) (GPWD) Grade. No doubt, this rather conservative approach has resulted in Australian milling wheats gaining an international reputation for soundness, which has been exploited for marketing purposes especially to those countries which mill Australian

wheat to produce flour for the manufacture of noodles. A
summary of the Australian Wheat Board's receival standards
for sprouted grain is shown in TABLE 3.

TABLE 3. Receival standards for sprouted wheat (1985/86)

Class	Tolerance for visibly sprouted grain	Falling Number requirement (sec)
Australian Standard White	nil	none
Australian Prime Hard	nil	none
Australian Hard No. 1	nil	none
Australian Hard No. 2	10%*	250
Australian General Purpose	0.1%*	200
Australian Feed	unlimited	none

* Variable visible sprouting tolerance depending on Falling
  Number of representative running samples at receival
  points.

        The Australian Wheat Board in its "Receival Standards
and Dockages Schedule" defines sprouted wheat as follows:

        Sprouted kernels are those in which the covering of the
        germ is split open and any further advanced stage of
        growth to the extent of the germ having grown or shot.

CLASSIFICATION AND ACCOUNTING FOR SPROUTED GRAIN

        The way in which sprouted wheat has been treated in the
Australian classification system and the way it has been
treated in the 'financial' sense has changed rapidly in
recent years. TABLE 4 shows the history of dockage levels
for sprouted grain.
        The way in which sprouted grain has been accounted for
is shown in TABLE 5.
        Prior to 1978/79, weather damage in wheat was generally
regarded among wheat growers as 'an Act of God' and a defect
over which the farmer had no control. Sprouted wheat was
classified as 'off-grade' and subject to a relatively small
fixed dockage. Any gain or loss on marketing was simply
absorbed by the 'Australian Standard White pool' which was a
national account. However, since then, the philosophy held
by the Australian Wheatgrowers Federation is that growers
should, as far as practicable, receive the actual market

6

TABLE 4. History of dockage levels for sprouted wheat

Year	Dockage scale
1947	3^d per bushell
1960	6^d per bushell
1971/72 - 1973/74	$1.10 per tonne
1974/75 - 1975/76	$2.20 per tonne
1976/77 - 1978/79	$3.00 per tonne
1979/80	$6.00 per tonne
1980/81 - 1982/83	$10.00 per tonne
1983/84	$20.00 per tonne (GPWD)
1984/85 *)	$22.00 per tonne (GP)
*)	$45.00 per tonne (Feed)
1985/86 *)	$24.00 per tonne (GP)
*)	$45.00 per tonne (Feed)

* Represent initial allowances and not final market returns

TABLE 5. Classification and accounting of sprouted grain

Year	Class	Accounting
71/72 - 73/74	'off grade'	Fixed dockage - loss or gain on marketing absorbed by ASW pool
74/75 - 82/83	'general purpose'	As above
83/84	'general purpose'	Provisional allowance - separate national account
	'Australian feed'	Provisional allowance - national accounting
84/85 - 85/86	'general purpose'	Separate guaranteed minimum price (GMP) - underwritten by Government - national accounting
	'Australian feed'	Separate guaranteed minimum price (GMP) - underwritten by Government - national accounting

value of their product. In effect the payment to growers of the market value for sprouted wheat and the requirement for the General Purpose and Feed Wheat Class to 'stand on its own" has withdrawn from the grower a form of 'insurance' previously held when market gain or loss was absorbed by the Australian Standard White pool.

With the advent of increased segregations a more appropriate level of dockage for sprouted wheat could be introduced by the Australian Wheat Board. Accordingly, a $20 per tonne 'provisional dockage' was introduced for the 1983/84 season. The philosophy of 'return of market value' which was incorporated in the Wheat Marketing Act of 1984/85, has resulted in the underwriting of various classes of wheat by the Australian Government. Consequently, sprouted grain can now be received into the General Purpose Class or the Feed Class which are separately underwritten and which have separate Guaranteed Minimum Prices (GMP). The GMP of the differential from ASW in the 1985/86 season was $15 per tonne for General Purpose wheat (plus a fixed dockage of $9 per tonne) and $45 for Feed wheat. The large financial burden attributed back to the individual wheat grower has resulted in an increased awareness of the detrimental effects of sprouted grain on the marketability of wheat. It has led growers to request more accurate objective assessment of sprouted wheat at the time of receival into the bulk handling system and the incorporation of sprouting tolerance in new wheat releases.

THE COMMERCIAL IMPLICATIONS OF PRE-HARVEST SPROUTING
IN THE AUSTRALIAN WHEAT CROP

1. Potential Economic Impact on the Industry

Given the long term averages for weather-damaged wheat, the resulting loss of value has impacted heavily on the wheat industry. Given todays values, the average per annum loss could be up to $25 million if the weather-damaged wheat was all classified as General Purpose or up to $47 million if it was all classified as Feed wheat. However, because significant weather damage occurs in particular seasons, potential losses in those years may be very high. For example, in a season such as 1983/84 where 4.2 million tonnes of wheat was damaged and marketed mainly as Feed wheat, the potential loss given todays market values could be as much as $190 million and approach 10 percent of the total market value of the Australian crop. Because of the importance of the wheat industry to the overall Australian economy, losses of such magnitude in certain years result in significant reductions in export income.

2. Economic Impact on the Individual Wheatgrower

In a period when the Australian wheatgrower faces difficult economic times, as is the case presently, the economic impact of pre-harvest sprouting on individual growers is most severe. To illustrate the effect of having wheat classified as General Purpose or Feed wheat on farm profitability, an example of a farm budget is described for the south west slopes of New South Wales. Variable costs estimated from the data of Godyn and Reilly (1985) are shown in TABLE 6.

TABLE 6. Farm budget for wheat - south west slopes of NSW
(Godyn and Reilly, 1985)

Variable costs per hectare	
Seed	$ 9.50
Tractor hours	$15.76
Implement repairs etc.	$ 1.25
Fertilizer	$18.55
Herbicide	$12.03
Harvesting (own plant)	$ 6.93
Cartage ($7 per tonne)	$14.70
Total variable costs (per hectare)	$78.93

Total variable costs are estimated at approximately $80 per hectare. Off-farm costs amounting to $43.20 for 1985/86 season are shown in TABLE 7.

The major component of these costs are rail freight (State Rail Authority) and the storage and handling charge (Grain Handling Authority of New South Wales).

A gross margin calculation has been performed using these costs, and an income based on the Preliminary Guaranteed Minimum Price (PGMP) established by the Australian Wheat Board and a yield level of 2.1 tonnes per hectare. The effect of downgrading wheat to the General Purpose or Feed grade is shown in TABLE 8.

TABLE 7. Off - farm costs (per tonne) (1985/86 estimates)

Rail freight	$24.50
Storage/handling (GHA)	$16.70
Wheat research tax	$ 0.35
Ceres house development fund	$ 0.10
Wharfage	$ 1.65
Off - farm costs (per tonne)	$43.30

TABLE 8. Gross margin calculation

Classification of wheat	PMPG 1985/86 $ per tonne	Income $ per hectare	Gross margin $ per hectare
Australian standard white	146.54	307.73	137.87
Australian general purpose*	131.54 (122.54)	276.23 (257.33)	106.37 (87.47)
Australian feed	101.54	213.23	43.37

* figures in brackets include fixed dockage for sprouted grain ($9.00) per tonne)

It is obvious from this Table that classification to the General Purpose or Feed grade removes the profitability in growing the crop. This is especially the case when additional fixed farm costs are taken into account. These costs include such things as general farm expenses (rates, electricity etc.), annual capital costs of farm structures, labour costs and interest on borrowed capital. Additional costs are generally assumed to be greater than variable costs - in this example we will take a conservative example of $80 per hectare. The 'break even' margin required is therefore around $80 per hectare. At a yield level of 2.1 tonnes per hectare the crop becomes unprofitable if wheat is sprouted and classified into the Australian Feed Class.

The effect of classification on profitability given various levels of yield is shown in TABLE 9.

TABLE 9. Effect of sprouted grain on gross margins at various levels of yield

Yield (tonnes per hectare)	Gross margin ($ per hectare) for class of wheat		
	ASW	GP*	FEED
1.0	32.01	8.01	-12.99
1.5	80.13	44.13	12.63
2.0	128.25	80.25	38.25
2.5	176.37	116.37	61.37
3.0	224.49	152.49	89.49

* Fixed dockage for sprouted grain included in calculation.

These calculations are based on the example from southern New South Wales. Taking additional costs to be in the vicinity of $80 per hectare the farmer needs a yield of 1.5 tonnes, 2.0 tonnes or 3.0 tonnes per hectare to 'break even' if the wheat is classified into the Australian Standard White, General Purpose or Australian Feed Class respectively. Average yield levels for the five years to 1983/84 are given for each State in TABLE 10.

TABLE 10. Yields of wheat in Australia (annual average for 1978/79 to 1983/84)

State	Area ('000 Ha)	Production ('000 tonnes)	Yield (t/Ha)
New South Wales	3,504	5,047	1.44
Victoria	1,430	2,524	1.76
Queensland	835	1,098	1.32
Western Australia	4,532	4,341	0.96
South Australia	1,452	1,795	1.24

Source: Australian Bureau of Statistics

Considering that the average yield levels in Australia are 1.5 tonnes per hectare or less, economic solvency is difficult to sustain. The advent of rainfall at harvest cannot be insured against and it is often 'the straw that breaks the camel's back'.

AUSTRALIAN WHEAT BOARD - NATIONAL FORUM 1984

As a result of concern expressed by wheatgrowers over the severe sprouting damage to wheat by rain in the 1983/84 season, the Australian Wheat Board sponsored a National forum to discuss all aspects of weather damage. Representations at this Forum came from all areas of the Wheat Industry and the following recommendations were made:

1. that an appropriate test be developed that could be used to objectively assess deliveries of wheat at receival,
2. that the training of receival agents (classifiers) be increased and formalized,
3. that a contingency plan be developed in each State to minimise the effects and minimise co-mingling of sprouted wheat with milling grades,
4. that research into the utilisation of damaged wheat for various end-products be sponsored,
5. that full support be given to breeding programmes for the development of varieties with sprouting tolerance.

Since this forum these recommendations have been partially implemented and, no doubt, a co-ordinated effort resulted in a better stage of preparedness by industry for the weather damage problems experienced in the 1985/86 season. A test suitable for use in the field at harvest has been developed and field-trialed in the past season by the Bread Research Institute of Australia and the CSIRO Wheat Research Unit. Given the economic loss incurred by the wheatgrower, objective testing for sprouting in the classification of wheat has become imperative.

OBJECTIVE ASSESSMENT GIVES ACCURATE SEGREGATION

With the availability of the falling number test the Australian Wheat Board in conjunction with the Prime Wheat Association introduced a pilot scheme in the 1982/83 season to allow a small percentage of sprouted grain into the Australian Prime Hard, Australian Hard No. 1 and No. 2 classes in northern New South Wales. Receival Standards were varied to allow 2.0 percent visibly sprouted grain into the Prime Hard and Hard No. 1 Classes provided that the falling number of truckloads delivered were above 350 and 300 seconds respectively. A limit of 10 percent sprouted grain was established for the Hard No. 2 class provided that the

site average falling number was above 250 seconds. The wet
harvest of 1983/84 provided an ideal situation for the operation and evaluation of the scheme. It was estimated that
objective assessment of wheat in this region returned
growers an additional 37-45 million dollars which would have
been otherwise lost due to inaccurate segregation (McMaster
and Quail 1984). In addition there is no doubt that the
uniformity of the grades improved dramatically and hence
improved the marketability of the wheat from this region.

Whilst the pilot scheme appears to have been most
successful, a sprouting tolerance has not been formalised in
the Australian Wheat Boards receival standards. Further ex-
tensions of objective assessment for weather damage can only
be justified by a full cost/benefit analysis to the wheatgrower and the industry.

RESEARCH INTO PRE-HARVEST SPROUTING DAMAGE

The incorporation of resistance/tolerance, especially into
white-grained wheats is proving to be very difficult.
Whilst some excellent advances have been made in Australia
incorporating tolerance to pre-harvest sprouting in
recently released cultivars (Mares 1984), the majority of
Australian wheats remain susceptible to sprouting.

Three important phases of research must be clearly
identified if significant advances are to be made on this
problem in cereals:

1. Basic research: the understanding of the underlying
   mechanisms of dormancy and germination.
2. Identification of useful tolerances/resistances to pre-
   harvest sprouting.
3. The successful incorporation of tolerance into new
   commercial cultivars.

Current work at the University of Sydney's Plant
Breeding Institute, by Dr. D. Mares points strongly to
potential tolerance to pre-harvest sprouting that may be incorporated in white wheats by wheatbreeders in the future.

TECHNICAL MARKETING STRATEGIES FOR SPROUTED WHEAT GRADES

In order that effective marketing strategies may be
formulated for wheat containing various levels of sprouted
grain, it is important to know the relative performance of

such wheats in the milling process and in the production of
various end-products. Generally weather damaged grain is
lower in test weight and the grain hardness is 'softened' to
some extent. The physical appearance of weather damage is
also very important as it is not attractive to prospective
purchasers. The wheat may be bleached, embryo sheaths may
be split, germs swollen and grains may be visibly sprouted.
Such an appearance detracts from the value of the grade.
Other factors combined with sprouting defects may further
affect the marketability of wheat. For example, weather
damaged wheat that is low in protein content is considerably less marketable than damaged wheat that is higher in
protein content.

Some end-products are affected more than others when
flour is milled from wheat containing sprouted grain. The
most sensitive of end-products to flour milled from sprouted
wheat are the various types of noodles made in South East
Asia, Japan, China and South Korea. In fact, small levels
of weather damage can result in major defects of these end-
products. Approximately one third of Australian wheat exports are used for noodles, and accurate segregation of
sound from unsound wheat is of paramount importance to these
types of markets. Other products such as the Middle-Eastern
flat breads, whilst not as sensitive as noodles to low
levels of sprouting, may be affected significantly. For the
'pocket' type of flat bread cooked at high temperature for
short periods, a high water absorption is required. Flour
milled from sprouted wheat however usually has a lower flour
water absorption value.

Other types of Middle Eastern flat breads exhibit a
reasonable tolerance to wheat with some weather damage.
Research into the potential uses of mildly sprouted wheat
for overseas markets is currently being conducted at the
Bread Research Institute of Australia. Studies in the
United States also provide encouragement for the use of
some sprouted wheats in flat bread products, especially if
blended with sound wheats (Finney, P.L. et al 1980). A
reduction in the quality of pan or volume breads may also
be observed if sprouting damage is present in wheat.
A sticky crumb, excessive crust colour and reduced loaf
volumes are evident in these breads and slicing problems may
arise in the modern automated bakery. The greatest tolerance to sprout damaged flour is exhibited by end-products
such as those of the Indian sub-continent (eg. chapattis and
rotis). Their organoleptic characteristics of colour, flavour and texture are still acceptable when prepared with
flour milled from wheat containing low levels of sprouted

grain.

The knowledge derived from research in this area has considerable implications for the development of appropriate marketing strategies for weather damaged wheat. Wheats containing various levels of sprouting can be further evaluated post-harvest, classified on their end-product making potential and allocated for sale to appropriate markets. Whilst such wheats cannot be expected to be priced similarly to those of sound wheats, the actual commercial value of sprouted wheat grades can be maximised and the wheatgrower is justified in receiving prices above that of Feed wheat.

REFERENCES

Finney, P. L., Morad, M. M., Patel, K., Choudhry, S. M., Ghiosi, K., Ranhotra, G., Seitz, L. M., and Sebti, S., 1980. Nine international breads from sound and highly field sprouted pacific northwest soft white wheat. Bakers Digest. June issue pp. 22-27.

Godyn, D. and Reilly, T., 1985. Farm budget handbook South West Slopes. New South Wales Department of Agriculture Agdex 815.

McMaster, G. J. and Quail, K. J. 1984. Pre-harvest sprouting in Wheat. An historical, current and commercial perspective. In: proceedings of the 34th Annual Conference of the Royal Australian Chemical Institute, Cereal Chemistry Division pp. 37-39.

Mares, D. J. 1984. Potential tolerance to pre-harvest sprouting in white wheats. In: Proceedings of the 34th Annual Conference of the Royal Australian Chemical Institute, Cereal Chemistry Division pp. 40-41.

Pre-Harvest Sprouting in Barley

Kåre Ringlund, Department of Crop Science,
Agricultural University of Norway, Aas,
Norway.

INTRODUCTION

The value of a barley crop is determined primarily by
the dry matter production. Price adjustments are generally
based on test weight, presence of impurities of different
kinds, and on protein content in certain cases. Sprout damage is only considered as one of the many quality deficien-
cies, such as the presence of green kernels, shrivelled kernels, insect damage, moulding etc. Against this background
pre-harvest sprouting is of very limited importance for barley growers and for the grain handling authorities. For the
end users of barley, however, sprouted kernels may represent
serious quality defects affecting the production of products. Based on the available literature on the subject, it
would appear that is is the physiologists, geneticists and
plant breeders together with a few brewers who are most concerned with the pre-harvest sprouting problem.

IMPORTANCE FOR GROWERS AND THE INDUSTRY

When pre-harvest sprouting occurs, it leads to a loss
of dry matter. Turcek et al. (1970) studied grain losses
during harvest over a period of three years. In one of
those years they found significant losses in dry matter
production from pre-harvest sprouting. This loss in yield
is probably the most important factor for the producers.
For the industry, sprout damage, strangely enough, has its
serious effects on the production of malt which is a product
of sprouted barley. Brookes (1980) states that a high incidence of pregerminated kernels, more than 5 percent, made
barley unsuitable for malting. A sprout-damaged kernel has

lost its ability to digest its own starch. Hence, a
sprouted kernel represents a lost kernel in the malting process. In addition, a dead kernel is very often attacked by
micro-organisms which represent quality hazards in beer production. In the production of barley for seed it is also
obvious that sprouting reduces quality. Fully sprouted kernels are not viable, whilst the germinative power is often
reduced before visible sprouting can be detected. Seed lots
containing sprouted kernels are likely to have a number of
kernels with non-visible sprout damage. For food and feed a
low incidence of pre-harvest sprouting may actually be
advantageous. Sundstøl (1970) reported an increased digestability in sprouted compared with non-sprouted barley when
fed to pigs (TABLE 1).

TABLE 1. Quality of barley protein from N-balance trials
with growing pigs (Sundstøl 1970).

	Protein digestability	Biological value	Net protein utilization
Normal barley	87.4	55.4	48.5
Sprouted 6 days	88.2	60.8	53.6
Sprouted 11 days	85.2	62.6	53.4

Sundstøl (1970) also found a significant increase in
the digestability of crude fibre when barley was sprouted.
This is an interesting result since plant physiologists have
found that cell wall degradation precedes starch degradation
during germination. The increased digestability, however,
was only large enough to balance out the loss of dry matter
caused by sprouting.

In experiments with rats the value of sprouted barley
was lower than for normal barley. Other experiments suggest
that more care is needed in protein supplementation when
weather damaged grain is used. No literature could be found
discussing the effect of sprouting on food products. In
practice the relatively small quantities of barley used for
food in industrialized countries can be carefully selected
and in any case barley for food is normally dehulled. In
the dehulling process most of the enzyme producing tissue is
removed and the amylolytic activity of the endosperm itself
is not sufficient to cause any problems for the technological properties of dehulled barley. In certain developing
countries naked barley is used directly for food. Pre-
harvest sprouting would be expected to reduce the technolog-

ical quality of naked barley for porridge, but since most of
the actual areas involved are very dry sprouting is normally
not a problem.

Since seed dormancy is the most important factor for
the control of pre-harvest sprouting, an evaluation of dormant grain should also be made. In the malting process a
dormant kernel has no more value than a sprouted kernel.
The difference is that a dormant kernel can be conditioned
to germinate. The conditioning can be either a storage
period or a heat treatment. Seed dormancy is due mainly to
inhibitors that supress biological activity in the grain.
It is highly probable that these same inhibitors will act as
antimetabolites when dormant barley is used for feed and
food. Thafvelin (1964) reported toxic effects of freshly
harvested barley used to feed pigs, but no clear evidence
was given to explain the nature of the toxicity. The author
refered to changes in the lipid fraction of the barley as a
possible cause for the problems encountered. Both the dormancy effects, and possibly other problems, associated with
freshly harvested grain disappear after a period of storage
at a reasonably high temperature.

THE GERMINATION PROCESS

In order to discuss the sprouting problem, some attention has to be paid to certain aspects of the germination
process. Three areas of research, namely determination of
amylase isoenzymes, hormonal effects on the germination process, and studies on endosperm degradation are of special
interest in the search for solutions to the sprouting problem.

Isoenzymes

Both alpha- and beta-amylases occur as different isoenzymes. The alpha-amylases have been grouped into three
different groups which can be characterized by their isoelectric points. There is some confusion on the numbering
systems since group I in Daussant's system (1980) corresponds to Sargent's group II (1980). Isoenzymes with isoelectric point at the lower PI are called "green' amylase
and the other groups are called "malt" amylases. Some
reports divided the malt amylases into two groups, II and
III, (Marchylo and Kruger 1983), but group III was later
found to be a complex of group II enzymes and an inhibitor

(Weselake et al., 1985). Marchylo et al., (1980) identified a total of 22 different isoenzymes in the three groups. One major difference between the green and the malt enzymes seems to be that the green enzymes cannot attach themselves to native starch granules (Sargent 1980). At the Plant Breeding Institute (PBI) in Cambridge a numbering system for the genes coding for the different isoenzymes has been developed. Alpha-Amy 1 codes for malt amylases or those with the highest PI, and Alpha-Amy 2 codes for the green amylases. Ainsworth et al., (1985) described a system of nomenclature based on gene mapping of these Alpha-Amy genes. The work on the inhibitor-enzyme complexes is also of particular interest to plant breeding. If different isoenzymes have different reactions to hormones and inhibitors, varieties with different sprouting "properties" might be produced through plant breeding.

Hormonal effects

In general enzyme activity is controlled by hormones. The role of gibberellic acid in the activation of the alpha-amylases has been clearly demonstrated by several authors. Nicholls (1983) has reported a deviation from normality that may be important to explain variations in sprout damage. He found that barley grown under specific environments may produce alpha-amylase in the aleurone layer without stimulation by gibberellic acid. The important implication of this finding is that the degradation of starch may start even if the embryo is still dormant. Corresponding results were reported by Gibbons (1983) who found that initial cell wall breakdown can proceed normally under conditions where rootlet and coleoptile growth are severely inhibited.

Endosperm breakdown

Gibbons (1980) demonstrated very clearly at the 2nd International Sprouting Symposium that the degradation of starch in the endosperm starts in the region adjacent to the scutellum and proceeds from the scutellum and into the endosperm. Only at later stages is alpha-amylase from the aleurone layer important in the degradation of endosperm starch. This has been confirmed by MacGregor et al. (1983). These workers also found that protein and cell wall degradation had proceeded further into the endosperm than starch degradation. This finding might be important in detecting sprout damage at an early stage. If methods for measuring the activity of proteases and cell wall degrading enzymes were developed, such methods should detect initial sprout damage before increased alpha-amylase activity could be measured.

SEED DORMANCY

The phenomenon of seed dormancy has been the subject of research since the beginning of this century and many different mechanisms for dormancy have been proposed. The early papers were mostly concerned with the seed coat impermeability theory, but today it is fairly well established that dormancy is due to inhibitory systems that act on the enzyme production. In general seed dormancy protects against pre-harvest sprouting. The relationship between the dormancy index as described by Strand (1965) and falling number was studied using varieties and breeding lines of barley harvested at different maturity stages (Ringlund 1970). The correlation coefficients between the dormancy index and falling number were low and non significant at an early harvest stage, but high and highly significant at the latest harvest stage when pre-harvest sprouting was most severe.

TABLE 2. Correlation coefficients between falling number and dormancy index in barley (Ringlund 1970).

Year	No. of varieties	Harvest time I	Harvest time II	Harvest time III
1960	49	-.008	.433**	.609***
1961	41	.186	.399*	.595***
1963	26	.513**	.403*	.627***
1964	19	.041	-.066	.752***

*P < 0.05, **P < 0.01, ***P < 0.001

As mentioned above some reports have indicated that there may be different control systems for certain types of starch degradation (Nicolls 1983, Gibbons 1983). Under specific environmental conditions these modified systems might be important and they might explain deviations from the normal germination pattern.

The build up and breakdown of dormancy have been studied by several authors. Reiner and Loch (1976) found two temperature dependent periods during the development of barley. Low temperatures 12-16 days after ear emergence and

high temperatures 30-41 days after ear emergence caused low
dormancy 3 weeks after harvest. Takahashi (1980) found that
low temperatures induce dormancy in barley whereas high
temperatures induce dormancy in rice. Strand (1983) studied
both the effect of temperature and the effect of rainfall
during the period from anthesis to yellow ripeness on dormancy. The effects of temperature were generally negative,
i.e. low temperatures caused high dormancy. The effects of
rainfall were positive, i.e. dormancy increased with increased rainfall. The effect of temperature on the breakdown of dormancy was studied by Strand (1965). If barley
grains are stored at very low temperatures, +3°C, the dormancy will last for many months. On the other hand, if the
storage temperature is 20-30°C, the dormancy will disappear
within a few weeks (FIGURE 1 and 2). Dormant barley will
show increased dormancy at higher germination temperatures.

FIGURE 1. Dormancy in barley stored at different temperatures and germinated at 12°C (Strand 1965).

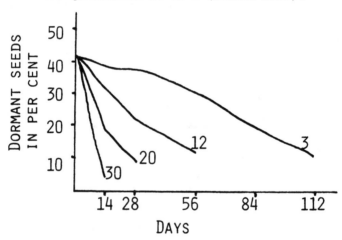

MEANS OF REDUCING PRE-HARVEST SPROUT DAMAGE

Several factors are potentially important for
achieving a reduction in the incidence of sprouting. The
usefulness of dormant varieties has already been discussed,
and this is the most important way of reducing pre-harvest
sprouting. Dormant varieties will however, never give complete resistance against sprouting and if the dormancy is
too severe, problems may arise in the seed production of
such varieties. One particular barley variety grown in
Norway during the 1950's, was still dormant in the spring

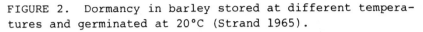

FIGURE 2. Dormancy in barley stored at different temperatures and germinated at 20°C (Strand 1965).

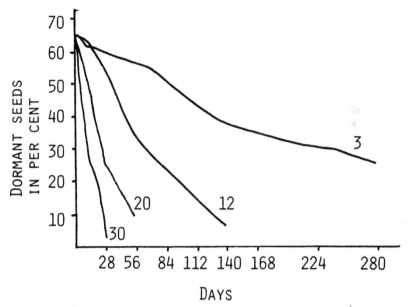

when stored at low temperatures. This can be avoided by
storage at a higher temperature, but heat treatment adds to
the price of the seed. In spite of these negative effects
of dormancy, the breeding of dormant varieties is still the
main factor for reducing pre-harvest sprouting.

In some areas where pre-harvest sprouting is a problem,
the rainfall increases throughout the harvest season. Early
maturing varieties will, therefore, avoid sprouting more
often than late varieties. Early varieties may have less
dormancy yet still be less affected by sprouting. Another
way of avoiding sprout damage is to harvest at an early maturity stage. Barley can be harvested, without severe
losses, with a moisture content as high as 25 percent, but
the drying costs are of course increased. If the risk of
sprouting is very high, such early harvests can be economical. Since sprouting is most severe in the presence of
lodging, limiting nitrogen application may also reduce
sprout damage. Finally sprouted kernels can be removed from
a bulk sample by the use of several screening methods.
Since sprouting reduces the density of the kernels, wind-
sifting or gravity table sifting are effective means of
removing sprouted kernels.

SUMMARY

Pre-harvest sprouting in barley is not a major problem for growers, grain handling authorities, and end users of barley in general, but for malt and seed production sprouting can cause important quality reductions in certain barley producing areas. A basic knowledge of the germination process and of the mechanisms controlling this process is a fundamental prerequisite to the solution of the sprouting problem. The use of dormant varieties, early harvesting, and cleaning by means of wind sifting or gravity tables can prevent or reduce the number of sprouted kernels in barley destined for malt and seed. In certain respects the quality of barley for feed and food may be improved by a low incidence of sprouting.

LITERATURE

Ainsworth, C. C., Doherty, P., Edwards, K. G. K., Martienssen, R. A. and Gale, M. D. Allelic variation at alpha-amylase loci in hexaploid wheat. Theor. Appl. Genet. 70, pp. 400-406.

Brookes, P. A. 1980. The significance of pre-harvest sprouting of barley in malting and brewing. Cereal Res. Comm. 8, pp. 29-38.

Daussant, J. and Mayer, C. 1980. Immunochemistry of cereal alpha-amylase in studies related to seed maturation and germination. Cereal Res. Comm. 8, pp. 39-48.

Gibbons, G. C. 1983. The action of plant hormones on endosperm breakdown and embryo growth during germination of barley. Third International Symposium on Pre-harvest Sprouting in Cereals. Kruger, J. E., LaBerge, D. E. Westview Press, Boulder, Co., USA. pp. 169-180.

Goldbach, H., Michael, G. 1976. Abscisic acid content of barley grains during ripening as affected by temperature and variety. Crop Science, 16, pp. 797-799.

MacGregor, A. W. 1983. Cereal endosperm degradation during initial stages of germination. Third International Symposium on Pre-harvest Sprouting in Cereals. Kruger, J. E., LaBerge, D. E. Westview Press, Boulder, Co., USA. pp. 162-168.

Nicholls, P. B. 1983. Environment, developing barley grain, and subsequent production of alpha-amylase by the aleurone layer. Third International Symposium on Pre-harvest Sprouting in Cereals. Kruger, J. E., LaBerge, D. E. Westview Press, Boulder, Co., USA. pp. 147-153.

Reiner, L. and Loch, V. 1976. Forcasting dormancy in
    barley. Ten years experience. Cereal Res. Comm.
    4, pp. 107-110.
Ringlund, K. 1969. Stivelseskvalitet hos bygg. Nord.
    Jordbr. forskn. pp. 303-308.
Strand, E. 1965. Studies on seed dormancy in barley.
    Meld. NLH. 44, pp. 1-23.
Strand, E. 1983. Effects of temperature and rainfall on
    seed dormancy of small grain cultivars. Third Inter-
    national Symposium on Pre-harvest Sprouting in Cereals.
    Kruger, J. E., LaBerge, D.E. Westview Press, Boulder,
    Co., USA. pp. 260-266.
Sundstøl, F. 1970. Undersøkelser over fôrverdien av bygg
    av ulik kvalitet. The Agricultural College of Norway,
    Institute of Animal Nutrition, Tech. Bull. No. 140,
    pp. 1-61.
Takahashi, N. 1980. Effect of environmental factors during
    seed formation on pre-harvest sprouting. Cereal Res.
    Comm. 8, pp. 175-183.
Thafvelin, B. 1964. Beakta riskerna vil utfodring av
    nyskärdat spannmål. Svinskjötsel 9, pp. 255-260.
Turcek, J., Supuka, S., Halabrin, M. 1975. Biological
    losses in spring barley. Acta Technologica Agricult-
    urae, 14, pp. 139-155.
Weselake, R. J., MacGregor, A. W. and Hill R. D. 1985.
    Endogenous alpha-amylase inhibitor in various cereals.
    Cereal Chem. 62, pp. 120-123.

Pre-Harvest Sprouting in Rye and Triticale

G. Oettler, Landessaatzuchtanstalt,
Universität Hohenheim, Postfach 70 05 62,
7000 Stuttgart 70, Germany F. R.

INTRODUCTION

In some years pre-harvest sprouting damage may cause considerable losses in cereals in various regions of the world. It occurs in well established crops such as rye (Secale cereale L.) and also affects the man-made cereal triticale (x Triticosecale Wittmack). Despite its global importance only limited research has been devoted to the problem in rye and research work in triticale is still in its infancy.

This paper gives an account of the current situation with regard to pre-harvest sprouting in rye and the wheat-rye hybrid triticale. Agronomical and technological measures to reduce the effects of sprouting damage will be examined but the main emphasis has been placed on breeding aspects since they appear to offer the most effective possibilities.

RYE

The main areas of rye production lie in central and eastern Europe (Germany, Poland, USSR) on poorer and sandy soils that generally are less suited to growing other cereals. Since substantial quantities of the rye produced are destined for human consumption high quality grain is required. The processing quality of rye is determined by the quantity and quality of the main grain components (starch, protein and pentosan), which exhibit special swelling properties, and by the level of enzyme activity, (Weipert, 1983). Wet conditions or rain prior to harvest can initiate germination and the production of enzymes and

cause the kernels to sprout in the head. Starch decomposition by alpha-amylase is the most important process.
Enzymatic degradation of the starch results in a decreased
swelling capacity and both processes ultimately lead to low
quality baked products.

Sprouted grain is of low quality and unsuitable for
food products such as rye bread (Drews and Seibel, 1975).
In countries where a grading system and a guaranteed fixed-
price system (intervention) exists, as for example in the
European Economic Community, rye has to meet certain requirements and intervention occurs if sprouted grain exceeds
8 percent (Drews and Seibel, 1976; Bunnies, 1982). Rye of
higher quality is classified as 'bread rye' with minimum
values for falling number of 75 seconds and for amylogram
(whole meal) of 200 AU at a gelatinizing temperature of
63°C. The proportion of sprouted grains should be below 2.5
percent. Good processing properties, however, are reached
only when the falling number is between 125 and 200 seconds.
Thus, securing optimum quality, even under adverse weather
conditions has to be the main aim for the rye producers, and
to this end various measures can be considered.

Agronomy and Technology

The processing quality of rye is determined, in part,
by precipitation and temperature during the grain filling
phase and harvest period. In the case of unfavourable weather conditions early or premature harvest can be recommended. However, high cost of drying the grain impose restrictions on this measure. Root rot and the excessive use
of nitrogen fertilizer can cause lodging. Lodged cereal is
attacked by micro-organisms which take part in the enzymatic
decomposition of the grain, leading to sprouting (Bolling,
1979). Thus balanced fertilizer application and a favourable rotation of crops are advisable. Lodging caused by
weak straw can be reduced by the application of growth
regulators.

The miller can mix flour of different qualities and upgrade lower quality flour milled from mildly weather damaged grain (Scharf, 1979). Milling intensity can also be
changed to reduce enzyme activity and improve quality
(Weipert, 1983). When using flour from sprout-damaged grain
the baker can lower the temperature during sour-dough fermentation to improve the quality. Increased baking time and
lower baking temperature may also be recommended (Bunnies,
1982).

Breeding

In designing a breeding programme to improve pre-harvest sprouting resistance, suitable testing methods for screening the material have to be applied. For the breeder's purpose the methods have to be quick, simple, inexpensive and reproducible in order to handle large numbers of genotypes. The most common test used to determine the level of sprouting potential in rye appears to be the Hagberg Falling Number Method, which measures the viscosity of a meal or flour suspension and reflects the level of alpha-amylase activity. Low falling numbers indicate high alpha-amylase activity, but the values are not linearly related. In the region of high alpha-amylase activity the falling number does not differentiate sufficiently (Kruger and Tipples, 1982; Fretzdorff and Weipert, 1983; Krey 1985).

The problem of sprouting damage in rye has been recognized for many years and as early as 1941 Popoff found genetic variability in Bulgarian rye populations. A successful breeding programme was conducted by Persson (1976) in Sweden, which lead to the release of the open-pollinated sprouting resistant cultivar 'Otello' in 1970-1971. Persson applied mass selection under provocative conditions in a moisture chamber and quality was determined by falling number and enzymatic test. Unfortunately, because of its low yield potential 'Otello' is of little importance in commercial production.

Plarre (1980) crossed the German cvs. 'Carstens Kurz' x 'Petkus Normal' and by sib-mating developed sprouting resistant inbred line which were crossed with 'Otello' to accumulate resistance genes. Crosses between the inbreds and S. silvestre also resulted in promising material which, however, did not surpass the population 'Otello'. Hence there seems to be little reason for using wild species to improve this character, particularly when the accompanying deterioration of other important traits is considered. It appears more profitable to exploit variation within the populations of adapted cultivated rye.

For a considerable period of time the variability in pre-harvest sprouting resistance was regarded as being predominantly due to environmental factors with little influence of genotype (Bolling, 1979; Weipert, 1981, 1983). This may be the case between particular populations, but within populations there exists a remarkable genetic variability, as has been conclusively established by Geiger (1975) and Geiger and Morgenstern (1979). These researchers showed that the range of variability within a source population became apparent in homozygous, genetically defined inbred lines and in their testcrosses. Whereas in populations only mean values can be utilized, in hybrids full use can be made of extreme or superior inbred genotypes. The above investigations also demonstrated that the main source of variation in the falling number was the genotype, i.e. inbred lines, and not locations and that through hybrid breeding a considerable improvement in quality characters could be expected in rye. Köhler et al. (1986) examined S_1 to S_4 generation inbred lines and showed the falling number to be a heterotic character. It was not adversely associated with other agronomic traits, which was consistent with the results of a two-year study of advanced homozygous inbred lines (Oettler, unpubl.). Moreover, the above authors found the falling number to be already determined in early selfing generations, which increases selection efficiency in population improvement and in developing inbreds for hybrid breeding.

TRITICALE

An estimated area of 650,000 ha was under triticale production worldwide in 1984 (Bernard and Bernard, 1985). The demand for high quality grain exists at present primarily for seed production, but not for human food. The crop is still not exploited commercially and is mainly grown for on-farm consumption as livestock feed (Skovmand et al., 1984). Triticale is highly prone to pre-harvest sprouting and this may be a limiting factor in triticale production in areas such as parts of Europe, Canada or Brazil where moist conditions prevail during the ripening and harvest period (Salmon and Helm, 1985; Baier and Nedel, 1985).

High levels of enzyme activity in the endosperm during the ripening period lead to the rapid conversion of starch to sugar, which is believed to result in shrinkage and partial collapse of the endosperm prior to the precocious germination (Klassen et al., 1971; Thomas et al., 1980; Barnlard et al., 1985). High levels of alpha-amylase activity, and corresponding low falling numbers, are found in triticale prior to harvest ripeness, in mature grain in the absence of visible sprouting (Baier and Nedel, 1985; Weipert, 1985; Oettler, unpubl.) and also in dry years. The enhanced enzyme activity, in particular alpha-amylase, in the mature triticale grain is one of the reasons for its low processing value with regard to baking quality. In general, the level of alpha-amylase activity in triticale is

substantially higher than in either wheat or rye (McEwan and
Haslemore, 1983; Branland et al., 1985), which indicates a
severe disorder in the grain.

Agronomy and Technology

The agronomic measures practised with rye to reduce the
incidence and the effects of pre-harvest sprouting can also
be applied to triticale. High levels of nitrogen fertilizer
were found to increase the amylase activity (Singh et al.,
1978), as has been reported for other crops (King, 1983).
With regard to technological measures which could be used to
improve the processing of sprouted grain, it has to be kept
in mind that triticale is a versatile food, which can be
used to make a range of products such as breads, breakfast
cereals, biscuits, cakes, crackers, pancakes and chapatis.
In contrast to rye, no classification system has been established for handling triticale as a food grain and it may
well prove difficult to do so in view of the wide range of
genomic constitutions in this crop. Thus any measures reported on should only be regarded as preliminary. Weipert
(1986) suggested that enzymatically damaged triticale grain
should be milled to a coarse meal if it is to be used for
bakery items. Amaya and Skovmand (1985) recommended a reduction in the fermentation period and the temperature and a
higher yeast concentration to improve the baking quality at
high levels of alpha-amylase. For sprout-damaged flour a
triticale-wheat mixture, using sour-dough for a rye-type
bread, has also been suggested (Drews et al., 1976).

Breeding

The Hagberg Falling Number test and colourimetric procedures for measuring alpha-amylase activity have also been
identified as useful testing methods to assess sprouting
potential in a triticale breeding programme (McEwan and
Haslemore, 1983; Amaya and Skovmand, 1985; Huskowska
et al., 1985). Chojnacki et al., (1976) and Krey (1985)
found falling number to be negatively correlated with alpha-
amylase activity in a curvilinear function, although some
samples with both a high falling number and a high enzyme
level were also identified. For low falling number values a
sample weight of 9g meal is recommended to allow better
differentiation between genotypes as is illustrated in
FIGURE 1. The year 1983 was unusually dry with excellent

29

FIGURE 1. Hagberg falling numbers of different sample weights for seven triticale lines grown at Hohenheim in 1983 and 1984.

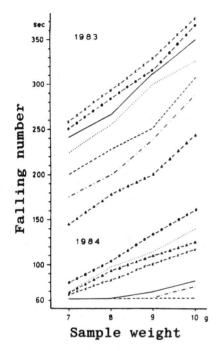

Sample weight

grain filling, whereas in 1984 humid and rainy conditions prevailed.

Improvement of pre-harvest sprouting resistance has been included in a number of breeding programmes (Baier and Nedel, 1985; Scarth et al., 1985, Huskowska et al., 1985). Working with genetically undefined materials of various origin and genomic constitutions some genetic variability was found, but the general level was lower than in the wheat and rye parents. Whether pre-harvest sprouting resistance can be incorporated into triticale at the required agronomic level by combining pre-selected wheat and rye parents, appears doubtful considering the information currently available. Despite considerable effort, Huskowska et al., (1985) were not able to introduce improved resistance, either from the rye cv. 'Otello' (representing the highest level of sprouting resistance at present), or from favourable wheats. Investigations with genetically defined triticale, which allowed the determination of the contribution of both its parental species, also indicated that the high falling numbers of the parents were not expressed in the derived hybrids (Oettler, 1986). Interaction effects of the parental components played a predominant role and apparently resulted in severe hormonal disorders.

Considering the importance of character associations in a breeding procedure, only scarce and contradictory information is available for triticale. Klassen et al. (1971) found a relationship between the degree of kernel shrivelling and alpha-amylase activity, with the poorest kernels having the highest activity. The triticale genotypes investigated by McEwan and Haslemore (1983), however, did not reveal a correlation between alpha-amylase level and test weight. Research with genetically defined primary

triticales confirmed the positive correlation between falling number and test weight (Oettler, 1986).

Chojnacki et al. (1976) reported on a tendency of dwarf
triticale to have higher falling numbers. In the defined
primary genotypes falling number and plant height were not
found to be correlated. Furthermore, this study did not
uncover any trait associations unfavourable for a breeding
programme.

To improve pre-harvest sprouting resistance in triticale the following points should be considered for further
research activities:

1. Since sprouting appears to be closely associated with
   grain shrivelling, which is still a most persistent
   problem of the new crop, improvement of both these
   characters should receive high priority. To overcome
   the obvious severe enzymatic disharmonies in the triticale grain, joint efforts of plant physiologists and
   breeders appear essential.
2. If falling number proves to be a heterotic character in
   triticale, advances in pre-harvest sprouting resistance
   can be expected through hybrid breeding.

CONCLUSION

Improvement in pre-harvest sprouting resistance in rye
appears possible both at the population and the hybrid
level, since considerable genetic variability is available
for exploitation. As for triticale, evidence suggests that
at present the potential variability and the level of pre-
harvest sprouting resistance which exists in both the parental species cannot be incorporated into the hybrid. More
intensive research on all aspects of sprouting in triticale
is needed to obtain reliable information.

REFERENCES

Amaya, A. and Skovmand, B. 1985. Current status of hexa-
    ploid triticale quality. In: Bernard, M. and Bernard
    S. (eds.): Genetics and breeding of triticale. INRA,
    Paris, pp. 604-606.
Baier, A. C. and Nedel, J. L. 1985. Triticale breeding in
    Brazil. In: Bernard, M. and Bernard, S. (eds.) Gene-
    tics and breeding of triticale. INRA, Paris. pp. 497-
    502.

Bernard, M. and Bernard, S. (eds.) 1985. Genetics and
    breeding of triticale. INRA, Paris, p. XVI.
Bolling, H. 1979. Qualitätskriterien für die Vermahlung von
    Roggen. Getreide, Mehl und Brot 33, pp. 216-218.
Branlard, G., Bernard, M., Antraygue, C., and Barloy, D.
    1985. Shrivelling and alpha-amylase activity in triti-
    cale and its parental species: preliminary studies.
    In: Genetics and breeding of triticale. Bernard, M.
    and Bernard, S. (eds.). INRA, Paris, pp. 617-631.
Bunnies, H. 1982. Mühlen- und bäckereitechnologische
    Anpassungen an unterschiedliche Roggenqualitäten und
    deren Kosten. Betriebs- u. marktwirtschaftliche
    Studien zur Ernährungswirtschaft, Kiel.
Chojnacki, G., Brykczyinski, J. and Tymieniecka, E. 1976.
    Preliminary information on sprouting in triticale.
    Cereal Res. Comm. 4, pp. 111-114.
Drews, E. und Seibel, W. 1975. Backtechnische Merkmale des
    Roggens. Getreide, Mehl und Brot 29, pp. 95-98.
Drews, E. and Seibel, W. 1976. Bread-baking and other uses
    around the world. In: Rye: Production, chemistry and
    technology. Bushuk, W. (ed.). Am. Ass. Cereal Chem.,
    St, Paul, Minn. pp. 127-178.
Drews, E., Weipert, D. und Meyer, D. 1976. Orientierende
    Untersuchungen über die Verarbeitungseigenschaften von
    inländischem Triticale. Getreide, Mehl und Brot 30,
    pp. 285-291.
Fretzdorff, B. und Weipert, D. 1983. Nephelometrische
    Methode zur Bestimmung der Alpha-Amylase-Aktivität bei
    Getreide. Z. Lebensm. Unters. Forsch. 177, pp. 167-
    172.
Geiger, H. H. 1975. Alternativen der heutigen Roggen-
    züchtung. Getreide, Mehl und Brot 29, pp. 197-202.
Geiger, H. H. und Morgenstern, K. 1979. Stand der Hybrid-
    züchtung bei Roggen. Getreide, Mehl und Brot 33, pp.
    225-231
Huskowska, T., Wolski, T. and Ceglinska, A. 1985. Breeding
    of winter triticale for improvement of grain char-
    acters. In: Genetics and breeding of triticale.
    Bernard, M. and Bernard, S. (eds.), INRA, Paris, pp.
    641-649.
King, R. W. 1983. The physiology of pre-harvest sprouting -
    a review. In: Proc. Third Int. Symp. on Pre-harvest
    Sprouting in Cereals. Kruger, J. E. and LaBerge, D. E.
    (eds.). Westview Press, Boulder, Co. pp. 11-21.
Klassen, A. J., Hill, R. D. and Larter, E. N. 1971. Alpha-
    amylase activity and carbohydrate content as related to
    kernel development in triticale. Crop Sci. 11, pp.
    265-267.

Köhler, K. L., Geiger, H. H. and Wilde, P. 1986. Analysis of inbreeding effects on falling number in diploid winter rye (Secale cereale L.). Proc. Eucarpia Meeting Cereal Sect. on Rye, June 11-13, 1985, Svalöf, Sweden (in press).

Krey, B. 1985. Methodische Untersuchungen zur rascheren Erfassung von Auswuchs in Durumweizen. Diplomarbeit Universität Hohenheim.

Kruger, J. E. and Tipples, K. H. 1982. Comparison of the Hagberg falling number and modified grain amylase analyzer method for estimating sprout damage in rye. Can. J. Plant Sci. 62, pp. 839-844.

McEwan, J. M. and Haslemore, R. M. 1983. The response of triticale and related cereals to conditions inducing pre-harvest sprouting. In: Proc. Third Int. Symp. on Pre-harvest Sprouting in Cereals. Kruger, J. E. and LaBerge, D. E. (eds.). Westview Press, Boulder, Co. USA. pp. 279-286.

Oettler, G. 1986. Variation of falling number in primary triticale and their wheat and rye parents. In: Proc. Fourth Int. Symp. on Pre-harvest Sprouting in Cereals. Mares, D. J. (ed.). Westview Press, Boulder, Co. USA. (in press).

Persson, E. 1976. Otello - a result of amylase selection for sprouting resistance. Cereal Res. Comm. 4, pp. 101-106.

Plarre, W. 1980. Breeding methodology for rye with resistance to sprouting. Cereal Res. Comm. 8, pp. 265-274.

Popoff, A. 1941. Über die Auswuchsneigung des Roggens. Z. Pflanzenzüchtg. 23, pp. 535-541.

Salmon, D. F. and Helm, J. H. 1985. Pre-harvest and post-harvest dormancy in spring triticale. Agron. J. 77, pp. 649-652.

Scarth, R., Larter, E. N., Helm, J., Salmon, D. and Townley-Smith, T. F. 1985. The triticale breeding program in western Canada. In: Genetics and breeding of triticale. Bernard, M. and Bernard, S. (eds.). IRNA, Paris, pp. 465-471.

Scharf, R. 1979. Qualitätskriterien für die Vermarktung von Brotroggen aus der Sicht der Müllerei. Gretreide, Mehl und Brot 33, pp. 220-221.

Singh, B., Patel, J. A. and Sapra, V. T. 1978. Amylase activity in triticale, x Triticosecale Wittmack. Euphytica 27, pp. 19-25.

Skovmand, B., Fox, P. N. and Villareal, R. L. 1984. Triticale in commercial agriculture: progress and promise. Adv. Agr. 37, pp. 1-45.

Thomas, J. B. Kaltsikes, P. J., Gustafson, J. P. and
   Roupakias, D. G. 1980. Development of kernel shrivel-
   ling in triticale. A. Pflanzenzüchtg. 85, pp. 1-27.
Weipert, D. 1981. Qualitätseigenschaften der in der
   Bundesrepublik Deutschland angebauten Roggensorten.
   Getreide, Mehl und Brot 35, pp. 199-204.
Weipert, D. 1983. Zur Beurteilung des Verarbeitungswertes
   von Roggen. Getreide, Mehl und Brot 37, pp. 229-234.
Weipert, D. 1985. Eigenschaften und Verwendungsmöglich-
   keiten von Triticale.   Getreide, Mehl und Brot 39, pp.
   291-298.
Weipert, D. 1986. Triticale processing in milling and
   baking. Proc. Int. Triticale Symposium, Sydney, 1986.
   Aust. Inst. Agr. Sci. Occasional Publication No. 24,
   pp. 402-411.

Pre-Harvest Sprouting in Rice

B. O. Juliano and Te-Tzu Chang,
International Rice Research Institute,
Los Banos, Languna, Philippines.

INTRODUCTION

Pre-harvest sprouting in rice frequently ranges from
0-20 percent, mainly in nondormant varieties exposed to prolonged rains and lodging during grain maturation particularly in japonica cultivars. Sprouted grain is used mainly
as poultry feed because the cracked caryopsis cannot be
properly milled except after parboiling. Full utilization
of rice grain requires appropriate grain size and shape, excellent milling quality (translucent, minimum brokens), and
the desired cooked rice texture (indexed mainly by amylose
content). Sprouting incidence may be effectively reduced by
incorporating moderate grain dormancy and lodging resistance
into cultivars, draining the fields after flowering whenever
possible, and timely harvesting.

NATURE AND EXTENT OF PRE-HARVEST SPROUTING

The rice grain (<u>Oryza sativa</u> L.) consists of 18-28 percent hull (mainly lemma and palea) that envelops the caryopsis (Juliano, 1985). The outer layers, pericarp, seed
coat, nucellus, and aleurone layer completely enclose the
endosperm and embryo.
"World production of rice was about 470 million tonnes
of rough rice in 1984 of which at least 90 percent was produced and consumed in Asia (FAO, 1985). Total area planted
was 147 million ha. corresponding to a mean yield of 3.18
tonnes/ha. The wet season crop is usually the main crop
as compared to the dry season, because of water availabil-

---

* Mailing address: P.O. Box 933, Manila, Philippines.

ity. The major post-harvest problem in the wet season is
the unreliable solar drying of grain, resulting in spoilage
and heat damage (stackburning), particularly of unthreshed
grain."

Pre-harvest sprouting is not a very common phenomenon
in rice. It correlates negatively with dormancy period
(days after heading for 50% germination) in both indica
(r = -0.85**, n = 98) and japonica (r = -0.74**, n = 84)
rices (Chen et al., 1980). Percentage of pre-harvest
sprouting was negatively correlated (r = -0.88**) with days
required after maturity to obtain 75% germination for 86 F_2
plants in the cross N22/Mahsuri (Seshu and Sorrells, 1985).
It is triggered by prolonged rains, delayed harvesting and
by lodging of plants in standing water during grain maturation. Pre-harvest sprouting (0-70%) was also found to be
positively correlated with the ability to germinate under
low temperature (14°C) (3-85%) (r = 0.92**) and under osmotic stress (-12 bars polyethylene glycol 6000 osmotic pressure) (0-85%) (r = 0.84**) among 22 cultivars (Seshu and
Sorrells, 1985).

Rice grain development in the tropics is usually completed 14-18 days after flowering and the remaining period
is for grain desiccation (Yoshida and Hara, 1977). The percentage of pre-harvest sprouting (harvested 25 days after
flowering and germinated 5 days) ranged from 0-70% in 22
cultivars (Seshu and Sorrells, 1985). Mahsuri, an indica x
japonica variety, is noted for its susceptibility to sprouting: levels of pre-harvest sprouting up to 62% have been
obtained. Among 20 lots of rice seeds of four japonica varieties with sprout damage in Aichi, Japan, in 1979 due to a
long spell of rain, percentage of sprouted seeds was 0-16%
for certified seed, 11-25% for quasi-certified seed, and 21-
44% for disqualified seed (Ito et al., 1980). In Taiwan,
Tang and Chiang (1955) reported a 10-20% yield loss due to
pre-harvest sprouting. In one rainy season crop in northern
Taiwan, mean percentage of panicle germination for 13 japonica varieties was 11% for 6 days continuous rain and 25%
for 10 rainy days (Lee, 1976). For varieties with no seed
dormancy, pre-harvest sprouting was 12-44% for 10 days continuous rain. In the Kada area of Malaysia, delayed harvest
during one wet season resulted in losses from pre-harvest
sprouting in 49% of farms: <5% loss in 16% of the farms:
5-10% loss in 20: and >10% loss in 13 (Mohamed, 1982).
During piling of cut and unthreshed panicles for 24 days in
the Philippines, the top layer showed 5% post-harvest germination, middle, 3%, and the bottom, 2%, whereas threshed
grains had 1% germination (Mendoza et al., 1982). Bulk

storage of rough rice at 23-26%-moisture with aeration for up to 11 days before drying, at two sites in the Philippines, resulted in 0.3% post-harvest sprouting (de Castro et al., 1980).

DORMANCY AND PRE-HARVEST SPROUTING

Dormancy in rice grain is not true embryo dormancy because excised embryos germinate readily (Navasero et al., 1975). The hull and, in some varieties, the caryopsis coat contribute to rice dormancy; caryopsis-coat dormancy is removed about 3 weeks ahead of hull dormancy (Seshu and Sorrells, 1985). The dormancy period tends to be shorter for japonica rices (0-30 days) than for indica rices (0-45 days) (Tang and Chiang, 1955). Pigmented rices had longer dormancy periods than nonpigmented rices; wild Oryza species also have longer dormancy periods than Oryza sativa (Tang and Chiang, 1955; Chen et al., 1980).

The mechanism of rice dormancy and its loss during storage at ambient temperature are not well understood. Presumably, diffusion of water (Martires, 1975) and of oxygen into the embryo (Navasero et al., 1975) are slower in dormant seeds than in nondormant seeds. Cracks on the hull of variety Ketaktara (Kataktara) dormant seeds, which may facilitate oxygen diffusion into the caryopsis, occur during high-temperature ripening or post-harvest storage (Hayashi and Hidaka, 1979). Peroxidase activity in the hull may also contribute to hull dormancy (Baun, 1972; Navasero et al., 1975). Dehulling increases germination percentage of dormant varieties in a variable manner depending on degree of caryopsis-coat dormancy (Nair and Sahadwan, 1962; Seshu and Sorrells, 1985). Scraping the pericarp over the embryo of brown rice completely breaks dormancy, but not of rough rice with the hull over the embryo removed (Navasero et al., 1975; Seshu and Sorrells, 1985).

UTILIZATION OF PRE-HARVEST SPROUTED AND NORMAL GRAINS

Sprouted seed cannot be used as certified seed because it is no longer viable. However, seeds pregerminated for 2 days in aerobic or anaerobic conditions and dried back can be sown again 1 week later, without any deleterious effect on germination and vigor (Krishnasamy V. and Seshu, D. V., unpublished IRRI data). Seeds with only the rudimentary roots showing may remain viable for a few days and can be used for seeding. In addition, the percentage of germinated grains is actually higher than percentage of sprouted rough rice due to grains with less developed sprouts that are seen only after dehulling (Ito et al., 1981a). However, percentage of emergence correlated well with germination percentage although increased seeding rate tended to reduce percentage of emergence and percentage of good seedlings (Ito et al., 1981b).

Sprouted grain will have a fissured endosperm as a result of water absorption before germination, because cracking occurs when relatively dry grain absorbs moisture (Kunze, 1985). Thus, sprouted rice will give rise mainly to brokens during dehulling and milling. Parboiling or steam treatment of steeped rough rice should salvage the crop by annealing the cracked grains together to produce whole-grain milled rice on processing (Juliano, 1985). Sprouted grain is used mainly as animal feed, particularly for poultry, because of their ability to ingest the siliceous hull.

Grain breakage during desiccation in the field and post-harvest drying should be minimized because rice is consumed mainly as whole-grain milled rice. Differences between varieties in their susceptibility to fissuring during moisture absorption may be related to differences in alkali spreading value (Srinivas and Bhashyam, 1985). Starch properties in addition to aroma, determine the cooking and eating quality of the grain. The compound 2-acetyl-1-pyrroline was found to be the principal aroma constituent of cooked rice of eight diverse aromatic rices (Buttery et al., 1983). Consumers prefer whole-grain, translucent, milled rice (Unnevehr et al., 1985). Amylose content is the major determinant of cooked rice texture, with waxy or glutinous rice being soft and sticky and high-amylose rices, flaky and nonsticky (Juliano, 1985). Among nonwaxy rices of similar amylose content, intermediate gelatinization temperature and soft gel consistancy are related to softer cooked rice texture. Amoung waxy rices, cooked high-gelatinization temperature rices have harder texture than low-gelatinization temperature rices. As the principal protein source in tropical Asia, rice must at least maintain, or preferably increase, its protein content of 7%.

POTENTIAL TO REDUCE PRE-HARVEST SPROUTING

Incorporating dormancy into rice varieties reduces preharvest sprouting. Recent genetic studies by Seshu and Sorrells (1985) under controlled greenhouse conditions indicated the following information:

1. Seed (hull) dormancy in rice is dominant, particularly
   if caryopsis-coat dormancy of N22 is taken into account, in the crosses N22/Mahsuri, N22/IR19735-5-2, and
   IR13429-287-3/Mahsuri.
2. The F_2 segregations for the hull dormancy confirmed a
   3:1 monogenic ratio in the two crosses involving N22
   dormant cultivar as parent and a 9:7 digenic complementary ratio in the cross with IR13429-287-3 dormant line
   as parent. The distribution in all three crosses was
   suggestive of modifier genes. Segregations at 10 days
   after harvest in the N22 crosses had the confounding
   effect of caryopsis-coat dormancy, which is absent in
   IR13429-287-3.
3. Hull and caryopsis-coat dormancies may be independently
   controlled.
4. Reciprocal crosses between N22 and Mahsuri indicated
   that cytoplasmic factors are not involved in the control of dormancy.
5. Dormancy intensity and duration are considered to be
   either independently controlled or both regulated by
   modifier genes.

Seshu and Sorrells (1985) concluded from these studies that
results of some earlier genetic studies conducted under
natural field conditions suggesting multigenic control of
dormancy might have been confounded by the effects of environmental factors and caryopsis-coat dormancy.

Seed dormancy has advantages and disadvantages. It
prevents pre-harvest sprouting during rainy weather or when
the crop lodges in standing water. Too strong a dormancy
prevents the use of seed from a crop to raise the succeeding
crop without pretreatment. The dormant seeds have to be
heat-treated at 45-50°C for 3-4 days or treated with ox-
idizing agent or dilute acid to break dormancy (Ellis et
al., 1983).

High temperature and low humidity during grain ripening, particularly 10-20 days after flowering, appear to reduce the dormancy period (Takahashi, 1980). Thus, the dry
season crop is generally considered to be less dormant than
the wet season crop in tropical Asia and is more susceptible
to pre-harvest sprouting. The early season crop in Japan
which matures in summer readily germinates within 2-3 days
after lodging of the plant. With late season culture, pre-
harvest sprouting seldom occurs due to low air temperature
during grain ripening (Hoshikawa, 1975). Thus, pre-harvest

sprouting is an important problem for early- and medium-maturity rices in southern Japan particularly with the advance of transplanting time (T. Ito, Aichi Agric. Res. Center., (pers. comm.). Draining fields as the grain matures not only allows hardening of paddy fields in time for harvest but also prevents grain wetting in case of lodging. Making the variety lodging resistant through breeding should also minimize pre-harvest sprouting. Lee (1976) reported that in 10 days continuous rain, pre-harvest sprouting was 15% for nonlodged and 26-28% for lodged japonica rices.

The relationship between growth duration and pre-harvest sprouting is complex: negative (r = -0.45*, n = 98) for indicas and positive (r = +0.20, n = 84) for japonicas (Chen et al., 1980). Similar correlations were obtained between growth duration and dormancy period: r = 0.51** for indicas and r = -0.37** for japonicas. In crosses involving indica varieties, no genotypic association between late maturity and strong dormancy was observed (Chang and Yen, 1969; Chang and Tagumpay, 1973), although the two traits were reported to be frequently associated in rices of India and Sri Lanka. In these studies rices should be harvested at the same time to minimize environmental effects.

REFERENCES

Baun, L. C. (1972). Biochemical studies on dormancy of the rice grain. M.S. thesis, Univ. Philippines Los Banos. p. 64.

Buttery, R. G., Ling, L. C., Juliano, B. O., and Turnbaugh, J. G. (1983). Cooked rice aroma and 2-acetyl-1-pyrroline. J. Agric. Food Chem. 31, pp. 823-826.

Chang, T. T. and Tagumpay, O. (1973). Inheritance of grain dormancy in relation to growth duration in 10 rice crosses. SABRAO Newslett. 5(2), pp. 87-94.

Chang, T. T., and Yen, S. T. 1969. Inheritance of grain dormancy in four rice crosses. Bot. Bull Acad. Sin. 10(1). pp. 1-9.

Chen, L. C., Chang, W. L. and Chiu, F. T. 1980. Studies on the varietal differences of pre-harvest sprouting and its relationship with grain dormancy in rice. Nat. Sci. Counc. ROC Monthly 8(2), pp. 151-161 (in Chinese).

de Castro, R. E., Andales, S.C., and de Padua, D. B. 1980. Field trials of pre-drying handling systems. Proc. 3rd Ann. Workshop Grain Post-Harvest Technol., Kuala Lumpur, 1980, pp. 126-145.

Ellis, R. H., Hong, T. D., and Roberts, E. H. 1983.

40

Procedures for the safe removal of dormancy from rice
seed. Seed Sci. Technol. 11, pp. 77-112.

FAO, 1985. Paddy Production in the Asia-Pacific region:
Past performance and future prospects, Food and Agricultural Organization of the United Nations, Rome.
Paper IRC/85/4 presented at International Rice Commission, 16th Session, Los Banos, Laguna, Philippines.
p. 46.

Hayashi, M., and Hidaka, Y. 1979. Studies on dormancy and
germination of rice seeds. 8. Temperature treatment
effects on rice ssed dormancy and hull tissue degenera-
tion in rice seed during the ripening period and after
harvest. Bull. Fac. Agric. Kagoshima Univ. 29, pp.
21-32(in Japanese).

Hoshikawa, K. 1975. Illustrated growth of rice plant.
Tokyo: Nosan Gyoson Bunka Kyokai. pp. 294-295 (in
Japanese).

Ito, T., Ito, Y., and Okawa, K. 1980. Seed technological
studies on sprouted grains on the panicle of rice. I.
Germinability of the sprouted seeds. Res. Bull. Aichi
Agric. Res. Cent. 12, pp. 37-44 (in Japanese).

Ito, T., Ito, Y., and Okawa, K. 1981a. Seed technological
studies on sprouted grains on the panicle of rice. II.
Selection measures of sprouted grains from seed rice.
Res. Bull. Aichi Agric. Res. Cent. 13, pp. 54-61 (in
Japanese).

Ito, T., Ito, Y., and Okawa, K. 1981b. Seed technological
studies on sprouted grains on the panicle of rice.
III. Raising of seedling with seed rice containing
grains. Res. Bull. Aichi Agric. Res. Cent. 13, pp.
62-67 (in Japanese).

Juliano, B. O., (ed.) 1985. Rice: chemeisty and technology, 2nd ed. Am. Assoc. Cereal Chemists, St. Paul, MN.
p. 774.

Kunze, O. R. 1985. Effect of environmental factors and
variety on milling qualities of rice. In: Rice grain
quality and marketing. Int. Rice Res. Inst., Los
Banos, Laguna, Philippines. pp. 37-47.

Lee, L. F. 1976. Studies on the ear-germination and dormancy of japonica rice. Nat. Sci. Counc. ROC Monthly
4(2), pp. 42-48 (in Chinese).

Martires, G. A. 1975. A rapid method of determining percentage germination of rice seeds (Oryza sativa L.) and
the factors affecting the accuracy of the method. Ph.
D. dissertation. Univ. Santo Tomas, Manila, Philippines. p. 72.

Mendoza, M. R., Rigor, A. C., and Abital, F. L. 1982.

Grain quality deterioration in on-farm level of operation. Proc. 5th Ann. Grains Post-Harvest Technol. Workshop. Chiangmai, Thailand, 1982, pp. 101-117.

Mohamed, N. H. 1981. Rice grain losses from Kada Area (Malaysia): and overview vis-a-vis traditional storage systems. Proc. 4th Workshop Grains Post-Harvest Technol., Manila, 1981, pp. 115-131.

Nair, N. R. and Sahadwan, P. C. 1962. An instance of bran layer influencing seed dormancy in rice. Curr. Sci. 31, pp. 72-73.

Navasero, E. P., Baun, L. C. and Juliano, B. O. 1975. Grain dormancy, peroxidase activity and oxygen uptake in *Oryza sativa*. Phytochemistry. 14, pp. 1899-1902.

Seshu, D. V. and Sorrells, M. E. 1985. Genetic studies on seed dormancy in rice. Proc. Int. Rice Genet. Symp. Los Banos, 1985.

Srinivas, T. and Bhashyam, M. K. 1985. Effect of variety on milling quality of rice. In: Rice grain quality and marketing. Int. Rice Res. Inst., Los Banos, Laguna, Philippines. pp. 49-59.

Takahashi, N. 1980. Effect of environmental factors during seed formation on pre-harvest sprouting. Cereals Res. Comm. 8, (1) pp. 175-183.

Tang, W. T. and Chiang, S. M. 1955. Studies on the dormancy of rice seed. Mem. Coll. Agric. Nat. Taiwan Univ. 4, (1) pp. 1-7. (in Chinese).

Unnevehr, L. J., Juliano, B. O. and Perez, C. M. 1985. Consumer demand for rice-grain quality in Southeast Asia. In: Rice grain quality and marketing. Int. Rice Res. Inst. Los Banos, Laguna, Philippines. pp. 15-23.

Yoshida, S. and Hara, T. 1977. Effects of air temperature and light on grain filling of an indica and a japonica rice (*Oryza sativa* L.) under controlled conditions. Soil Sci. Plant Nutr. 23, pp. 93-107.

Section II

Breeding and Selection for Pre-Harvest Sprouting Tolerance

Breeding and Selection for Pre-Harvest Sprouting Resistance in Red Wheats

E. Czarnecki, Research Station,
Agriculture Canada, Winnipeg,
Manitoba, Canada.

INTRODUCTION

Sprouting damage in cereal crops can occur in many different parts of the world where the unharvested grain is
subjected to adverse weather (Mac Key 1976). In 1968 export
shipments of high grade Canadian hard red spring wheat to
Japan showed elevated levels of alpha-amylase activity.
This indicated a potential problem in commercial markets requiring sound, high quality grain and lead to pre-harvest
sprouting resistance being incorporated into the breeding
program at Winnipeg.

Testing methods.

Barley malting equipment was originally used to establish a screening program to identify potential sources of
good resistance. Early attemps at screening and backcrossing followed a steeping and germination process of intact heads similar to the procedure for barley malting.
Steeping at 11°C and germination at 13°C was successful in
separating susceptible and highly resistant cultivars.
Since 1978, a rain simulator similar to the type described
by McMaster and Derera (1976) has been used in assessing
heads for visual sprouting resistance and falling number
determinations.

Head samples of cultivars or lines are taken as they
mature (<25% moisture determined by seed hardness and
chlorophyll loss) along with 10cm or more of stem. Samples
consist of 10 heads for visual sprouting assessment (>15
heads for segregating lines) and 15-20 heads for falling
number tests. After indoor drying to 13.5% or less moisture

45

(about 5 days at 20°C) samples may be tested or stored at
-15°C to maintain seed dormancy until tests are undertaken
(Noll and Czarnecki 1982;   Mares 1983).   Samples should be
taken at the same stage of maturity (Weilenmann 1975;
Strand 1979) as immature or overipe material may result in
misclassification, especially in comparing selections taken
from segregating populations.   Under Manitoba conditions
late maturing cultivars often ripen under cooler fall temperatures and develop a long dormancy period.   Samples in
the rain simulator receive about 60-70mm of water during an
initial three-hour interval and are then maintained at 100%
relative humidity and a temperature of 15-17°C for a pre-
determined time.   Each year early testing of standard resistant and susceptible cutivars establishes the temperature
and length of treatment required for visual or falling number determinations.   The force of the overhead water spray
may pry the chaff open in shatter-prone material which is
useful in screening out this characteristic.

Sources of sprouting resistance in red wheats.

Between 1970 and 1975, wheats from many different
sources were tested and evaluated on the basis of sprouting
in the ear but only a small number showed promise.   Wheats
tested included collections from central Asia, historic and
recent cultivars representing many different countries.   The
best of these were RL4137, Chris, Frontana, Exchange,
Sommerweizen 8793, Chinese Spring and Park.   A number of
these had fair to moderate resistance but few equalled the
resistance of RL4137.   More recently, in co-operation with
Dr. John Noll, falling number determinations of rain simulator treated samples have established the relative dormancy
of RL4137 compared to other resistant cultivars (TABLE 1).
Qualitative tests also indicated that RL4137 maintained low
levels of alpha-amylase, a high falling number and amylograph viscosity compared to Neepawa (TABLE 2).   The well
adapted cultivar, RL4137 has therefore been used extensively
in backcrossing and in the pedigree breeding program at
Winnipeg.
The highly resistant cultivars RL4137 and Chris have
Frontana in their pedigree and the dormancy may have been
derived from attempts to transfer leaf rust resistance
(DePauw and McCaig, 1983).   Columbus (Neepawa*6/RL4137) however, inherited the leaf rust gene Lrl6 (Samborski and Dyck
1982) which has not been found in Frontana.   Frontana has
also been used as a source of sprouting resistance in Europe

(Olsson 1975). Mac Key (1975) indicated some wild diploid wheats possess dormancy adapted for short and long term survival.

TABLE 1. Relative comparison of falling number and pedigree of some red spring wheats resistant to sprouting.

Cultivar	Falling Number[1] Range (Seconds)	Pedigree
RL4137	400-450	RL2520//Thatcher*6/ Kenya Farmer
Columbus[2]	400-450	Neepawa*6/RL4137
Leader[3]	400-450	Fortuna/Chris
Chris	400-450	Frontana/3*Thatcher/ 3/Kenya58/New-thatch //2*Thatcher
Frontana	350-400	Fronteira/Mentana
Exchange	350-400	Warden/Hybrid English
Sommerweizen 8793	350-400	Unknown
Park	300-350	Mida/Cadet//Thatcher
RL2520	250-300	Frontana/3/McMurachy /Exchange//2*Redman
Thatcher	200-250	Marquis/Iumillo// Marquis/Kanred
Neepawa	150-200	RL4125/RL4008

[1]simulator tested.   [2]licensed in 1980.   [3]licenced in 1981.

TABLE 2. Alpha-amylase activity, falling number and amylograph viscosity on non-weathered and weathered samples of Neepawa and RL4137.

	Alpha-amylase activity[1]		Falling Number (seconds)		Amylograph Viscosity (Brabender)	
	NW[2]	W	NW	W	NW	W
Neepawa	5	17	545	359	720	450
RL4137	2	1	610	597	985	1020

[1]one amylase unit = 1mg glucose/100g grain.
[2]NW = non-weathered, W = weathered.

Inheritance of sprouting resistance.

Post-harvest dormancy has been reported to be closely associated with the genes for red grain colour with the most resistant cultivars having up to 3 loci acting in an additive manner (Gfeller and Svejda 1960; Freed et al., 1976). The Canadian cultivars RL4137, Park and Neepawa have 3 genes for seed-coat colour (Baker 1981) but show different levels of sprouting resistance (TABLE 1). Allelic differences at the red seed colour loci may be responsible for the variation in sprouting resistance of cultivars with equal numbers of genes. DePauw and McCaig (1983) reported that RL4137 had two or more genes controlling seed dormancy, one gene associated with seed colour and another one or more independent of seed colour. It is not known if the genes for dormancy and seedcoat colour are pleiotrophic, linked or whether one may be required for expression of the other. Gordon (1978) suggested a link between resistance to sprouting and a chemical precursor of the seed coat pigment. The maternal seedcoat and/or endosperm appear to be required for the full expression of dormancy and, depending on the parents used in crosses, the embryo component of seed dormancy may show partial dominance (Noll et al., 1982) or none at all (Mueller 1963). Estimates of heritability range from moderate (Piech et al., 1970; Gordon 1978) to relatively high values (Gfeller and Svejda, 1960). A recent study of the inheritance of pre-harvest sprouting resistance obtained heritability estimates from crosses RL4137/Timgalen and RL4137/Neepawa of 0.74 and 0.78, respectively (Dyck, P.L., Noll, J. S. and Czarnecki, E. unpublished data). Analysis of the cross RL4137/Neepawa suggested at least a two gene difference independent of red seed colour. King (1984) found cultivars differed widely in water uptake by the grain with awns accounting for some of the differences in ear water uptake (King and Richards 1984). Interestingly, some sources of pre-harvest sprouting resistance (RL4137, Chris, Columbus and Leader) are significantly more susceptible to common root rot (Cochliobulus sativus) than is the cultivar Neepawa. Columbus (Neepawa*6/RL4137) must have acquired its susceptibility from the donor parent RL4137.

Breeding and selection.

Both the backcross and pedigree methods of breeding for pre-harvest sprouting resistance are used at the Winnipeg Research Station. A bulk method of screening has been applied in Europe (Svensson 1975).

The backcrossing procedure to transfer RL4137 sprouting resistance to Neepawa was started in 1969 and took about 2 years. The recurrent parent Neepawa was used as the female parent in all backcrosses making it possible to eliminate any selfs during testing. Sprouting tests of the pollen sources used for backcrossing was started after the second backcross and was continued for the remaining backcrosses. Threshed seeds were tested in petri dishes at 20°C and only crossed seed orginating from the most resistant backcross F_1 plants were used for the next backcross. After the second backcross, at least 25 or more plants (representing five or more crosses) were grown at each stage of backcrossing. Each plant was backcrossed to the recurrent parent. Germination tests on seeds from backcross F_1 plants were compared to the recurrent and donor parents. According to Noll et al., (1982) the crossed F_1 seeds may express dominance and early germinating ones may be eliminated, however this technique was not applied.

After backcrossing was completed, an F_2 selection from the most resistant cross was increased and F_3 lines planted in the field nursery. The F_3 through to the F_5 were tested visually for sprouting resistance with the most resistant selections being retained (limited alpha-amylase tests were conducted on F_3 lines). Increasing was done in the F_5 generation. The first year yield test showed considerable genetic variation for different agronomic and quality traits indicating that probably several genes for dormancy were involved in the transfer (TABLE 3). Nineteen lines were reduced to 7 for the "B" test and only 2 entered the Coop (3 year prelicensing trials) before resulting in the release of Columbus (BW37) in 1980.

Similar programs to transfer improved sprouting resistance were started with Glenlea, a utility class wheat, Chester, a sawfly resistant cultivar, Benito, an early maturing wheat, and Katepwa, a recently licensed cultivar. RL4137 was used as a donor source for the Glenlea, Chester and Katepwa recurrent parents and Columbus and Takahe for the Benito backcrossing.

The procedures used in the Glenlea and Chester backcrosses were similar to those used for Columbus except that unthreshed rather than threshed heads from each donor male parent were tested for visual sprouting resistance. Upon completion of backcrossing, the F_3 to F_5 generations were initially screened by visual sprouting. Then, the most resistant lines were further tested for falling number in the rain simulator. Increasing again occured in F_5.

TABLE 3. Variability of Neepawa*6/RL4137 backcross lines.[1]

Characteristic	Neepawa	Neepawa*6/RL4137 range	LSD.05
Yield, kg/ha	39.2	29.3 - 39.4	3.9
Maturity, days	93.5	93.7 - 97.9	2.1
Lodging, 1-9	1.6	1.5 - 2.5	0.5
Height, cm	83	86 - 94	4.0
Sprouting %	32.0	4 - 28	
Leaf rust reaction	MS	R-MS	
Stem rust reaction	TR	TM-MS	
Loose smut reaction	R	R-MS	

[1]19 lines.

TABLE 4. Field weathered falling number analysis of variance on Glenlea backcross lines.

Source of variation	DF	SS	MS	F
Glenlea vs. backcrosses	1	48001.3	48001.3	14.40**
Columbus vs. backcrosses	1	18249.9	18249.9	5.48*
Among backcrosses	16	29441.0	1840.0	<1 NS
Error	18	59992.2	3332.9	

*, ** significant at the 5 and 1% levels.
NS, not significant.

In 1982 the best of the Glenlea and Chester backcross lines were assessed by field weathered and simulator derived falling numbers. The Glenlea backcrosses showed a signficant difference over the recurrent parent but were not equal to Columbus (TABLE 4).

The attempt to improve Chester by backcrossing was not successful. Rather than visual screening for sprouting resistance during backcrossing, improvement of Chester may be possible by testing for alpha-amylase activity (falling number tests require a larger sample size than is available from individual plants).

In the Benito back crossing, F_5 lines from the fourth backcross were increased in 1985. Falling number tests on Benito*4/Columbus and Benito*3/Columbus lines have confirmed that early maturing selections are similar to Columbus in pre-harvest sprouting resistance. Takahe showed a high level of visual sprouting resistance and had been used in backcrossing however falling number values were lower in comparison to Columbus or RL4137. The Katepwa backcrossing program is not completed.

Pedigree screening of F_2 rust resistant selections was started in 1976 and is now conducted routinely on most crosses. Some crosses are screened lightly to eliminate the most susceptible plants. The lightly screened ones do not have a highly resistant parent. All tests include Neepawa, RL4137 and Columbus as standards. In some years up to 8000 heads from rust resistant plants have been tested. Scoring has been on the basis of the intact spike with further selection made on threshed grain however recently most selections are done on unthreshed spikes. No further selection is made until the F_7 increases, which are then tested visually.

TABLE 5. Comparison of falling number on pedigree material in 1978 and 1985[1].

1978 Cultivar Key No.	Falling Number (seconds)	1985 Cultivar Key No.	Falling Number (seconds)
Neepawa	405	Neepawa	235
Columbus	438	Columbus	386
39	137	90	239
40	101	92	117
41	97	96	349
42	153	97	207
44	286	98	359
47	268	100	268
48	394	101	300
Mean (n=20)	285	Mean (n=20)	289
LSD .05	138	LSD .05	109

[1]Simulator tested.

All entries are evaluated for falling number (2 reps) in the final 4 years of yield testing. The most advanced yield tests are further assessed by amylograph viscosity and alpha-amylase activity (Canadian Grain Commission GRL) on untreated composited samples representing many locations. The screening of pedigree material has resulted in a marked improvement of lines entering yield tests (TABLE 5).

Breeding and selection for pre-harvest sprouting resistance is an on-going program and the release of Columbus has received a positive response both from the producer and industry. At 20.7% of the seeded area, Columbus ranked second to Neepawa (41.1%) and was planted on 2.3 million hectares in 1985. Columbus is taller, later maturing and more susceptible to common root rot than Neepawa, thus, a short, early maturing, high yielding cultivar with good quality, disease and sprouting resistance, is still highly desirable.

REFERENCES

Baker, R. J. 1981. Inheritance of seedcoat colour in eight spring wheat cultivars. Can. J. Plant Sci. 61, pp. 719-721.

de Pauw, R. M. and McCaig, T. N. 1983. Recombining dormancy from R14137 with white seed colour. In: Proc. Third Int. Symp. on Pre-harvest Sprouting in Cereals. Kruger, J. E. and LaBerge, D. E. (eds.). Westview Press, Boulder, Co. USA. pp. 251-259.

Freed, R. D., Emerson, E. H., Ringlund, K. and Gullord, M. 1976. Seedcoat colour in wheat and the relationship to seed dormancy at maturity. Cereal. Res. Commun. 4, pp. 147-149.

Gfeller, F. and Svejda, F. 1960. Inheritance of post-harvest seed dormancy and kernel colour in spring wheat lines. Can. J. Plant Sci. 40, pp. 1-6.

Gordon, I. L. 1978. Selection against sprouting damage in wheat. A synopsis. Proc. 5th Int. Wheat Gen. Symp. pp. 954-962.

King, R. W. 1984. Water uptake in relation to pre-harvest sprouting damage in wheat: Grain characteristics. Aust. J. Agric. Res. 35, pp. 337-345.

King, R. W. and Richards, R. A. 1984. Water uptake in relation to pre-harvest sprouting damage in wheat: Ear characteristics. Aust. J. Agric. Res. 35, pp. 327-336.

Mac Key, J. 1975. Seed dormancy in nature and agriculture. Cereal Res. Commun. 4, pp. 83-91.

Mares, D. J. 1983. Preservation of dormancy in freshly harvested wheat grain. Aust. J. Agric. Res. 34,

pp. 33-38.

McMaster, G. J. and Derera, N. F. 1976. Methodology and sample preparation when screening for sprouting damage in cereals. Cereal Res. Commun. 4, pp. 251-254.

Mueller, H. N. 1963. About breeding of sprouting resistant cereals. Genet. Today. 1, p. 238.

Noll, J. S. and Czarnecki, E. 1979. Methods of extending the testing period for harvest-time dormancy in wheat. Cereal Res. Commun. 8, pp. 233-238.

Noll, J. S., Dyck, P. L. and Czarnecki, E. 1982. Expression of RL4137 type of dormancy in Fl seeds of reciprocal crosses in common wheat. Can. J. Plant Sci. 62, pp. 345-349.

Olsson, G. 1975. Breeding for sprouting resistance in wheat. Proc. 2nd Int. Winter Wheat Conference. pp. 108-113.

Piech, J., Ruszkowski, M. and Jaworska, K. 1970. Inheritance of the seed dormancy stage duration in winter wheat (Triticum aestivum L.). Genet. Pol. 11, pp. 227-240.

Samborski, D. J. and Dyck, P. L. 1982. Enhancement of resistance to Puccinia recondita by interactions of resistance genes in wheat. Can. J. Plant Pathol. 4, pp. 152-156.

Strand, E. 1979. A seed dormancy index for selection of cultivars of cereals resistant to pre-harvest sprouting. Cereal Res. Commun. 8, pp. 219-223.

Svensson, G. 1975. Screening methods for sprouting resistance in wheat. Cereal Res. Commun. 4, pp. 263-266.

Recovery of Sprouting Resistance from Red-Kernelled Wheats in White-Kernelled Segregates

R. M. De Pauw and T. N. McCaig,
Research Branch, Agriculture Canada,
Swift Current, Saskatchewan, Canada,
S9H 3X2.

Wheat with white kernel coat colour has traditionally been considered to have a short dormancy period and therefore to be susceptible to sprouting in intact spikes prior to threshing. Reports of a positive relationship between red kernel coat colour and resistance to sprouting include Gfeller and Svedja (1960); Khan and Strand (1977); Wellington and Durham (1958).

Recent research, perhaps stimulated by the International Symposia on Pre-harvest Sprouting in Cereals, has provided evidence that there are exceptions to this generalized relationship. Gordon (1983) detected only a moderate relationship between kernel coat colour and resistance to sprouting in 97 genotypes of several Triticum species. Clarke et al., (1984) studies weathering damage, as reflected by changes in Hagberg falling number value in 14 cultivars. Because the mean falling number for the weathering resistant white kernel coat class exceeded that of the weathering susceptible red kernel coat class in both years, they concluded that a range of genotypic expression of weathering resistance existed within both red and white wheats.

Reitan (1980) analyzed an 8 x 8 diallel for kernel coat colour and dormancy and detected two mechanisms controlling dormancy, one associated with and one not associated with kernel coat colour. De Pauw and McCaig (1984) provided evidence that RL4137, a red-kernelled wheat with a long stable dormancy period, had two or more mechanisms controlling resistance to sprouting in intact spikes. One of the mechanims was not associated with kernel colour. They produced Losprout, a sprouting resistant white-kernelled wheat, which derived its sprouting resistance from RL4137 (De Pauw et al., 1985).

54

This paper reports on the techniques used to screen for kernel colour and sprouting resistance, and to verify the recovery of sprouting resistance.

MATERIALS AND METHODS

Five parental genotypes were used to make seven hybrid populations within which selection was exercised to recover white-kernelled segregates with resistance to sprouting (SR) of kernels in intact spikes (TABLE 1). Classification of kernel colour was based on the reaction of kernel pigment to sodium hydroxide (De Pauw and McCaig 1983).

TABLE 1. Resistance of kernels in intact spikes to sprouting, length of sprouting resistance period of five genotypes with 0-3 genes for red kernel colour, and the hybrid populations studied.

Genotype	No. of genes for red colour	SR of kernels in intact spikes	Length of SR period
RL4137	3	very good	very good
HY320	2	moderate	moderate
Takahe	1	-	-
Kenya 321BT.1.B1	0	moderate	moderate
7722	0	poor	short

Hybrid populations studied: 7722/RL4137; 7722*3/RL4137; 7722/3*RL4137; HY320/K321; HY320*3/K321; K321/Takahe.

Experiment I. Assessment of techniques to identify sprouting resistance.

Twenty-three genotypes differing in number of genes for kernel colour, maturity, sprouting resistance and alpha-amylase activity were studied to evaluate the effectiveness of using evidence of visible sprouting of intact spikes to select for sprouting resistance. The genotypes were grown in plots four rows wide and 3m long that were replicated four times. Days to heading were scored and used as a preliminary indicator of relative maturity. When genotypes reached about 18% moisture content (T1), a sample of ten random healthy primary spikes was collected and stored at -23°C. Seven days later (T2) another sample of ten random

healthy spikes was collected. Spikes from both times of
sampling were placed in a controlled environment cabinet and
subjected to 135mm of simulated rain over a 3-hour period.
Each day an additional 15mm of water was misted onto the
spikes. After seven days at 20°C and about 100% relative
humidity, the number of spikes with visible sprouts per sample were counted (TABLE 2). Each sample was dried down at
50°C and threshed by hand. The percentage of kernels which
germinated was determined. Alpha-amylase activity was measured using a gel-diffusion technique described by Hejgaard
and Gibbons (1979). Alpha-amylase activity was also measured on non-weathered grain from the field using both gel-
diffusion and Hagberg falling number techniques. The data
were subjected to an angular transformation, where appropriate, prior to analysis of variance. Regression and correlation analysis was performed on untransformed data.

TABLE 2. Glossary of abbreviations.

Abbreviation	Variable
HDAT	days to heading
T1HS and T2HS	no. of spikes sprouted/sample from T1 and T2.
T1GM and T2GM	% germination from samples T1 and T2.
T1WD and T2WD	diameter of alpha-amylase ring from samples T1 and T2.
T1T2	average No. of spikes sprouted from T1 and T2.
NGWD	alpha-amylase activity on non-weathered grain sample.
FNO	Hagberg falling number values for non-weathered sample.

Experiment II. Screening for white-kernelled sprouting
resistant segregates.

Selection for white-kernelled sprouting resistant
segregates was exercised using kernel pigment reaction to
sodium hydroxide and evidence of visible sprouting in intact
spikes for four cycles in all populations (TABLE 1) except
7722/3* RL4137, because effort was devoted to identifying
white-kernelled segregates for backcrossing to RL4137. In
1985 86 lines, parents, and check cultivars were grown in
plots four rows wide and 3m long. The procedure for

sampling spikes and response to simulated rain was similar
to that for Experiment I.

RESULTS AND DISCUSSION

Selecting White-kernelled Segregates from Red by White
Hybrids

An unequivocal determination of kernel coat colour is
required to select for a recombination of SR from red-
kernelled wheats in white-kernelled segregates. Within a
"red" by "white" population red kernel colour has been
demonstrated to have a positive effect on SR (De Pauw and
McCaig 1983; Gjeller and Svedja 1960; Khan and Strand
1977). RL4137 has been demonstrated to have at least two
mechanisms controlling SR, one associated with colour and
one not associated with colour (De Pauw and McCaig 1984).
Kernel colour is controlled by up to three recessive
genes. White-kernelled segregates occur at a low frequency
(TABLE 3). The frequency at which they are recovered can
be increased by identifying those types which have a low
frequency of genes for red kernel colour. Since the genes
will be in either a homozygous or heterozygous condition,
some white-kernelled types will be recovered the following
generation from the heterozygotes.

TABLE 3. Comparison of segregation patterns for two parents
differing by 1-3 gene pairs for kernel colour.

No. of gene pairs	F2 (W/R)			BC1F1 (W*2/R)			Random inbreeding (W/R)		
	Homo white	Seg	Homo red	Homo white	Seg	Homo red	Homo white	Seg	Homo red
1	1	2	1	1	1	0	1	0	1
2	1	8	7	1	3	0	1	0	3
3	1	26	37	1	7	0	1	0	7

De Pauw and McCaig (1983) demonstrated that the intensity of kernel colour to sodium hydroxide was influenced
more by the number of genes for kernel colour than by environmental factors.

Experiment I. Assessment of Techniques for Identifying
Sprouting Resistance

Analysis of variance detected highly significant differences among genotypes for each of the ten variables
(TABLE 4). The regression of each of the nine variables as
a function of HDAT was non-significant (TABLE 5). The correlation values were low negative and not significant (data
not shown) indicating that maturity differences could influence each variable. Maturity differences, however, were
adequately accounted for by using days to head as a preliminary indicator of relative maturity and then making
the final determination of sampling date by estimation of
kernel moisture content.

TABLE 4. Analysis of variance for ten variables measured
on 23 genotypes.

	F-value		F-value		F-value
HDAT	*				
T1HS	*	T2HS	*	T1T2	*
T1GM	*	T2GM	*	FNO	*
T1WD	*	T2WD	*	NGWD	*

* mean squares significant at p = 0.01.

TABLE 5. Regression of row variable as a function of column
variable.

| | HDAT | T1GM | T2HS | T2WD | NGWD | |
	T1HS	T1WD	T2GM	T1T2	FNO

HDAT
T1HS   ns
T1GM   sig
T1WD   sig
T2HS   sig
T2GM   sig
T2WD   sig
T1T2   sig
NGWD   ns
FNO   ns

Neither measure of alpha-amylase activity (NGWD and FNO) based on non-weathered grain samples from the field were related to the variables used to measure sprouting resistance and alpha-amylase activity on samples subjected to a simulated rain (TABLE 5). The simulated rain treatment, therefore, prompted the genotypes to respond to the favourable germination conditions. The differential response to the favourable germination conditions was used to determine the sprouting response of the genotypes.

The number of spikes with visible evidence of sprouting per sample of ten spikes, percentage of kernels germinated, and diameter of the alpha-amylase ring at both T1 and T2 were significantly related to one another (TABLE 5). Larger coefficients of determination obtained for the regression of T1GM, T1WD, and T1T2 as a function of T1HS than to T2 variables (74%-88% vs. 19%-62%). Similarly larger coefficients of determination obtained for the regression of T2GM, T2WD, and T1T2 as a function of T2HS than to T1 variables (59%-76% vs. 29%-40%).

Visible evidence of sprouting in intact spikes related quite well to the percentage of germinated kernels and alpha-amylase activity. Counting spikes with visible evidence of sprouting requires about 1/100 of the time to determine percentage germination or alpha-amylase activity. Using evidence of visible sprouting on a sample of 10 spikes per line, permits a rapid effective separation of a population into sprouting susceptible and sprouting resistant categories.

Experiment II. Screening for White-kernelled Sprouting Resistant Segregates

At T1 the mean number of spikes with visible evidence of sprouting was smaller for the white-kernelled checks than that for the red-kernelled checks (TABLE 6). At T2, however, the mean for the red-kernelled checks was smaller than that for white-kernelled checks. The range for number of spikes sprouted/sample at both T1 and T2 was larger for the red-kernelled checks than the white-kernelled checks.

At T1 the mean for each population in which RL4137 was used as the source of SR was lower than the T1 mean for both the white-kernelled checks and red-kernelled checks. The T1 "Lo-values" for these populations were comparable to the T1 "Lo-values" for the white-kernelled checks. Also, the T1 "Hi-values" were less than the T1 "Hi-values" for both white-kernelled checks and red-kernelled checks.

TABLE 6. Mean number of spikes sprouted/sample for white-kernelled checks, red-kernelled checks, and for six populations at T1 and T2.

	No. of families	No. of lines	T1 $\bar{x}$	T1 Lo+	T1 Hi	T2 $\bar{x}$	T2 Lo	T2 Hi
7722/ RL4137	19	32	3.7	2.0	6.0	7.7	6.0	10
7722*3/ RL4137	7	21	5.8	2.6	8.3	9.0	7.3	10
7722/ 3*RL4137	-	3	0.7	0	2	6.3	4	8
HY320/ K321	1	6	1.8	0	4	3.7	0	7
HY320*3/ K321	4	12	7.4	6.3	8.5	10.0	10.0	10.0
K321/ Takahe	4	12	1.8	0.5	4.0	3.7	2.8	5.5
Red-kernelled checks	5[+]		8.0	1.0	10.0	8.0	0	10.0
White-kernelled checks	7[+]		6.7	2.0	10.0	8.7	4	10.0

+ Lo and Hi values represent the mean of 2-5 lines within a family or the mean of a check cultivar within a colour-type.

The T2 mean for two populations with RL4137 was equal to or less than the mean for both colour types of checks. The T2 "Lo-values" for all three populations with RL4137 were less than the T2 "Hi-values" for red-kernelled checks. Therefore, some white-kernelled SR segregates were recovered with better SR than some red-kernelled checks in all three populations involving RL4137. The best level of SR was expressed in a population which had been backcrossed twice to RL4137. This would lead one to tentativly suggest that the SR mechanism that is independent of kernel colour in RL4137 might be controlled by more than one gene.

The T1 mean number of spikes sprouted/sample for two populations involving K321 was less than the mean for both colour types of checks. The T1 "Lo-value" for two populations was less than the T1 "Lo-value" for the checks. Also, the T1 "Hi-value" for three populations was less than the T1 "Hi-value" for the checks. The T2 mean for two populations

was less than the T2 mean for both colour types of checks.
The T2 "Hi-value" for three populations was less than the T2
"Hi-value" for both sets of checks.

White-kernelled SR segregates were recovered in two
populations involving K321. The response to sprouting at
both T1 and T2 indicate that the lines from the population
HY320*3/K321 had a low level of SR which lasted a short
period. Recovering white-kernelled sprouting resistant
segregates using a partial backcross strategy with RL4137 as
a donor was more successful than transferring both white-
kernel colour and SR with K321 as a donor.

The white-kernelled checks included Losprout, a sib to
Losprout and K321 whose SR was comparable to some red-
kernelled types (TABLE 7). The inclusion of three white-
kernelled SR checks accounted for the mean performance of
the white-kernelled checks being comparable to the red-
kernelled checks.

Verifying the Recovery of White-kernelled Sprouting Resistant Segregates

The mean number of spikes sprouted per ten spike sample
at T1 for all genotypes, except RL4137, varied from year to
year (Table 7). Most genotypes expressed a short SR period.
RL4137, however, consistently expressed a high level of SR
and a long SR period. Losprout, derived from the cross
7722/RL4137, expressed a higher level of SR and a longer SR
period than other genotypes except RL4137.

TABLE 7. Mean number of spikes sprouted per sample of ten
spikes for seven genotypes at two times of sampling in each
of three years.

Genotype	No. of genes for red colour	n	Mean no. of spikes sprouted/sample							
			1983			1984			1985	
			T1	T2	n	T1	T2	n	T1	T2
Losprout	0	4	3.5	4.5	4	7.5	9.8	4	0	1.0
7722	0	6	10.0	9.8	4	8.5	10.0	4	2.0	7.8
K321	0	7	4.4	8.6	4	7.3	9.8	4	0	2.8
Pitic	1	10	9.9	9.9	4	6.8	10.0	4	0.5	3.8
Glenlea	2	8	5.3	9.3	4	3.0	10.0	4	0.3	6.8
Neepawa	3	9	4.2	9.6	4	2.3	9.8	4	3.0	9.0
RL4137	3	9	0.2	0.3	4	0	0	4	0	0

62

In 1984 both Losprout and K321 expressed a poorer level
of SR than in 1983 and 1985. The primary weather-variable
that differed among years was temperature, which was above
normal during the ripening phase. Belderok (1961) and
Takahashi (1980) both reported that temperature can affect
the development of dormancy.

Obviously in 1984 the response to selection for SR
would have been less effective because of a reduced level
of phenotypic variation for SR which points to the need for
using multi-environments to verify detection of SR. Sampling at two times permits selection of genotypes which have
a long SR period. In the early generations, however, this
might be difficult given the large populations that are
being evaluated for many traits. It may be more efficient
to delay sampling to perhaps a few days after 18% moisture
and using a single time of sampling. After several cycles
of selection for SR, elite inbred lines could be evaluated
in greater detail by using two sampling times.

REFERENCES

Belderok, B. 1961. Studies on dormancy in wheat. Proc.
    Int. Seed Test. Assoc. 26, pp. 697-760.
Clarke, J. M., Christensen, J. V. and De Pauw, R. M. 1984.
    Effect of weathering on falling numbers of standing and
    windrowed wheat. Can. J. Plant Sci. 64, pp. 457-463.
De Pauw, R. M. and McCaig, T. N. 1983. Recombining dormancy
    and white seed colour in a spring wheat cross. Can. J.
    Plant Sci. 63, pp.581-589.
De Pauw, R. M. and McCaig, T. N. 1983. Evidence for a
    genetic mechanism controlling seed dormancy independent
    of seed colour. In: Proc. 6th Int. Wheat Genetics
    Symp. Kyoto, Japan. 1984, pp. 629-633.
De Pauw, R. M., McCaig, T. N. and Townley-Smith, T. F. 1985.
    Registration of a sprouting resistant white-seeded
    spring wheat germplasm line. Crop Sci. 25, pp.
    577-578.
Gfeller, F. and Svejda, F. 1960. Inheritance of post-
    harvest seed dormancy and kernel colour in spring wheat
    lines. Can. J. Plant Sci. 40, pp. 1-6.
Gordon, I. L. 1983. Sprouting variability in diverse
    Triticum spp. germplasm. In: Proc. 3rd Int. Symp. on
    Pre-harvest Sprouting in Cereals. Kruger, J. E. and
    LaBerge, D. E. (eds.). Westview Press, Boulder, Co.
    USA. pp. 221-230.

Hejgaard, J. and Gibbons, G. C. 1979. Screening for alpha-
amylase in cereals. Improved gel-diffusion assay using
dye-labelled starch substrate. Carlsberg Res. Commun.
44, pp, 21-25.

Khan, F. N. and Strand, E. A. 1977. Investigations into the
genetics of kernel colour and dormancy in wheat (Triticum aestivum L.). Meld. Nor. Landbrukshogk 56, pp.
1-12.

Reitan, L. 1980. Genetical aspects of seed dormancy in
wheat related to seed coat colour in an 8 x 8 diallel
cross. Cereal Res. Com. 8, pp. 275-282.

Takahashi, N. 1980. Effects of the environmental factors
during seed formation on the pre-harvest sprouting.
Cereal Res. Com. 8, pp. 175-183.

Wellington, P. S. and Durham, V. M. 1958. Varietal differences in the tendency of wheat to sprout in the ear.
Emp. J. Exp. Agric. 26, pp. 47-54.

Pre-Harvest Sprouting Tolerance in White Grained Wheat

D. J. Mares, The University of Sydney,
Plant Breeding Institute, Narrabri,
NSW., Australia.

INTRODUCTION

Australia, as a wheat producing nation, is unique in that it traditionally produces only white-grained wheat and exports a large proportion of the annual crop. Interest in white-grained wheat has increased significantly in recent years following a recognition of some of its inherent advantages, in particular milling quality, over red wheat in the international market place. Until recently it was generally accepted that white grained cultivars were all susceptible to pre-harvest sprouting and that this rep-resented a considerable constraint to their successful cul-tivation in many wheats growing regions of the world. Con-sequently the increased interest in white wheats has been paralleled by the establishment of research programs at several centres throughout Canada and USA aimed at improving sprouting tolerance. Similar programs have been in progress for many years in Northern NSW, Australia, and more recently Queensland, where weather damage or pre-harvest sprouting damage affects, on-average, 15-20% of the wheat harvest annually.

In this review a brief historical account of the changes which have occured in the tolerance to sprouting of Australian wheats will be presented leading up to the estab-lishment of sprouting tolerance screening programs, the release of varieties deliberately selected for their toler-ance, and finally, the identification of new and better sources of sprouting tolerance in white-seeded genotypes. In addition these new sources will be compared with existing varieties and sprouting tolerant red wheats and the mech-anisms of tolerance and durability or stability of tolerance will be discussed. The importance of mechanisms other than

grain dormancy will be examined with special reference to
the effects of environmental factors, particularly temperature and precipitation, or sprouting tolerance.

MATERIAL AND METHODS

Seed used in this study was obtained from the Australian Winter Cereals Collection and from individual breeders
through the agency of Interstate Variety Trials and Northern
Cooperative Trials. Field trials were conducted at Narrabri
in Northern NSW. Trials were sown in early June (early
winter) and normally matured in October/November (late
spring/summer). Average maximum temperature for October and
November were approximately 27°C and 32°C respectively.
1. Assessment of Sprouting Tolerance

At eosin ripeness (no further movement of water, labelled with the red dye eosin, into the upper stem or ear) ears
were harvested at random from field plots and stored either
at ambient temperature under cover prior to testing or in
deep freeze storage (Noll and Czarnecki, 1979; Mares,
1983a). Tolerance to pre-harvest sprouting was assessed
using an artificial rain simulation facility maintained at
20°C and high relative humidity (McMaster and Derera, 1976).

(a) Visual Sprouting. Ten ears per variety were treated
    with 50mm of rain and maintained in the rain simulator
    until all ears had sprouted. Ears were examined at
    daily intervals for visual evidence of sprouting and
    discarded when rootlets protruded from one or more
    spikelets. The standard varieties SUN 56A, Songlen,
    Cook, Kite, Suneca, Kenya 321 and RL 4317 were used to
    establish a scale of increasing tolerance to pre-
    harvest sprouting.

(b) Falling number. Thirty ears per variety were subjected
    to a standard weather treatment (50mm of rain in the
    first 2h followed by 58h of high relative humidity at
    20°C) followed by air-drying at 40°C. Sprout damage
    was estimated using the Falling Number method or a
    direct assay of alpha-amylase activity (Barnes and
    Blakeney, 1974).

2. Grain Germination

Grain was separated from mature ears by hand and stored

in a deep freeze at -20°C until required. Germination tests were conducted over a range of temperatures between 5°C and 30°C using duplicate samples of 50 grains on filter paper moistened with 4ml deionized water as previously described (Mares, 1984). Percent germination was recorded at daily intervals for seven days.

3. Movement of water into ears and grains

Intact ears were treated in the artificial rain simulator and samples taken at intervals over the first 30h, prior to the onset of germination. Ear and grain moisture contents were determined as previously described (Mares, 1983b) and expressed as percent dry weight.

RESULTS AND DISCUSSION

1. Comparison of current NSW wheat varieties, older Australian varieties and advanced breeding lines

The relative tolerance to pre-harvest sprouting of a number of white wheats compared with the tolerant red wheat, RL 4237, is illustrated in FIGURE 1. Many early Australian wheat varieties were found to be similar to Kite and Kenya 321 with respect to sprouting tolerance. More recently developed varieties, up to 1980, which form the bulk of current commercial cultures in Australia tended to be less tolerant, ranging from Kite to Songlen with the majority being similar to Cook. This trend towards greater susceptibility is further substantiated by a comparison of recent advanced breeding lines originating from programs where there has been no selection for sprouting tolerance. Of the 45 lines in this category which were examined in 1983 and 1984, over 75% were equal to or more susceptible than songlen and in general, the remainder, were only similar to Cook. The reasons for this continuing decline in sprouting tolerance are not clear although rapid, mechanized plot harvesting and the use of several generations per year to speed up variety development are seen as practices which either eliminate natural selection or disadvantage segregates with reasonable levels of dormancy at maturity. The lack of sprouting tolerance in advanced breeding lines and in many introductions from international nurseries poses considerable problems for Australian breeders facing increasing pressure for higher yields and better disease resistance.

In stark contrast to this trend outlined above are two varieties released by The University of Sydney. These lines were developed from crosses involving one sprouting tolerant parent by subsequent selection within the populations for tolerant segregates. Suneca, derived from Spica, and Sunelg, derived from Kite, (FIGURE 1) retain the tolerance of the sprouting tolerant parent in high yielding, disease resistant, good quality backgrounds. The release of these varieties was important, not only because these varieties filled the immediate needs of local wheat growers, but also because it clearly demonstrated that sprouting tolerance could be transferred by breeding into varieties which were acceptable to both growers and grain marketing authorities.

FIGURE 1. Relative pre-harvest sprouting tolerance of some white-grained wheats compared with the tolerant red wheat, RL 4137.

Standards	Recent varieties and breeding lines	New sources of tolerance
RL 4137		
		Aus 1408, Aus 1490
		Saberbeg, Hellas
		Sth. African 1166, Homera, Chile 59
		Kenya 59
		Mexican 852, Pakistan C228
		Egypt 1167-109
Kenya 321		Bihar 124, Spoetnik, Lerma 52
	Suneca**	
Kite* Spica**	Sunelg*	
Cook		
Songlen	Most advanced breeding lines 1983, 1984,	
Sun 56A	and new varieties.	

(left margin: ↑ Increasing Tolerance)

2. New, white-seeded sources of tolerance to pre-harvest sprouting

In an attempt to identify better sources of sprouting tolerance, over 3000 white-grained bread wheats from the Australian Winter Cereals Collection were screened over a period of three years. Approximately 20 accessions were recovered with tolerance equal to, or better than, Kite.

Subsequently, three years of intensive evaluation have highlighted 14 apparently distinct genotypes (TABLE 1) with
sprouting tolerance consistently equal to or better than
Kite and Kenya 321 (FIGURE 1). The best of these lines
approach the level of tolerance characteristic of RL 4137
grown under environmental conditions typical of Northern NSW.

TABLE 1. New sources of tolerance to pre-harvest sprouting.

Aus 109 - Chile 59	Aus 1408 - (From) Transvaal
Aus 320 - Hellas	Aus 1490 - Unknown
Aus 471 - Lerma 52	Aus 5820 - Kenya 59
Aus 633 - Mexican 852	Aus 7122 - Bihar 124
Aus 758 - Pakistan C228	Aus 13054 - Egypt 1167-109
Aus 1293 - South African 1166	Aus 16115 - Saberbeg
Aus 1305 - Spoetnik	Aus 16116 - Homera

(Aus number = Australian Winter Cereals Collection code
number).

3. Analysis of new sources of tolerance

Sprouting tolerant varieties, grown in field trials in
1983, 1984 and 1985, were compared with the range of standards described earlier. With respect to the time required
for mature ears to sprout, the new lines were ranked between
Suneca and RL 4137 (FIGURE 2). Similarly when falling numbers or alpha-amylase activities were used as a measure of
the weather damage resulting from a standard weather treatment, applied to samples of ears 15 and 25 days after eosin
 ripeness, the superiority of the new lines over existing
white grained varieties was clearly demonstrated (TABLE 2).
With the exception of Lerma 52 all of the sprouting
tolerant lines had control falling numbers in the range 350-
500sec. By contrast Lerma 52 was characterized by a low
falling number (240sec) and a high alpha-amylase activity
($84mUg^{-1}$) at maturity in the absence of rain. The artifical
weather treatment, however, did not induce any marked change
in these parameters. Low falling numbers in apparently
sound, mature grain do not necessarily preclude the use of
such varieties in a variety improvement program. For
example, the sprouting tolerance of Suneca was derived from
Spica which normally has a low falling number (200sec) and a
high alpha-amylase activity at maturity. This indicates
that the factors which control grain germinability and
sprouting in the ear can be separated from those which condition low falling numbers and high alpha-amylase activity
in mature grain of particular varieties in the absence of
sprouting.

FIGURE 2. Visual sprouting in heads of new sources of tolerance (hatched area) compared with Sun 56A (●) (very susceptible), Suneca (▲) (moderately tolerant) and RL 4137 (✦) (tolerant). Samples were grown at Narrabri, NSW in 1985 and tested at 5 days after eosin ripeness.

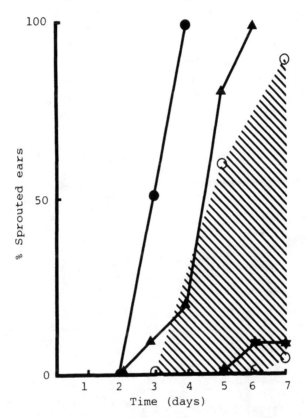

high alpha-amylase activity in mature grain of particular varieties in the absence of sprouting.

4. Mechanisms of tolerance to pre-harvest sprouting

The sequence of events which occur during pre-harvest sprouting can be simply summarized as shown in FIGURE 3. By definition any factor(s) which reduces or inhibits one or more of the steps prior to embryo germination and the production of germinative enzymes should result in some degree of sprouting tolerance. There has been a tendancy in the past to equate sprouting tolerance with dormancy, however, whilst it is clear that dormancy is a very important mechanism, theoretically there are many other possibil-

TABLE 2. Changes in falling number and alpha-amylase activity induced by a standard artificial weather treatment which was applied at 15 days and 25 days after eosin maturity.

	Time after eosin maturity		
Variety	15 day	25 day	
	ΔFN* (sec)	ΔFN* (sec)	ΔAlpha-amylase (mUg^{-1})
New sources of tolerance			
(Mean of 14 lines)	-6	14	9
RL 4137	-38	-39	3
Suneca	31	59	17
Kite	182	280	144
SUN 56A	356	391	614

Δ*FN and alpha-amylase = (control falling number - falling number after weather treatment) and (control alpha-amylase activity - alpha-amylase activity after weather treatment) respectively.

FIGURE 3. Sequence of events in pre-harvest sprouting.

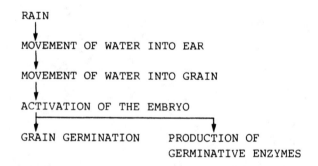

ities. Mechanisms can be divided into factors associated with the intact ear for example, movement of water and gases to the surface of the grain, evaporation of water and the presence of germination inhibitors in the vegetative parts of the ear; or factors determined by the grains themselves for example, water and gas movement into the grains, grain germinability or dormancy, production of germinative enzymes. In order to locate tolerance mechanisms it is necessary to have methods for estimating the rates of these individual steps in isolation.

As a first approximation it was found useful to compare

rates of ear sprouting with rates of germination of isolated grains under identical conditions. For varieties where tolerance is primarily determined by the level of grain dormancy there should be a simple linear relationship between the two measurements. By contrast, any departure from the straight line for example, rate of ear sprouting much slower than predicted on the basis of grain germinability, should indicate the presence of additional factors associated with the ears. Of the new sources of tolerance, eight fitted closely to the linear relationship outlined above whilst six showed varying degrees of deviation from the line. Aus 1408, Aus 1490 and Hellas exhibited high levels of dormancy, approaching RL 4137, whilst Saberbeg showed the most marked deviation from the sprouting/germination regression line. King and Richards (1984) have suggested that the physical structure of the ear, in particular the presence or absence of awns, may substantially effect the water trapping capacity of the wheat ear. Similarly, Mares (1983b) demonstrated that there was significant variation in the rates of movement of water into grains in intact ears which could not be readily explained in terms of differences in either total ear water uptake or water uptake by isolated grains. This factor, i.e. reduced movement of water into grains in intact ears, seemed to be responsible for a large part of the tolerance of varieties such as Suneca. The rate of movement of water into grains in ears of Saberbeg was not particularly slow suggesting that someother factor(s) was responsible for the reduced rate of sprouting in this line. Varieties were compared with respect to (a) the time taken for grains in intact ears to accumulate 35% of their dry weight in water, and (b) the time required for 50% of isolated grains to germinate at 20°C (TABLE 3).

TABLE 3. Variability in rate of movement of water into grains and in grain germinability at maturity.

Variety	Time to 50% germination (days)	Time to reach a grain moisture of 35% (dry wt.) (h)
SUN 56A	2.5	12
Suneca	4	30
RL 4137	>7	25
New sources of tolerance	4->7	8-28

Clearly the variation is much greater for dormancy than for rate of grain water intake. Rates of production of alpha-amylase were also examined, under conditions where differences in grain germinability were minimized, but here again the extent of variability was markedly less than for grain dormancy.

5. Stability and duration of tolerance to pre-harvest sprouting

Sprouting tolerance is a dynamic phenomenon and for accurate characterization genotype, environmental conditions and stage of maturity are all required. The level of tolerance or grain dormancy which is present at grain maturity is strongly influenced by both genotype and environmental factors such as temperature and rainfall. Low temperatures during grain ripening, for example, favour the development of dormancy. Higher temperatures are less favourable and also increase the rate of decay of dormancy during post-maturity after-ripening. For imbibed, mature grain the response to temperature is rather different. Dormant grains germinate readily at low temperatures (10-15°C) but germinability decreases with increasing temperatures i.e. dormancy is enhanced or expressed more strongly. Short periods of cool treatment (18-48h) of imbibed grains lead to the abolition of dormancy and the acquisition of the capacity to germinate readily at higher temperatures (Mares, 1984). The loss of dormancy resulting from short periods of cold treatment of imbibed grain is not reversed by redrying. George (1967) showed that the germination rate of non-dormant grains increased with increasing temperature up to 35°C.

Thus it is clear that there are environmental conditions which may either
(a) reduce the level of dormancy which develops in maturing grains, or
(b) abolish dormancy either slowly (normal temperature dependant after-ripening) or rapidly (cool temperature treatment of wet grain), and thereby dramatically increase the susceptibility of the crop to pre-harvest sprouting. Some insurance, i.e. other less environmentally sensitive tolerance mechanisms, in these situations would be of great advantage. Mechanisms such as the reduced rate of water movement into grains in intact ears which do not appear to be significantly affected by after-ripening, may assume considerable importance where dormancy has been greatly reduced.

Further studies of the interaction of this and other similar tolerance mechanisms with the environment are required.

SUMMARY

New sources of tolerance to pre-harvest sprouting have been identified which have the potential to dramatically improve the tolerance of white grained, bread wheats. Some of these lines developed a strong dormancy at maturity, approaching that seen in some of the better red-grained wheats, under the environmental conditions typical of Northern NSW and Queensland. In other varieties there was some evidence that mechanisms associated with the non-grain components of the ear were acting in conjunction with dormancy and it is suggested that such mechanisms may provide insurance in those instances where dormancy has declined or has been destroyed.

REFERNCES

Barnes, W. C. and Blakeney, A. B. 1974. Determination of cereal alpha-amylase using a commercially available dye labelled substrate. Staerke. 26, pp. 193-197.

George, D. W. 1967. High temperature seed dormancy in wheat (Triticum aestivum L.). Crop Sci. 7, pp. 249-253.

King, R. W. and Richards, R. A. 1984. Water uptake in relation to pre-harvest sprouting damage in wheat : ear characteristics. Aust. J. Agric. Res. 35, pp. 327-336.

McMaster, G. J. and Derera, N. F. 1976. Methodology and sample preparation when screening for sprouting damage in cereals. Cereal Res. Comm. 4, pp. 251-254.

Mares, D. J. 1983a. Preservation of dormancy in freshly harvested wheat grain. Aust. J. Agric. Res. 34, pp. 33-38.

Mares, D. J. 1983b. Investigation of the pre-harvest sprouting damage resistance mechanisms in some Australian white wheats. In: Proc. Third Inter. Symp. on Pre-harvest Sprouting in Cereals. Kruger, J. E. and LaBerge, D. E. (eds.). Westview Press, Boulder, Co. USA. pp. 59-65.

Mares, D. J. 1984. Temperature dependence of germinability of wheat (Triticum aestivum L.) grain in relation to pre-harvest sprouting. Aust. J. Agric. Res. 35, pp. 115-128.

Noll, J. S. and Czarnecki, E. 1980. Methods of extending
the testing period for harvest time dormancy in wheat.
Cereal Res. Commun. 8, pp. 233-238.

Breeding and Selection for Pre-Harvest Sprouting Tolerance in Barley

L. Reitan, Kvithamar Agricultural
Research Station, N-7500 Stjørdal, Norway.

INTRODUCTION

Pre-harvest sprouting in barley occurs in harvesting seasons with much rain and dewfall. Severe damage may develop in both malting barley and fodder barley crops in years with conditions favouring general, low levels of dormancy and when harvesting is delayed by bad weather.

Pre-harvest sprouting tolerance is based upon different mechanisms, which alone, or in concert lead to varietal or genotypical protection against sprouting. In barley there is disagreement as to which mechanisms are the most significant and both the seed coverings and the embryo are claimed to be important. Chaff has less effect.

Whilst true seed dormancy is the major factor giving protection, other factors such as straw-stiffness, earliness, ear density, ear shape and angle, awns and other "water-holding" properties affecting drying after rainfall all affect the tendency to sprout in the ear. Belderok (1968) classified three phases of dormancy in barley :

1. An immediate post-harvest phase in which practically no grains germinate in germination tests.
2. A subsequent transitional phase following the after-ripened stage in which dormancy terminates in gradually increasing numbers of grains.
3. Water sensitivity; a phase in which the grains are capable of full germination at optimum moisture, but cannot do so if excess water is present.

Water sensitivity is important for malsters. Interruption of the steeping process and provision of an "air rest" overcomes water sensitivity (Briggs 1978). Malting barley, furthermore, requires rapid and uniform germination.

Brookes (1980) discussed "germinative energy" and "germinative capacity" which reflect viability and dormancy, respectively and stated that for UK malsters, a delay until at
least October is necessary to overcome dormancy in season-
harvested barley crops.

Concerning pre-harvest sprouting tolerance there is a
need to collect exact information on the mechanisms that are
important at different times and which interact with different factors outside the grain :

1. How these mechanisms are inherited; how many loci are
    involved and where are they situated on the chromo-
    somes.
2. Are there simple methods for screening for these genes
    in a selection programme.
3. What methods can select for the important mechanisms
    involved.

Not one of these ambitious questions can be answered fully.
Pre-harvest sprouting tolerance is difficult to work with
because of the continuous variation with time. The only way
to fix the actual level is to freeze the dry ear or grain.
Simple measurements which distinguish between dormant and
non-dormant grain at a particular after-ripened stage are
important.

BASIC KNOWLEDGE ON SPROUTING AND DORMANCY

Different climatic conditions interact with genotype
and the expression of sprouting fluctuates in different
years and at different sites. The major climatic factor is
the temperature from anthesis and during the ripening stage,
but rainfall also has a great influence in certain periods
(Belderock 1968; Reiner and Loch 1976; Strand 1983). In
general the higher the temperature during the ripening
period, the lower the seed dormancy level. Strand (1983)
found that precipitation in a 10 day period after anthesis
had a marked effect on seed dormancy level. For late harvest, rain in a period just prior to and during yellow
ripeness had a great influence. Dormant barley grains
sprout less with increasing temperature but germinate well
under low temperatures 4-10°C. [In Avena fatua L., the
opposite response has been observed in some genotypes,
although the overall rule applies even here (Sawhney and
Quick 1985)].

As a rule dormancy declines for some point after ripening. It can be raised temporarily under certain conditions

of grain and low temperatures (Wellington and Bradnock 1964; Strand 1983). Significant interactions between havest time and year, and between harvest time and genotype for seed dormancy have been reported (Ringlund 1965; Lallukka 1971; Strand 1983). This is of great importance in selection programs. Seed dormancy is also thermo-unstable. The higher the storage temperature, the faster the decrease in dormancy.

Permeability barriers for water uptake and gas exchange may occur (Woodbury and Weibe 1983). Imbibition is a "passive" process, but it can induce processes that depress gas diffusion. Strongly dormant grains require much more oxygen than non-dormant and the latter can germinate with as little as 0.3-0.5% of atmospheric oxygen (Griggs 1978). Oxygen - requiring mitochondrial activity for repair-processes is present both in dormant and non-dormant grains in the first hours after imbibition (Taylorson and Hendricks 1977). A water film on "semi-dormant" grain, which drastically decreases the diffusion of oxygen causing an oxygen demand, may lead to "water-sensitivity". This effect is enhanced if the temperature is high, due to a lower oxygen concentration in the water (Briggs 1978).

Oxygen-absorbing, enzymic reactions may also create an oxygen barrier to the embryo and depress gibberellic acid synthesis which regulates mRNA production coding for alpha-amylases in aleurone layers.

Takahashi (1985) found that the germination of japonica rice (Oryza sativa) was inhibited by excess amounts of oxygen i.e. the opposite of the normal situation.

As will be discussed later the sprouting tolerance character is highly genetically expressed, and for barley there seems to be a relatively small number of genes involved per se. Expression of dormancy is modified by a number of factors which act on biochemical processes during growth and development, and during storage and germination of the grain. Interactions between genotype and milieu are possibly controlled by biochemical processes with different temperature- and pH-optima and a manipulation of the balance between promoting and inhibiting factors. Physiological patterns for development of the outer layers, the oxygen uptake, "weather resistance" against the rupture of these layers etc., are all to some extent at least, genetically based. The accumulation of all these factors make the "sprouting-tolerance factor" a complex phenomenon.

TESTING METHODS

   Sprouting scores of certain batches of grain are the
simplest, but also the most unreliable test of resistance to
pre-harvest sprouting. Measurements of starch-related
enzymatic processes and levels such as alpha-amylase activity are much more precise. The Falling Number method, which
measures a complex of starch quality and alpha-amylase activity, is effective in barley for distinguishing between
levels of sprouting tolerance. Moist chambers or rain simulators, alone or in combination with other methods, take into account not only dormancy but also effects due to other
"head" properties. A disadvantage here is the free water
control and the "water sensitivity" of barley.

GENETICAL ASPECTS OF SPROUTING TOLERANCE

   Obviously the variation in sprouting tolerance is
partly due to genetic variation among barley genotypes. A
number of reports show clear varietal differences in sprouting, alpha-amylase activity, Falling Number, and seed dormancy levels in response to equal treatments. (Strand 1965,
1983; Harvey et al., 1983; Reitan 1983; Buraas and
Skinnes 1985). There is good documentation of genetical
aspects of wheat (Nilsson-Ehle, 1914; Freed, 1972; Reitan,
1980; Derera and Bhatt, 1980) and the major dominant genes
related to seedcoat colour. In oats (Avena sativa L.)
Johnson (1935) found that dormancy is inherited recessively,
and that a small number of polymeric factors were present.
These factors were based on properties of the embryo and not
on those of the coat layers and grains, since clear segregation ratios were found in F_2 grains (F_1 coat layers). In
barley, little work has been reported on genetical studies
of dormancy. Freistedt (1935) studied four spring barley
crosses and found seed dormancy to be recessive with 1 - 2
genes involved. Moorman (1942) stated that dormancy was
located partially in the seed coat and partially in the
embryo, and found reciprocal differences due to maternal
effects of the seed coat.
   Buraas and Skinnes, (1984) found, in their investigation on simple and double crosses in barley :

1. No cytoplasmic effects on seed dormancy
2. Seed dormancy was a recessive character with almost
   complete dominance for low dormancy
3. Several genes governed seed dormancy
4. The same genes were expressed at different harvests as

well as at different recording stages
5. No close association between seed dormancy and other agronomic characters were found
6. The heritability of seed dormancy was high
7. Mass selection for seed dormancy of single seeds was efficient as early as the F_3 generation
8. The selection response in the field was lower than in the glass house
9. The highest response to selection for seed dormancy was combined with an early recording stage.

They found, by means of covariance analysis, that the test for dormancy at five weeks after yellow ripeness was more effective compared with one at three weeks; and a germination duration of seven days compared with eleven. In addition they made selections with single seeds in the phytotron in F_3, and an F_8 progeny-selection in the field. They found heritability values of about h2 = 0.8. They also found no relationship between germination rate and sprouting resistance among parents in the crosses, but due to differences in germination rate it was deemed important to overcome rate differences by delaying seed counts in germination test until the 7th day. They made selections in both high (H) and low (L) directions and found marked responses to selection. In the double cross (two F_1's crossed) the selection lines did deviate more from the most extreme parents than in the single cross.

Harvey et al., (1983) found rather low heritabilities in studies of crosses between dormant and non-dormant genotypes in F_2-F_3, but year differences confounded some of the heritability. They also found significant effects due to hull extracts from dormant grain. All naked barley genotypes sprouted well and they concluded that dormancy factors are located in the grain covering. The germination of non-dormant seed was not influenced by mixing dormant seed in the same petri dish. Storing grain for 8 months at 18°C was sufficient to break the dormancy in the progeny. They found a correlation (r = 0.485**) between germination at harvest and alpha-amylase activity on malted grain after storage, but it was possible to detect progeny with both high dormancy and high amylase activity.

SELECTION METHODS

In cereal breeding, crosses are seldom made with the sole aim of obtaining genotypes with a specific dormancy period or resistance to sprouting. More commonly sprouting tolerance is only one of many characters which must be considered. Breeders therefore need a rapid screening test to handle large numbers of samples and in this situation ear germination tests are often preferred. It is very important that tests on genotypes are repeated after harvest ripeness since protection has to be effective for some weeks after ripeness (Belderok 1968; Strand 1965; Reitan 1983; Larsson 1985). Both harvest x genotype and genotype x year interactions are significant and selection in more than one year is necessary.

Decision-making on parent-combinations should include the sprouting resistance factor. If not known, parental screening tests for this character are necessary. Double-crosses, composite crosses etc. may be necessary to increase genotypical variation in a certain population. At least some Scandinavian and European barley lines and varieties possess early maturing and high dormancy properties.

Many more or less simple selection methods are used for cereals although most of them were developed for wheat. Since the aim is generally similar in other cereals, a lot of the tests are also suitable for barley. Depending on the stage in a breeding program, and of precision required, different techniques are possible on the same segregating material at different stages.

Svensson (1976), McMaster and Derera (1976), Strand (1980), Reitan (1983) and Buraas and Skinnes (1984) described different mass selection "bulk" methods for screening for sprouting resistance. Some of the tests are based on moist chambers for intact ears, others select for dormant grain in germination tests under controlled conditions of water and temperature. Building up sub-populations of more dormant grain in early generations is possible in barley because of the recessive nature of the dormancy inheritance, and leads to a higher frequency of positive genotypes in later generations. For malting barleys, intact ears should be included in the test in order that the effects of awns, glume and grain attachments are also taken into account. (See TABLES 1 and 2.)

Other, more sophisticated methods are suitable for advanced progeny- and variety testing (Belderok 1968; Weilenmann 1976, 1980; Strand 1965; Svensson 1976; and Reddy 1985, among others.). These are characterized by precision but are laborious. In between these two extremes there are screening tests more suited for individual testing of large numbers of genotypes in early generations. Alpha-amylase and Falling number tests for exposed material are

also effective in barley. However controlled conditions are
required, otherwise valuable material can be lost due to
excessive sprouting. Simple germination tests are often
more convenient because of the dormancy which is almost
independant of weather or water stress. Simple germination
tests are described by Jonansson (1976) and Larsson (1985)
using petri dishes, and by Reitan (1979) who used paper
towels and two different germination temperatures to com-
pute a "dormancy index". This index, worked out by Strand
(1965, 1980) takes into account the fact that dormancy is
expressed more strongly at higher temperature, and dormancy
at lower temperature will count more in the index.

Different breeding methods may have some influence on
when and how the sprouting tolerance selection may take
place. The well-known pedigree method has considerable
disadvantages. It requires quite a lot of book-keeping and
the material is bound to increase in volume. Another difficulty is that the effectiveness of early generation plant
selection may be hampered by intergenotypic competition
(Powell et al., 1985). Thus random single plant selections
in early generations are recommended. Heritability of
yield and other quantitatively inherited characters are very
low on a single plant basis. In the pure pedigree method,
mass selection in barley should take place in F_2, and head
sprouting tests during some of the next generations, followed by exact tests in the later steps.

TABLE 1. Examples of variation for sprouting resistance in
barley populations

	Population Number									
	1	2	3	4	5	6	7	8	9	10
% Dormant grain	0.8	6	10	16	24	29	34	47	58	73
Found SD for 10 tests	1.2	2.9	3.9	2.7	4.4	5.9	5.6	4.9	3.0	5.0
Expected SD	0.9	2.4	3.3	3.7	4.3	4.7	4.7	4.9	4.9	4.5

SD = standard deviation.

Tests were made following storing at low temperature
(10-1°C) for 5 months, using 10 x 100 grains on moist paper
at 22°C for 7 days.

In the "Bulk method" one or two early mass selections are recommended to increase the frequency of resistant geno- types. During the "homozygosity-process" over generations, simple germination tests on a population basis may give valuable information about different population levels. A random head selection for testing could give an indication of the variation within the populations.

The F_2-progeny method (tandem selection method, Frey 1968) is a powerful method in the sense that it is possible to work with a lot of crosses and simultaneously have vari- ation between lines for quite early generation yield selec- tion etc; avoid strong intergenotypical competition caused for, example by, straw length and strength; and the possi- bility of screening for recessive characters in F_2, such as sprouting resistance in barley. Rosielle (1984) used com- puter simulation techniques to compare the efficiency of bulk yield F_2 testing (in F_4-F_5) with direct-line yield testing, and found F_2 bulk-yield testing to be almost always more efficient than direct-line testing.

In F_2, head tests in moist chambers or individual heads or plants in paper towels should be effective, after selec- tion for factors such as fungal resistance and earliness. Selection among F_2 progeny in later generations, followed by retesting of dormancy for 1 - 2 generations is necessary. Reselections from F_2 bulks should enter a new cycle of sprouting selection, either with heads in a moist chamber or individual germination tests the following year.

TABLE 2. Examples of positive progeny from a cross between dormant (Agneta) and non-dormant (Yrjar) parents, bulk selected and reselected on individual tests. (expressed as Dormancy Index)

| Line | YEAR | | | | | |
	1981	1982	1983	1984	1985	$\bar{X}$
H3550	28	91	70	69	26	56.8
H3553	75	91	73	65	34	67.6
H3554	63	92	82	66	34	67.4
H3556	35	88	21	42	28	42.8
H3556	57	64	50	53	28	59.4
H3576*	2	28	1	27	5	12.6

* Low-selection line. (In 1981 on single rows)

New time-saving breeding methods such as single seed descent (SSD) should commence with a mass-selection step, followed by appropriate grain treatments in order to prevent the loss of the dormant ones in the next generation.

Back-crossing methods for transferring sprouting tolerance in established varieties are possible. As mentioned above, dormancy in barley is recessive. In contrast to readily detectable recessive disease resistance characters, dormancy requires a lot of individual testing, and this, together with crossing work, makes it expensive. (It is not possible to make crosses in the same generation as the testing unless late tillers can be used. In the meantime better breeding material is delayed.

Double haploid techniques such as H. bulbosum and anther-culture show potential for rapid advance.

New Recombinant DNA-techniques may solve the problems in future.

In a practical breeding programme it is necessary to run a number of different populations rather than putting maximum effort into one single cross. For that reason it is of great value to make early bulk selections on most crosses. As the material advances, it becomes more important to run individual tests together with advanced yield tests. If the sprouting resistance of a line is too low at this stage, there is no point in continuing yield trials. (In Norway there is an official requirement for dormancy of new feed barley varieties to be comparable with the known varieties.)

CORRELATED RESPONSE

In "mass" or "bulk" selections for sprouting resistance, correlation with other characters may lead to a parallel shift in the population means of these characters. For example a negative correlation between earliness and sprouting tolerance has been reported and Larsson (1976) observed a 1-4 day delay in ripening with strong selection in wheat. In contrast, Buraas and Skinnes (1984) and Reitan (1979) found no such correlation in segregating populations of barley and wheat if samples were tested at similar ripening stages and they suggested that differences in after-ripening conditions could lead to a spurious correlation. Stepwise selections at fixed ripening stages could eliminate this effect.

REFERENCES

Belderok, B. 1968. Field Crops Abs. 21, pp. 203-211.

Briggs, D. E. 1978. Barley, Chapman and Hall, London.

Brookes, P. A. 1980. Cereal Res. Comm. 8, pp. 29-38.

Buraas, T. and Skinnes, H. 1984. Hereditas, 101, pp. 235-244.

Buraas, T. and Skinnes, H. 1985. Acta Agric. Scand. 35, pp. 233-244.

Derera, N. F. and Bhatt, G. M. 1980. Cereal Res. Comm. 8, pp. 193-198.

Freed, R. D. 1972. Ph. D. Thesis, Dept. of Crop and Soil Sci, Michigan State Univ., E. Lansing MI.

Frei, M. 1968. Crop Sci. 8, pp. 235-238.

Freistedt, P. 1935. Z. Zücht. Reihe A. Pflanzenzücht. 20, pp. 169-209.

Harvey, B. L., Rossnagel, B. G. and Muderewich, R. P. 1983. In: Proc. Third Int. Symp. on Pre-harvest Sprouting in Cereals. Kruger, J. E. and LaBerge, D. E. (eds.). Westview Press. Boulder, Co. USA. pp. 239-243.

Johansson, N. E. 1977. Thesis, Uppsala, Sweden.

Johnson, L. P. V. 1935. Can. J. Res. 13, pp. 367-387.

Lallukka, U. 1971. J. Sci. Agr. Finl. 43, pp. 167-177.

Larsson, S. 1985. Sveriges Uts. fören. Tidssk. 95, pp. 87-98.

McMaster, G. J. and Derera, N. F. 1976. Cereal Res. Comm. 4, pp. 251-255.

Moorman, B. 1942. Kühn-Arch. 56, pp. 41-79.

Nilsson-Ehle, H. 1914. Zeitschr. Pfl. Zücht. 2, pp. 153-187.

Powell, W., Caligari, P. D. S., Goudappel, P. H. and Thomas, W. T. B. 1985. Theor. Appl. Genet. 71, pp. 443-450.

Reiner, L. and Loch, V. 1976. Cereal Res. Comm. 4, pp. 107-110.

Reddy, L. V., Metzger, R. J. and Ching, T. M. 1985. Crop Sci. 25, pp. 455-458.

Reitan, L. 1979. Thesis Dr. Sci. Agr. Univ. Norway. N120p.

Reitan, L. 1980. Cereal Res. Comm. 8, pp. 265-274.

Reitan, L. 1983. In: Proc. Third Int. Symp. on Pre-harvest Sprouting in Cereals. Kruger, J. E. and LaBerge, D. E. (eds.). Westview Press, Boulder, Co. USA. pp. 244-250.

Ringlund, K. 1965. Sci. Res. Agr. Coll. Norway. 44, (23).

Rosielle, A. A. 1984. Euphytica. 33, pp. 153-159.

Sawhney, R. and Quick, W. A. 1985. Ann. Bot. 55, pp. 25-28

Strand, E. 1965. Sci. Res. Agr. Coll. Norway. 44, (7) pp. 1-22.

Strand, E. 1983. In: Proc. Third Int. Symp. on Pre-harvest

Sprouting in Cereals. Kruger, J. E. and LaBerge, D. E. (eds.). Westview Press, Boulder, Co. USA. pp. 260-266.

Svensson, G. 1976. Cereal Res. Comm. 4, pp. 263-266.

Takahashi, N. 1985. Ann. Bot. 55, pp. 597-600.

Taylorson, R. B. and Hendricks, S. B. 1977. Ann. Review of Pl. Physiol. 28, pp. 331-354.

Weilenmann, F. 1976. Cereal Res. Comm. 4, pp. 267-274.

Weilenmann, F. 1980. Cereal Res. Comm. 8, pp. 209-218.

Wellington, P. S. and Bradnock, W. T. 1964. Agr. Bot. 10, p. 129.

Woodbury, W. 1983. In: Proc. Third Int. Symp. on Pre-harvest Sprouting. Kruger, J. E. and LaBerge, D. E. (eds.). Westview Press, Boulder, Co. USA. pp. 51-58.

Germless Grains — The Ultimate Answer to Pre-Harvest Sprouting

D. R. Marshall, *The University of Sydney,
Plant Breeding Institute, Narrabri, NSW.,
Australia.*

SUMMARY

The use of cultivars which produce germless grains is
proposed as a means of radically reducing weather damage in
winter cereals. The sorts of systems and genetic stocks
requird to produce such cultivars are outlined along with
their advantages and disadvantages.

INTRODUCTION

Pre-harvest sprouting remains a serious problem
periodically causing substantial economic losses in many
cereal producing areas, despite the considerable effort that
has gone into the breeding of resistant cultivars.
Moreover, resolution of this problem via the traditional
approach of breeding for seed dormancy is, in many cases,
likely to be a long and difficult task. The reasons for
this are many and varied. They include a lack of suitable
germplasm, particularly in white wheats, rye and triticale,
the complex genetic control of sprouting resistance and the
problems this poses in its manipulation in breeding programs, and difficulties in reliably screening early generation breeding lines for sprouting resistance. As a consequence there is a need for more radical and innovative
approaches to the development of sprouting resistant cultivars. This need is underlined by the increased interest
in F_1 hybrids in winter cereals sparked by recent improvements in the effectiveness of the cytoplasmic male sterility
system in wheat and the development of male gametocides or
chemical hybridizing agents. As hybrids increase in pop-
ularity they may pose new problems with respect to sprout-

ing susceptibility (Jönsson, 1976) but they will also open
the way for new approaches to the breeding of sprouting
resistant cultivars. One such approach which appears to
merit serious consideration is the use of cultivars which
produce germless or embryoless grains. I consider this
approach in detail here.

GERMLESS GRAINS

Occurrence

Germless or embryoless grains occur rarely but
routinely in most cereals - apparently due to the single
fertilization of the polar nuclei (Harlan and Pope, 1925;
Lyon, 1928). The frequency of embryoless grains depends on
both the genotype and the environment and some genotypes
produce elevated levels of germless grains in particular
environments. An excellent case in point is the wheat cultivar 'Yorkstar', which in New York State, produces 10%
germless grains (Khan et al., 1973).
Genetic mutants which cause the development of germless
grains are also known. Such mutants, both natural and
artificial, have been commonly recorded in maize. They are
generally simple recessive genes which act as embryo lethals
and which may occur at several independent loci (Coe and
Neuffer, 1977).

Response to pre-harvest rain

Clearly germless grains cannot sprout regardless of
pre-harvest weather conditions. We would expect, therefore,
germless grains to react to pre-harvest rains in a similar
way to fully dormant grains. That is, we would expect damage to be largely restricted to bleaching and/or discolour-
ation of the seed coat and to loss of bushel weight. We
would not expect significant starch damage due to the
production of hydrolytic enzymes by either the scutellum or
aleurone.
There is substantial experimental support for these
expectations from studies of embryoless or de-embryonated
grains (e.g. Khan et al., 1973; Gilmour and MacMillan,
1984). The bulk of these studies suggest that only limited
amounts of alpha-amylase are produced in embryoless grains
compared to normal grains or embryoless grains treated with
GA_3. Nevertheless, in some genotypes of barley in some

environments, substantial amounts of alpha-amylase can be produced in embryoless grains in the absence of GA_3 (Nicholls, 1983). The practical significance of this finding remains unclear. However, it would be prudent for breeders to avoid such genotypes in developing cultivars with germless grains to reduce sprouting.

SEED PRODUCTION SYSTEMS

There are several potential ways of developing cultivars which produce germless grains. I describe three below.

1. Complementary genes

The first system involves a pair of dominant complementary genes which cause embryo lethality in the progeny of their carriers. This system, if the appropriate genetic stocks were developed, could be used in hybrid production systems based on either cytoplasmic male sterility or chemical hybridising agents. It is not applicable to pure line breeding programs.

If the complementary lethal genes are designated L_1 and L_2, respectively, then to produce F_1 hybrids which yield germless grains we require:

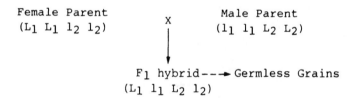

$$
\begin{array}{ccc}
\text{Female Parent} & & \text{Male Parent} \\
(L_1\ L_1\ l_2\ l_2) & \text{X} & (l_1\ l_1\ L_2\ L_2)
\end{array}
$$

$$
F_1 \text{ hybrid} \dashrightarrow \text{Germless Grains}
$$
$$
(L_1\ l_1\ L_2\ l_2)
$$

It should be stressed that the lethal genes must not act in the hybrid embryos otherwise the result would be germless F_1 hybrid seed which is obviously of no value.

2. Embryo lethal linked to a gametophytic factor

The second system involves a recessive embryo lethal closely linked to a gametophytic factor or gamete eliminator. This system is similar in principle to previously proposed systems for the use of a gametophytic factor in the exploitation of nuclear male sterility in the production of F_1 hybrid varieties (e.g. Marshall and Ellison, 1986). It is also only applicable to hybrid production systems.

The development of an effective system requires:

(a) a dominant gametophytic factor which acts in both male and female gametophytes. The gametophytic factor (Ga) must act as a gamete eliminator and cause abortion of gametes carrying the recessive (ga) allele in heterozygous (Ga/ga) individuals,

(b) the Ga locus must show complete or nearly complete penetrance (preferably 98%) so that Ga/ga genotypes produce only, or largely, Ga/Ga progeny,

(c) the Ga locus be tightly linked to a recessive embryo lethal (L/l locus so that crossovers between the two loci are rare (preferably less than 1-2%).

Given these requirements hybrids which produce germless grains could be derived as follows:

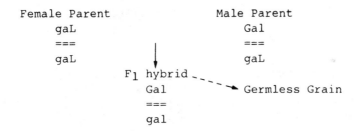

Since the male parent would also yield germless grains it would have to be produced each generation from isogenic lines of the appropriate genotype which differed only at the Ga and l loci:

Male parent production

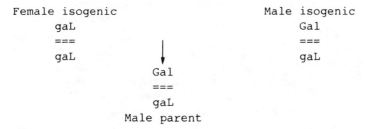

It would not be possible to adapt this system for use with cytoplamic male sterile based hybrid production systems because of the need to produce the male parent each generation and the difficulties this would pose in manipulating restorer genes. However, it could be readily adapted for use in hybrid production systems based on male sterilising chemicals.

3. Conditional embryo lethals

The third system involves the use of embryo lethals
whose expression is environmentally (either physically or
chemically) dependent. For this system to function mutants
must be found which cause embryo lethality in normal grain
production environments but not in others. Plants would be
grown in the environment where lethality was not expressed
to produce seed with embryos for sale. Farmers would grow
the same seed in an environment where lethality was
expressed to produce sprouting resistant germless grains.
This system could be readily incorporated in hybrid production systems based on both cytoplasmic male sterility and
chemical hybridizing agents. It would also be feasible to
apply this system to pure-line cultivars.

PRACTICAL LIMITATIONS

It is clear that, in theory at least, there are several
ways to create commercial cultivars which produce germless
grains. However, it is also clear that there are obvious
limitations to their exploitation in pratice.

With respect to the first system the major practical
problems appear to be ascertaining whether appropriate
variants are likely to exist and designing procedures for
their isolation. Complementary lethal or sublethal genes
are known in several crops. In wheat, for example, hybrid
necrosis, hybrid chlorosis and grass clump dwarfing are all
controlled by such genes. Consequently, it may be possible
to generate appropriate variants by mutation in some crops.
However, even if one accepts this premise, the identification and isolation of such variants would be a very onerous
task since all possible mutants at each locus would have to
be tested in hybrid combination. As a result, although this
system has the obvious advantage of simplicity once suitable
mutants are available, the prospects of its development in
any crop in the immediate future seems remote.

The second approach is clearly more complex than the
first. However, it has the advantage that suitable mutants
are known in some crops. As noted earlier, germless grains
occur routinely in mutation programmes, at least in the
diploid crops. They are less common in polyploid species
such as bread wheat but would undoubtedly be found if
seriously sought. Gametophytic factors, especially those
that act in both male and female gametes, may be less com-

mon. Nevertheless, the 'cuckoo' chromosome derived from
Aegilops sharonensis and transferred to wheat appears to
have all the characteristics required to develop a successful system (Law, 1983). The critical problem in this case
is likely to be the need for tight linkage between the loci
governing the gametocidal gene and embryo lethality.

The third approach is the simplest of the proposed
schemes. However suitable conditional mutants are likely to
occur rarely and to be difficult to isolate and identify.
For example, it has repeatedly been suggested that chromo-
somal male sterility genes which are conditional in their
expression on a specific environment or chemical could be
used to produce F_1 hybrid crop plants (Driscoll, 1986).
Yet, such a system has not been developed in any crop -
primarily because of the lack of suitable mutants. However,
because conditional mutants offer the only means of developing pure line varieties which produce germless grains,
this proposal still merits consideration.

DISCUSSION

The use of varieties which produce germless grains
should substantially reduce weather damage in cereals.
Since the lack of an embryo is likely to be controlled by
one or two genes this trait would be relatively easy to
manipulate in breeding programs. Further, since germless
grains are readily detected, use of this character would
eliminate entirely the need for costly testing of breeding
lines for tolerance to sprouting.

Germless grains would also offer the cereal breeder
other advantages. For example, in wheat the germ (embryo
plus scutellum) usually accounts for 2.5-3.5% of the grain
weight and this goes mainly into the milling offal. Consequently, we would expect germless grains to have substantial advantage over their normal counterparts in milling
yield. Further, if germless grains rather than prolonged
dormancy associated with red seed coat colour were to become
the main mechanism of sprouting resistance in commercial
cultivars this would allow the use of white wheats and the
advantages they offer in terms of milling yield in all wheat
producing areas. Finally, the development of F_1 hybrids
which produce germless grains so that their F_2 progenies
cannot be grown, may be of advantage to seed companies which
produce hybrid cultivars using a chemical hybridising agent.

The use of germless grains would also have its disadvantages. First, it would force farmers to buy planting

seed each year. However, if F_1 hybrid cultivars come to
dominate winter cereal production as now seems possible,
then this problem will diminish in time. Second, it would
be incompatible with grain end uses where germinative capacity is important e.g. in malting barley. Nevertheless, the
potential advantages of germless grains would appear to substantially outweigh the disadvantages in most winter cereal
crops.

REFERENCES

Coe, E. H. and Neuffer, M. G. 1977. The genetics of corn.
    In: Corn and corn improvement. Sprague, G. F. (ed.).
    Amer. Soc. Agron. Inc., Madison, Wisc. pp. 111-223.
Driscoll, C. J. 1986. Nuclear male sterility systems in
    seed production of hybrid varieties. Critical Reviews
    in Plant Sciences (in press). pp.
Gilmour, S. J. and MacMillan, J. 1984. Effect of inhibitors
    of gibberellin biosynthesis on the induction of alpha-
    amylase in embryoless caryopses of Hordeum vulgare cv.
    Himalaya. Planta 162, pp. 89-90.
Harlen, H. V. and Pope, M. N. 1925. Some cases of apparent
    single fertilization in barley. Amer. J. Bot. 12, pp.
    50-53.
Jönsson, J. P. 1976. Pre-harvest sprouting problems in
    hybrid wheat breeding. Cereal Res. Commun. 4, pp.
    115-119.
Khan, A. A., Verbeek, R., Waters, E. C. and Van Onckelen, H.
    A. 1973. Embryoless wheat grain. A natural system for
    the study of gibberellin-induced enzyme formation.
    Plant Physiol. 51, pp. 641-645.
Law, C. M. 1983. Prospects for directed genetic manipula-
    tion in wheat. II Miglioramento Genetico dei Cereali.
    Acad. Nazional dei Lincei, Roma. pp. 13-33.
Lyon, M. E. 1928b. The occurrence and behaviour of embryo-
    less wheat seeds. J. Agr. Res. 36, pp. 631-637.
Marshall, D. R. and Ellison, F. W. 1986. The potential use
    of gametophytic factors in the production of F_1 hybrid
    varieties. Euphytica (in press).
Nicholls, P. B. 1983. Environment, developing barley grain,
    and subsequent production of alpha-amylase by the aleu-
    rone layer. In: Third Int. Symp. on Pre-Harvest
    Sprouting in Cereals. Kruger J. E. and LaBerge D. E.,
    (eds.). Westview Press, Boulder, Col. USA. pp.
    147-153.

Ten Years Experience with using Sprouting Index in Wheat Breeding

F. Weilenmann, *Swiss Federal Research Station for Agronomy, CH-8046 Zurich-Reckenholz, Switzerland.*

SUMMARY

The production of high quality wheat is one of the most important factors in Swiss agriculture and includes the prevention of sprouted wheat. In the winter and spring wheat breeding program, the so-called sprouting index, calculated from the falling number dry (directly from the field) and the falling number wet (after 3 days treatment in the moist chamber) has been used to classify varieties and breeding lines. At different sampling times the sprouting index is a more accurate way of estimating the sprouting resistance of new varieties than are the falling numbers.

Ecoanalytical studies were used to measure the phenotypic stability of sprouting resistance of varieties in different environments. There were great differences in the ecostability between varieties and years. The ecovalues of the sprouting indices were more meaningful for varieties and for years as compared to the ecovalues of the falling numbers alone.

INTRODUCTION

Wheat production in Switzerland is used exclusively for human consumption. Every year the Swiss farmers are constrained to produce high quality bread wheat. In the past great losses due to sprouting occurred repeatedly. In 1976 42.8% and in 1982 43.9% of the wheat production could not be used for baking purposes. In these years losses due to sprouting reached several million Swiss francs.

To reduce these enormous losses the Swiss Federal Research Station for Agronomy at Zurich-Reckenholz initiated a breeding program for sprouting resistance. Thus, for more than ten years the selection of varieties for sprouting tolerance together with high yielding ability and disease resistance have been the major goals of the program. Specific crosses were made and a screening method has been developed for the selection of breeding lines (Weilenmann, 1976). Experiments to determine sprouting resistance have been carried out systematically in specific sprouting trials (including a range of standard varieties) as well as in yield trials (new breeding lines) of the official variety testing. In this report experience with these tests is discussed and some ecoanalytical studies of the results are presented.

MATERIAL AND METHODS

Specific sprouting trials were sown in the field at the Swiss Federal Research Station. Varieties tested are summarized in TABLE 1. The trials were arranged as randomized complete block designs. From each plot, 30 to 40 ears were sampled at weekly intervals after they had reached the eosin ripening stage. Depending on the year, 3 to 5 samples were taken during that period. The ear samples were divided into two equal parts for either drying or treatment in a moist chamber for 3 days. The sprouting test was based on the falling number method and on the calculation of a sprouting index (Weilenmann, 1976).

Apart form specific sprouting trials, spring and winter wheat variety yield trials were also analyzed. These trials included breeding lines as well as standard varieties, arranged as balanced lattice squares. Of the hundreds of lines tested so far a limited range of cultivars orthogonally chosen over years are outlined in TABLE 1. The results for some representative cultivars are discussed below.

Trials were located at Ellighausen, Reckenholz, Worben/ Oensingen and in some years also Delley. These locations are identical with those of the official variety testing and are uniformly distributed over the whole wheat growing area of the northern part of Switzerland. In these trials, samples of 40 ears were taken just before combine harvesting and were analyzed as mentioned above. Due to geographical and technical reasons, it was not possible to carry out a specific ripening test with this material.

For testing the ecostability of the sprouting tolerance the method described by Wricke (1962, 1965) was applied. Ecovalence values (= ecovalues) for the falling numbers dry

TABLE 1. Varieties in Test in different Years

Varieties Years 19..	76	77	78	79	80	81	82	83	84	85
Sprouting Trial-Spring Wheat										
No. of samples	3	3	4	-	3	3	3	3	3	4
Tano	x	x	x		x	x	x	x	x	x
Calanda					x	x	x	x	x	x
Walter					x	x	x	x	x	x
Hermes					x	x	x	x	x	x
Sprouting Trial-Winter Wheat										
No. of samples		5	5	-	5	4	5	5	5	5
Zenith		x	x		x	x	x	x	x	x
Zenta					x	x	x	x	x	x
Eiger						x	x	x	x	x
Sardona						x	x	x	x	x
Arina						x	x	x	x	x
Yield Trial-Winter Wheat										
No. of Locations	2	4	4	3	4	3	3	3	3	3
Probus	x	x	x	x	x	x	x	x	x	x
Zenith	x	x	x	x	x	x	x	x	x	x
Eiger	x	x	x	x	x	x	x	x	x	x
Arina		x	x	x	x	x	x	x	x	x
Sardona		x	x	x	x	x	x	x	x	x
Partizanka			x	x	x	x	x	x	x	x
Bernina		x	x	x	x	x	x	x	x	x

and wet as well as for the sprouting indices were calculated. The smaller the calculated ecovalues the more stable
and therefore the less environmentally influenced are the
characters under consideration.

RESULTS AND DISCUSSION

     The averages of the falling numbers and of the calcu-
lated sprouting indices obtained in different years are
presented in TABLE 2. A large variation between years is
obvious. In 1982, which has already been mentioned as a

TABLE 2. Average Falling-Numbers and Sprouting Indices in different years of Winter Wheat Yield Trials

Year	Falling number dry (sec)	Falling number wet (sec)	Sprouting index
1978	292.4	80.1	6.3
1979	318.9	261.0	3.3
1980	314.1	237.1	3.5
1981	290.2	212.0	4.0
1982	147.2	98.3	7.2
1983	296.1	158.3	5.1
1984	252.5	102.5	6.1
1985	307.5	126.8	5.3

sprouting year, falling numbers were very low. Little or no sprouting was observed in 1985, 1979, 1980 and 1981. Consequently the average falling numbers were high and sprouting indices low.

1. Tendency to sprout in the ripening phase

In FIGURES 1-4 the progress of both falling numbers (dry and wet) and of the sprouting indices are shown for the cultivars Zenith, Arina, Tano and Hermes for the years 1982 and 1985. In the sprouting year 1982 the falling number wet indicated no varietal differences for either spring and winter wheat. Therefore, a selection based on this criterion only would not be successful.

In winter wheat the falling number dry was fairly specific for each variety at the first two sampling dates. For spring wheat, only the variety Hermes showed a distinctly superior falling number.

It can also be seen that the calculated sprouting index gave a good indication of the tendency of a variety to sprout. The scale of 1 to 9 was empirically chosen in such a way that the score of 9 corresponds to the minimum Falling Number of 60 seconds, and the best score of 1 corresponds to a Falling Number of 300 seconds.

With winter wheat in particular, the sprouting index allowed a good differentiation of the varieties; with spring wheat, however, the differentiation is minimal. Nevertheless the index gave a better differentiation between varieties than the falling number wet.

The average values for sprouting characters over the

FIGURE 1. Sprouting tendency in the ripening phase for the winter wheat varieties Zenith and Arina in 1982.

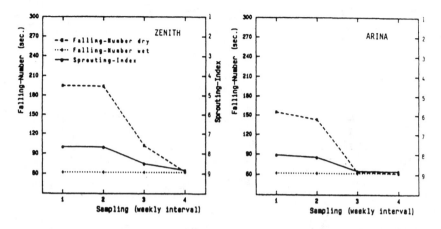

FIGURE 2. Sprouting tendency in the ripening phase for the spring wheat varieties Tano and Hermes in 1982.

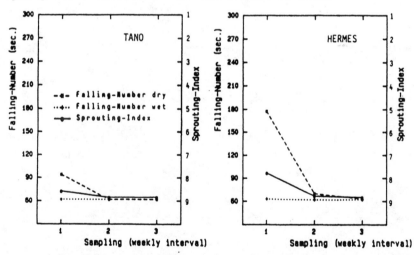

years are presented in TABLE 3 for both wheat types. These figures include results over 5 years for winter wheat (1981 to 1985) and over 6 years (1980 to 1985) for spring wheat. The winter wheat varieties Zenith and Eiger with a sprouting index of 5.8 clearly have better sprouting resistance than the varieties Sardona (6.9) and Zenta (6.7), respectively. Concerning the spring wheats, the varieties Walter (6.6) and Hermes (6.5) have lower sprouting indices than the varieties Tano (7.7) and Calanda (7.6).

FIGURE 3. Sprouting tendency in the ripening phase for the winter varieties Zenith and Arina in 1985.

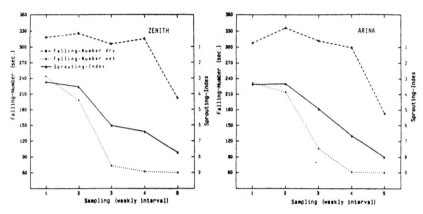

FIGURE 4. Sprouting tendency in the ripening phase for the spring varieties Tano and Hermes in 1985.

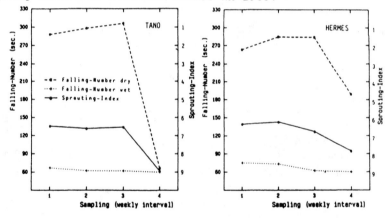

2. Sprouting tendency at different locations

In the yield trials every year, samples from 2 to 4 locations were collected to determine the sprouting tendency. FIGURES 5 and 6 show the results of two selected years, 1982 and 1985, respectively.

In 1982 the falling numbers as well as the sprouting indices varied considerably from location to location. At Reckenholz, a differentiation of the varieties was observed if the falling number dry was considered but not with the falling number wet. At Worben, the process of sprouting was already advanced to a point where neither the falling numbers nor the sprouting indices could differentiate the cultivar.

TABLE 3. Falling numbers and Sprouting Indices for Spring
and Winter Wheat Varieties in the Sprouting Trials

Variety	Falling number (sec)		Sprouting Index
	dry	wet	
Spring wheat (1980-1985)			
Tano	170	67	7.7
Walter	213	95	6.6
Hermes	234	94	6.5
Calanda	159	72	7.6
Winter wheat (1981-1985)			
Zenith	255	127	5.8
Zenta	212	95	6.7
Eiger	245	125	5.8
Sardona	197	87	6.9
Arina	230	113	6.2

FIGURE 5. Falling numbers (F.N.) and sprouting indices
(Spr. Ind.) for winter wheat varieties at different locations in 1982.

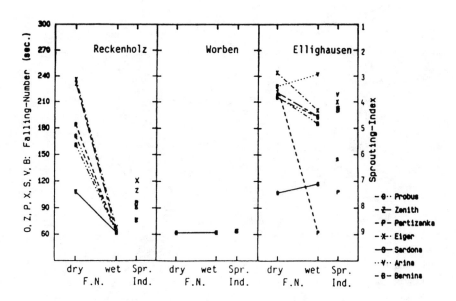

FIGURE 6. Falling numbers (F.N.) and sprouting indices
(Spr. Ind.) for winter wheat varieties at different locations in 1985.

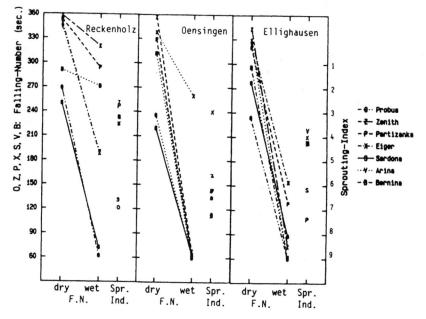

At the third location, Ellighausen, both analyses of the
falling numbers and sprouting indices gave large differences
between varieties.

In 1985 a clearly different reaction of the varieties
was observed. Only at Oensigen were the falling numbers
wet very low, except for the variety Eiger. The differentiation between varieties was therefore only approximately
the same at all locations.

The reason for the lack of correspondence in the
results from different locations in some years is due to the
fact that sampling of the ears were not adjusted to the
specific ripening stage.

3. Sprouting tendency over locations and years

The mean of the sprouting values over all locations
within a single year gave clearer results. As can be seen
from FIGURES 7 and 8 the varieties react typically under
the moist chamber stress. In particular the varieties
Probus and Partizanka show a more pronounced depression of
falling numbers than do Zenith and Eiger. In addition,

FIGURE 7. Falling numbers (F.N.) and sprouting indices (Spr. Ind.) for winter wheat varieties over locations in the years 1979, 1980 and 1981.

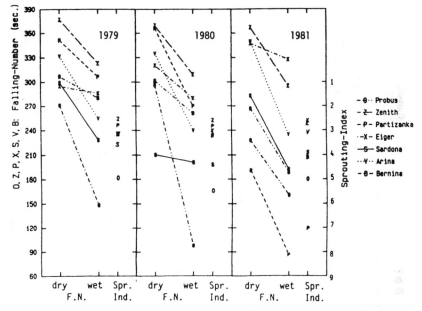

FIGURE 8. Falling numbers (F.N.) and sprouting indices (Spr. Ind.) for winter wheat varieties over locations in the years 1982, 1983 and 1985.

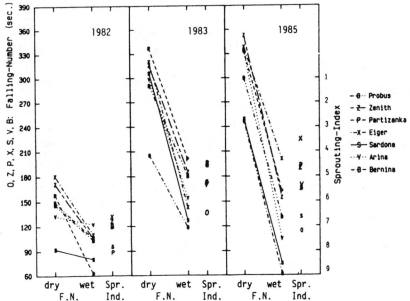

Partizanka demonstrates a higher variation for this character, e.g. in the years 1979, 1980 and 1983 this variety
indicated a good sprouting resistance whereas in the years
1981 and 1982 the resistance was rather low.

From these figures it can be deduced that the sprouting resistance of a cultivar is not the result of either the
falling number dry or wet alone but rather a combination of
the two as indicated by the overlapping curves. A variety
with a high initial falling number dry can sustain a
stronger depression before it reaches an unacceptable level.
The method described for calculating the sprouting index
takes account of this fact.

In FIGURE 9, the average indices calculated over all
years (1978-1985) correspond well with the observed values
of the varieties for each year. Therefore, a sprouting
analysis including 3 locations can be considered adequate
for determining the varietal differentiation.

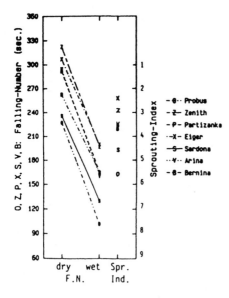

FIGURE 9. Falling numbers
(F.N.) and sprouting indices
(Spr. Ind.) for winter wheat
varieties over locations in
the years 1978 to 1985.

4. Ecoanalytical Studies

The calculated eco-
values are considered to be
a measure of the stability
of the characters, falling
number and sprouting index,
over varieties and years.
The absolute magnitude of
these indices depend
strongly on the magnitude of
the initial values. A
transformation of these values was therefore necessary and
are shown together with the ranks in the TABLES 4 and 5.

The varieties Zenta, Walter and to certain extent
Calanda, show higher ecovalues and therefore lower phenotypic stability than do the other varieties tested. The
first cultivars mentioned are therefore more difficult to
classify than e.g. the varieties Zenith and Hermes.
Furthermore, it is evident that some varieties can be described as stable on the basis of falling number dry but on

TABLE 4. Eco-values, relative and Ranks, for varieties and years of the Spring Wheat Sprouting Trial

Variety/ Year	Falling number				Sprouting Index	
	dry Eco-value rel. %	Rank	wet Eco-value rel. %	Rank	Eco-value rel. %	Rank
Tano	45.3	2	105.9	3	58.3	1
Walter	197.2	4	145.4	4	163.1	4
Hermes	38.9	1	93.4	2	79.1	2
Calanda	118.4	3	55.0	1	98.6	3
1980	176.2	5	65.1	4	184.3	6
1981	31.8	1	340.4	6	47.9	1
1982	52.7	3	82.2	5	98.9	4
1983	199.1	6	36.3	2	132.2	5
1984	105.3	4	33.6	1	86.4	3
1985	34.6	2	42.1	3	48.9	2

TABLE 5. Eco-values, relative and ranks for varieties and years of the Winter Wheat Sprouting Trial

Variety/ Year	Falling number				Sprouting Index	
	dry Eco-value rel. %	Rank	wet Eco-value rel. %	Rank	Eco-value rel. %	Rank
Zenith	39.2	1	54.5	1	61.8	2
Zenta	245.5	5	169.1	5	211.5	5
Eiger	49.9	2	102.1	4	84.7	3
Sardona	81.8	3	79.2	2	86.6	4
Arina	83.6	4	95.1	3	56.7	1
1981	294.3	5	216.0	5	205.1	5
1982	25.1	1	53.7	2	31.8	1
1983	46.8	2	66.1	3	79.6	3
1984	65.7	3	139.9	4	147.1	4
1985	68.1	4	24.2	1	37.6	2

the basis of falling number dry but not on falling number wet. Specifically this is the case for the cultivars Eiger, Hermes and Tano. On the other hand the varieties Calanda, Walter, Zenta, and to some extent Sardona, can be considered as more stable for the falling number wet. With this classification, care should be taken because the more sprouting susceptible varieties show a false stability due to the very low falling number wet.

The ranks of the ecovalues for falling numbers dry and wet do not fully correspond for all varieties. Where large differences exist, the sprouting indices lead to a certain adjustment.

The annual ecovalues for winter wheat in 1981 were exceptional. In TABLE 4 and 5 the high ecovalues and thus the unstable behaviour of the cultivars in that year can be seen. In 1982 ver low ecovalues were obtained for both types of wheat. These however were due to very low falling numbers dry and wet. The year 1985 showed low ecovalues with both falling numbers. In that year a very good rating of a variety specific sprouting resistance was achieved for both spring and winter wheat. In the years 1980 and 1983 the rating was lowest.

CONCLUSIONS

The sprouting index has been successfully applied for the assessment of sprouting resistance for the last ten years. A classification of different varieties is made possible by calculating means over locations where even specific location differences in falling numbers occur. By this method it is possible to include simultaneously, morphological influences of ears, e.g. awns, angle of the ears etc. (King and Chadim, 1983).

By calculating the ecovalues for varieties and years the sprouting differentiation can be improved. In doing so absolute values of the falling numbers also have to be considered. The ecovalues of the sprouting indices compared with those of the falling numbers were more evident for the varieties as well as for the years.

ACKNOWLEDGEMENTS

Thanks are due to the collaborators Dr. P. M. Fried,
Dr. W. Saurer and Dr. H. Winzeler for their kind help during
the preparation of the manuscript. The author wishes to
thank also all cooperators Miss M. Joerger, Miss V. Boesch,
Mr. H. Schneebeli, Mr. M. Anders, Mr. N. Stucki for technical assistance as well as Mrs. K. Schmic for typing the
manuscript.

REFERENCES

King, R. W. and Chadim, H. 1983. Ear wetting and pre-
    harvest sprouting of wheat. Proc. Third Int. Symp. on
    Pre-harvest Sprouting in Cereals. Kruger, J. E. and
    LaBerge, D. E. (eds.). Westview Press, Boulder, Co.
    U.S.A. pp. 36-50.
Weilenmann, F. 1976. A selection method to test the sprout-
    ing resistance in wheat. Cereal Res. Com. 4, pp. 267-
    273.
Weilenmann, F. 1979. Plant breeding aspects of sprouting
    resistance and experience with the selection technique
    using sprouting index. Cereal Res. Com. 8, pp. 209-
    218.
Wricke, G. 1962. Ueber eine Methode zur Erfassung der
    oekologischen Streubreite in Feldversuchen. Zeit-
    schrift fuer Pflanzenzuechtung. 46, pp. 92-96.
Wricke, G. 1965. Die Erfassung der Wechselwirkung zwischen
    Genotyp und Umwelt bei quantitativen Eigenschaften.
    Zeitschrift fuer Pflanzenzuechtung. 53, pp. 266-343.

Effects of Intermating on Kernel Colour and Pre-Harvest Dormancy in Durum Wheat

J. F. Soper, R. G. Cantrell, and J. W. Dick,
North Dakota State University, Fargo,
North Dakota, USA.

SUMMARY

An attempt is being made to incorporate pre-harvest dormancy from RL4137 (Triticum aestivum L. em. Thell) into durum wheat (T. turgidum L. var. durum) germplasm. An experiment was conducted to determine if random intermating could be utilized to reduce the genetic associations between kernel colour and pre-harvest dormancy in populations derived from the hybridization of 'Vic' durum and PD#44, a red-seeded tetraploid experimental line possessing pre-harvest dormancy derived from RL4137. Results indicated that one cycle of random intermating did not reduce the genetic associations between kernel colour and dormancy (visual sprouting scored using intact spikes in a rain simulation chamber) ranged from -0.46 to -0.59. Following random intermating, the genotypic variance for sprouting was significantly higher, and small changes in the population means were detected for sprouting and kernel colour. Fortunately, significant variation exists among amber-seeded lines derived from the experimental populations. Several lines appear to have dormancy levels similar to PD#44. In a backcrossing program using RL4137 and Kenya 321 sib as dormant parents, and Vic as the recurrent parent, we have obtained BC_3F_2 progeny which appear to have good levels of dormancy. Further research is being conducted to evaluate the alpha-amylase activity of experimental lines using a gel-diffusion technique (amylopectin azure blue). Preliminary results indicate that relative alpha-amylase activity, as determined by the area of gel clearance zones, is moderately correlated with sprouting and kernel colour.

INTRODUCTION

Durum wheat, Triticum turgidum L. var. durum, is
extremely susceptible to sprouting with the onset of
harvest-time rains. Sprouting may reduce semolina yield,
lower pasta colour scores, increase speck count in pasta
products, and reduce product shelf life (Dick et al., 1974;
Donnelly, 1979; Harris et al., 1943; Leu, 1969).
Additionally, growers suffer an economic loss resulting from
a reduction in the U.S. numerical grade and durum subclass-
ification due to the presence of sprouted and non-vitreous
kernels. In 1980, 75-80% of North Dakota's durum crop
suffered some degree of sprouting, resulting in an economic
loss to the state estimated at $80 million (Dick et al.,
1980; Maier, 1980). Attempts to identify stable sources of
pre-harvest dormancy in durum breeding material have been
unsuccessful. For this reason, an attempt is being made to
incorporate pre-harvest dormancy (sprouting resistance) from
RL4137 (T. aestivum L. em. Thell) into durum breeding germ-
plasm. RL4137 has a long, stable dormancy period and is the
source of sprouting resistance in 'Columbus' hard red spring
wheat (De Pauw and McCaig, 1983). The pre-harvest dormancy
in RL4137 is strongly associated with the presence of red
kernel colour, however, a partial transfer of RL4137's
dormancy to white-seeded spring wheat lines has been
reported by De Pauw, and McCaig (1983). Their results
suggest that it may be possible to incorporate pre-harvest
dormancy into amber-seeded durum germplasm if the genetic
factors involved are associated with the A or B genomes.
The probability of obtaining amber-seeded lines with
adequate levels of pre-harvest dormancy may be increased by
utilizing random intermating prior to selection if the
association between kernel colour and pre-harvest dormancy
is due to genetic linkage. The primary objective of this
study was to determine the effects of random intermating on
pre-harvest dormancy, kernel colour, and their inter-
relationship in populations derived from a cross between
'Vic' durum and PD#44.

MATERIALS AND METHODS

Two populations derived from the cross Vic/PD#44 were
used in the experiment. Vic is a high yielding, strong
gluten, amber durum cultivar which occupied 53% of North
Dakota's durum acreage in 1985. PD#44 is a red-seeded
tetraploid experimental line possessing pre-harvest dormancy

derived from RL4137 and having the pedigree [Coulter/RL4137/
/Vic, F5 (#44)]. It was obtained from D. Leisle at Agriculture Canada. F$_2$ data from this cross indicate that PD#44
possesses a single dominant gene for red kernel colour. The
base population (C0S1) consisted of 188 S0-derived S1 families (analogous to F2-derived F3 families). C0S1 family
seed was harvested from C0S0 plants grown in 1983 at Fargo.
A second population (C1S1) consisted of 188 S0-derived S1
families following one cycle of random intermating. Intermating was conducted in the greenhouse using hand emascula-
tion. Two seeds were planted, one at each of two dates,
from each of 200 C0S1 families, thus giving rise to approximately 400 plants in the crossing block. By randomly
utilizing each of these plants as a parent in one cross,
approximately 200 intermatings were obtained. Hybrid (C1S0)
seeds from the crosses were planted in the greenhouse during
the winter of 1983-84 and plants were allowed to self-
pollinate for production of C1S1 seed.

The experiment was conducted as a sets/rep design (2
reps, 4 sets) in Fargo and Langdon, North Dakota in 1984.
Each set consisted of 47 C0S1 families, 47 C1S1 families,
and five check entries (Vic, PD#44, PD#45, RL4137, and
D7075). PD#45 is a sister line of PD#44 and D7075 is a
North Dakota durum experimental line which has exhibited
moderate dormancy in Canadian experiments (D. Leisle,
personal communication). Twenty to thirty seeds from each
family were planted in a 2m row within each replication.
Agronomic field evaluation was limited to the measurement of
days to heading and plant height. Five individual plants
were harvested from each family row at approximately 5-15
days after harvest ripeness (HD1 and HD2, respectively).
Spikes from harvested plants were stored at 21-24°C for 10
days to reduce grain moisture to safe storage levels and
then placed in a walk-in freezer (-15 to -20°C) to maintain
dormancy until sprouting tests could be conducted. Dormancy
was assessed by placing intact spikes (one/harvested plant)
into a rain-simulation chamber. The spikes were exposed to
a fine water spray (simulating 6cm of rainfall) for 4 hours.
Spikes remained in the chamber for 5 days at 20°C and 100%
relative humidity. Following this treatment, individual
spikes were visually scored for sprouting on a scale from
0-9 [0 = no visible sprouting, 9 = severe sprouting (uniform
sprouting over the spike with coleoptiles exceeding 2cm in
length)]. Sound seed samples from each harvested plant were
visually rated for kernel colour on a scale from 0-6 (0-3 =
light to dark amber yellow, 4-6 = light to dark red).

Statistical analyses were conducted using family-row

means for all traits. All sources of variation in the
model were assumed random. Variance and covariance
estimates were obtained for each population using analyses
of variance and covariance similar to the form used for the
presentation of mean squares in TABLE 1. The components
were utilized to estimate genotypic variances, heritabilities, and genotypic correlations on an S1 family mean basis
for the ith and jth traits as indicated:

Genotypic variance $\hat{o}g_i$ = (MS fam$_i$ - MS fam*loc$_i$)/4

Genotypic covariance og_{ij} = (MSCP fam$_{ij}$ - MSCP fam*loc$_{ij}$)/4

Heritability H = (MS fam - MS loc*fam)/(MS fam)

Genotypic correlation rg_{ij} = og_{ij} /(og_i og_i)

Population mean differences were tested by using an F-test
involving partitioned mean squares [MS (C0 vs. C1)/ MS (loc
* C0 vs. C1)]. The "C0 vs. C1" error was pooled with "loc
* C0 vs. C1" and used as the denominator in this test if
"loc * C0 vs. C1" was not significant at the 0.20 level.
F-tests also were used to determine if the genotypic
variance estimates differed significantly between the two
populations.

TABLE 1. Mean squares for sprouting and kernel colour.

		Mean squares			
		Sprouting		Kernel colour	
Source	df	HD1°	HD2	HD1	HD2
Families/Sets (F/S)	368	3.84**	3.79**	4.33**	4.66**
Among C0S1	184	3.54**	3.39**	4.68**	4.72**
Among C1S1	184	3.95**	4.04**	4.04**	4.64**
C0S1 vs. C1S1	4	12.29**	10.34	1.70*	2.87**
Locations * F/S	368	1.24*	1.91**	0.31	0.29
Locations*C0S1	184	1.44**	2.01**	0.31	0.33
Locations*C1S1	184	1.02	1.85**	0.31	0.25
Locations* (C0S1 vs. C1S1)	4	2.01	0.71	0.11	0.59
Error	732	0.95	1.21	0.29	0.31
Within C0S1	366	0.92	1.24	0.28	0.32
Within C1S1	358	0.96	1.10	0.29	0.31
C0S1 vs. C1S1	8	1.82	5.17	0.48	0.16

*,** Significant at the 0.05 and 0.01 levels, respectively.
° HD1 and HD2 refer to harvest dates at 5 and 15 days after
harvest ripeness, respectively.

RESULTS AND DISCUSSION

Small but significant population mean differences were detected for all traits except plant height (TABLES 1 and 2). The C1S1 population exhibited less sprouting, higher kernel colour score, and later maturity. The change in population means could indicate the presence of epistasis involving linkage or may be due to random drift or inadvertant selection. The changes which occurred would appear to have little benefit to durum breeding efforts since higher colour scores (indicating a higher proportion of red seeded lines) and later maturity accompanied the desirable increase in pre-harvest dormancy following random intermating.

TABLE 2. Population means for the COS1 and C1S1 families, and checks.

	Sprouting HD1°	Sprouting HD2	Kernel Colour HD1	Kernel Colour HD2	Heading days	Height cm
	---score¶---		---score§---		days	cm
COS1						
mean	2.59**	4.19	4.28*	4.25*	59.5**	79.4
range	0.2-5.6	1.7-6.5	2.0-5.9	2.1-6.0	57-64	66-88
C1S1						
mean	2.31	3.93	4.37	4.37	59.9	79.1
range	0.2-5.4	1.5-6.4	2.2-5.8	2.0-6.0	57-64	68-90
Vic	3.23	4.08	2.15	2.06	58.6	86.4
D7075	2.66	4.05	2.11	2.12	56.8	78.6
PD#44	1.21	3.29	5.66	5.74	57.8	75.5
PD#45	0.66	2.26	5.18	5.18	58.7	76.1
RL4137	0.18	1.30	5.55	5.48	59.4	85.6

*, ** Significant difference between the population means at the 0.05 and 0.01 levels, respectively.
°HD1 and HD2 refer to harvest dates at 5 and 15 days after harvest ripeness, respectively.
¶Scale 0-9 (0 = no visible sprouting, 9 = severe sprouting).
§Scale 1-6 (1-3 = light to dark amber, 4-6 = light to dark red).

Significant variation was found among families in both populations for all traits examined (TABLE 1). Location by family interactions were significant for sprouting and heading date. The genotypic variance for sprouting was significantly higher in the C1S1 than in the COS1 population (TABLE 3).

TABLE 3.   Genotypic variance and heritability estimates for sprouting, kernel colour, and agronomic traits.

Trait	Population	Genotypic Variance	Heritability
Sprouting (HD1)°	COS1	0.53*	0.60
	ClS1	0.73	0.74
Sprouting (HD2)	COS1	0.35*	0.41
	ClS1	0.55	0.54
Kernel Colour (HD1)	COS1	1.09	0.93
	ClS1	0.93	0.92
Kernel Colour (HD2)	COS1	1.10	0.93
	ClS1	1.10	0.95
Days to Heading	COS1	1.31	0.67
	ClS1	1.38	0.73
Plant Height	COS1	7.14	0.59
	ClS1	8.83	0.63

* Significant difference between the population variance estimates at the 0.05 level.
° HD1 and HD2 refer to harvest dates at 5 and 15 days after harvest ripeness, respectively.

Corresponding increases in the heritability estimates for sprouting also were obtained. The increase in genotypic variability following intermating would suggest that some of the principal genes controlling pre-harvest dormancy in this population may be associated in a repulsion-phase linkage. Vic may be contributing a favorable dormancy gene to the COS1 and ClS1 populations, despite its relatively low dormancy level. Linkage and an epistatic association between this gene and the dormancy gene(s) contributed by PD#44 could explain the increased genotypic variance for sprouting and decreased sprouting mean following intermating. Despite the increased genotypic variance in the ClS1 population, no significant changes in the genotypic variance and heritability estimates for sprouting were detected between the populations when only considering the amber-seeded families on which selection will be practiced. Fortunately, there is significant variation for pre-harvest dormancy among amber-seeded families within both populations. Several families appear to possess pre-harvest dormancy levels similar to PD#44 but none approach the level of RL4137 (TABLE 4). Variability among these families suggests a partial transfer of dormancy, not associated with kernel colour, from PD#44 and/or recombination between minor genes controlling pre-

TABLE 4. Sprouting means for checks and amber-seeded, red-seeded, and segregating families within the COS1 and C1S1 populations

	N	HD1°		HD2	
		mean	range	mean	range
		---------------score¶---------------			
Amber-seeded families	87	3.28	0.95-5.55	4.66	2.46-6.35
Red-seeded families	83	2.06	0.15-4.50	3.76	1.45-6.05
Segregating families	206	2.25	0.45-4.70	3.93	1.80-6.45
Vic		3.23		4.08	
D7075		2.66		4.05	
PH#44		1.21		3.29	
PD#45		0.66		2.26	
RL4137		0.18		1.30	

°HD1 and HD2 refer to harvest dates at 5 and 15 days after ripeness, respectively.
¶Scale 0-9, (0 = no visible sprouting, 9 = severe sprouting)

harvest dormancy from the Vic and PD#44 backgrounds. A partial transfer of dormancy from RL4137 to white-seeded spring wheat lines has been reported by De Pauw and McCaig (1983). Genotypic variance and heritability estimates for other traits were similar in both populations. Heritability estimates for all traits were in the moderate to high range.

Genotypic correlation estimates between sprouting and kernel colour were moderate and did not decrease following random intermating (TABLE 5). This indicates that the genetic association between red kernel colour and pre-harvest dormancy is either pleiotropic or due to an extremely tight linkage. The strong association between red kernel colour and pre-harvest dormancy will continue to hamper efforts to obtain amber-seeded lines with high levels of pre-harvest dormancy. Maturity was not correlated with sprouting within either population. There was a weak positive correlation between plant height and sprouting at the first harvest date in the COS1 population while no correlation between these traits was detected in the C1S1 population, possibly indicating a weak coupling-phase linkage between these traits which approached linkage equilibrium with random inter-

TABLE 5. Genotypic (above the diagonal) and phenotypic (below the diagonal) correlation estimates between the sprouting, kernel colour, and agronomic traits.

		S1	S2	C1	C2	DH	HT
Sprouting-HD1°							
(S1)	COS1		0.96	-0.56	-0.53	-0.05	0.22
	C1S1		1.04	-0.59	-0.60	-0.14	0.10
Sprouting-HD2							
(S2)	COS1	0.72**		-0.46	-0.48	-0.05	0.27
	C1S1	0.80**		-0.52	-0.59	-0.12	-0.10
Kernel Colour-HD1							
(C1)	COS1	-0.42**	-0.28**		1.00	0.05	-0.05
	C1S1	-0.49**	-0.35**		1.00	0.04	-0.17
Kernel Colour-HD2							
(C2)	COS1	-0.39**	-0.30**	0.94**		0.02	-0.08
	C1S1	-0.48**	-0.37**	0.94**		0.00	-0.12
Days to Heading							
(DH)	COS1	-0.09	-0.01	0.05	0.02		0.12
	C1S1	-0.10	-0.05	0.03	0.02		-0.07
Plant Height							
(HT)	COS1	0.23**	0.13	-0.02	-0.05	-0.08	
	C1S1	0.00	0.08	-0.13	-0.09	-0.18**	

** Correlation coefficient significantly different from zero at the 0.01 level.
°HD1 and HD2 refer to harvest dates at 5 and 15 days after harvest ripeness, respectively.

mating.

Another approach to recombining dormancy with amber kernel colour is backcrossing. Preliminary data from a backcrossing program with RL4137 and Kenya 321 sib as dormant parents, and Vic as the recurrent parent, are presented in TABLE 6. Three generations of backcrosses have been completed with selection for dormancy practiced within each cycle. The BC_3F_2 progeny distribution for sprouting indicates that 73% of the amber-seeded progeny were superior to Vic and 14% approach RL4137. The level of dormancy in the population derived from Kenya 321 sib was lower yet some progeny demonstrated strong dormancy. Further selection and evaluation within each of these populations should produce useful durum germplasm with improved levels of pre-harvest dormancy.

Further research is being conducted to evaluate entries for alpha-amylase activity following rain-simulation.

TABLE 6. Mean sprouting scores from BC_3F_2 progeny from two different sources of dormancy evaluated at Langdon, ND in 1985.

Population	Kernel Colour	N	Sprouting Score¶					Check Mean
			0-1	2-3	4-5	6-7	8-9	
			--------% of progeny-------					score
RL4137/ Vic*3								
	Red	66	20	65	15	0	0	
	Amber	137	14	59	27	0	0	
Kenya 321 sib/Vic*3	Amber- white	200	7	55	38	0	0	
Vic								5.4
RL4137								0.5

¶ Determined on a 5 spike sample harvested from each plant approximately 7 days after harvest ripeness and then placed in the rain-simulation chamber for 5 days.

We are utilizing a gel-diffusion technique (amylopectin azure blue, Calbiochem) as described by Fox and Eslick (1980). Preliminary results indicate that relative alpha-amylase activity, as measured by the area of gel clearance zones, is moderately correlated with sprouting [$r = 0.46**$ (HD1), $0.61**$ (HD2)] and kernel colour [$r = -0.41**$ (HD1), $-0.52**$ (HD2)]. In an additional study concerning the effects of sprouting on pasta quality, relative alpha-amylase activity (gel-diffusion) has correlated well with falling number ($r = -0.85**$) and alpha-amylase activity (log 10 transformed) as measured by the Perkin-Elmer Grain Amylase Analyzer ($r = 0.91**$). The gel diffusion technique appears promising for large scale screening due to its rapidity, simplicity, and relatively low cost.

CONCLUSIONS

It appears that a single cycle of random intermating did not improve the probability of obtaining amber-seeded lines with suitable levels of pre-harvest dormancy in the populations examined. This is apparently due to a pleiotropic or tight linkage relationship between red kernel colour and pre-harvest dormancy. Fortunately, there is significant variation among amber-seeded families within

the COS1 and C1S1 populations. Several families appear to
have dormancy levels similar to PD#44. Backcrossing efforts
have produced BC_3F_2 progeny with amber colour and good
levels of pre-harvest dormancy. Further testing will be
conducted to evaluate the level and stability of dormancy in
promising amber lines and families. Evaluation of a second
cycle of random intermating is underway.

REFERENCES

De Pauw, R. M., and McCaig., T. N. 1983. Recombining
    dormancy and white seed colour in a spring wheat cross.
    Can. J. Plant Sci. 63, pp. 581-589.
Dick, J. W., Banasik, O. J. and Vasiljevic, S. 1980.
    Quality of 1980 durum wheat crop. Agr. Exp. Sta.,
    North Dakota State Univ., Fargo.
Dick, J. W., Walsh, D. E. Walsh, and Gilles, K. A. 1974.
    The effect of field sprouting on the quality of durum
    wheat. Cereal Chemistry. 50, pp. 180-188.
Donnelly, B. J. 1979. Effect of sprout damage on durum
    wheat quality. Trans. 8th Joint Conference of the
    Association of Operative Millers District no. 13 and
    the Am. Assoc. of Cereal Chemists, Canadian section
    no. 14. Winnipeg, Manitoba, Canada.
Fox, G., Eslick, R. F. 1980. Rapid determination of cereal
    alpha-amylase using amylose azure blue dyed starch
    suspended in agar medium. Agron. Abstr., Am. Soc.
    Agronomy, Madison, WI. p. 124.
Harris, R. H., Smith, G. S. and Sibbitt L. D. 1943. The
    effect of sprout damage on the quality of durum wheat,
    semolina, and macaroni. Cereal Chemistry. 20, pp.
    333-345.
Leu, R. 1969. The effect of wheat sprouting on the quality
    of spaghetti. M. S. Thesis, North Dakota State Univ.,
    Fargo.
Maier, M. G. 1980. Widespread sprout damage. Macaroni J.
    62, pp. 20-21.

Differences in Pre-Harvest Sprouting and Alpha-Amylase Activity among Wheat Cultivars

K. Fukunaga,* T. Hoshino,* U. Matsukura,**
H. Taira** and S. Oda.*
*National Agriculture Research Centre,
** National Food Research Institute,
Yatabe, Tsukaba, Japan 305.

SUMMARY

    Changes in sprouting percentage and alpha-amylase
activity with days after heading (DAH) were investigated in
20 wheat cultivars. The relationship between sprouting and
alpha-amylase activity was analyzed in connection with the
selection of breeding lines with sprouting resistance and
low alpha-amylase activity. Sprouting percentage was high
at 35 DAH and at 60 DAH, while low at 45 DAH or 50 DAH.
Alpha-amylase activity was high before maturity and continued to decrease until 45 DAH when the lowest activity was
recorded. The relationship between sprouting and alpha-
amylase activity was significant after maturity but not
significant when cultivars with a high sprouting percentage
and/or high alpha-amylase activity were omitted. These
results suggest that some of the cultivars with a low
sprouting percentage displayed a high alpha-amylase activity
at 50-60 DAH. For the selection of breeding lines with
sprouting resistance and low alpha-amylase activity, pre-
harvest sprouting should be tested twice at 35 and 60 DAH,
while alpha-amylase activity should be measured at 50-60
DAH.

INTRODUCTION

    One of the most important breeding objectives in Japan
is the release of cultivars with a high flour quality for
noodle production. Several leading cultivars in our country are susceptible to pre-harvest sprouting when subjected
to rain during the ripening stage. Sprouting is closely
related to alpha-amylase activity (Bhatt et al., 1981;

116

Huang et al., 1980). Alpha-amylase consists of multiple
forms, and alpha-amylase-1 and -2 behave differently with
grain maturation (Gale et al., 1983). Alpha-amylase of
sprouted grain is different from the alpha-amylase that
operates during the maturation of the developing grain
(Meredith 1980-1982).

The objectives of our studies were to investigate the
changes in pre-harvest sprouting percentage and alpha-
amylase activity with days after heading (DAH), and to analyze the inter-relation between sprouting and the enzyme
activity during the ripening stage of wheat.

MATERIALS AND METHODS

Twenty wheat cultivars and breeding lines with varying
percentages of pre-harvest sprouting (TABLE 1) were grown at
the National Agriculture Research Centre from the end of
October 1982 to the end of June 1983 (1983) and from the end
of October 1983 to the beginning of July 1984 (1984).
Approximately 20 spikes were randomly harvested from drill
strips, 8 meter long, at 5 day intervals from 30 DAH in 1983
and 20 DAH in 1984 to 60 DAH (anthesis occurred 5-7 days
later than heading). A subsample of 7 spikes per cultivar
was used for the pre-harvest sprouting test and the remaining spikes, wrapped with a nylon bag, were stored at -20°C
after freeze-drying for assay of alpha-amylase activity.

Winter was very warm in 1983, but very cold in 1984
(mean temperature in February: 3.1°C in 1983, 0°C in 1984).
Heading of cultivars occurred 12-16 days later in 1984 than
in 1983 (heading date of Norin 61 was April 26 in 1983 and
May 9 in 1984). The rainy season started at similar times
in both years, however the cultivars received more rain in
1984 than in 1983 during the late ripening stage (rainfall:
157.2mm in 1983 and 221.3mm in 1984 per cultivar from 20-59
DAH).

1. Sprouting test: Seven spikes per cultivar were placed
   in a rain simulator maintained at 16°C, in a standing
   position immediately after harvest. About 85mm of
   "rain" per day was applied to the spikes for 2 weeks.
   Germinated grains on the spikes were removed from the
   simulator and the grains were considered to have germinated when the pericarp over the embryo had ruptured.
   Pre-harvest sprouting percentage (%) was calculated on
   the basis of the sprouting number of 5 spikes; omit-
   ting the maximum and minimum values recorded in the

TABLE 1. Change of pre-harvest sprouting percentage with the days after heading in 1984.

Cultivar	Heading day	Days after heading								
		20	25	30	35	40	45	50	55	60
Toyohokomugi	May 10	0	0	22.5	9.8	4.9	4.1	6.1	5.7	40.0
Ejimashinriki	May 9	0	0	35.2	55.2	16.7	8.7	23.3	21.5	53.7
Sakigakekomugi	May 7	0	0	3.5	25.2	5.4	14.8	20.4	34.8	32.2
Norin 61	May 9	0	0	15.8	15.8	8.8	9.8	22.1	20.0	27.9
Ushiokomugi	May 9	0	0	4.2	16.6	7.2	3.8	15.8	9.4	40.0
Fukuhokomugi	May 7	0	0	9.6	73.6	21.6	34.8	51.6	40.0	74.4
Junreikomugi	May 7	0	0	2.0	50.0	8.6	24.0	40.4	44.8	49.6
Shinchunaga	May 9	0	0	13.2	40.0	23.2	29.8	40.4	48.6	77.1
Kanto 100	May 5	0	0	0	40.4	33.3	12.1	21.7	12.9	18.8
Shirasagikomugi	May 9	0	0	12.1	50.6	22.6	12.6	38.1	21.5	48.7
Fujimikomugi	May 10	0	0	22.8	80.0	47.6	30.8	42.4	22.0	37.2
Zenkojikomugi	May 10	0	0	9.1	42.7	23.6	17.0	20.9	18.2	20.0
Kanto 85	May 9	0	0	24.5	74.5	59.7	24.2	38.4	29.7	51.9
Norin 67	May 11	0	0	48.9	86.4	47.5	34.3	22.9	63.2	62.5
Aobakomugi	May 11	0	0	0.5	23.2	59.1	40.0	40.5	87.3	96.4
Saitama 27	May 12	0	0	25.0	53.1	41.9	45.8	65.4	48.7	70.8
Omasekomugi	May 8	0	0	22.9	70.2	17.6	44.9	21.2	39.6	66.9
Ichigoukumamotokomugi	May 12	0	1.9	52.2	73.0	85.1	64.6	69.7	70.0	59.2
Jessore	May 12	0	0	11.2	48.5	64.1	57.6	50.5	38.3	36.6
Kanto 99	May 9	0	0	42.3	61.0	76.8	95.2	92.3	72.9	71.9

subsample of 7 spikes.

2. Alpha-amylase activity: The method of Mathewson and
   Pomeranz (1977) was modified to obtain stable data
   (Matsukura et al., 1984).   Activity was expressed as
   the absorbance per 1 gram of wholemeal (Unit).

RESULTS AND DISCUSSION

1. Pre-harvest sprouting

     All the cultivars except Ichigoukumamotokomugi failed
to sprout at 20 and 25 DAH (TABLE 1). Sprouting tendency
appeared suddenly at 30 or 35 DAH (milky stage) and
decreased until 45 or 50 DAH (maturity) in most cultivars.
After maturity, sprouting continued to increase until 55 to
60 DAH (dead ripening stage). Kanto 100 and Zenkojikomugi
showed a relatively high percentage of sprouting at 35 DAH
but a low response from 40-60 DAH. Kanto 99 and Ichigou-
kumamotokomugi exhibited a very high percentage of sprouting
from 30-60 DAH.

     The large amount of rain in the early to middle part of
June resulted in sporadic pre-harvest sprouting prior to
grain maturity. Wheat often stands for a long period of
time in the field after the dead ripening stage as harvesters cannot be operated in the field with abundant rain.
Pre-harvest sprouting should be tested at 35 and 60 DAH when
the cultivars are susceptible to sprouting to avoid damage
caused by early and late sprouting under the climatic conditions prevailing in Japan.

2. Alpha-amylase activity

     Kanto 100 showed the highest activity (49.9 unit in
1983 and 68.2 unit in 1984) at 30 DAH, and a low activity
(1.6 unit in 1983 and 1.1 unit in 1984) at 50 DAH compared
with the other cultivars. In general, the alpha-amylase
activity of all the cultivars exhibited the highest value on
the first sampling day (30 DAH in 1983 and 25 DAH in 1984),
and declined rapidly until 35 DAH. The activity continued
to decrease gradually until 45 or 50 DAH (maturity). Most
of the cultivars showed a higher activity at 30 and 35 DAH
in 1984 than in 1983 and almost the same activity at 40, 45,
50 DAH in both years. After maturity, the activity increased gradually in 1983 and rapidly in 1984. The pattern
of change of the alpha-amylase activity with days after

heading was similar to that reported by Gale et al., 1983;
and Marchylo et al., 1980.

3. Relationship between pre-harvest sprouting and alpha-amylase activity

The relationship between sprouting percentage and
alpha-amylase activity was similar in 1983 and 1984. Correlation coefficients between sprouting at any time and
alpha-amylase activity at 30, 35, 40, 45 DAH was not
significant (TABLE 2).

TABLE 2. Relationship between pre-harvest sprouting and
alpha-amylase activity in 1983.

Alpha-amylase activity	Pre-harvest sprouting						
	30	35	40	45	50	55	60#
30#	-0.038	0.178	0.066	0.215	0.362	0.371	0.123
35	-0.332	0.189	-0.145	-0.046	0.112	0.135	-0.212
40	-0.223	0.257	0.002	0.095	0.229	0.254	-0.053
45	-0.007	0.226	0.135	0.273	0.426	0.366	0.111
50	0.237	0.524*	0.726***	0.652**	0.754***	0.602**	0.562**
55	0.198	0.588**	0.672**	0.581**	0.604**	0.617**	0.612**
60	0.230	0.590**	0.754***	0.661**	0.731***	0.657**	0.645**

Days after heading
*, **, ***, significant at the 5, 1 and 0.1% level,
respectively.

A significant correlation was observed between sprouting
from 35 DAH onward and alpha-amylase activity at 50, 55, 60,
DAH. The results obtained in this study may support
indirectly those reported by Meredith (1980-1982) and Gale
et al., 1983. However no significant correlation between
sprouting and alpha-amylase activity was observed when the
cultivars with a high percentage of sprouting (>10% in 1983,
>50% in 1984) and/or a high alpha-amylase activity (>1.0
unit in 1983, >10.0 unit in 1984) were omitted for the correlation analysis (r = 0.127 in 1983, r = 0.407 in 1984)
(FIGURE 1). The propensity of some wheat cultivars to have
unacceptable high levels of alpha-amylase activity at maturity even in the absence of visible sprouting has been recognized for some years (Gale et al., 1983). Selection for
both sprouting resistance and low alpha-amylase activity is
recommended because the two traits are not mutually

FIGURE 1.   Relationship between alpha-amylase activity and sprouting percentage at 55 days after heading in 1983.

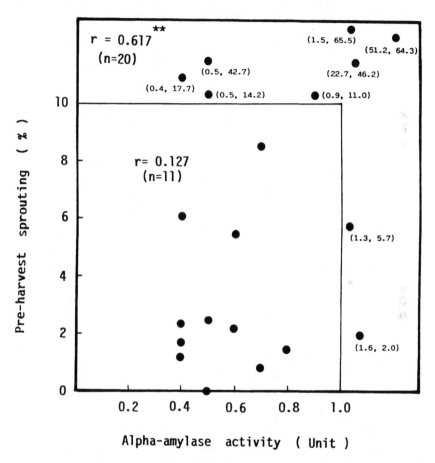

inclusive (McCrate et al., 1981).
    It is suggested that the alpha-amylase activity should be measured at 50-60 DAH for the selection of breeding lines with sprouting resistance and low alpha-amylase activity.

REFERENCES

Bhatt, G. M., Pausen, G. M., Kulp, K. and Heyne, E. G. 1981. Pre-harvest sprouting in hard winter wheats: Assessment of methods to detect genotype and nitrogen effects and interaction. Cereal Chem. 58, pp. 300-302.
Gale, M. D., Flintham, J. E. and Arthur, E. D. 1983. Alpha-amylase production in the late stages of grain develop-

ment - An early sprouting damage risk period? In: Third Int. Symp. on Pre-harvest Sprouting in Cereals. Kruger, J. E. and LaBerge, D. E., (eds.). Westview Press, Boulder, Co., USA. pp. 29-35.

Hung, G. and Varriano-Marston E. 1980. Alpha-amylase activity and pre-harvest sprouting damage in Kansas hard white. J. Agric. Food Chem. 28, pp. 509-512.

Marchylo, B. A., LaCroix L. J. and Kruger J. E. 1980. Alpha-amylase isoenzymes in Canadian wheat cultivars during kernel growth and maturation. Can. J. Plant Sci. 60, pp. 433-443.

Mathewson, P. R. and Pomeranz Y. 1977. Detection of sprouted wheat by a rapid colourimetric determination of alpha-amylase. Journal of AOAC. 60, pp. 16-20.

Matsukura, U., Kato, I., Taira, H. and Imai, T. 1984. Quality of Japanese wheat. Rept. Natl. Food Res. Inst. 44, pp. 97-100.

McCrate, A. J., Nielsen, M. T., Paulsen, G. M. and Heyne, E. G. 1981. Pre-harvest sprouting and alpha-amylase activity in hard red and hard white winter wheat cultivars. Cereal Chem. 58, pp. 424-428.

Meredith, P. 1980-1982. Sprouting damage. New Zealand wheat review. 15, pp. 78-83.

Variation of Falling Number in Primary Triticale and their Wheat and Rye Parents

G. Oettler, Landessaatzuchtanstalt,
Universität Hohenheim, Postfach 70 05 62,
7000 Stuttgart 70, Germany F. R.

SUMMARY

Pre-harvest sprouting potential was assessed by the Hagberg Falling Number Method in genetically defined octoploid and hexaploid primary triticale and their constituent wheat and rye parents. Consistently low falling number values were found in triticale at both ploidy levels, despite the high per se performance of the parental wheats and/or rye inbred lines. The genetic variability of the parents was not expressed in their hybrids, indicating severe hormonal disorders in the amphiploid. Analysis of variance revealed significant paternal and interaction effects for falling number. Significant positive correlations were found in hexaploid triticale between falling number and both grain yield and test weight. Thus, in triticale breeding programmes more attention should be devoted to the improvement of fertility and grain filling.

INTRODUCTION

Susceptibility to pre-harvest sprouting received consideration in triticale breeding as early as two decades ago (Müntzing, 1963). Unfortunately, very limited research activity has been devoted to overcoming this problem, although it was recognized as being a major defect of the new crop.

Investigations relating the alpha-amylase activity, mainly assessed by falling number, of various triticale to the values for standard wheat and rye cultivars (Chojnacki et al., 1976; McEwan and Haslemore, 1983) revealed that the hybrid compared very unfavourably with both parental species.

For an effective breeding and selection programme which is aimed at improving pre-harvest sprouting behaviour in the intergeneric hybrid triticale (x Triticosecale Wittmack), the following aspects should be considered: (1) relationship between per se performance of the wheat and rye parents and their synthesized hybrids, (2) parental and interaction effects in the hybrid, and (3) associations between pre-harvest sprouting and other plant and seed characteristics in the amphiploid. Definitive information on these three subjects will only be obtained from studies of genetically defined primary triticale and comparisons with the homozygous wheat and rye inbred lines from which they are derived.

With respect to the first point, preliminary results (Oettler, 1985) indicated that the falling number of triticale was low and that there was little variation, despite the considerable variability observed among the rye inbred lines. Some aspects of the second point have recently been addressed in a study of yield characters in defined triticale (Oettler, 1986). Information regarding the association of pre-harvest sprouting with other characters, is however, only available for undefined materials. Hill et al., (1974) found a significant negative correlation between test weight and alpha-amylase activity and the latter trait showed a close negative curvilinear association with the falling number as documented by Krey (1985). Chojnacki et al., (1976) also noted that shrivelled grain had higher levels of alpha-amylase and they found a tendency for dwarf triticale to have high falling numbers.

In the work reported here the three aspects, cited above as being important with regard to the development of pre-harvest sprouting tolerant triticale, were studied using a large number of genetically defined octoploid and hexaploid primary triticale.

MATERIALS AND METHODS

The experimental material, all of winter habit, comprised three Triticum aestivum L. cultivars, five lines of T. durum L., one T. turgidum L., 17 Secale cereale L. inbreds and selfed seed of 71 primary tricticale newly synthesized from these parents. Seed of eight primary triticale was kindly provided by Dr. T. Lelley, Göttingen. Wheat, rye and triticale were tested in 1984 at Hohenheim, in randomized block designs with three replications, 2m² plots and a seeding rate of 200 grains per m².

The Hagberg Falling Number test was used to assess pre-harvest sprouting potential. At harvest ripeness grain samples of 25g were ground and portions of 9g (triticale and rye) and 7g (wheat) were tested.

To lessen unorthogonality, a reduced and nearly balanced set of 38 triticale, produced from two T. aestivum (Kormoran, Jubilar), three T. durum (D4, D30, D40), one T. turgidum (T 34) and eight rye inbred parents, was used for obtaining mean values as least squares estimates (LSE). For further evaluation octoploid triticale were disregarded since their falling numbers did not vary. To calculate the analysis of variance the falling number values were transformed to logarithms. Trait correlations were computed for all 48 hexaploid triticale of the original set.

RESULTS

The falling numbers of 38 triticale from a factorial crossing plan together with the values of wheat and rye parents are shown in TABLE 1.

TABLE 1. Falling number values (secs) of defined octoploid and hexaploid primary triticale and their wheat and rye parents.

Designation			Hexaploid Jub	Kor	WHEAT Tetraploid D4	D30	D40	T34	Mean (LSE)
			348	322	62	65	70	200	178
	RYE								
L	25	119	62	62	78			72	71
L	37	131		62	62		66	62	62
L	70	64	62		62	62	62	62	62
L	145	172	62		62	62		84	67
L	150	68	62	62	62	62	78	95	70
L	155	174	62	62	62			62	65
L	185	269	62		62	81	70	72	68
L	301	158	62	86	90	120	102	89	92
Mean (LSE)		144	62	65	68	75	74	75	70

126

In general, the individual triticale values and the means were at a low level and varied little, despite the large differences in per se performance of their parents: 62 to 348 seconds for wheat and 64 to 269 seconds for rye. With all but one octoploid triticale the lowest falling number of 62 seconds was obtained, whereas in hexaploids the values ranged from 62 to 120 seconds.

The analysis of variance for falling number of hexaploid triticale (TABLE 2) revealed highly significant effects due to the rye parents and parental interactions.

TABLE 2. Mean squares of defined hexaploid primary triticale for falling number (logarithmic transformation).

Source	d.f.	MS
Wheat	3	0.058
Rye	7	0.250**
Wheat x Rye	15	0.049**

** Significant at P = 0.01.

The correlation coefficients between falling number and various agronomic and seed characters in hexaploid triticale (TABLE 3) were small to moderate. Significant positive correlations existed with grain yield and test weight, and a significant negative value was found for grain protein content.

TABLE 3. Correlations between falling number and other plant and seed characters of defined hexaploid primary triticale.

Character correlated with falling number	r (N = 48)
Heading (days from May 1st)	0.10
Plant height (cm)	-0.04
Spikes per m²	0.14
Grain yield (g/m²)	0.49**
1000-grain weight (g)	0.22
Test weight (kg/hl)	0.56**
Grain protein content (%)	-0.37*

*, ** Significant at P = 0.05 and 0.01, respectively.

DISCUSSION

During the 1984 season conditions were conducive to pre-harvest sprouting, which is evidenced by the low individual and mean falling numbers of octoploid and hexaploid triticale. Low falling numbers have also been reported for genetically undefined materials (Tymieniecka and Wolski, 1980; McEwan and Haslemore, 1983).

The very high values characteristic of the common wheat cultivars 'Jubilar' and 'Kormoran' were not reflected in their octoploid hybrids, despite the high genomic contribution of the wheat parents. The results were in accord with previous experience with conditions unfavourable to pre-harvest sprouting (Oettler, 1985). In addition, octoploids had even lower falling numbers than hexaploids and completely lacked variability. At the hexaploid level also, a high falling number of one wheat parent (T 34) was not expressed in its hybrids.

The performance of the rye inbreds also varied considerably, but appeared to show little relationship to the variation and level of falling number in the hybrids. One rye inbred (L 301) of only moderate performance made a favourable contribution to the pre-harvest sprouting behaviour of its hybrids.

Obviously, the low falling numbers in triticale were not primarily due to the rye parents, a view held by many authors (Müntzing, 1963; Hill et al., 1974; Huskowska et al., 1985). Likewise, with regard to the female parent, it was not the tetraploid wheat which caused susceptibility to pre-harvest sprouting, since hexaploid triticale tended to have higher falling numbers. High parental performance generally resulted in hybrids where the character was strongly inhibited in its expression. Thus, the unfavourable condition in particular in octoploid triticale, appeared to be the effect of severe enzymatic disorders, caused by the combination of alien genomes.

The analysis of variance also demonstrated the small effect due to the wheat parents. This result was unexpected, since it was known that most production characters in triticale were equally affected by both parents (Oettler, 1986). The highly significant interaction effects will partly impede the successful use of rye inbreds pre-selected for high falling numbers.

With respect to character associations, the relationship between pre-harvest sprouting resistance and the dwarf character reported by Chojnacki et al., (1976) was not found to be valid in the present study. It may, however, still be valid for genotypes containing the Rht 3 height reducing gene.

The highly significant correlation between falling number and grain yield in hexaploid triticale emphasized the importance of fertility for pre-harvest sprouting. A relationship between falling number and seed set was also detected in rye inbred lines by Köhler et al., (1986).

The significant positive association between test weight and falling number contradicted results previously reported for undefined materials by Huskowska et al., (1985), where negative coefficients were disclosed. Also McEwan and Haslemore (1983) could not find a correlation between test weight and alpha-amylase activity. The significant value obtained in this experiment indicated endosperm and starch anomalies and clearly pointed to the importance of improving seed quality in triticale.

The negative correlation between falling number and protein content in hexaploid tritical agreed with the fact that shrivelled kernels had higher protein.

CONCLUSIONS

The results indicated that the high falling numbers of wheat and/or rye parental genotypes were not expressed in triticale at an acceptable level. The genetic variability in falling number of the parents could not be exploited in the intergeneric hybrid, leaving very limited scope for selection. Since falling number in triticale was decisively determined by interaction effects between the parental genomes, there is little possibility of making predictions of the hybrid performance by means of the parents. The correlations between falling number and other characters suggested that progress in pre-harvest sprouting resistance of triticale can be accomplished only by accompanied improvement of fertility and grain filling.

The combination of alien parental genomes apparently caused severe hormonal imbalances, resulting in various defects in the hybrid. Thus, susceptibility to pre-harvest sprouting seemed to be primarily a physiological rather than a genetic problem.

ACKNOWLEDGEMENTS

The author wishes to thank Prof. Dr. H. H. Geiger for helpful discussions and to acknowledge the practical contribution made by Ms. E. Schwarz and Mr. J. Häge in this experimental work.

REFERENCES

Chojnacki, G., Brykczyinski, J. and Tymieniecka, E. 1976. Preliminary information on sprouting in triticale. Cereal Res. Comm. 4, pp. 111-114.

Hill, R. D., Klassen, A. J. and Dedio, W. 1974. Metabolic factors influencing kernel development in triticale. In: Triticale Proc. Int. Symp. MacIntyre, R. and Campbell, M. (eds.). El Batan, Mexico, Oct. 1-3, 1973, pp. 149-154.

Huskowska, T., Wolski, T. and Ceglinska, A. 1985. Breeding of winter triticale for improvement of grain characters. In: Genetics and Breeding of Triticale. Bernard, M. and Bernard S. (eds.). INRA, Paris, pp. 641-649.

Köhler, K. L., Geiger H. H. and Wilde, P. 1986. Analysis of inbreeding effects on falling number in diploid winter rye (Secale cereale L.). Proc. Eucarpia Meeting Cereal Section on Rye, June 11-13, 1985, Svalöf, Sweden (in press).

Krey, B. 1985. Methodische Untersuchung zur rascheren Erfassung von Auswuchs in Durumweizen. Diplomarbeit, Universität Hohenheim.

McEwan, J. M. and Haslemore, R. M. 1983. The response of triticale and related cereals to conditions inducing pre-harvest sprouting. In: Third Int. Symp. on Pre-harvest sprouting. Kruger, J. E. and LaBerge, D. E. (eds.). Westview Press, Boulder, Co., pp. 279-286.

Müntzing, A. 1963. Some recent results from breeding work with rye and wheat. In: Recent Plant Breeding Research, Svalöf 1946-1961. Åkerberg, E., Hagberg, A., Olsson, G. and Tedin, O. (eds.). John Wiley, New York, pp. 167-178.

Oettler, G. 1985. The influence of the wheat and rye genome on the performance of primary triticale. In: Genetics and Breeding of Triticale. Bernard, M. and Bernard, S. (eds.). INRA, Paris, pp. 125-134.

Oettler, G. 1986. Variation and covariation of agronomic characters in primary triticale and their wheat and rye parents. Proc. Int. Triticale Symp. Sydney, 1986, Aust. Inst. Agric. Sci. Occasional Publication No. 24, pp. 120-123.

Tymieniecka, E. and Wolski, T. 1980. Performance of some foreign triticale strains in Central Poland. Hod. Rosl. Aklim. Nasiennictwo 24, pp. 583-592.

Relative Rates of Sprouting, Alpha-Amylase Production and Endosperm Breakdown during Intact Head Wetting at Different Temperatures

R. Henry and B. McLean, Queensland
Wheat Research Insitiute, Toowoomba,
Australia.

SUMMARY

The sprouting of intact heads of three varieties of
wheat was studied under controlled conditions. Three
approaches to assessing sprouting damage; visual assessment
of sproutinq, estimation of alpha-amylase and determination
of endosperm modification were compared. Sprouting assessed
visually, increased with increasing temperature, while the
rate of alpha-amylase production or endosperm modification
was not always increased by higher temperatures. The relationship between alpha-amylase and sprouting or endosperm
modification was dependent upon the wheat variety and the
temperature indicating that determination of alpha-amylase
was the only reliable method for assessment of sprouting
damage.

INTRODUCTION

Pre-harvest sprouting leads to a loss of quality in
wheat. A small amount of sprouted wheat may substantially
lower the value of large quantities of sound wheat if
admixed. The segregation of sprouted wheat requires the
reliable evaluation of sprouting damage.
The assessment of pre-harvest sprouting damage in
cereals may be approached in several ways. The extent of
sprouting may be assessed visually. However, this is not an
objective test and is usually considered unreliable (Jensen
et al., 1984). Measurement of alpha-amylase levels is an
objective method with the further advantage that an increased alpha-amylase level is directly associated with a loss in
quality while sprouting (growth of the embryo) may not it-

131

self be the cause of a serious loss of quality. Methods for assessing sprouting damage based upon endosperm modification have also been proposed (Jensen and Law, 1983; Jensen et al., 1984).

As the grain sprouts the embyro grows, alpha-amylase is produced and the endosperm is broken down (Lukow and Bushuk, 1984). In this paper, the relationship between these three temperatures has been examined. This allowed an evaluation of the influence of genetic and possible environmental factors on the relative rates of these processes, and the value of measuring these processes for assessment of pre-harvest sprouting damage.

MATERIALS AND METHODS

Source of wheat: Heads from three cultivars of wheat (Cook, Kite and Timgalen) were supplied by Dr. G. Wildermuth. The wheat had been grown in the field, removed from the field at maturity and stored as whole plants for six weeks prior to this study. This length of storage minimized effects due to differences in maturity or dormancy.

Wetting wheat: A total of 45 replicate samples of each variety were sprouted in the steeping cabinet of a Seeger micro-malting plant. Each replicate included ten heads held upright in a mesh support and irrigated from above with a fine spray of water for 10s every 30min. Three replicate samples of ten heads of each variety were removed after 8, 24, 48, 72 and 96h at 10°C, 20°C and 30°C. All heads were scored for sprouting (MacMaster and Derera, 1975) and dried overnight in the kiln with forced air heated to 30°C.

Alpha-amylase determination: Grain was ground to pass an 0.8mm sieve in a laboratory mill 3100 (Falling number) and alpha-amylase was determined in duplicate using a dye-labelled substrate in a procedure described previously (Henry, 1985a).

Assessment of endosperm modification: Fluorescein dibutyrate is sensitive to the presence of lipases or esterases that are present in those parts of the endosperm modified by sprouting. Grain was sectioned and stained with fluorescein dibutyrate (Jensen and Heltved, 1982; Helved et al., 1982). All samples were counterstained with Fast Green prior to examination with a Carlsberg macrofluorescence microscope. The percentage modification calculated was the average for

100 grains (Henry, 1985a).

RESULTS

All three indicators of sprouting damage; visual sprouting, alpha-amylase and endosperm modification increased with increasing wetting time.

Growth of embryo: Visual sprouting increased with increasing temperature. Sprouting was observed in Cook and Timgalen earlier than in Kite. However, after prolonged wetting, all varieties had similar sprouting scores.

Alpha-amylase: Alpha-amylase production at 30°C was lower than alpha-amylase production at 10°C and 20°C. Alpha-amylase concentrations in Cook and Timgalen were higher than those in Kite. Highest concentrations of alpha-amylase were found in Timgalen wet for more than 48h at 20°C. Cook also produced most alpha-amylase at 20°C after these longer periods of wetting.

Endosperm modification: Endosperm modification was also influenced by variety and temperature.

Relationship between sprouting indicators: The relationship between alpha-amylase, visual sprouting and endosperm modification were examined in pairs using a linear model. Results for Kite are shown in FIGURES 1, 2 and 3. The overall correlations were all significant ($P = 0.01$, $n = 162$); alpha-amylase versus visual sprouting ($r^2 = 0.69$); alpha-amylase versus endosperm modification ($r^2 = 0.72$) and endosperm modification versus visual sprouting ($r^2 = 0.74$). However, the slopes of the lines were significantly different ($P = 0.01$) for variety x temperature for all three relationships. The correlations for individual varieties and temperatures were usually much better than the overall correlations.

All three varieties produced less alpha-amylase in relation to the amount of visible sprouting and endosperm modification at 30°C than at lower temperatures. Highest levels of endosperm modification in relation to visible sprouting were at 20°C for all three varieties.

FIGURE 1. Alpha-amylase versus endosperm modification (Kite)

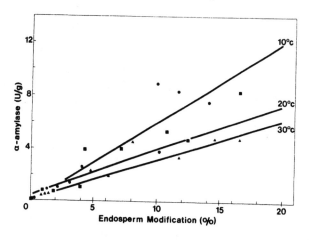

FIGURE 2. Alpha-amylase versus sprouting score (Kite)

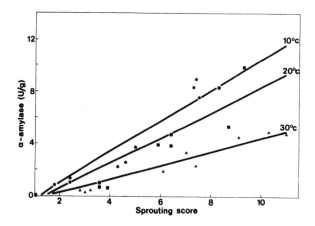

DISCUSSION

These results indicate that measurement of any of these three sprouting processes cannot be used to reliably predict the other two.

The use of measures of endosperm modification to predict sprouting damage was proposed by Jensen et al. (1984) for the segregation of grain into different silos as it is delivered from the field. Henry and McLean (1985) suggested that differences in the relationship between endosperm modification and alpha-amylase levels for different varieties would prevent the widespread use of this technique. The present study has indicated the potential for environmental

FIGURE 3. Endosperm modification versus sprouting score (Kite)

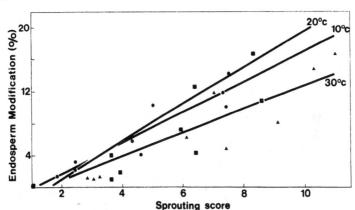

factors such as temperature to prevent the reliable prediction of alpha-amylase levels from endosperm modification.

Alpha-amylase concentrations in the grain are correlated with visual estimates of sprouting. However, both genetic and environmental factors influence the relationship. This led McCrate et al., (1981) to suggest that both low visual sprouting and low alpha-amylase should be selected for in breeding for resistance to pre-harvest sprouting, since the two traits were not mutually inclusive. The present study indicates that a third factor endosperm modification could be included.

The influence of temperature on the germination of wheat (Strand, 1983; Mares, 1984; Reddy et al., 1985) and the production of alpha-amylase, especially in response to gibberellic acid (Nicholls, 1980), has been studied extensively. However, the influence of temperature on the relative rates of these processes during sprouting in the head has received little attention, although Reddy et al., (1983) reported that the influence of temperature on alpha-amylase concentrations during sprouting was also dependant upon the temperature under which the wheat had been grown and matured. The results of the present study show that while growth of the embryo (visual sprouting) may increase with increasing temperature other processes may not. These results indicate that the relative rates of synthesis of alpha-amylase and other hydrolytic enzymes and their movement into the endosperm may be influenced by temperature. Both the production and activity of hydrolytic enzymes may be influenced by temperature, but not necessarily in the same way. The extent of endosperm breakdown is, therefore, not a reliable indicator of the concentration of alpha-amylase in the grain. Protein and cell wall degradation may progress much further into the endosperm than starch degradation (MacGregor, 1983).

The relative importance of measurements of sprouting, endosperm breakdown and alpha-amylase concentrations in assessing loss of wheat quality due to pre-harvest rain must relate to the relative impact of these changes on the utility of the grain. The part of the grain most altered by visual sprouting, the embryo may be removed in milling. Starch damage due to sprouting or protein modification, both related to endosperm modification, may adversely affect the quality of the wheat. However, the factor limiting the use of sprouted grain for breadmaking and other purposes is probably the alpha-amylase level. Excessive breakdown of starch during processing may result from high alpha-amylase concentrations.

Any segregation of sprouted wheat requires the establishment of categories with definite limits. This study has demonstrated that limits based upon visual assessment of sprouting, alpha-amylase concentration or endosperm modification may each result in dramatically different segregations. Major considerations in the choice of a basis for segregation are, the ease and speed of assessment, the reproducability of that assessment and the extent to which the resulting segregation separates wheat into classes with defined quality.

These results indicate the importance of basing methods for assessment of sprouting damage on the measurement of alpha-amylase concentrations and suggest that research should be concentrated upon the improvement of the efficiency of methods for alpha-amylase determination rather than development of tests for other quality characters that may be associated with sprouting.

REFERENCES

Heltved, F., Aastrup, S., Jensen, O., Gibbons, G. and Munck, L. (1982). Preparation of seeds for mass screening. Carlsberg Res. Commun. 47, pp. 291-296.

Henry, R. J. 1985a. Evaluation of methods for the assessment of malting quality in barley breeding. Euphytica. 34, pp. 135-145.

Henry, R. J. 1985b. Rates of endosperm modification during micro-malting of some Australian barleys. J. Inst. Brew. 91, pp. 318-320.

Henry, R. J. and McLean, B. T. 1985. Re-evaluation of
    fluorescein dibutyrate staining as a method for assess-
    ment of pre-harvest sprouting. J. Cereal Sci. 4, pp.
    51-56.

Jensen, S. A. and Heltved, F. 1982. Visualization of enzyme
    activity in germinating cereal seeds using a lipase
    sensitive fluorochrome. Carlsberg Res. Commun. 47,
    pp. 297-303.

Jensen, S.A. and Law, D. P. 1983. A rapid method for the
    detection of sprouting in populations of wheat kernels.
    Cereal Chem. 60, pp. 406-407.

Jensen, S. A., Munck, L. And Kruger, J. E. 1984. A rapid
    fluorescence method for assessment of pre-harvest
    sprouting in cereal grains. J. Cereal Sci. 2, pp.
    187-201.

Lukow, O. M. and Bushuk, W. 1984. Influence of germination
    on wheat quality. II. Modification of endosperm pro-
    tein. Cereal Chem. 61, pp. 340-344.

McCrate, A. J., Nielsen, M. T., Paulsen, G. M. and Heyne,
    E. G. 1981. Pre-harvest sprouting and alpha-amylase
    activity in hard red and hard white winter wheat cul-
    tivars. Cereal Chem. 58, pp. 424-428.

MacGregor, A. W. 1983. Cereal endosperm degradation during
    the initial stages of germination. In: Third Int.
    Symp. on Pre-harvest Sprouting in Cereals. Kruger, J.
    E. and LaBerge, D. E., (eds.). Westview Press, Boulder
    Co. USA. pp. 162-168.

McMaster, G. J. and Derera, N. R. 1975. Methodology and
    sample preparation when screening for sprouting damage
    in cereals. Cereal Res. Commun. 4, pp. 251-254.

Mares, D. J. 1984. Temperature dependence of germinability
    of wheat (Triticum aestivum L.) grain in relation to
    pre-harvest sprouting. Aust. J. Agric. Res. 35, pp.
    115-128.

Nicholls, P. B. 1980. Development of responsiveness to gib-
    berellic acid in the aleurone layer of immature wheat
    and barley caryopses: Effect of temperature. Aust. J.
    Plant Physiol. 7, pp. 645-653.

Reddy, L. V., Ching, T. M. and Metzger, R. J. 1984. Alpha-
    amylase activity in wheat kernels matured and germina-
    ted under different temperature conditions. Cereal
    Chem. 61, pp. 228-231.

Reddy, L. V., Metzger, R. J. and Ching, T. M. 1985. Effect
    of temperature on seed dormancy of wheat. Crop Sci.
    25, pp. 455-458.

138

Strand, E. 1983.   Effects of temperature and rainfall on
     seed dormancy of small grain cultivars.   In: Proc.
     Third Int. Symp. on Pre-harvest Sprouting in Cereals.
     Kruger, J. E. and LaBerge, D.E. (eds.).   Westview Press
     Boulder, Co., USA.   pp. 267-273.

Progress in Pre-Harvest Sprouting Resistance of Winter Triticale in Poland

T. Wolski, Ewa Tymieniecka,
Teresa Huskowska, Laski Plant Breeding
Station, Poland.

SUMMARY

Results and observations made in the course of a 17 year breeding programme we presented. Sprouting resistance in fodder triticale is required to ensure stable seed production. Determination of dormancy period and exact evaluation of sprouting in a breeding programme is difficult. For this reason the falling number method is used for comparison of single plant progenies, from the F_4 generation onward, and advanced strains. No interdependence is found between test weight and falling number in advanced generations. Although the falling number may not show a good correlation with sprouting, strains with a relatively high value in this test appear to be rather resistant to pre-harvest germination. They are, however, much more subject to environmental variation than sprouting resistant wheats. This variation is caused not only directly by weather conditions, but also by lodging. A marked progress in falling number represented by recent varieties and by some new crossbreds is noted, but it seems that so far only the level of rye has been reached. For further progress research on the dormancy period of triticales of different origin is necessary as well as new germplasm.

INTRODUCTION

Winter triticale has become an important crop in Poland as a substitute for rye and an acreage of 1 million ha is foreseen in the near future. The grain is used as fodder for swine and poultry.

One of the main and most difficult problems in the

breeding of triticale is the improvement of it's sprouting
resistance. This susceptibility, inherited from tetraploid
wheats and rye, is a serious drawback for the new crop. In
a grain fodder cereal, however, the importance of this trait
is restricted to ensurance of a safe and stable seed production.

In Central Europe weather conditions inducing pre-
harvest sprouting occurred every few years. In Poland this
occured as recently as 1985.

BREEDING AND TESTING PROCEDURES

The improvement of resistance to sprouting in triticale
is difficult and results are less notable than with such
traits as resistance to lodging and winterhardiness.
Judging from observations on the octoploid offspring of
crosses between sprouting resistant wheat and rye strains,
the improvement of triticale by this means seems doubtful
(Huskowska et al., 1985). This is confirmed in hexaploid
offspring of triticale-wheat and triticale-rye crosses.
Oettler (1985) found, in cross-progenies of hexa- and tetraploid wheats with rye, a lower average falling number than
the parental means. The same phenomenon is observed in the
inheritance of some other characters. It seems therefore,
that intergenome interactions play a more important role
in triticale than in other allopolyploid cereals. This is
the reason for the transgression observed in triticale
crosses for falling number as well as for some other characters (Tymieniecka and Wolski 1986). In this case results
are unpredictable.

Over the course of 17 years in the triticale programme
at the Laski Plant Breeding Station, rigorous visual selection against sprouting was applied in the years when it
appeared. As single plants are selected in the field at
some stage after the optimal harvest time, such selection
is effective, although restricted to some years. A rigorous
selection for plump grain is practiced at the same time.
Thomas et al., (1980) found that early alpha-amylase
inactivation tends to be a side effect rather than a cause
of grain shrivelling. In the Laski programme no relationship was found between falling number and test weight
(TABLE 1).

The choice of selection criteria in breeding for
resistance to sprouting is not easy. Different methods of
provoking germination in heads or in threshed grain may be
helpful. The counting of sprouted seeds is very labour-

consuming and errors may arise due to removal of sprouts in the course of threshing. Branlard et al., (1985) demonstrated that a the genetical differentiation of the material tested is more pronounced when seeds are germinated. This data favoured the use of direct alpha-amylase evaluation however, in our experience this is less reliable than the falling number method, displaying more fluctuation from year to year. This latter, simple method has shown good repeatability from year to year, which enables effective selection (TABLE 2). The falling number test is applied to single plant progenies tested in microtrials beginning with F_4 and to bulked strains tested in trials at 2 locations beginning from the F_5 or F_6.

It is assumed that the selection for high falling number does not only affect alpha-amylase activity and resistance to sprouting but also starch quality/starch value (Ringlund 1983).

TABLE 1. Interdependence of falling number and test weight/ correlation coefficients/ for triticale offsprings tested in microtrials in Laski.

Year	Correlation coefficient	No. of offsprings	Falling number (sec)		Test weight (kg/hl)	
			mean	c.v. %	mean	c.v. %
1981	-0.089	182	83	31.1	68.2	3.0
1982	-0.100	239	160	38.9	72.6	2.8
1983	-0.183**	392	130	49.3	72.8	2.9
1984	-0.224**	342	145	42.2	69.5	3.8
1985	-0.044	398	96	52.4	70.2	2.7

** significance at 0.01 level

TABLE 2. Repeatability of falling number/correlation coefficients in offsprings tested in microtrials in successive years.

Years	No. of offsprings	Correlation coefficient	Falling number			
			first year		second year	
			mean	c.v. %	mean	c.v. %
1981-1982	82	0.521**	86	31	184	32
1982-1983	102	0.686**	153	44	129	48
1983-1984	79	0.470**	132	53	158	39
1984-1985	137	0.632**	143	44	92	46

** significance at 0.01 level

RESULTS

The first Polish released triticale cultivar Lasko, proved to be relatively resistant to sprouting, as manifested by a rather high falling number and low alpha-amylase activity (Beaux and Martin, 1985; Merker, 1985).

A comparison of the falling numbers of single plant offspring tested in microtrials in Laski in 1979-1980 and later in 1984-1985, indicated progress in the number of progenies superior to Lasko (14 and 19% in the early years and 39 and 36% 4 years later). All 4 years, taken into consideration in FIGURE 1, were favourable to pre-harvest germination.

Another aspect of progress in falling number is presented in FIGURE 2. The falling number values of two new triticale varieties entered in official trials, Largo and LAD 285, were compared in Laski in 1983-1985 with Lasko and 3 rye cultivars. Largo occupies around 75% of the Polish rye growing area, and Otello and LAD 185 from Svalof, Sweden, represent a significant improvement in quality. Both new varieties are superior to Lasko, equal or even superior (in 1985) to Dańkowskie Złote rye with the comparisons with the sprouting resistant ryes being more variable.

Resistance to lodging is an important trait affecting grain quality and test weight and this may partly explain the good performance of LAD 285 triticale. Lodging, in addition to the direct effects of the conditions, are responsible for observed differences in falling number and other grain characteristics of Lasko over the years.

The relatively high falling number of LAD 285 has been confirmed by data from the Eucarpia trial in Laski in 1985 (TABLE 3). Here a comparison of Lasko, Grado and its sister strain Dagro with 3 semi-dwarf varieties, carrying the dominant wheat dwarfing gene of Tom Thumb (Rht 3) is presented. Chojnacki et al., (1976) demonstrated the superiority of this type of triticale with regard to lower alpha-amylase activity. This is confirmed by the falling numbers presented in TABLE 3. The tall, but lodging resistant variety LAD 285, is in this case still inferior to the semi-dwarf type, but much superior to Lasko and Dagro.

As mentioned previously, 1985 was an exceptionally bad year for sprouting. The falling number and germination capacity of eight varieties, was tested in Laski 10 days before harvest and again at harvest time, with and without cooling of the grain for 1 day in order to gain some information on the dormancy period (TABLE 4).

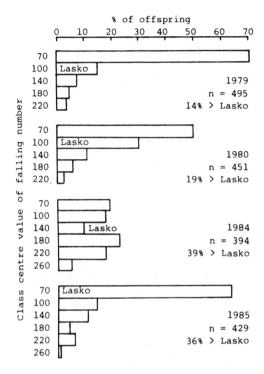

FIGURE 1. Falling number of triticale offspring tested in microtrials in Laski in the years 1979, 1980, 1984 and 1985.

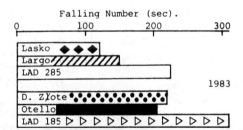

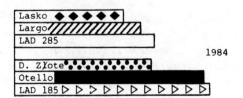

FIGURE 2. Falling number in offspring of 3 triticale tested in microtrials in Laski in 1983-1985 as compared with 3 rye varieties.

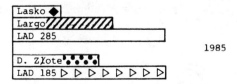

TABLE 3. Falling number and plant height of dominant dwarfs as compared with tall varieties tested in Laski in the Eucarpia trial in 1982-1985.

Variety	Country	1982		1984		1985	
		Falling number (sec)	Plant height (cm)	Falling number (sec)	Plant height (cm)	Falling number (sec)	Plant height (cm)
Lo 415	G.F.R.	211	74	-	-	-	-
BR 2	Czecho-slovakia	-	-	243	97	-	-
Local	G.F.R.	-	-	-	-	177	75
Lasko	Poland	203	105	148	145	64	116
Grado	"	96	99	64	130	-	-
Dagro	"	-	-	-	-	64	115
Largo	"	-	-	-	-	83	119
LAD 285	"	-	-	-	-	130	126

The germination capacity of the 8 varieties was also determined 4 months later when it was assumed that slightly sprouted seed, which could still germinate at the time of harvest, would no longer have such a capacity. Despite the very low falling number values, all tested strains showed a good germination in the early period high lighting the advantages of early harvesting and drying the seeds. Only one variety, Presto, showed a lack of dormancy although Largo, with a superior falling number had only a short dormancy period. Most strains tested had some dormancy period with Salvo, a variety with low falling number, possessing the strongest dormancy. These preliminary results need further confirmantion.

The year 1985, with a rainy harvest period, was the first year of relatively high certified seed production of Grado, reaching around 4,000 t. Nevertheless, the greater part of the produce was qualified, and thus triticale passed a severe test with success. This is rather encouraging since Grado is inferior in sprouting resistance to Lasko. In general, however a higher proportion of seed had to be discarded in the course of seed processing than in the case of rye.

TABLE 4. Falling number and germination capacity of triticale varieties tested in Laski in 1985.

Variety	10 days before harvest 5th August			at harvest time 15th August			4 months later 5th December
	Falling number (sec)	% Germination after 7 days		Falling number (sec)	% Germination after 7 days		% Germination
		without cooling	with 1 day cooling		without cooling	with 1 day cooling	
Grado	64	85	100	62	56	89	95
Dagro	121	86	100	63	62	94	97
Bolero	71	78	98	67	67	93	96
Salvo	77	74	96	64	56	91	92
Presto	62	98	100	62	94	98	97
Lasko	73	89	100	63	77	91	94
Largo	159	93	100	73	85	96	91
LAD 285	120	86	99	63	65	86	91

CONCLUSIONS

It is clear, on the basis of the presented observations, that the procedures applied at the Laski Plant Breeding Station achieved an acceptable resistance to pre-harvest germination. The recent improvement, compared with the first Polish triticale Lasko, is not very significant, but much better combinations of falling number with yield, test weight and lodging resistance were obtained. In order to achieve a further progress a better understanding of seed dormancy would be necessary. Probably a provocation of sprouting will be necessary and as a consequence direct testing of alpha-amylase activity (Branlard et al., 1985) may be introduced. New germ-plasm, if available, could make further improvement easier and more effective.

REFERENCES

Beaux, Y. and Martin, G. 1985. Bread-making aptitude in hexaploid triticale. Genetics and breeding of Triticale. EUCARPIA meeting, Clermont-Ferrand, France, INRA, Paris. pp. 651-655.

Branlard, G., Bernard, M., Antraygue, C., and Barloy, D. 1985. Shrivelling and alpha-amylase activity in triticale and its parental species: preliminary studies. Genetics and breeding of Triticale. EUCARPIA meeting, Clermont-Ferrand, France, INRA, Paris. pp. 617-631.

Chojnacki, G., Brykczyński, J. and Tymieniecka, E. 1976. Preliminary information on sprouting in Triticale. Proc. First Inter. Symp. on Pre-harvest Sprouting in Cereals. Sweden, 1975. Cereal Research Communications 4, pp. 111-114.

Huskowska, T., Wolski, T. and Ceglińska, A. 1985. Breeding of winter triticale for improvement of grain characters. Genetics and breeding of Triticale. EUCARPIA meeting, Clermont-Ferrand, France, INRA, Paris. pp. 641-649.

Merker, A. 1985. Triticale breeding in Svalöv. Genetics and breeding of Triticale. EUCARPIA meeting, Clermont-Ferrand, France, INRA, Paris. pp. 429-433.

Oettler, G. 1985. The influence of the wheat and rye genome on the performance of primary triticale. Genetics and breeding of Triticale. EUCARPIA meeting. Clermont-Ferrand, France, INRA, Paris. pp. 125-134.

Ringlund, K. 1983. Differences in Falling Number at constant alpha-amylase activity. Proc. Third Inter. Symp.

on Pre-harvest Sprouting in Cereals. Kruger, J. E. and LaBerge, D. E. (eds.). Westview Press, Boulder, Co., U.S.A. pp. 111-118.

Thomas, J. B., Kaltsikes, P. J., Gustafson, J.P. and Roupakias, D. G. 1980. Development of kernel shrivelling in Triticale. Review. Z. Pflanzenzuchtg. 85, pp. 1-27.

Tymieniecka, E. and Wolski, T. 1986. Utilization of diverse germ-plasm in the Laski and Choryń breeding programme. Inter. Triticale Symp., Sydney. Australia. AIAS Occasional Publications No. 24, pp. 469-472.

Comparisons of Tolerance to Pre-Harvest Sprouting within Triticum Species, Triticale and Barley

D. J. Mares, The University of Sydney,
Plant Breeding Institute, Narrabri,
NSW., Australia.

INTRODUCTION

Pre-harvest sprouting is widely recognized as being a significant problem in the production of quality cereal grain in many areas of the world. Consequently it is not surprising that many breeders are trying to improve the tolerance of their varieties either through deliberate selection of more tolerant parents or by selection within breeding populations. In general, the source of tolerance has been selected from within the existing cultivated germ-plasm for the species in question, unlike the situation in disease resistance breeding where erosion of resistance genes has forced breeders to look at wild species and related genera. In wheat there are a considerable number of recognized species and races, many of which have not been subjected to intense selection or domestication, which represent a potential pool of new and different sources of tolerance to sprouting. MacKey (1976) discussed the property of seed dormancy and noted that wild cereals where all characterized by levels of dormancy which enabled them to survive and adapt to particular environments. MacKey (1976) also discussed the loss, during domestication, of some mechanisms which prevented precocious germination. Whilst some of these are likely to be incompatible with the agronomic and quality requirements of modern cereals, others may well be of considerable use. Gordon (1983) examined a range of wheat varieties and species and noted considerable variation in sprouting tolerance. The aim of the present investigation was to extend this study to include members of all the commonly recognized species and races of wheat (Peterson, 1965), to characterize tolerance in relation to the range already established in the hexaploid bread wheats and, in addition, to compare wheat with the alternate winter cereal crops such as barley and triticale which are currently cultivated in Australia.

MATERIALS AND METHODS

Seed of several accessions of each of the commonly recognized species and races of wheat (TABLE 1) were obtained from the Australian Winter Cereals Collection. Triticale varieties were obtained from Dr. N. Darvey and Dr. F. Ellison of the University of Sydney and included five selections from the 12th. International Triticale Screening Nursery which had shown high falling numbers. One hundred barleys of diverse origin, together with current varieties and advanced breeding lines, were supplied by Dr. D. Sparrow of the Waite Agricultural Research Institute in South Australia.

The cereals were grown in twin row plots (5m in length) in the field at Narrabri, N.S.W., in 1985 along with different levels of tolerance (SUN 56A - very susceptible white wheat; Suneca - moderately tolerant white wheat; RL 4137 - tolerant red wheat). Ripening occurred in a warm to hot environment and there was no significant rainfall in the 3-4 weeks prior to grain maturity. Data was also drawn from trials conducted under similar conditions in previous years where applicable. Ear samples were harvested at random at eosin maturity and stored under cover at ambient temperature until required. At 5 days, 10 ears per accession were treated in an artificial rain simulator (McMaster and Derera, 1976) and scored for visual sprouting at daily intervals. Varieties were ranked according to the number of days required for the first ears to sprout. Ears were considered to have sprouted when the seminal roots protruded from one or more of the spikelets. At 25 days after eosin maturity 30 ears per line were subjected to a standard weather treatment (50mm of rain followed by 58h at high relative humidity and 20°C) and the extent of weather damage assessed by the falling number method. Falling numbers for non-weathered samples were used as controls.

RESULTS AND DISCUSSION

The relative tolerance to sprouting of the wheat species are compared in TABLE 1. The wild diploid wheat (T. dicoccoides) together with T. monococcum (diploid) were the

TABLE 1. Distribution of wheat, barley and triticale accessions according to the tendency to sprout in the ear at maturity.

	Time required for the first ears to sprout (days)								Total number of accessions tested
	1	2	3	4	5	6	7	>7	
Triticum									
(a) diploid									
T. boeticum							1	3	4
T. monococcum			1	4			1	2	8
(b) tetraploid									
T. dicoccoides	1						1	3	5
T. timopheevi	2	2							4
T. dicoccum	3	3	3	1	1				11
T. durum		1	3	9	1				14
T. turanicum	1	3	4	1					9
T. turgidum	3	5	4				1		13
T. polonicum		2	4	1			1		8
T. carthlicum			1	10	4		1		16
(c) hexaploid									
T. macha	2				1				3
T. spelta	3	2	1						6
T. compactum		2	4		1				7
T. sphaerococcum			1	2	5	2	2		12
T. aestivum									
(i) Best white wheats				4	4	3	3		14
(ii) Standards	SUN 56A			SUNECA			RL 4137		
Barley	26	47	11	2			7		93
Triticale			1	2	3	2			8

only groups in which there were individuals with more tolerance than the red wheat RL 4137, or the best white bread wheats (T. aestivum). Within species and between species there was considerable variation and in most groups there were accessions equivalent to or better than the moderately tolerant white-grained commercial wheat variety, Suneca.

TABLE 2. Distribution of Australian commercial wheat and
barley varieties according to the tendency to sprout in the
ear at 10 days after maturity (1984).

	Time required for first ears to sprout (days)					Total number of varieties tested
	1	2	3	4	5	
WHEAT						
(a) Commercial varieties	-	3	13	4	-	20
(b) Advanced lines	-	29	15	1	-	45
BARLEY						
varieties and advanced lines	-	9	26	2		35

The most susceptible groups appeared to be T. timopheevi, T.
macha and T. spelta, however, the number of accessions
tested in each case was limited. Within the tetraploid
wheat T. durum showed the most tolerance whilst in the hexaploids there were some T. sphaerococcum lines which were as
good as some of the best white-grained bread wheats. A
large proportion of the lines in each of these species were
ranked between the moderately tolerant Suneca and the
sprouting tolerant RL 4137. By comparison, current commercial bread wheat varieties and breeding lines from Australian programs were ranked between SUN 56A and Suneca with a
strong skew towards the susceptible end (TABLE 2).

A large proportion of the barleys sprouted more readily
than the better bread wheats, although there were some barleys with good tolerance to sprouting. Current Australian
malting barleys tended to be more susceptible to sprouting
than the popular wheat varieties (TABLE 2) in contrast to
the reasonable levels of tolerance observed in some 6 row
feed barleys.

Tetraploid and hexaploid species were further compared
on the basis of falling number before and after an artificial weather treatment (FIGURE 1a, b). In general the
effect of the weather treatment on falling number was consistent with the visual sprouting patterns. Although not
shown here,the falling number of the best white bread wheats
ranged from Suneca up to RL 4137 (Mares, 1980). Of special
note were the consistently high falling numbers of the
durums and the sphaerococcums (FIGURE 1). This confirmed
data collected over a number of seasons for the local durum

FIGURE 1. Falling number values for hexaploid and tetraploid wheats, barley and triticale lines following an
artificial rain treatment applied 25 days after eosin
maturity.

FIGURE 1a. Tetraploid wheat species

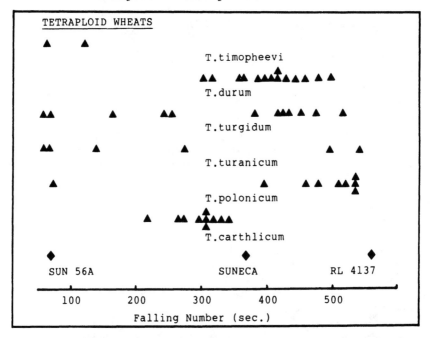

variety, Kamilaroi, and the sphaerococcums N72.72.b.CIII
(derived from the cross Spica x Indian Dwarf and selected
for sphaerococcum grain and head type), India 37, India 100,
India 137 and Punjab 7, which indicated that the tolerance
was relatively stable over the range of environmental conditions experienced in northern N.S.W. In both instances
grain dormancy appeared to be the major component of tolerance although, in the case of N72.72.b.CIII, the rate of
water movement into grains in intact ears was also reduced.
Dormancy in the durums and sphaerococcums responded to different germination temperatures and dormancy breaking treatments (e.g. temperature stratification, exogenous GA_3,
removal of the pericarp over the embryo) in similar manner
to the bread wheats.

These observations raise some interesting questions
concerning possible associations between sprouting tolerance
and species specific characters. Certainly in the case of
the Australian durum varieties there has been no deliberate
selection for sprouting tolerance. In addition there does

FIGURE 1b. Hexaploid wheat species

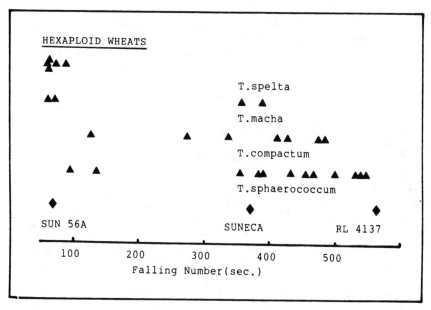

FIGURE 1c. Barley and triticales

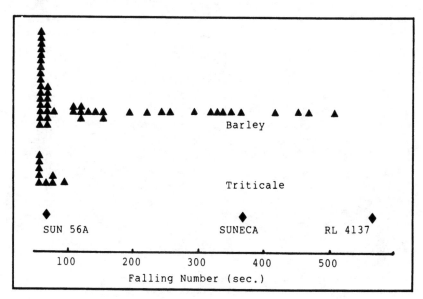

not appear to have been a continued decline in tolerance in
recent years in durum wheats to parallel the observed
decline in the bread wheats (mares, 1986). Clearly, to be
of practical use the tolerance mechanisms must be capable of

being transferred from the particular species into commercial varieties. Investigations are underway with a number
of T. sphaerococcum x bread wheat crosses to determine whether tolerance can be separated from the typical sphaerococcum head and grain characteristics. Preliminary data
suggests that some bread wheat type segregates may have
levels of tolerance similar to that of the sphaerococcum
parent (Mares and Marshall, unpublished data).

Also of interest were the high falling numbers of the
majority of T. polonicum accessions compared with the T.
carthlicum lines. This would not have been anticipated on
the basis of similar visual sprouting patterns and deserves
further investigation.

Falling number values gave a better differentiation
within the barley samples (FIGURE 1c) than did visual
sprouting. For most of the barley varieties tested the
standard weather treatment, which gave good discrimination
between bread wheat cultivars, was too severe. For the
triticale lines there was little or no relationship between
the visual sprouting data and the very low falling numbers,
either before or after the artificial weather treatment
(FIGURE 1c). Even in the absence of weathering the falling
numbers for triticales in 1985 were of the order of 60-100
sec; the one exception being a CIMMYT line which had a control falling number of approximately 180sec. Paradoxically
this line had one of the highest visual sprouting scores
which was reflected in the low falling number after weathering (79sec). Trials conducted over several seasons at
Narrabri have, on occasion, produced mature triticale grain
with high falling numbers (TABLE 3). Subsequent weather
treatments resulted in a decrease in falling number but this
was not markedly different from the response of some standard wheat varieties. The results suggest that the propensity of triticale to develop high levels of enzyme activity
at maturity is not necessarily related to the ability of
the embryo to germinate when incubated under favourable
conditions.

SUMMARY

Several accessions of each of the commonly recognised
species and races of Triticum were compared with respect to
pre-harvest sprouting tolerance. Lines with better tolerance than RL 4137, a tolerant red wheat, or the best white-
grained wheats were only found in the diploid species and
the wild tetraploid (T. dicoccoides). Of the other species

TABLE 3. Effect of a standard weather treatment on falling number of some Triticale lines and standard bread wheat varieties, 5 and 15 days after eosin maturity. Samples taken from field trials grown in 1982.

| Line or variety | Falling number (sec). | | | |
| | 5 days | | 15 days | |
	Control	Weathered	Control	Weathered
1. Triticale				
Ningadhu	125	103	130	79
Sam 81	284	250	255	162
Sam 145	326	272	387	182
Sam 175	317	257	345	210
2. Wheat				
SUN 56A	530	308	545	143
Suneca	567	473	565	343

T. durum, T. sphaerococcum, and perhaps T. polonicum, appeared to be the most promising. In each of these groups there were lines which deserve further investigation to determine the exact nature of the tolerance.

The barley lines examined tended to be more susceptible to sprouting than the wheats although there were several six row barleys with good levels of tolerance. Triticale lines showed reasonable tolerance to visual sprouting in the ear but, in general, were characterized by low falling numbers even in the absence of an artificial treatment or adverse weather conditions in the field during the later stages of grain maturation. Falling numbers of triticale samples at maturity did not appear to be related to grain germinability or the tendency to sprout in the ear.

REFERENCES

Gordon, I. L. 1983. Sprouting variability in diverse Triticum spp. germplasms. In: Proc. Third Int. Symp. on Pre-harvest Sprouting in Cereals. Kruger, J. E. and LaBerge, D. E. (eds.). Westview Press, Boulder, Co. USA. pp. 221-230.
MacKey, J. 1976. Seed dormancy in nature and agriculture.

156

Cereal Res. Commun. 4, pp. 83-91.

McMaster, G. M. and Derera, N.R. 1975. Methodology and sample preparation when screening for sprouting damage in cereals. Cereal Res. Commun. 4, pp. 251-254.

Mares, D. J. 1986. Pre-harvest sprouting tolerance in white-grained wheats. In: Proc. Fourth Int. Symp. on Pre-harvest Sprouting in Cereals. Mares D. J. (ed.). Westview Press, Boulder, Co. USA. pp.

Peterson, R. F. 1965. Wheat, botany, cultivation and utilization. Hill, London.

Combining Ability Analysis of Wheat Grain Ripening and Germinability in Two Ripening Environments

I. L. Gordon, Massey University,
Palmerston North, New Zealand.

SUMMARY

Five wheat cultivars, representing variability in dormancy, grain coat colour, and germinative alpha-amylase, were used as parents in a combining-ability diallel analysis. Two ripening environments were provided by repeating the experiment at a later sowing-date. Regression fits of harvest-ripeness (HR), HR germinative maturity, HR dormancy and HR germination provided data for the diallel analysis.

Dormancy and germination had specific combining-ability (sca) as their main source of variance. In contrast, the main variance for maturity was sca-environment interaction, with general combining-ability (gca) as the only non-trivial genetic component. Bunker-germination was influenced most by environmental variance, with notable-contributions also from gca and sca. The environmental variance was substantial as well for sprouting-index (for which gca also was non-trivial), alpha-amylase (where gca-environment interaction was substantial too), and test-weight (for which sca variance also was significant). Grain coat colour had large gca and sca variances, with a small non-trivial environmental variance. The flavanol pigment precursors were affected substantially by environmental variance, with a smaller notable genetic effect from sca only.

These results revealed substantial heterosis for dormancy and suggested that hybrid-cultivars may be a viable method of obtaining non-sprouting wheats, rather than pure-lines. However, selection against alpha-amylase production was not indicated as useful in this experiment. Selection using bunker-germination relied on different genetic variability to that for dormancy. Sprout-index was not indicated as a useful selection criterion.

INTRODUCTION

Descriptions of grain development and ripening have been undertaken to try to establish useful selection criteria for breeding against sprouting damage (Gordon et al., 1977, 1979; Gordon 1979). It is common experience in breeder's plots that simple characters, such as grain coat redness or gibberellin insensitivity, are not consistently useful in obtaining dormancy (loc. cit., Gordon 1983). Review of the physiology of germinability and dormancy reveals that many complex processes affect the end-result during both grain development and at germination time (e.g. Gordon 1980). However, physiological and biochemical information is insufficient to evaluate the efficacy as a selection criterion of any character, for the plant breeder also requires genetic information such as heritability and genetic correlations before it is possible to decide which are useful criteria.

The present paper describes a pilot study to estimate such genetic information for the ripening attributes reported in the earlier work.

MATERIALS AND METHODS

All crosses, including reciprocals, were made amongst five inbred lines representing various physiological extremes in the wheat germinability spectrum. The parents were: Gamut (white-grained, highly germinable, Australian), Sonora 64A (red-grained, highly germinable, Mexican), Pembina (red, dormant, Canadian), Timgalen (white, reduced alpha-amylase production, Australian) and Thatcher (red, dormant, Canadian). These also represented a diverse genetic base.

Hybrid plants were transplanted at deci-meter spacing into single-row three-meter plots, and one Timgalen seedling was planted in between each. Timgalen plants were used also at plot ends and in guard-rows. Three randomised complete blocks were used in each of two experiments planted two weeks apart. These were intended to provide two different warm ripening environments. Ears were tagged at anthesis, and weekly serial samples were taken as described by Gordon et al., (1979). Attributes were measured as described in that paper. Subsequently, plot by plot regressions were fitted (Gordon et al., 1979; Gordon 1979), and estimates of key values were made at harvest ripeness. These regression fits provided the data for the attributes harvest ripeness

(HR), embryo maturity and dormancy. Harvest ripeness was defined as that time (days from anthesis) when moisture content first reached 12.5%, based on the regression fit. The maturity property involved a special dormancy-bypassing germination test (Gordon et al., 1979). Other attributes were obtained form plot-bulk grains at two to three weeks after harvest-ripeness. These included grain coat colour, flavanols, sprouted alpha-amylase, bunker germination and sprouting score. Details of all of these are given in Gordon et al., 1977; Gordon et al., 1979; Gordon 1979; Gordon 1983(b).

Data were analysed using a combining-ability diallel model (Griffing 1956), with reciprocal crosses included but parents excluded. A pooled experiment version of the analysis was established by the author (unpublished), enabling the various genetic-environment interactions to be extracted. Variance components were estimated, and various heritability estimates were made from these, in particular, of general and specific combining abilities (gca and sca). In obtaining these heritability estimates, the phenotypic variance used was the usual one consisting of the sum of all variance components except environment and blocks. From these, it would be possible to decide which attributes would be useful as selection criteria, and to discuss any special breeding strategy which might be indicated.

RESULTS

The means for each attribute are presented in TABLE 1, together with their units of measurement and error variance components. The magnitudes of the sources of variation are given in TABLE 2, which shows the ratios of each variance component to its respective error component. The significances of the components are given also.

Heritability estimates (for gca and sca) are given in TABLE 3, together with their standard errors. Finally, the genotypic and phenotypic correlations between these attributes are given in TABLE 4. The latter correlations are the usual ones provided in many reports.

DISCUSSION

The main source of genetic variation in harvest ripeness and maturity was general combining ability. This formed a marked contrast with dormancy and germination,

TABLE 1. Means, and error variances of each germinability attribute.

Attribute	Unit	Mean	Hybrids Range	Error Variance	(*)
HR	days	45.5	40.3-50.6	7.7861	1.7410
Maturity	%	80.4	49.4-98.1	209.2343	46.7862
Dormancy	%#	74.6	1.1-99.8	382.5200	85.5341
Germination	%	23.8	0.1-91.6	377.2457	84.3547
Bunker Germ'n	%	61.9	22.5-92.5	447.3026	100.0199
Sprout Score	Rankit[2]	2.62	2.00-3.28	0.1021	0.0228
Amylase	EV	88.2	12.58-205.65	2385.17	533.3405
Colour	Rankit[3]	2.38	1.24-2.85	0.0421	0.0094
Flavanols	ug cg^{-1}	89.7	72.18-139.68	386.8305	86.4979

(*) Standard error of estimate
Dormancy of mature embryos
[2] Increasing score means more damage
[3] Increasing score means greater redness.

TABLE 2. Relative magnitudes[1] and significances[2] of variance components

Attribute	gca	sca	Reciprocal	gca-E	sca-E	Rec-E	Env.	Blks
HR	0.98	-0.14	0.01	-0.11	0.31	-0.15	0.17	-0.01
	**	NS	NS	NS	(NS)	NS	(NS)	NS
Maturity	0.60	-0.23	0.02	-0.13	1.10	-0.16	0.18	-0.05
	(NS)	NS	NS	NS	**	NS	NS	NS
Dormancy	0.07	1.97	0.04	0.30	0.13	-0.16	0.24	-0.02
	NS	*	NS	NS	NS	NS	NS	NS
Germination	0.07	1.75	0.04	0.28	0.13	-0.16	0.28	-0.03
	NS	*	NS	NS	NS	NS	NS	NS
Bunker germ'n	0.67	0.26	0.10	0.02	-0.09	-0.21	1.10	-0.02
	*	*	NS	NS	NS	NS	**	NS
Sprout score	0.69	0.27	0.10	0.16	0.05	-0.16	3.23	0.03
	(NS)	NS	NS	NS	NS	NS	**	NS
Alpha-amylase	0.15	0.05	-0.02	0.84	0.14	0.14	4.63	-0.04
	NS	NS	NS	*	NS	NS	*	NS
Colour	2.34	1.64	0.12	-0.07	0	-0.12	0.11	-0.01
	(NS)	**	NS	NS	NS	NS	(NS)	NS
Flavanols	-0.21	0.65	0.04	0.11	0.12	-0.20	0.80	0.09
	NS	(NS)	NS	NS	NS	NS	*	(NS)

(1) Estimated error variances used as ratio basis.
(2) NS = not significant at 10%; (NS) = not significant at 5% (but significant at 10%); * = significant at 5%; ** = significant at 1%.

TABLE 3. Heritabilities for general and specific combining abilities[1].

Attribute	gca heritability[2]		sca heritability[3]	
HR	0.515	(0.364)	0	(0.054)
Maturity	0.273	(0.005)	0	(0.002)
Dormancy	0.020	(0.001)	0.588	(0.003)
Germination	0.023	(0.001)	0.561	(0.003)
Bunker Germination	0.383	(0.004)	0.147	(0.002)
Sprout Score	0.328	(10.308)	0.129	(8.353)
Alpha-amylase	0.067	(0.0002)	0.022	(0.0002)
Colour	0.477	(9.529)	0.334	(21.338)
Flavanols	0	(0.002)	0.428	(0.004)

[1] Standard errors in parentheses (Gordon, unpublished).
[2] gca heritability = o^2gca/o^2p•[4]
[3] sca heritability = o^2sca/o^2p•[4]
[4] $o^2p = o^2 + o^2gca\text{-}E + o^2sca\text{-}E + o^2Rec\text{-}E + o^2gca + o^2sca + o^2Recip$•

where specific combining ability predominated. The gca is composed mainly of genetic variance associated with average effects of alleles. Conventional wheat breeding exploits this form of genetic variation in the breeding of pure-lines or composites of them. The sca, on the other hand, represents "dominance" deviations derived from the departure of the heterozygote from the mid-point between the two opposing homozygotes. It is often considered as a component of hybrid-vigour. These results therefore suggest that hybrid cultivars could be useful in obtaining sprouting-resistant wheats! If we examine the heritabilities of these attributes, it is apparent that moderate genetic advance could be expected in them. Germination and dormancy would probably be the most directly useful ones to use.

Of the other three "sprouting" selection criteria, alpha-amylase showed hardly any genetic variation, despite the presence of a parent which was putatively diminished in its amylase generation. There was nothing in these results

162

TABLE 4. Genotypic correlations (upper triangle) and phenotypic correlations amongst germinability attributes.

Attribute	1	2	3	4	5	6	7	8	9
1. HR	1 ...	+.81 **	-.42 (NS)	+.46 *	+.80 **	+.79 **	+.65 **	-.10 NS	-.37 NS
2. Maturity	+.66 **	1 ...	-.65 **	+.67 **	+.87 **	+.87 **	+.56 **	-.36 NS	-.25 NS
3. Dormancy	-.29 **	-.43 **	1 ...	-.99 **	-.61 **	-.73 **	-.57 **	+.79 **	+.21 NS
4. Germination	+.31 **	+.46 **	-.99 **	1 ...	+.63 **	+.75 **	+.58 **	-.77 **	-.21 NS
5. Bunker	+.36 **	+.54 **	-.40 **	+.42 **	1 ...	+.81 **	+.95 **	-.54 *	-.31 NS
6. Score	+.26 *	+.47 **	-.52 **	+.54 **	+.34 **	1 ...	+.67 **	-.63 **	-.33 NS
7. Amylase	+.11 NS	+.31 **	-.45 **	+.47 **	+.85 **	+.75 **	1 ...	-.39 (NS)	-.10 NS
8. Colour	-.12 NS	-.34 **	+.62 **	-.60 **	-.38 **	-.42 **	-.21 (NS)	1 ...	+.16 NS
9. Flavanols	-.08 NS	-.18 NS	+.13 NS	-.14 NS	-.37 **	-.43 **	-.42 **	+.13 NS	1 ...

NS = not significant at P = 0.10; (NS) = not significant at P = 0.05 (but significant at P = 0.01); * significant at P = 0.05; ** = significant at P = 0.01.

to encourage its use as a selection criterion against sprouting damage. This goes against general inclinations based on cereal technology, and is particularly disturbing in the light of the diversity in the parental germplasm. Bunker germination was revealed to be a particularly useful selection criterion, having a moderate gca heritability. The sprouting score also had a moderate gca heritability, but it also had a substantial environmental component which made it less useful than the bunker germination. In using this criterion, the emphasis would be on low germinations, of course. Pure-line breeding is indicated in both cases.

Grain coat colour had substantial gca and sca components, and moderate heritability for each. The relative sizes of gca and sca were similarly large. This is typical of cases where the heterozygote approaches one of the homozygotes in phenotypic expression: that is where classical dominance occurs. It is already well known that grain coat redness is inherited in this manner (Nilssohn-Ehle 1914). It may be somewhat surprising to find that the combined heritability for colour was not higher than 0.81. The flavanols are putative precursors of the pigment, and have been proposed as germination inhibitors in the grain. Gordon (1979) found that all lines examined, irrespective of their dormancy, had similar levels of flavanols, and did not consider them as in vivo inhibitors. In this study, however, the flavanols had high levels of sca, in parallel with dormancy. This should not be interpreted immediately as evidence of cause and effect, however. Pure lines should fix grain coat colour (gca), while hybrid varieties again are suggested as alternatives for flavanol selection.

There are no contrary correlations, either genotypic or phenotypic, amongst these selection criteria. The differences between the genotypic and phenotypic correlations are noteworthy, emphasizing the need for caution when considering only phenotypic results. It should be possible to estimate a selection index to optimise the combined genetic advance of all the useful attributes. This has not been done with the present data.

REFERENCES

Gordon, I. L., 1979. Selection against sprouting damage in wheat. III. Dormancy, germinative alpha-amylase, grain redness and flavanols. Aust. J. Res. 30, pp. 387-402.

Gordon, I. L., 1980. Germinability, dormancy and grain development. Cereal Res. Commun. 8, pp. 115-129.

Gordon, I. L., 1983. Factor analyses of characters useful in screening wheat for sprouting damage. In: Third Int. Symp. on Pre-harvest Sprouting in Cereals. Kruger, J. E. and LaBerge, D. E. (eds.). Westview Press, Boulder. Co. USA. pp. 231-238.

Gordon, I. L., 1983. Sprouting variability in diverse Triticum spp. germplasms. In: Third Int. Symp. on Pre-harvest Sprouting in Cereals. Kruger J. E. and LaBerge D. E., (eds.). Westview Press, Boulder. Co. USA. pp. 221-230.

Gordon I. L., Balaam, L. N. and Derera, N. F., 1979. Selection against sprouting damage in wheat. II. Harvest ripeness, grain maturity and germinability. Aust. J. Agric. Res. 30, pp. 1-17.

Gordon, I. L., Derera, N. F. and Balaam, L, N., 1977.
Selection against sprouting damage in wheat. I. Germination of unthreshed grain with a standard wetting
procedure. Aust. J. Agric. Res. 28, pp. 583-596.

Griffing, B., 1956. Concept of general and specific combining ability in relation to diallel crossing systems.
Aust. J. Biol. Sci. 9, pp. 463-493.

Nilssohn-Ehle, H., 1914. Zur Kenntnis der mit der Keimungs-
physiologie des Weizens in Zusammenhang stenhenden
inneren Faktoren. Z. Pflanzensuchtung. 2, pp.
153-187.

Section III
Commercial Utilization of Sprouted Grain

The Sensitivity of Various Products to Sprouted Wheat

R. A. Orth and H. J. Moss,
Bread Research Institute of Australia.

SUMMARY

Various foods were made from rain-damaged wheat from
the 1983-1984 Australian harvest and some were compared with
those made from corresponding sound wheat. This indicated
that Cantonese-style noodles were the most sensitive of the
products considered, followed by pan bread, flat bread and
chapattis.

Factors other than alpha-amylase were important. Enzyme
systems causing darkening were important in loss of noodle
quality; softening of the grain was the most serious wheat
quality effect in flat breads and chapattis, and the results
indicated that factors other than alpha-amylase were affecting bread quality.

INTRODUCTION

In the 1983-1984 season in Australia, unprecedented
quantities of rain-affected wheat were produced. Although
markets were available to dispose of this for animal feeding, it was thought worthwhile that the suitability of this
wheat for manufacture of food products of different types
be investigated. Accordingly, representative samples of
rain-affected wheat from six areas were collected, along
with corresponding samples of unaffected wheat. Chinese-
type noodles and pan bread (from 74% extraction flour),
pocket bread (both from flour and 90% atta) and chapattis
(from 81% extraction flour) were prepared and examined for
ease of processing and product quality.

EXPERIMENTAL

The samples were as shown in the accompanying TABLE 1.
Corresponding sound samples from the same general areas were
also selected. The falling numbers of these are shown in
parentheses in TABLE 1. None showed any sprouting, and the
Phadebas alpha-amylase units were in the range 30-40. Each
sample was milled to a straight run flour, and flours of 81%
and 90% extraction were also prepared. The following TABLES
2-5 show the comparison in each case of the affected versus
the unaffected sample. The figures for the latter are in
parentheses. Some of the variation was, of course, attib-
utable to factors other than pre-harvest rain.

TABLE 1. Salient features of wheat samples examined in this
work.

Wheat	Sprouted (%)	Falling No. (Secs.)	Phadebas alpha-amylase	Control grade	
NSW GP	2.4	239 (314)	94	NSW	ASW
WA GP	5.8	94 (375)	428	WA	ASW
NSW LW	6.2	152 (396)	308	NSW	HARD
NSW WD	24.4	69 (396)	776	NSW	HARD
QLD GP	29.8	72 (416)	838	QLD	HARD
VIC GP	30.5	62 (414)	909	VIC	ASW

TABLE 2. Milling value.

Wheat	Flour Yield %		Colour Grade		Flour Protein %	
NSW GP	74	(74)	-0.3	(-0.7)	9.6	(8.9)
WA GP	72	(74)	-0.3	(-0.5)	8.8	(9.2)
NSW LW	74	(75)	0.3	(-0.4)	11.7	(11.5)
NSW WD	74	(75)	0.3	(-0.4)	11.2	(11.5)
QLD GP	75	(74)	1.4	(-1.1)	11.9	(11.3)
VIC GP	73	(74)	-1.1	(-0.7)	8.9	(8.7)

Products were then prepared from the various flours.
Cantonese (alkaline) noodles were prepared by the method of
Miskelly and Moss (1985) using straight run flour. Pocket
bread was made by a modification of the method of Faridi et
al., (1983). This product was made from both straight run

TABLE 3.   Farinograph data.

Wheat	(%) Water absorption		Development time (min.)	
NSW GP	58.2	(57.5)	3.5	(3.5)
WA GP	55.1	(57.8)	2.1	(3.4)
NSW LW	61.2	(61.0)	4.0	(5.6)
NSW WD	56.8	(61.0)	3.9	(5.6)
QLD GP	58.4	(61.9)	4.2	(6.5)
VIC GP	51.1	(56.9)	1.7	(2.3)

TABLE 4.   Extensograph data.

Wheat	Resistance (E.U.)		Extensibility (cm.)	
NSW GP	295	(375)	24.4	(20.8)
WA GP	275	(280)	22.2	(20.4)
NSW LW	315	(440)	27.5	(24.3)
NSW WD	345	(440)	28.2	(24.3)
QLD GP	310	(460)	27.8	(23.1)
VIC GP	215	(235)	20.5	(18.6)

TABLE 5.   Pasting Peak (viscograph)
               Amylograph units

Wheat	Rain damaged	Equivalent unaffected grade
NSW GP	120	220
WA GP	35	560
VIC GP	0	305
QLD GP	20	650
ANW WD	15	295
NSW LW	45	295

flour and the 90% extraction atta. 1kg flour was taken so
that a number of specimens could be inspected. The breads
were baked at approximately 400°C. Chapattis were made by a
method similar to that of Ram and Nigam (1981). Pan bread
was made by the rapid dough method described by Marston
(1966) and by a conventional lean formula 2.5 hour fermentation method.

     In each case the products were evaluated according to
an established scale. It soon became apparent that scoring

systems appropriate to evaluating products made from sound wheat were not always adequate when applied to those from unsound wheats. Furthermore, actual scores allotted would convey little meaning to the general reader.

The effect on each product was as follows.

Cantonese Noodles

Cantonese noodles were made from the four flours with the highest Viscograph peak, i.e. NSW LW, NSW GP, Queensland GP and WA GP. Colour was evaluated subjectively with the raw fresh noodle, the cooked noodle and the raw noodle after 24 hours. The colour of the freshly prepared dough sheet, and the dough sheet after 24 hours was measured in the Hunterlab Colour Difference Meter. Points were allotted for colour (maximum 6 in each case) and eating quality (maximum 24). Mean figures for each of the four flours along with mean figures for the control samples are shown in TABLE 6.

TABLE 6. Cantonese noodles - Mean figures for four samples in each category

	Rain affected	Unaffected
Subjective tests (maximum)		
Raw noodle colour (6)	3.8	5.4
24 hour noodle colour (6)	3.6	5.2
Cooked noodle colour (6)	4.1	5.4
Eating quality (24)	14	18
Hunterlab colour difference meter readings		
Raw noodle brightness %	73	76
yellowness %	48	48
24 hour noodle brightness %	65	68
yellowness %	53	56

The Hunterlab figures did not indicate a great change in colour attributable to rain damage, but panel members noted a significant darkening, especially in the case of the samples from Queensland and Western Australia. Eating quality was more or less acceptable in the Queensland sample, but markedly inferior to that of the Hard Grade sample of equal protein content. The other three rain-damaged samples furnished soft, mushy noodles, difficult to bite cleanly and quite unacceptable. We expected the NSW WD sample and that from Victoria to be even worse.

Chapattis

It was anticipated that the nature of the chapatti-making procedure would make this product somewhat immune from the effects of pre-harvest rain. The dough processing time is relatively short and, during baking, the heat penetrates the dough piece so quickly that alpha-amylse activity ceases within seconds of the commencement of starch gelatinisation.

Chapattis were made by the normal procedure from a flour prepared by adding bran dustings then pollard to a straight run flour from the Buhler experimental mill until an extraction rate of 81% had been achieved. Notwithstanding the Farinograph water absorption figures, the doughs made from these flours actually had very low absorptions (only 48% in the case of the Victorian sample). Only the NSW GP approached normal absorption.

In four cases, chapattis made at these low absorptions were of fair quality, but inferior to those made from equivalent sound wheats. The Victorian GP sample gave a very poor product. Doughs of this wheat were sticky and difficult to handle. The Queensland sample gave doughs which were tough and difficult to sheet, but very sensitive to increased water.

With reasonable protein content and a Falling Number over, say, 120, it is felt that chapattis of adequate quality could be made from a flour of 80-85% extraction, normal on the Indian sub-continent. However, sharp changes in the water absorption of successive deliveries of flour, and a generally low absorption are likely to cause difficulties, both with housewives and manufacturers who are not advised of discontinuities in quality.

Pocket Breads

Flour of 74% extraction and atta of 90% extraction (prepared by combining products from the Buhler mill after sieving the coarsest bran particles away from the rest) were prepared from four samples, viz. Queensland and West Australian GP, New South Wales WD and LW. Pocket breads were made from each, by processes normal in the Middle-East, baking in each case in the high temperature arabic bread oven.

The flours and attas each had extremely low water absorptions which would be regarded with disfavour.

The Queensland flour gave pocket breads which had unattractive blisters, poor crumb colour and a leathery over-all texture. They were difficult to fold or roll. Atta from the same wheat gave pocket breads which tended to tear, allowing steam to escape in the oven, and giving a product of poor texture.

Flour milled from the West Australian wheat gave a fair product at the low water absorption, but atta from this wheat gave breads of poor texture which were difficult to fold or roll.

White flour milled from NSW LW wheat gave pocket breads which developed cracks on the surface and showed unattractive blisters. The breads were easily torn and had uneven crumb texture. Equivalent bread from LW atta was quite acceptable.

WD wheat from NSW gave the best all-round result, but even so a reduction in water addition was required to achieve a product of reasonable quality. Crust colour, crumb texture and resistance to tearing were below standard.

Of the four wheats evaluated for pocket bread, three gave a better product at 90% extraction flour (atta) than at 74% extraction. These were NSW WD, NSW LW and Queensland GP. The fourth, the West Australian sample produced a better product from the 74% extraction flour than from the atta. It appears from these findings that the higher protein, stronger wheats from Queensland and NSW react well to inclusion of bran and germ in the flour, whereas the weaker wheats may not.

Although none of the flat breads from weather damaged wheats were of equal quality to the control, atta flours produced flat bread of acceptable quality for two of the wheats. However each flour and atta exhibited such low water absorptions that product yields would be much lower than for sound wheat.

TABLE 7 shows the scores allotted to pocket breads made from these flours and attas. The NSW Hard sample was used as a control in each case.

Pan Bread (sandwich or condensed loaves)

Pan bread was made from each flour. Bread (900g or 2lb. loaves) was made both by the "rapid" or "no-time" and conventional 2.5 hour fermentation methods. Once again, water absorptions were low, extending from 50% (Victoria) to

TABLE 7.   Pocket Bread (Loaf score %)

Wheat	Flour	Atta
WA	62	57
NSW LW	57	78
NSW WD	60	78
QLD	50	65
NSW HARD 1 (Control)	75	90

53% (Queensland and Southern NSW).

Western Australia, Victoria and Queensland GP wheats gave a loaf by each method which was quite unacceptable. The Queensland one had a very dark crust and crumb, and all were irregularly shaped, the crust separated from the crumb in the slicer and slices adhered together and tore when attempts at separation were made. The Victorian "no-time" loaf collapsed under its own weight while cooling, and the Western Australian loaf showed a similar tendency. WA and Victorian loaves had unacceptably open, coarse crumb. The NSW WD rapid dough loaf was also unacceptable.

The NSW LW loaves were barely acceptable, and the GP (NSW) gave acceptable loaves, but with little margin of tolerance. We consider that NSW wheat with adequate protein content for breadmaking and a Falling Number over 150 seconds can make reasonable pan bread, but the low water absorption would not be acceptable.

TABLE 8.   Pan bread.

Wheat	Bakery absorption %	Loaf score %	
		Rapid dough	Fermented
NSW GP	53	76	77
WA GP	52	62*	55*
NSW LW	52	71	73
NSW WD	51	70*	75
QLD	53	67*	72*
VIC	50	49*	42*
NSW HARD 1 *	58	78	81

* See text.

TABLE 8 shows the points allocated to the respective loaves. The stronger flours tended to give better loaves by the fermented method and the weaker flours vice versa.

The asterisks indicate loaves with slices adhering to each other, crust separating from crumb, excessive pressure in the lidded tins, partial collapse of the loaf on cooling, and dark crust. It should be noted that not all the variation in loaf scores is attributable to sprout damage.

DISCUSSION

Different degrees of loss of quality were suffered by wheats in different areas, but some general conclusions may be drawn. So far as the flours milled from the wheats are concerned, there was a decline in milling yield - colour grade profile, a distinct reduction in flour water absorption, and marked weakening of the dough.

Chapattis were the least affected, suffering more by the lower water absorption than from manifestations of enzyme activity. Pocket or Middle Eastern breads were less affected than some other products. Once again, low bakery absorptions were of great concern, but blisters, poor crumb colour, surface cracks and difficulties in rolling and folding were apparent to varying degrees in the rain damaged samples. The higher protein samples tended to be better than those of lower protein, but still suffered by comparison with the sound sample.

Pan bread, whether high top or condensed, rapid dough or fermented, was more sensitive to rain damage than were flat breads. As a generalisation, pan breads made from affected flour had irregular shape, dark crust, crust separating from crumb on slicing, and the slices adhering to each other. Low water absorption was also a commercial disadvantage, both in itself and in difficulties which would be caused by varying quantities of affected wheat in successive deliveries. It is possible that enzymes other than alpha-amylase were responsible for some of the effects.

Of the products considered, Cantonese noodles were by far the most sensitive. Even the least rain-affected samples had a dark, unattractive appearance; the surface of the cooked noodle was slimy and the noodles had a soft, mushy eating quality. No tolerance to pre-harvest rain damage was shown by this product. We would expect salted or Japanese-type noodles to be even more sensitive as the high pH of Cantonese noodles restricts some enzyme activity. One might assume that proteases and oxidases may be more deleterious than amylases in this product.

CONCLUSION

   None of the food products considered here was found to
have a tolerance to pre-harvest sprout damage. Softening of
the grain, and its consequent effect on flour water absorption, was at least as disadvantageous as generation of enzymes in loss of flour quality.
   Protein content and dough strength also govern product
quality. Where doughs are weak the sensitivity to rain damage is greater.

REFERENCES

Miskelly, D. M. and Moss H. J. 1985. J. Cereal Sci. 3,
     p. 379.
Faridi, H. A., Finney, P. L. and Rubenthaler, G. L. 1983.
     J. Fd. Sci. 48, p. 1654.
Ram, B. P. and Nigam, S. G. 1981. J. Fd. Sci. 47, p. 231.
Marston, P. E. 1966. Cereal Science Today. 11, p. 530.

The Control of Pre-Harvest Sprouting in Cereals for Seed, Malting and Milling

L. Munck, Department of Biotechnology,
Carlsberg Research Laboratory,
10 Gamle Carlsberg Vej, DK-2500 Valby
Copenhagen, Denmark.

INTRODUCTION

Although pre-harvest sprouting in cereals has a common biological background, the consequences of such damage have vastly different effects in the various cereal industries depending on use and on technology. The seed and the malting industries depend not only on the germination rate per se but as much on the necessity that the seeds which germinate produce vigorously growing seedlings. The key character is vitality. Obviously pre-harvest sprouted kernels tend to give less percentage of seeds which germinate - it is difficult for a seed to germinate twice with an intercepted desiccation - but what is equally important, the character of sprouted kernels might be indirectly correlated to weather conditions, microbiological infection and /or variety which might be important for seed viability and storage life of the remaining seeds in the batch which do not show any sign of sprouting.

The seed and malting industry should, at the time of raw material purchase, be able not only to analyse germination and vitality but also to forecast these parameters up to 12 months ahead under the given storage conditions. This has hitherto been an empirical undertaking, sometimes, however, spiced with expensive surprises.

On the other hand, the situation in the milling and baking industries is quite different from that of the seed and malting professions. In one aspect the task is easier to solve because the germination quality character is of less direct importance. However, it is evident that the minute amounts of sprouted kernels which are needed to totally spoil a batch of premium wheat make growing and seed handling a demanding undertaking. In Canada, for example, max. 0.5% sprouted kernels are allowed in grade No. 1 for Canadian winter red wheat.

The crucial questions to be asked about pre-harvest sprouting are:

CAN IT BE PREVENTED? This question has obviously already been answered with a partial yes by the recent sprouting resistant varieties which have been obtained through the dedicated work of several plant breeders.

CAN IT BE TOLERATED? The question is intimately related to the reliability of screening analyses which could be used in seed handling and which could implement changes in technology such as milling and baking. The flour analyses of choice are here the Falling Number method and alpha-amylase analysis. However, milling a wheat batch with several slightly sprouted wheat seeds moves relatively less alpha-amylase activity to the endosperm fraction compared to milling a wheat sample having the same level of alpha-amylase activity but with a few highly sprouted seeds (Kruger and Tipples 1983). Thus there is a need to quantify the degree of sprouting of individual seeds. Acidification of wheat and rye flours which inhibits alpha-amylase has been used for centuries to make an acceptable bread out of a flour from strongly sprouted kernels. The recently defined alpha-amylase subtilisin enzyme inhibitor (ASI) in barley and wheat (Mundy and Hejgaard, 1983) points at the theoretical possibility of adding a very effective alpha-amylase inhibitor to wheat flours to avoid starch breakdown during dough treatment prior to baking. The ASI inhibitor is effectively destroyed at temperatures above 50°C (Munck et al., 1985).

CAN IT BE CURED? It is a tendency to forget that a batch of wheat seeds from even the same variety and the same field is not a homogeneous sample but consists of a complex population of wheat seed individuals with often very different properties. Pre-harvest sprouting does not strike randomly among seeds in a wheat field. It is mostly confined to the late ripening spikes which seem to have a relaxed dormancy defense. Those are softer and less dense than seeds from other spikes. Added to this fact comes the effect of the pregermination process per se which tends to make the seed additionally softer and lighter.

There are technological possibilities to grade seeds according to density and thus remove the sprouted grain fraction. Such equipment - pneumatic density gradient tables - have been available on the market for processing in large scale for more than half a century. However, these tables are difficult to tune-in due to the many parameters (slope in three dimensions, vibration and air) which must be regulated properly. The convenient and rapid FDB-method which is discussed in the following is ideal for analysing populations of grains to follow the grading of presprouted kernels on a density gradient table. The lack of a suitable analysis method has hampered the use of appropriate grading technology in the elevator industry.

Obviously, two kinds of analytical methods for assessment of pre-harvest sprouting are needed:

(a) SEED TESTS: The assessment of visual sprouting has been employed for centuries. Seed inspectors in eg. Canada the U.S.A. for example still employ this simple but demanding test, often with good result (Kruger and Tipples 1983). For instance, in a test with 138 wheat samples graded to be sprouted less than 0.5% by the inspectors, only two misgraded samples were found. However, the visual method does work well with rye and not at all with triticale. In most countries visual sprouting assessment has now been exchanged for more rapid and objective screening methods mostly based on analyses of a milled flour like the Falling Number and the alpha-amylase methods which are less time-consuming but demands less skill by the operator.

In the seed industry, seed tests for vitality e.g. the Vitascope have been employed with varying success. In the malting industry, an alpha-amylase method employing incubation of half seeds on a starch gel followed by development of the hydrolyzed zones with iodine has been used (Huygers and Van der Beken 1976). We have developed a single seed test - the FDB test for pregermination - by directly treating half seeds with a fluorochrome causing strongly fluorescent spots marking enzyme activity in pregerminated seeds (Jensen et al., 1984).

(b) FLOUR TESTS like the Falling Number method, the amylograph method and the alpha-amylase analysis (Kruger and Tipples 1983) are a matter of choice in the selection of wheats to mill an acceptable flour for baking, but they do not substitute for the information obtained from a seed test like the FDB-analysis.

In the following, the advantages and limits of single seeds and flour tests as well as their relationships will be discussed.

1. THE FDB-TEST AND THE MORPHOLOGICAL AND BIOCHEMICAL BASIS
FOR THE EVALUATION OF SPROUTED KERNELS IN SEED TESTS

The modern enzymatic techniques employing fluorochromes
are easy and fast to perform as well as being relatively
specific. In order to obtain a convenient seed analysis we
have developed a "seed fixation system" in which 50-300
seeds (1 to 6 grams of seeds) are fixed in a block of
plastic clay and later halved by abrasion in a rotary
grinder (FIGURE 1, a-f). This process takes about 2 minutes.

FIG. la. Plastic block with
holder.

FIG. lb. Multi-seed matrix
(300 seeds) with seeds.

FIG. lc. Seed fixation press
pressing la and lb together.

FIG. ld. Seeds pressed into
the plastic block.

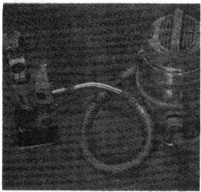 

FIG. 1e. Rotary grinder to   FIG. 1f. Treatment of seeds
halve the seeds in the plastic with FDB-solution.
block (d) with dust remover.

The technique can be used to study vitreousity in seeds as
well as to perform analyses where various stains are
employed for e.g. detection of cell wall breakdown in barley
by Calcofluor (Munck et al., 1981) or pregerminated seeds in
barley (Jensen and Heltved 1983) and in wheat, rye and triticale (Jensen et al., 1984) by fluorescein dibutyrate.
This is followed by inspection in UV-light in the seed
quality analyser, system Carlsberg (FIGURE 2) or in a
fluorescence microscope. In FIGURE 3, half seeds of different degrees of germination are treated with FDB and
photographed in the seed quality analyser.
    FDB is decomposed by esterase activity to form the
strongly fluorescent compound, fluorescein. Fluorescent
techniques are flexible and convenient to work with in both
research and in practice at several different levels: (a) at
the morphological level by the seed quality analyser or at
higher magnification in a fluorescence microscope, (b) the
visible fluorescence can be quantified in a TLC fluorescence
scanner. By taking the other half of the seed which has
been investigated in (a) and (b) and milling it in a ball
mill, its reaction with a fluorochrome can (c) be quantified
in suspension in a spectrofluorimeter and/or another enzyme
analysis e.g. alpha-amylase could be made.
    The purpose of using specific fluorochromes in seed
analyses is to study the location and dissemination area of
enzymes which are synthesized de novo during germination.
It is shown that the area of alpha-amylase dissemination as
detected by a specific florescent antibody is comparable

FIGURE 2. Seed quality analyser with accessories.

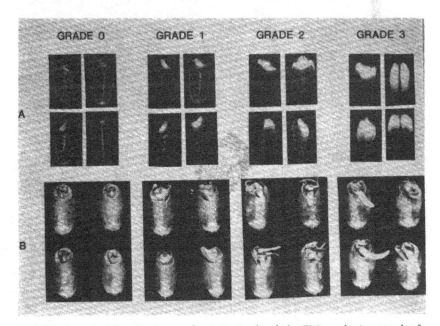

FIGURE 3. Half wheat seeds treated with FDB, photographed as seen in the seed quality analyser (Jensen et al., 1984).

with the area of cell wall breakdown with Calcofluor in the
same sections of barley seeds malted for various times
(Gibbons 1979, 1981). Enzyme dissemination and cell wall
breakdown start in a zone near to and parallel with the
crushed cell layer of scutellum at 2-3 days germination.
Hydrolytic activity is likewise seen near to the aleurone
layer. It is important to note from these investigations
that the patterns of alpha-amylase and cell wall breakdown
in practice are identical. No alpha-amylase activity is
seen to diffuse through intact cell walls. We have shown
that areas of FDB hydrolysis activity in germinating barley
also coincide with the areas of cell wall breakdown as measured by Calcofluor. (Jensen and Heltved, 1982).

Thus, the area of dissemination of the de novo synthesized enzymes in the germinating seeds is our prime target.
The rate of activity during sprouting of the enzyme system
used as a tracer (e.g. FDB-esterases) need not be identical
with that of alpha-amylase.

The rate of modification - cell wall breakdown and concomitant enzyme dissemination - during germination is dependent on the thickness of the endosperm cell walls and the
compactness (vitreousity) of the starch/protein matrix as
well as on the activity and amounts of enzymes secreted from
the aleurone and the scutellum. This can be lucidly demonstrated by our thin cell wall, soft endosperm textured mutant M-737 from Minerva barley (Aastrup 1983). In spite of
having identical levels of alpha-amylase and beta-glucanase
with the mother variety, this mutant modifies faster because
of less resistance against hydrolytic breakdown due to the
thinner cell walls and the softer endosperm texture. We can
thus conclude that the area of dissemination of enzymes
represented by the yellow zone in the FDB-technique is
limited to a semi-quantitative estimate of alpha-amylase
activity because this area is not only dependent on alpha-
amylase activity per se but also on the activity of cell
wall degrading enzymes and finally also on the endosperm
texture and cell wall thickness.

The FDB-technique should thus be regarded as a complement to the flour analyses such as Falling Number and alpha-
amylase analysis applicable as the seed analysis of choice
for pre-harvest sprouting - a modernized visual seed inspection method.

2. ESTIMATION OF SPROUT DAMAGE IN BARLEY FOR MALT AND IN
WHEAT, RYE AND TRITICALE FOR BREAD WITH SEED AND FLOUR
ANALYSES

In the malting industry, alpha-amylase in single seeds
has been studied by the iodine method. Barley seeds are
halved by hand and incubated on a starch gel for a few
hours, after which the zones of hydrolytic activity are
determined by staining the starch with iodine. Jensen and
Heltved (1983) have compared this method with the FDB-analysis and have found in the image analyser that the two meth-
odds measure essentially the same area, the FDB-method, however, being superior with regard to precision, rapidity and
easy of handling, making analyses possible within 10 minutes. In 1985, the FDB-analysis replaced the iodine method
as the official analysis of the European Brewery Convention for sprouted seeds in malting barley. Jensen and
Heltved studied the effects of storage on germination in
barley material with the FDB-method. Three weeks after
harvest, low germination ability was due to dormancy in a
few varieties while the other varieties germinated 88-97%.
After storage for one year there was a strong indirect correlation between percentage of sprouted seeds with FDB-test
and germination percentage in wet sand (FIGURE 4). It is
apparent that the FDB-test indirectly gives an estimate of
those barley samples which are at risk with regard to losses
in germination after a long time of storage.

FIGURE 4. Correlation between % sprouted kernels, FDB and
germination after one year of storage in a barley material
(Jensen and Heltved 1983).

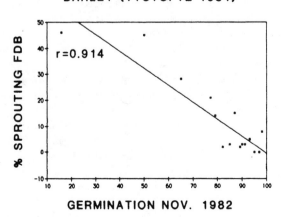

BARLEY (TYSTOFTE 1981)

$r=0.914$

% SPROUTING FDB

GERMINATION NOV. 1982

In another study Jensen et al., (1984) compared the
FDB-method with alpha-amylase and Falling Number in Canadian material consisting of wheat, rye and triticale which
had been checked independently by two grain inspectors. The
investigation was performed in cooperation with Dr. J.
Kruger at the Canadian Grain Commission laboratory in
Winnipeg.

TABLE 1. Correlation between the seed analyses, visible
sprouting, % sprouting FDB and the flour analyses Log Falling Number and alpha-amylase activity in wheat, rye and
triticale (Jensen et al., 1984).

			Correlation	
X	Y	Wheat	Rye	Triticale
% Visible sprouting	Log falling number	-0.88	-0.86	-0.37
% Visible sprouting	% Sprouting FDB	0.98	0.95	0.40
% Sprouting FDB	Alpha-amylase activity	0.97	0.96	0.98
% Sprouting FDB	Log falling number	-0.91	-0.93	-0.89

TABLE 1 shows correlation data recalculated from this paper
and demonstrates that the grain inspectors are not able to
determine sprouting in triticale due to the very high level
of incipient sprouting. The FDB-method is, however, highly
correlated with alpha-amylase activity (r = 0.98) and with
Falling Number (r = 0.89) in triticale as well as in the
wheat and rye samples, where the grain inspectors were more
successful. As seen in FIGURES 5 and 6, there is a good
correlation between alpha-amylase activity over the whole
range and % sprouting FDB in wheat, rye and triticale, but
it is also obvious that the FDB-method is not well correlated with Falling Numbers above 150 s (log Falling
Number = 5.0).

The seed quality analyser system with the FDB fluorochrome does not need wet laboratory facilities as do the
Falling Number and the alpha-amylase methods and is therefore well suited to give separate estimates of percentage of
sprouted kernels at the elevator within 5 minutes. Being a
seed analysis, the FDB-method can give an estimate of percentage of slightly sprouted and of severely sprouted seeds,

185

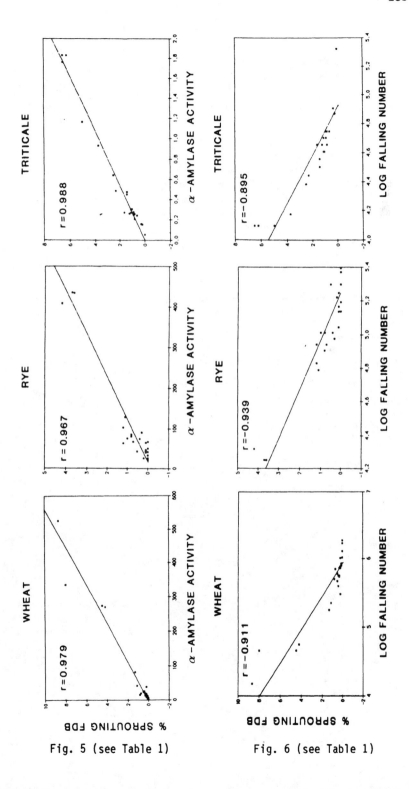

Fig. 5 (see Table 1)

Fig. 6 (see Table 1)

which is of interest in wheat milling when it comes to distribution of alpha-amylase in the bran and flour streams.

Finally, the FDB-method can also be used for controlling density gradient tables to sort out light, softly
sprouted seeds, a technology which is applicable both in the
seed and in the bread cereal industries. We have thus
sorted a feed wheat sample with a Falling Number of 70s into
two fractions, 1/3 with a Falling Number of 60s and 2/3 with
a Falling number of 200s, using the FDB-method to tune-in
the density gradient table during running.

It is evident that further practical studies are desirable to exploit the obvious benefits of seed analyses as a
compliment to the present flour analyses, Falling Number
and alpha-amylase.

REFERENCES

Aastrup, S. 1983. Selection and characterization of low
    beta-glucan mutants from barley. Carlsberg Res. Comm.
    48, pp. 307-316.
Gibbons, G. C. 1979. On the localisation and transport of
    alpha-amylase during germination and early seedling
    growth of Hordeum Vulgare. Carlsberg Res. Comm. 44,
    pp. 353-366.
Gibbons, G. C. 1981. Visualization of alpha-amylase trans-
    port and cell wall breakdown during malting of barley.
    Ann. Meeting of the American Society of Brewing Chem-
    ists, Minneapolis. J. Am. Soc. Brewing Chemists, ASBC-
    Journal. 39, pp. 55-59.
Huygers, R. and Van der Beken, R. 1976. Le Petit Journal du
    Brasseur. 84. (Referred to in European Brewery Con-
    vention. Analytica III. Method 2.6A).
Jensen, S. Aa. and Heltved, F. 1982. Visualization of en-
    zyme activity in germinating cereal seeds using a
    lipase sensitive fluorochrome. Carlsberg Res. Comm.
    47, 297-303.
Jensen, S. Aa. and Heltved, F. 1983. An improved method for
    the determination of pregerminated grains in barley.
    Carlsberg Res. Comm. 48, pp. 1-8.
Jensen, S. Aa., Munck, L. and Kruger, J. E. 1984. A rapid
    fluorescence method for assessment of pre-harvest
    sprouting of cereal grains. J. Cereal Sci. 1, pp.
    187-201.
Kruger, J. E. and Tipples, K. H. 1983. Some experiences
    with monitoring alpha-amylase levels in Canadian wheat.
    In: Proc. Third Int. Symp. on Pre-harvest Sprouting

in Cereals. Kruger, J. E. and LaBerge, D. E. (eds.).
Westview Press, Boulder, Co. USA. pp. 125-131.

Munck, L., Gibbons, G. C. and Aastrup, S. 1981. Chemical
and structural changes during malting. Proc. 18th
European Brewery Convention Congress, Copenhagen, pp.
11-30.

Munck, L., Mundy, J. and Vaag, P. 1985. Characterization of
enzyme inhibitors in barley and their tentative role in
malting and brewing. Proc. 50th Annual Meeting of Ame-
ican Society of Brewing Chemists, St. Louis, MO (1984).
ASBC Journal, 43, pp. 35-38.

Mundy, J., Hejgaard, H. and Svendsen, I. 1984. Characterization of a bifunctional wheat inhibitor of endogenous
alpha-amylase and subtilisin. FEBS Letters 167, pp.
210-214.

Effects of Sprouting on Wheat Proteins and Baking Properties

W. Bushuk and O. M. Lukow[1], University
of Manitoba, Winnipeg, MB, Canada, R3T 2N2.

SUMMARY

Changes in the protease-gluten protein system of wheat
endosperm during post-maturity germination (and sprouting)
were investigated in relation to "sprouting damage" of
breadmaking potential. Protease activity, which is negligible in sound wheat, increased substantially during germination under laboratory conditions. Endoprotease activity
increased more than exoprotease activity. Concomitant
changes in the flour proteins were consistent with the
observed increase in proteolytic activity. The most
significant change, in relation to breadmaking potential,
was the gradual decrease in the proportion of insoluble
glutenin and an equivalent increase in the proportion of
soluble glutenin plus low MW. protein components. The
observed changes in breadmaking potential were consistent
with the changes in protein solubility. In the case of the
standard quality Canadian bread wheat variety (cv. Neepawa),
breadmaking potential (loaf volume) deteriorated with degree
of sprouting damage. On the other hand, the variety with
very strong gluten (cv. Glenlea) showed an improvement in
breadmaking potential with limited sprouting damage and a
subsequent deterioration with additional sprouting damage.
It is concluded that changes in the protease-gluten proteins
system contribute significantly to the loss of breadmaking
potential during sprouting and that wheat varieties can be
selected, in terms of gluten strength, for improved tolerance of sprouting damage.

[1]. Present address: Agriculture Canada Research Station,
195 Dafoe Road, Winnipeg, MB R3T 2M9, Canada.

INTRODUCTION

The well known detrimental effect of post-maturity germination and sprouting of wheat on its breadmaking potential is generally attributed to the changes in the alpha-amylase-starch system (Meredith and Pomeranz, 1985 and references therein). The large increase in alpha-amylase activity during sprouting has been substantiated by many studies. Much of the detrimental effect on breadmaking performance occurs subsequently during the breadmaking process and derives from excessive degradation of the starch with the production of low molecular weight starch fragments (dextrins) and concomitant loss of water-holding capacity and development of sticky doughs and crumb. The overall detrimental effect of "sprouting damage" on baking potential also includes changes in the starch granules which occur in the wheat kernel prior to milling. These changes are generally referred to as "biological starch damage". Subsequently, during milling, the starch granules which had suffered biological damage are more susceptible than sound starch to further damage by the physical action of the milling process. Effects of increased alpha-amylase activity and starch damage in the breadmaking process appear to be synergistic.

While there is some evidence that sprouting damage also includes the effects of changes in the protease-gluten proteins system, relatively little research attention has been paid to this system. The effective palliative modifications of the baking process, such as lowering of water absorption and shortening of the mixing time, suggest that effects of changes in the protease-gluten system are probably greater than believed. In technological tests, including the baking test, it is not easy to separate the effects of the amylase-starch and protease-gluten systems.

In this article I will examine sprouting damage on the basis of the protease-gluten proteins system. My comments will be based primarily on results of two separate studies from our laboratory (Hwang and Bushuk, 1973; Lukow and Bushuk, 1984a,b).

PROTEOLYTIC ACTIVITY

Our earlier study (Hwang and Bushuk, 1973), with one variety of Canadian hard red spring wheat (cv. Neepawa), showed that the proteolytic activity of flour milled from partially sprouted wheat increased 17-fold during eight days of laboratory sprouting. While most of this activity was associated with the water-soluble protein fraction there was substantial activity associated with the acetic acid-soluble fraction (i.e. soluble glutenin).

The later study (Lukow and Bushuk, 1984a,b), on two wheat varieties of widely different dough mixing properties (cvs. Neepawa and Glenlea), showed that both exo- and endo-proteolytic activities increased approximately two-fold during 54 hours of sprouting under laboratory conditions. While it is not possible to translate explicitly these results to field-sprouted samples, it is obvious that wheat samples that had suffered sprouting damage will have an abnormally high proteolytic activity along with the increased alpha-amylase activity. In the context of the contribution of sprouting damage to breadmaking potential, it is important to note that during the early period of sprouting, the exoproteolytic activity showed little increase (6% after 18 hr of sprouting) while the endoproteolytic activity showed a significant increase (38%). In relation to gluten degradation, endoproteolytic activity is analogous to alpha-amylase activity in relation to starch degradation. In both cases, a small increase in activity translates into a major effect on the physical properties of the substrate because of the drastic decrease in molecular weight of the substrate resulting from the enzymic action.

EFFECTS ON PROTEINS

Our first study (Hwang and Bushuk, 1973) showed that the major effect of germination and sprouting on endosperm proteins was the transformation of insoluble glutenin (residue protein) into soluble glutenin (FIGURE 1). The concomitant loss of baking potential that was observed is consistent with our earlier findings (Orth et al., 1972) which showed that baking performance, as reflected by loaf volume, is inversely related to the proportion of soluble glutenin in the gluten and directly to the proportion of insoluble glutenin (residue protein).

In addition to the increase in the proportion of the soluble glutenin fraction, sprouting produced a significant increase in the proportion of the low molecular weight protein nitrogen (see FIGURE 1). This nitrogen is lost in the fractionation process during the dialysis step used to separate the albumins and globulins. It appears that the low molecular weight substances are formed primarily from the degradation of the glutenin proteins. In addition, a small

FIGURE 1. Effect of sprouting treatment on distribution of flour proteins among six fractions for bread wheat Neepawa.

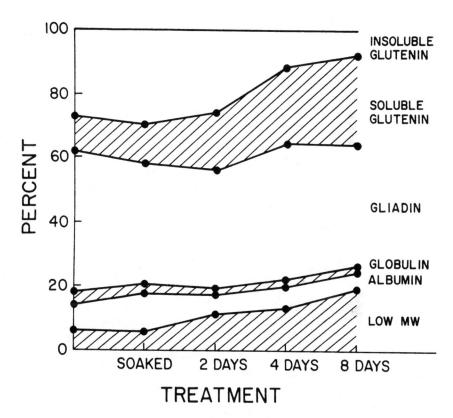

proportion may be derived from the albumins.

In general, an analogous effect of sprouting on the solubility of flour proteins was obtained from gel-filtration chromatography results (FIGURE 2) for the fractionation of the AUC (0.1M acetic acid, 3M urea and 0.01M cetyltrimethylammonium bromide) soluble proteins. Under the conditions used in that study, AUC dissolved 98% of flour protein. Our results showed that during sprouting, there occurs a gradual decrease in the proportion of high molecular weight gluten proteins with the formation of low molecular weight, more soluble, fragments. Such a change would have a drastic effect on the rheological properties of doughs (e.g. elasticity) that are particularly important to breadmaking potential.

The increase in protein solubility during sprouting

FIGURE 2. Gel filtration chromatograms (Sephadex G-150) of AUC-soluble proteins of flour milled from cv. Neepawa that had been subjected to various sprouting treatments (Hwang and Bushuk, 1973).

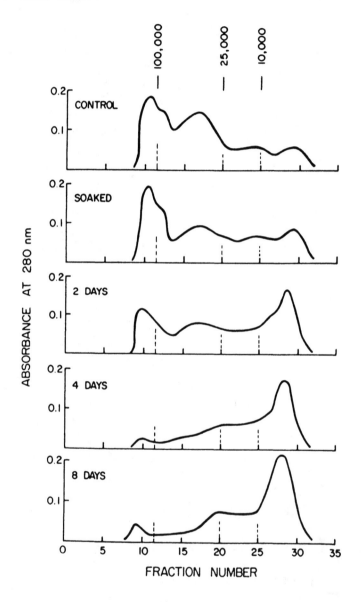

appears to result mainly from the degradation of peptide bonds by the increased proteolytic activity. Concomitant increases in the content of free amino groups (Hwang and

Bushuk, 1973) and in free amino acids (Lukow and Bushuk, 1984a) is cited as evidence of peptide bond cleavage.

Results on the changes during sprouting obtained in our two studies are generally consistent. However, they do show that even in the same laboratory it is difficult to reproduce quantitatively the results of sprouting experiments.

EFFECTS ON MILLING AND BREADMAKING PROPERTIES

Milling properties of wheat are markedly affected by wetting and sprouting (TABLE 1). Samples of both wheat varieties used in our second study (Lukow and Bushuk, 1984a) showed that soaking and to a lesser extent limited germination, produced a marked increase in the yield of flour during the break stages of the milling process. The yield of reduction flours decreased by a similar amount so that total flour yield remained virtually unaffected by soaking and germination. These results indicated that soaking in water alters the milling properties of hard wheat endosperm in a way that makes it similar to soft wheat endosperm.

TABLE 1.   Milling results[1]

Sample	Break flour	Reduction flour	Total flour	Ash content
	%	%	%	% (14% mb)
GLENLEA				
sound	14.7	59.0	73.7	0.52
soaked	38.6	37.0	75.6	0.34
germinated-18h	48.6	24.0	72.6	0.32
-35h	48.4	23.8	72.2	0.28
-54h	50.1	23.6	73.7	0.28
NEEPAWA				
sound	18.6	55.9	74.5	0.37
soaked	37.1	40.1	77.2	0.34
germinated-18h	48.3	26.0	74.3	0.29
-35h	46.8	28.7	75.5	0.28
-54	46.6	28.6	75.2	0.27

[1]From Lukow and Bushuk, 1984a.

While this may be a desired effect under some milling conditions, pre-milling treatment of wheat by soaking has not been used in commercial milling because of added costs and

194

potential sanitation problems.

The extensive lowering of ash content during soaking is attributed mainly to leaching of water-soluble ions.

FIGURE 3. Farinograms of flours milled from wheat that had been subjected to various sprouting treatments (Lukow and Bushuk, 1984a).

GLENLEA

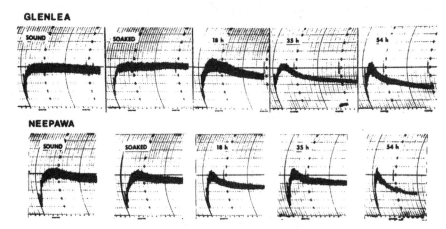

NEEPAWA

Farinograms (FIGURE 3) indicate a progressive "weakening" of dough with increasing germination period. Along with a gradual decrease in dough development time and an increase in mixing tolerance index, the samples showed a substantial decrease in water absorption (TABLE 2). The large loss of water absorption during soaking is probably due to the decrease in starch damage. While the flours from the germinated wheat samples showed a gradual increase in starch damage, water absorption continued to decrease. This decrease is probably due to loss of water holding capacity suffered by the gluten proteins resulting from degradation during sprouting. Damaged starch and protein "quality" along with quantity, are the key factors that contribute to the water absorbing capacity of flour (Greer and Stewart, 1959). The gradual decline in the Sedimentation Test value with germination obtained in our study (TABLE 2) is consistent with the farinograph results, both sets of results are consistent with the protein solubility results presented above.

All of the indices of breadmaking potential described thus far showed a unidirectional change with degree of germination. In contrast, loaf volume data showed a different

TABLE 2.   Breadmaking quality data[1]

Sample	Farinograph absorption	Sedimentation value	Loaf volume
	%	cc	cc
GLENLEA			
sound	62.9	72.5	600
soaked	54.2	67.0	630
germinated-18h	53.5	60.5	813
-35h	53.1	57.0	603
-54h	52.6	42.0	433
NEEPAWA			
sound	66.3	50.4	728
soaked	60.3	47.4	438
germinated-18h	58.0	31.0	615
-35h	57.7	24.9	460
-54h	55.4	11.0	395

[1]From Lukow and Bushuk, 1984a.

effect for the two varieties used in our second study (TABLE 2). For the variety Neepawa, loaf volume decreased with increasing degree of germination, in accord with common knowledge. For this variety, soaking produced an unexpectedly large drop in loaf volume. This could result from a loss of gas producing ability caused by the lowering of starch damage and loss of fermentable sugars and minerals. Some of the latter may be important as yeast nutrients. The weakening of the gluten that was observed would, in this variety, cause a further loss of loaf volume.

The baking results for the very strong variety, Glenlea, are technologically interesting in that soaking and limited germination produced an improvement in loaf volume. It appears that in this case, the combined effect of the changes in flour constituents, which leads to overall weakening of the dough, produced a beneficial effect on loaf volume. A major deficiency of overly strong varieties like Glenlea grown under normal climatic conditions is the inability of the constant mixing time used in the baking test to fully develop the dough. Apparently, development of doughs of such wheat varieties improved when the gluten is weakened by germination as shown here. Furthermore, doughs from such varieties can be fully developed by increasing mixing time (Bushuk et al., 1969).

Our studies have shown that changes in the protease-gluten proteins system that occur during "sprouting damage" are detrimental to breadmaking potential of standard bread wheat varieties. However, varieties with very strong glutens can be selected for improved tolerance to sprouting damage. Such varieties show an improvement in breadmaking potential during limited sprouting.

REFERENCES

Bushuk, W., Briggs, K. G., and Shebeski, L. H. 1969. Protein quantity and quality in the evaluation of bread wheats. Can. J. Plant Sci. 49, p. 133.

Greer, E. N., and Stewart, B. A. 1959. The water absorption of wheat flour: Relative effects of protein and starch. J. Sci. Food Agric. 10, p. 248.

Hwang, P., and Bushuk, W. 1973. Some changes in the endosperm proteins during sprouting of wheat. Cereal Chem. 50, p. 147.

Lukow, O. M., and Bushuk, W. 1984a. Influence of germination on wheat quality. I. Functional (breadmaking) and biochemical properties. Cereal Chem. 61, p. 336.

Lukow, O. M. and Bushuk, W. 1984b. Influence of germination on wheat quality. II. Modification of endosperm protein. Cereal Chem. 61, p. 340.

Meredith, P., and Pomeranz, Y. 1985. Sprouted grain. Pages 239-320. In: Advances in Cereal Science and Technology, Vol. VII. Pomeranz, Y. (ed.) American Association of Cereal Chemists, Inc. St. Paul, MN.

Orth, R. A., Baker, R. J. and Bushuk, W. 1972. Statistical evaluation of techniques for predicting baking quality of wheat cultivars. Can J. Plant Sci. 52, p. 139.

Contribution No. 735 of the Department of Plant Science, University of Manitoba, Winnipeg, MB R3T 2N2, Canada with financial assistance of the Natural Sciences and Engineering Research Council of Canada.

Processing of Grain according to Alpha-Amylase Activity and Preparing Optimal Blends from Sound and Damaged Grain

D. Weipert, Federal Research Centre of Cereal and Potato Processing, Detmold, Federal Republic of Germany.

SUMMARY

Blends of wheat, rye, and triticale of high and low alpha-amylase activity were prepared and falling number and amylograph measurements carried out. The relationship of these results to blend ratio proved to be cuvilinear. The regression equation which was nearest to the obtained curves, was of a reciprocal type. The reciprocal transformation of quality criteria seemed to be very helpful in predicting the blend ratio required to achieve particular characteristics. Only small quantities of sprout damaged grain can be utilized in blends for bread processing and only the breeding of sprout resistant varieties can solve this technological problem.

INTRODUCTION

Grain quality in relation to baking has been the subject of many studies. The amount and the properties of proteins and pentosans are considered to be the main determinants of baking quality. Amylolytic activity and the gelatinization properties of starch are the next most important factors governing the processing value of the bread grains, wheat and rye. In this context, the susceptibility to sprouting as a function of genus (wheat and rye) or variety is well known to grain breeders, growers and processors. Problems concerning sprout damage and its measurement were recently discussed by Meredith and Pomeranz (1985).

In countries with unfavourable climates sprout damage is constantly evident and unfortunately there are few practicable and efficient remedies which enable acceptable

197

bread to be produced from sprout damaged grain. The preparation of blends from sound and damaged grains is perhaps
one of the simplest, cheapest and most lucrative ways
utilizing damaged grain for human food but this is still
limited mostly by the severity of sprout damage of blend
components.

Besides the visual determination of sprout damage a
number of objective methods, including the determination of
alpha-amylase activity are used for scientific and practical
purposes. Of these Falling number and the Amylograph have
found the widest acceptance in the grain trade and bread
processing industry. Both are viscometric devices and thus
record the water binding capacity of other grain constituents (proteins, pentosans), besides gelatinized and disag-
gregated starch. The relationships between alpha-amylase
activity and falling number and amylograph viscosity data
are described by curvilinear equations. This was confirmed
by blending grains of different qualities. Transformation
of falling number figures into liquefaction numbers (LN) is
used to determine the blend ratio of two grain batches
required to give a particular falling number (Perten, 1964;
Weipert and Bolling, 1979). Attempts to linearise the amylogram viscosity of slurries with added malt have been made by
Hlynka (1968) and Paterson and Crandall (1967).

Although both the falling number apparatus and the
amylograph record viscosity of flour-water slurries, they
are assessed differently by wheat and rye processors.
Whilst the rapid falling number offers a satisfactory evaluation of enzymatic status in wheat processing, the baking
quality and behaviour of rye is considered to be described
more comprehensively by the peak viscosity and peak temperature readings of an amylogram (Drews, 1966; Drews and
Seibel, 1975; Weipert, 1983). In this paper the blending
behaviour of grain and the possibilities of mathematical
evaluation of blend ratio will be discussed.

MATERIAL AND METHODS

A number of wheat, triticale, and rye lots from several
years and locations were chosen to provide samples with low,
medium and high alpha-amylase activity (TABLE 1). In turn
these gave samples with high and low falling number and
amylogram data.

The grain was ground in a laboratory mill and blends,
in increments of 10% were prepared. Falling number and
amylograms of blends were determined according to standard

methods (ICC-107 and ICC-126). The results were analyzed
with the aid of a graphical relationship obtained by computer plotting of regression lines.

TABLE 1. Characteristics of samples used for blends

Sample (grain)		Alpha-amy. activity ICC-U	Falling number sec.	Amylogram peak	
				Visc. BU	Temp. °C
Wheat	1	0.56	374	990	88.0
	2	0.38	366	765	88.5
	3	5.19	161	100	65.5
	4	13.36	99	60	68.0
Triticale	1	0.26	347	585	87.0
	2	1.50	191	175	71.5
	3	19.98	62	75	61.0
	4	61.56	62	30	62.5
Rye	1	0.25	284	820	75.5
	2	1.81	189	870	68.5
	3	9.80	88	210	63.0
	4	12.31	70	230	62.0
Rye (sprout damage)	2-s	-	71	260	60.0

RESULTS AND DISCUSSION

In previous papers the presence of an optimal alpha-
amylase activity for good baking results has been discussed
for wheat.

This optimum is equated with a falling number of 250
seconds; the higher ones have to be optimized by addition of
malt, (Petzold and Pusch, 1977; Seibel, Stephan 1973). Due
to the specific properties of rye bread processing a falling
number in the range of 100-120 seconds is considered optimal. Because of swelling substances in rye (proteins and
pentosans) and the distinctly lower gelatinization temperature of rye starch, it was established that the peak temperature provides a more reliable evaluation of rye quality
than the peak viscosity and the falling number (Drews, 1966;
Drews, Seibel, 1975). An optimum was found in the vicinity
of 65°C and the value of 63°C established as a limiting
quality factor in EEC-standards (Seibel et al., 1983).

Though the viscosity of hot slurries is a function of
alpha-amylase activity as well as the water binding capacity

FIGURE 1. Relationship between falling number and amylogram peak temperature with rye from different crops.

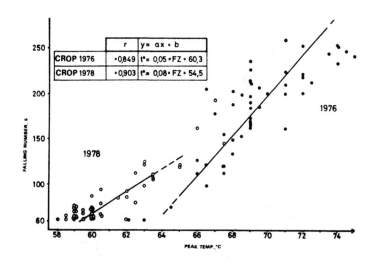

of swelling substances, significant differences between these two methods were observed, due to the relative rates of swelling and gelatinization in the course of the test procedures. Swelling and gelatinization properties are influenced by climatic conditions during grain ripening, and this may explain why in dry and moderate rainy years a falling number in the range of 80-90 seconds describes the same baking properties of rye as a falling number of 100-120 seconds in a rainy year; the peak termperature being 64°C in both cases (FIGURE 1) (Weipert and Bolling, 1979). This observation highlights the value of the peak temperature as a reliable criterion and indicates that it may be possible to prepare rye blends on the basis of peak temperature. This phenomenon was also observed in wheat by Ringlund (1983) where it was termed starch value.

Wheat

Due to the inherent gelatinizing properties of starch, sound wheat is characterized by a high falling number, a high peak viscosity and a high peak temperature of the amylogram. Sprout damaged wheat shows a lower viscosity but still higher than the sprout damaged rye. The falling number and the amylogram data of blends with sound and

sprout damaged wheat proved to be of a curvilinear type
(FIGURE 2). The influence of the sprout damaged wheat with
a high amylase activity is evident, as a low blend ratio of
damaged wheat has decreased the figures distinctly. This
influence is more evident in amylogram data than in falling
number - the upper curve is less curvilinear than the one
for wheat blends with higher amylase activity.

FIGURE 2. Blending behaviour of wheat of different qualities.

WHEAT

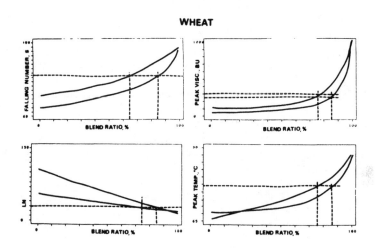

If a batch of wheat with a particular falling number
is required the blend ratio can be predicted by using the
liquefaction number equation. The relationship, liquefaction number versus blend ratio, is expected to be linear
(Perten, 1964). A batch with the falling number of 250
seconds or liquefaction number of 30 can be obtained by
blending 65% sound wheat with 35% of wheat with a lower, or
by blending 85% sound with 15% of wheat of the higher amylase activity respectively. The blends with this ratio will
have similar amylogram data, with the peak viscosity in the
range of 200-300 BU and the peak temperature in the range of
75-77°C.

Rye

The different origin of rye baking functions is evident

202

in a lower falling number and peak temperature. Besides the
starch-alpha-amylase interaction the content and properties
of pentosans and proteins are responsible for viscosity of
rye slurries (Drews, 1966; Drews and Seibel, 1975; Weipert
1981). Rye with a falling number of approximately 120 seconds and a peak temperature of 65°C is required to ensure
safe processing of different bakery items (Petzold and
Pusch, 1977; Seibel and Stephan, 1973; Seibel et al.,
1983; Stephan, 1970). As rye is known to be highly susceptible to sprouting rye batches from different provenances
and years are very often blended. The linear transformation
of falling number is of considerable use. (Hlynka, 1968;
Patterson and Crandall, 1967; Weipert and Bolling, 1979).
The relationship between liquefaction number and the blend
ratio is not always linear, but this did not lessen the usefulness of the LN as a means of predicting the blend ratio
(FIGURE 3). A blend ratio of rye, which provides a falling
number (120 seconds), also seems to ensure a safe range of
peak temperature above 63°C. As for wheat, a small proportion of enzyme rich rye decreases the quality significantly
and an optimal blend, generally contains a much greater part
of the rye with better functional properties.

FIGURE 3. Blending behaviour of rye of different qualities
(s = sprout damaged rye).

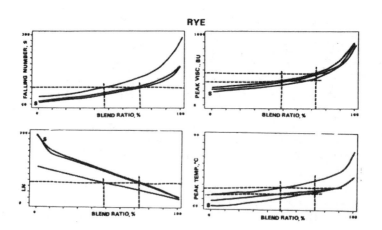

Triticale

Generally, the gelatinizing properties of triticale
starch are characterized by a high peak temperature, but
lower peak viscosity than wheat and rye. Due to the severe
susceptability to sprouting, those characteristics and the
falling number can be severely decreased.

FIGURE 4. Blending behaviour of triticale of different
quality.

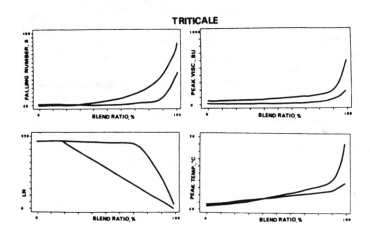

TRITICALE

Following the preparation of triticale blends from a
sound (FN 191 and 347) and a damaged triticale (FN 62) from
rainy years, a dominant influence of damaged triticale was
observed (FIGURE 4). Extremely low falling numbers and
amylogram data were found at every blend ratio showing that
proportions as low as 10% of sprouted triticale had spoiled
the blend for the bread processing. The standard viscometric methods did not appear to reflect the severity of
sprout damage in this instance. For better differentation
modified methods using concentrated or acidulated slurries
are helpful.

Rye-wheat blends

Wheat and rye mixed breads are very popular in Germany
and in some other European countries. The advantages of

these breads are firstly the flavour and secondly the longer
shelf life. In addition to this, it enables grain and flour
of different qualitits to be used in bread processing. As a
rule a wheat flour with a high falling number will be blended with a rye flour with a lower falling number, to increase
the baking properties of the flour blend, although the
reverse situation is also conceivable (Seibel et al., 1983;
Stephan and Morgenstern, 1984). The blending behaviour of
the components is best described by falling number and peak
temperature rather than by peak viscosity (FIGURE 5). The
phenomenon, that the peak viscosity, and sometimes the falling number of blends does not change significantly by
comparison with the starting flours, is due to interaction
and counteracting changes in soluble swelling substances and
enzymatic activity, both of which influence the water binding capacity of slurries (Drews, 1966). Usually, a wheat
flour with a high falling number is required to improve the
properties of a mixture with rye flour with a lower falling
number; the blend ratio for a particular bread is specified
by the recipe.

FIGURE 5. Blending behaviour of rye and wheat of different
qualities.

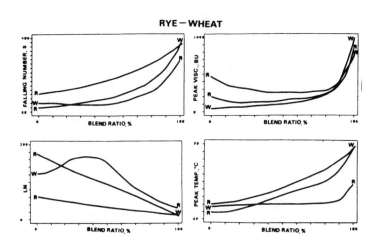

CONCLUSIONS

The falling number and amylogram data were used in
evaluating wheat and rye quality. One of the aims is to
prepare an optimal blend of flour from sound and sprout damaged grains. As a rule the viscometric data for the blend
were curvilinear. The curves of blending behaviour can be
described by the mathematical equation:

$$Y = \frac{1}{a + b^x},$$

where Y is falling number, peak viscosity, or peak temperature, and x is the blend ratio of the poor enzymatic component. Reciprocal transformation of falling number, peak
viscosity, and to a lesser extent peak temperature, yielded
linear relationships. Although the lines did not touch
every blend-ratio-value and in some cases the initial values are not included, the reciprocal transformation of
amylogram data seems to be useful for predicting the blend
ratio of a grain or flour batch which is required to give
particular properties. The adequacy of this procedure will
be examined by the use of baking tests. The preparation of
blends from sound and sprout damaged, grains and flours, is
certainly one way of utilizing grain of insufficient processing value in bread processing. The quantity of sprout
damaged grain which can be blended will always be low so
that the development of sprout resistant varieties is still
the best solution in the long term.

REFERENCES

Drews, E., 1965. Amylogramm und Fallzahl bei Mehlmi-
    schungen. Brot und Gebäck. 19, pp. 110-113.
Drews, E., 1966. Die Bedeutung der Schleimstoffe für die
    Bewertung der Roggenqualität. Getreide und Mehl.
    16, pp. 21-26.
Drews, E., and Seibel, W. 1975. Backtechnische Merkmale des
    Roggens. Getreide, Mehl und Brot. 29, p. 95.
Hlynka, I., 1968. Amylograph mobility and liquefaction num-
    ber. Cereal Sci. Today. 13, pp. 245 and 254.
Meredith, P. and Pomeranz, Y. 1985. Sprouted grain. Y.
    Pomeranz (ed.). Advances in Cereal Science and Techno-
    logy. VIII pp. 239-320.
Patterson, B. E. and Crandall, L. G. 1967. Amylograph vs.
    Falling number values compared. Cereal Sci. Today.
    12, pp. 332-335.

Perten, H., 1964. Application of the falling number method for evaluating alpha-amylase activity. Cereal Chem. 41, pp. 127-140.

Petzold, H. and Pusch, H. 1977. Ermittlung von Qualitätsschäden bei Roggen- und Weizenmehlen und Verarbeitung dieser Mehle zu Roggenmisch- und Mischbrot. Bäcker und Konditor. 25, pp. 321-322.

Ringlund, K. 1983. Differences in Falling Number at constant alpha-amylase activity. In: Third Int. Symp. on Pre-harvest Sprouting in Cereals (eds.) Kruger, J. E. and La Berge, D. E., Westview Press, Boulder, Co. USA. pp. 111-118.

Seibel, W. and Stephan, H. 1973. Mehlbeurteilung mit rheologischen Methoden. Getreide, Mehl und Brot. 27, pp. 206-213.

Seibel, W., Brümmer and Stephan, H. 1983. Influence of flour from sprout-damaged rye on the quality of rye and rye-mixed bread. Cereal Foods World. 28, pp. 503-505.

Stephan, H. 1970. Die Aussagekraft von Laborergebnissen über den Verarbeitungswert von Mehl und Schrot. Brot und Gebäck. 24, pp. 231-239.

Stephan, H. and Morgenstern, G. 1984. Wechselseitige Beeinflussung der Mischbrotqualität durch verschiedene Roggen- und Weizenmehle. Getreide, Mehl und Brot. 38, pp. 369-372.

Weipert, D. 1983. Zur Beurteilung des Verarbeitungswertes von Roggen. Getreide, Mehl und Brot. 37, pp. 229-234.

Weipert, D. 1985. Eigenschaften und Verwendungsmöglichkeiten von Triticale. Getreide, Mehl und Brot. 39, pp. 291-298.

Weipert, D. and Bolling, H. 1979. Zur Beziehung Fallzahl-Amylogrammdaten bei Roggen. Die Mühle und Mischfuttertechnik. 116, pp. 485-487.

Publication No. 5418.

Pearling and Milling as Techniques for the Improvement of the Quality of Sprouted Wheat

R. J. Henry, D. J. Martin and A. B. Blakeney,*
Queensland Wheat Research Institute, Toowoomba,
Australia.
*Yanco Agricultural Institute, Yanco, Australia.

SUMMARY

The quality of flour from sprouted wheat may be
improved by pearling the wheat prior to milling. This procedure may extend the limits of sprouting acceptable in
milling wheats. The poor quality of sprouted wheat may be
largely attributed to the presence of high levels of alpha-
amylase and other hydrolytic enzymes present in parts of the
grain removed by pearling. Pearling greatly reduced flour
alpha-amylase in field sprouted wheat, making sprouted wheat
sound. Falling numbers as low as 62 seconds were improved
to over 200 seconds by this process, with a more than 10-
fold decrease in alpha-amylase in some samples.

INTRODUCTION

The quality of sprouted wheat is reduced because of the
presence of hydrolytic enzymes, especially alpha-amylase.
High concentrations of alpha-amylase can result in excessive
breakdown of starch during processing. Treatments designed
to remove or inactivate alpha-amylase may therefore improve
the utility of sprouted wheat.

Studies of laboratory sprouted wheat samples have indicated the value of pearling in improving the quality of
sprouted wheat (Henry, Martin and Blakeney, unpublished).
Pearling prior to milling helps to avoid contamination of
sound endosperm material in the centre of the grain with
alpha-amylase from the outer layers of the grain.

This paper reports a study of the influence of pearling on alpha-amylase concentrations in field sprouted wheat.

MATERIALS AND METHODS

Source of field sprouted wheat: Wheat was obtained from
experimental field plots grown in Queensland Department of
Primary Industry trials at two sites, Bauhinea Downs and
Mt Murcheson in Queensland in 1985. Wheats from four
replicate plots of 22 varieties and experimental lines were
tested from each site.

Pearling of wheat: Wheat was pearled using a Satake TM-05
grain testing mill (Satake Engineering Co., Tokyo, Japan).
Pearlings were removed using a Bates laboratory aspirator
(H. T. McGill Inc., Houston, Texas, U.S.A.).

Determination of alpha-amylase: Alpha-amylase was determined in duplicate by the method of Barnes and Blakeney
(1974) using 0.25g of grain ground to pass through a 0.8mm
sieve in a laboratory mill 3100 (Falling number).

Determination of Falling Number: Falling numbers were
determined on samples equivalent to 7g on a 15% moisture
basis.

RESULTS AND DISCUSSION

Influence of pearling on alpha-amylase: Pearling to remove
about 20% of the grain weight reduced alpha-amylase in all
sprouted samples. Wheat with falling numbers as low as 62
seconds had falling numbers of over 200 seconds after
pearling. Alpha-maylase levels were reduced up to 10-fold
from unacceptable levels to levels similar to those in sound
wheat.

Relationship between alpha-maylse levels in pearled and
untreated grain: The relationship between the falling numbers of pearled and unpearled wheat is shown in FIGURE 1.
Alpha-amylase levels are compared in FIGURE 2. Both
relationships were linear. A greater proportion of the
total alpha-amylase was removed from more severely sprouted
samples.

Varietal Differences: Varietal differences in response to
pearling were not observed. Wheat varieties that could be
efficiently pearled to remove damaged parts of the grain
with maximum yield of sound flour could be pursued as a
breeding objective. Factors such as grain size and shape

FIGURE 1. Relationship between falling number of sprouted wheat before and after pearling. (r^2 = 0.86, n = 176). Falling number after pearling = 0.76 x Falling number before pearling + 164.

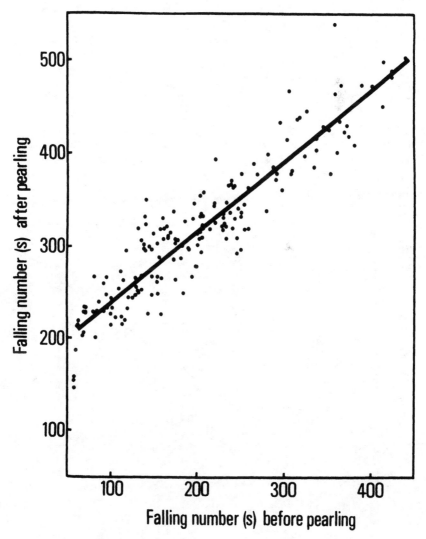

are likely to be important.

<u>Milling of pearled wheat</u>: Milling, especially at low extraction, reduces flour alpha-amylase. Alpha-amylase is not distributed evenly in flour fractions, allowing selec-

tion of fractions low in alpha-amylase. Pearling prior to milling further reduces alpha-amylase levels, especially in those flour fractions containing the highest alpha-amylase concentrations (Henry, Martin and Blakeney, unpublished). The pearling procedure used in this study should allow milling to give high overall yields of flour (>70%) with greatly improved quality. The success of pearling may be attributed to the removal of those parts of the grain containing high concentrations of alpha-amylase which would normally contaminate the flour when it is milled directly without prior pearling.

FIGURE 2. Relationship between alpha-amylase in sprouted wheat before and after pearling (r^2 = 0.71, n = 176). Alpha-amylase after pearling = 0.18 x alpha-amylase before pearling + 27.

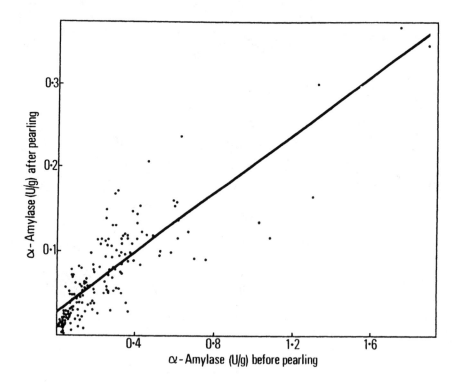

CONCLUSIONS

Pearling in combination with selection of specific mill-stream fractions offers the potential to produce acceptable flour from sprouted wheat in high yield. The removal of alpha-amylase in this way, alone or in combination with enzyme inactivation (Meredith and Pomeranz, 1985) could allow the successful utilization of severely sprouted wheat.

REFERENCES

Barnes, W. C. and Blakeney, A. B. 1974. Determination of cereal alpha-amylase using a commercially available dye-labelled substrate. Die Staerke. 26, pp. 193-197.
Meredith, P. and Pomeranz, Y. 1985. Sprouted grain. In: Advances in Cereal Science and Technology Volume VII. Pomeranz, Y. (ed.). American Association of Cereal Chemists, St. Paul. pp. 239-320.

Problems encountered in Storage of Sprout Damaged Barley

T. J. Moor, Barley Marketing Board,
Toowoomba, Queensland, Australia.

SUMMARY

As the malting process involves the germinating of
barley grains, it follows that malting quality barley must
possess a high germination at time of processing. Pre-
harvest sprouting is one of the factors which contributes to
the loss of germinative capacity in bulk grain storage. The
link between visibly sprouted grain and storage longevity is
well evidenced in Queensland barley. The link between non-
visibly sprouted grain and storage longevity is not so
clear. Presented in this paper are, the current methods
used to segregate sprout damaged grain, the developed storage quality monitoring systems, historical commercial data,
varietal differences, and the economics of sprout damage in
Queensland barley crops.

INTRODUCTION

Barley in Queensland, although grown in a temperate
climate, mainly on the fertile soils of the Darling Downs,
is grown in a predominately summer rainfall area as shown by
average rainfall data shown in FIGURE 1. Queensland malting
barley production has averaged some 200,000 mt for the past
three seasons with 40,000 mt of this being used by local
maltsters annually. Normally export malting barley is exported soon after harvest with only local malting supplies
being stored for extended periods. This pattern, however,
is changing, with some export shipments of malting being
made some months after harvest. Barley supplied as malting
barley must possess a germinative energy in excess of 95%
(Institute of Brewing BIRF Method). To achieve this malt-

ing barley is carefully segregated at intake, specialized
storage techniques of refrigerated and ambient aeration are
employed, and storage quality is regularly monitored. We
see the issue of storage longevity as falling into three
broad areas. Firstly, the variety - some varieties store
better than others and it has been suggested this may be
related to their relative dormancies and hence pre-harvest
sprouting. Secondly, the condition of the grain to be
stored - i.e. harvest damage (the germ may be physically
damaged), moisture level, drier damage (too high a temperature may have been used to dry the grain), and finally weather damage (pre-harvest sprouting). The third area is the
storage regime - moisture and temperature levels as well as
chemical protectant and fumigant usage may all contribute to
germination decline in storage if not carefully managed.

SEGREGATION METHODS

A clear link has been established in our laboratory
between visibly sprouted grain and germination decline of
malting barley in refrigerated, aerated and non-aerated
storage. FIGURES 2 and 3 show germination and sprout damage
profiles of two bins, clearly demonstrating the relationship
between the two parameters. FIGURE 4 shows this relationship plus data from a third bin in a scatter diagram. FIGURE
5 shows a similar relationship for a larger number of samples stored in the laboratory for four months.
Barley classifiers are shown how to visually detect
sprouted barley at annual workshops on barley classification. Sprouted Grain is defined in the Boards Intake Specifications as 'grains with rootlet growth evident on the germ
end of the kernel' and includes those grains where, although
the rootlet has shrunk back inside and is not visible, the
husk has been ruptured by the growing rootlet node. Since
the advent of an extensive classifier education campaign on
this subject there has been a marked decline in the amount
of malting barley loosing germination capacity in Queensland
storages. A number of laboratory methods for detecting
sprout damage have been proposed but at this stage such
methods have only been used in trial work and have not been
implemented as routine measures because of the poor correlation between the laboratory data and storage longevity
as described below. Certainly methods, such as the Falling
Number Method, measuring absolute levels of alpha-amylase
have been rejected on theoretical grounds as not providing
information on the actual percentage of seeds affected.

STORAGE QUALITY MONITORING SYSTEMS

All full storages of malting barley are sampled at least every six weeks using a Probe-A-Vac sampling device. This device allows a sample to be withdrawn every 1.2 m (4') down to maximum of 24m (80') in vertical silos which are the predominant storage type for malting barley in Queensland. Temperature of the sample is measured immediately by an electronic thermometer and the sample placed in a sealed plastic bag. On reaching the Board's laboratory, moisture and germination tests (both 4 ml and 8 ml IOB) are carried out. It was also planned when designing this monitoring system that delivery sample composites as well as samples from the first inspection would also be tested for pre-germination level by some laboratory method.

All the data collected is stored in the Board's computer with detail, summary and graphical reports automatically printed on a regular basis so as to aid the effective management of malting barley storages. FIGURE 6 is an example of such a graphical report for Dalby West SA1-05.

HISTORICAL COMMERCIAL DATA

As can be seen from FIGURE 6 of SA1-04 Dalby West 1985, germination declined steadily throughout the year as did the temperature despite the absence of visibly sprouted grains. FIGURE 7 shows the germination at various times during the year together with pregermination (EBC) levels. Similarly FIGURE 8 shows Bin SA1-06 and FIGURE 9 Bin SA1-07. Where both pre-germination tests, EBC (Anon., 1979) and Carlsberg FDB (Jensen and Heltved, 1982), were used, both results are shown. Not only is the correlation between storage longevity and pre-germination bad but also the correlation between the two methods.

In 1980/81 there was a drought in Queensland with only some 90,000 tonne of barley being produced. In order to get the 40,000 tonne of malting barley required by the local malting industry the Board took the decision to accept up to 1% sprout damaged grain into malting receivals. FIGURE 10 shows rather high pregermination levels in barley held in the refrigerated squat silo at Brookstead for 7 months. The barley stored suprisingly well given the observed levels of pre-germination.

In 1980 a storage trial (Moor, 1983) was carried out to emulate the refrigerated bin storage results using low oxygen atmospheres. Germination declined in parts of the various bins in the trial and pre-germination was cited as an explanation for this. A poor correlation, similar to that already shown, was obtained for the three bins of this trial. However, by comparing the "Mean Loss Rate" (i.e. % germination loss per month) with mean pre-germination levels for each bin, a more definite relationship can be seen in FIGURE 11 and TABLE 1.

TABLE 1.

Bin	Germination Jan '80	Germination Mar '81	Loss rate % per mth	% Pre-germination
A	96.4	77.9	1.23	7.2
B	94.7	79.5	1.01	3.4
C	97.3	93.8	0.23	0.8

VARIETAL DIFFERENCES

Although Grimmett has been the dominant malting variety grown in Queensland for the past three seasons, Clipper was the dominant variety from the early seventies until replaced by Grimmett. Clipper possessed little dormancy, probably in the order of two weeks, in direct contrast to Prior, the malting variety which Clipper replaced, which had a dormancy period of some three months. The introduction of Clipper was obviously of great advantage to the maltsters as they were able to malt new seasons grain some two months earlier. Although Clipper was higher yielding than Prior there are some disadvantages to the farmer. As a result of the lower dormancy there was an increased risk of sprouting, leading to down grading to Feed grade. Grimmett possesses slightly more dormancy than Clipper as shown in FIGURE 12. Although dormancy is difficult to establish accurately and also appears to change with the season, it is currently accepted that relative to a dormancy period of two weeks for Clipper, Grimmett has a dormancy of approximately four weeks. This extra dormancy has afforded the farmer some protection against sprout damage except in extremely wet harvests.

A new variety, Schooner, has been introduced into Queensland in the past two years and it would appear at this stage to have a similar dormancy to that of Clipper. The advantage of Grimmett's extra dormancy was well evident in Queensland this past season, where only 1.2% of the Grimmett sprouted but where some 42% of the Schooner crop was sprout-

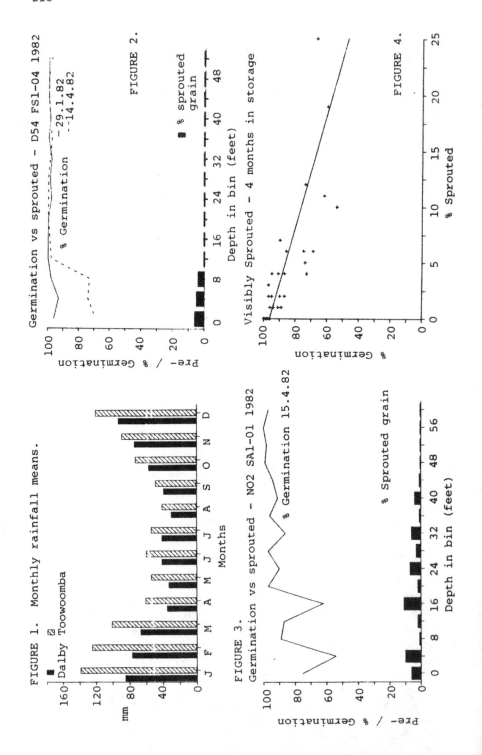

FIGURE 1. Monthly rainfall means.

FIGURE 2.
Germination vs sprouted - D54 FS1-04 1982

FIGURE 3.
Germination vs sprouted - NO2 SA1-01 1982

FIGURE 4.

217

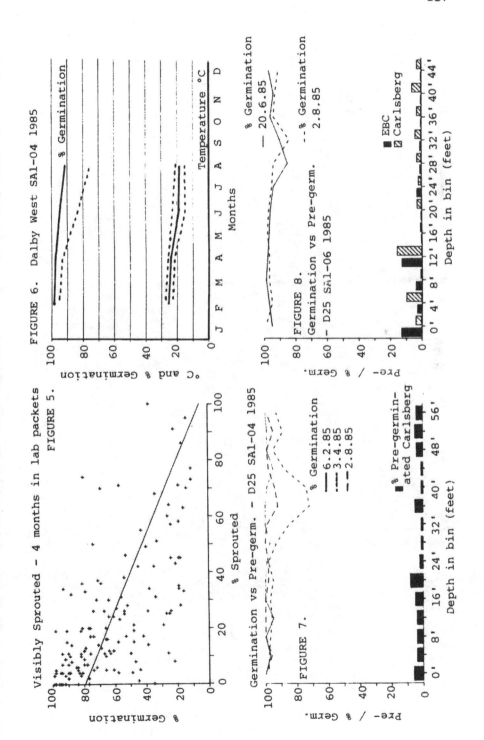

FIGURE 6. Dalby West SA1-04 1985

% Germination

Temperature °C

Months

FIGURE 5.

Visibly Sprouted - 4 months in lab packets

% Sprouted

% Germination

% Germination
—— 20.6.85
--- % Germination
     2.8.85

FIGURE 8.
Germination vs Pre-germ.
- D25 SA1-06 1985

Pre- / % Germ.

Depth in bin (feet)

EBC
Carlsberg

Germination vs Pre-germ. - D25 SA1-04 1985

% Germination
—— 6.2.85
--- 3.4.85
—— 2.8.85

FIGURE 7.

Pre- / % Germ.

Depth in bin (feet)

% Pre-germin-
ated Carlsberg

218

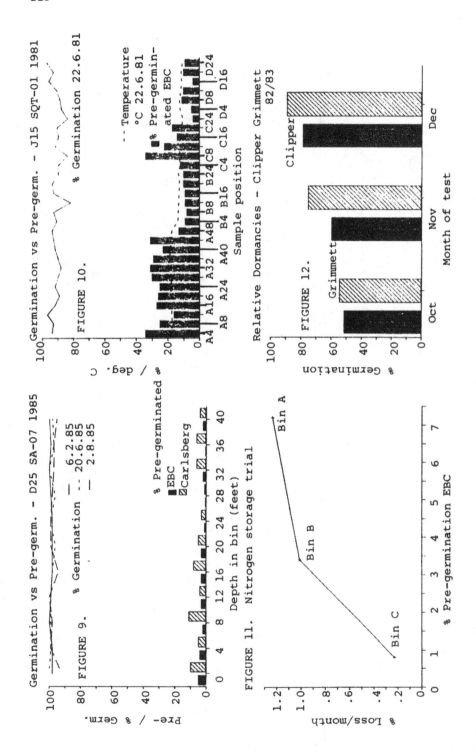

FIGURE 9. Germination vs Pre-germ. - D25 SA-07 1985

FIGURE 10. Germination vs Pre-germ. - J15 SQT-01 1981

FIGURE 11. Nitrogen storage trial

FIGURE 12. Relative Dormancies - Clipper Grimmett 82/83

FIGURE 13.

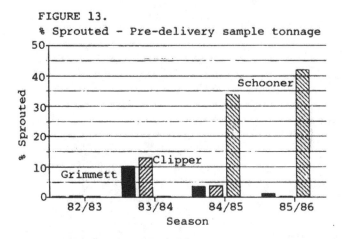

% Sprouted - Pre-delivery sample tonnage

It would also appear that this increased dormancy of Grimmett has also increased the storage longevity of this variety relative to Clipper. Certainly there has not been the storage problems with Grimmett which were encountered with Clipper. To date there has been insufficient experience with the storage of Schooner malting barley.

TABLE 2. % Tonnage sprouted, pre-delivery sample tonnages.

Season	Grimmett	Schooner	Clipper
1982/83	0.07	–	0.41
1983/84	10.20	–	12.96
1984/85	3.55	33.67	3.75
1985/86	1.16	42.00	–

ECONOMICS OF SPROUT DAMAGE

When considering the economics of pre-harvest sprout damage in barley in Queensland, there are two areas to be considered, namely costs that are directly attributable to sprout damage and those which are indirectly and only partly attributable to pre-harvest sprouting :-

Direct Costs

The exclusion of sprout damaged barley from the malting grade, and hence the foregoing of the malting premium, is main direct cost associated with pre-harvest sprouting. The Board's Intake Specifications also provides for a dockage on feed grade barley which exceeds 10% visibly sprouted grain, however dollar dockage has been minimal over the seasons. TABLE 3 shows details on the amount of barley in the past three years which did not make malting grade because of sprout damage, details on tonnage docked and their respective dollar amounts.

TABLE 3.   Sprout damage details.

	1983/84	1984/85	1985/86
Tonnage downgraded	47,867	23,664	22,563
Malting premium	$15	$7	$5
Dollar loss to Grower	$718,005	$165,648	$112,815
Total production (tonne)	446,725	665,255	730,000
Downgrading cost/tonne	$1.61	$0.25	$0.15
Tonnage docked	1,359	1,114	431
Total dockage	$2,980	$741	$1,248

The weighted average loss in downgrading over the past three seasons is therefore calculated at $0.54/t or an absolute annual cost of $330,000.

Indirect Costs

Expenses which are indirectly and only partly attributable are shown in TABLE 4 together with estimates of annual costs.

TABLE 4. Annual indirect cost estimates.

(a)	Training of classification staff	$10,000
(b)	Laboratory testing	20,000
(c)	Specialised storage techniques	20,000
(d)	Monitoring systems	20,000
(e)	Research funding	5,000
(f)	Cost of holding buffer stocks	
	i) Interest foregone	162,000
	ii) Premium loss	50,000
	Total	$287,000

Only 50% of the annual indirect cost could reasonably be attributed to pre-harvest sprout damage, i.e. $143,500. This brings the estimated cost of pre-harvest damage in round figures to $0.5 million.

REFERENCES

Jensen, S. A. and Heltved, F. 1982. Carlsberg Res. Comm. 47, p. 297.
Anon. 1979. Analytica - EBC, 3rd Edition. European Brewing Convention. Schweizer Brauerei-Rundschau: Zurich, E20/1.
Moor, Tl J. 1984. Controlled atmosphere and fumigation in grain storages. Ripp, B. E. (ed.). Elsevier, Amsterdam. pp. 105-122.

The Feeding Quality of Sprouted Wheat for Poultry and Pigs

R. J. Johnson and M. R. Taverner,
Victorian Department of Agriculture and
Rural Affairs Animal Research Institute,
Werribee, Victoria, 3030, Australia.

INTRODUCTION

Cereal grain provides the major energy source in diets for poultry and pigs. Feed formulators require detailed information on the nutrient content and relative feeding value of ingredients for proper diet formulations. Major economic appraisal of ingredients is by price relative to nutrient content and availability and feeding value. Seasonal conditions which cause moisture damage to cereal grains often lead to some reductions in price and, as such, have considerable ramifications for stockfeed manufacturers and producers. In Australia this is particularly relevant for wheat as it is the major cereal grain used in poultry and pig diets. Large quantities of sprouted wheat may be available for animal diets on an irregular basis. For example, in Australia approximately 5 million tonnes of wheat were weather-damaged during the 1983-84 harvest season.

Although there are studies on the effects of sprouting on the nutrient content of wheat (e.g. Taverner, 1975) and on animal performance (e.g. King, 1976), there remains uncertainty regarding the changes in important nutrients such as metabolisable energy and in feeding value. The reason for this is that in previous studies sound wheats used as controls for comparison with sprouted wheats were not of the same origin as the sprouted wheats examined. It is well known that many factors can influence the feeding value of sound cereal grains (Taverner et al., 1981), so such studies cannot give an accurate estimate of changes due to sprouting.

The present paper reports our preliminary findings on feeding experiments with poultry and pigs in which wheat was

sprouted under controlled conditions designed to simulate
the field sprouting of wheat.

MATERIALS AND METHODS

Sprouting procedure

Two wheat cultivars each of 20 tonnes (wheats 1 and 2)
were used. For each wheat, six tonnes were retained as the
sound control (A) and the remaining twelve tonnes were
sprouted at a commercial malting facility. Grain was
immersed in water at 18°C for 4h, drained then maintained at
95% relative humidity at 18°C for either 22h (Wheat 1) and
31h (Wheat 2) (B) or 38h (Wheat 1) and 45h (Wheat 2) (C)
prior to drying at 70°C for 16h to a grain moisture content
of approximately 10%. All grains were subsequently stored
in silos. Therefore, each of the wheat cultivars had three
sub-samples, namely (A) sound control, (B) moderately
sprouted, and (C) severely sprouted. Three other wheats
used for the feeding trial evaluations included a sound
wheat (cv. Millewa, Wheat 3) and two natural pre-harvest
sprouted wheats (Wheats 4 and 5) obtained commercially.
Physical and chemical composition of the wheats are given
in TABLE 1.

Animal Experiments

The wheat samples were incorporated into diets for
weaner pigs and meat-type (broiler) chicks. Diets for
chickens were formulated to provide equal levels of apparent metabolisable energy (AME) and essential amino acids
and other major nutrients, and each diet was fed to four
pens each of fifty chickens from hatching to 42d of age.
Diets contained approximately 64% cereal grain in conjuction
with 17% soyabean oil meal and 9% meat and bone meal.
Tallow was used to equalize the level of AME. Nutrient
levels were as recommended by the SCA (1983). Birds were
housed in a controlled environment deep-litter shed. Feed
intake, liveweight and mortality were measured.
The grains were incorporated into diets for weaner pigs
at approximately the same level (TABLE 2) to give the following expected nutrient levels: digestible energy (DE,
MJ/kg), 15.1; crude protein (%), 22; available lysine (%),
1.02. Minimum dietary levels of other amino acids were set
according to standard recommendations (ARC, 1981).

TABLE 1. Physical characteristics and chemical composition of sound and sprouted wheats and corn

	1 A	1 B	1 C	2 A	2 B	2 C	3	4	5
Falling number (secs)	336	62	62	435	81	62	467	62	111
Test weight (kg/hl)	81	73	73	81	74	74	82	76	76
Dry matter (%)	90.3	90.3	90.5	89.3	90.2	90.5	88.7	90.1	90.0
Crude protein (NX6.25%)	10.9	11.1	11.5	13.4	13.8	13.4	11.7	10.9	13.1
Starch (%)	52.0	56.3	55.9	32.1	37.5	ND	48.4	48.7	58.6
Gross energy (MJ/kg)	16.5	16.7	16.8	16.4	16.8	16.8	16.6	15.8	16.1
AME (MJ/kg)	13.7	13.9	14.3	13.6	15.0	14.6	13.5	13.8	13.4

ND is not determined

TABLE 2. Composition of diets fed to pigs from 10 to 20 kg liveweight

		Diet				
Ingredient	WHEAT	1 1A sound	2 1C CS+	3 2A sound	4 2C CS	5 3 sound
Cereal grain		66.2	66.2	69.2	69.2	66.1
Meat and bone meal		12.0	12.0	11.0	11.0	12.0
Soyabean meal		15.0	15.0	13.0	13.0	15.0
Common*		6.8	6.8	6.8	6.8	6.8

+Controlled sprouted.
*Common ingredients were blood meal 2%, Lysine 0.2%, Methionine 0.1% Vitamins and minerals 0.5% and fat 4%.

Each diet was fed ad libitum to eight weaner pigs housed individually from 10 to 20 kg liveweight in an environment-ally-controlled shed. Feed intake and rate of liveweight gain were measured.

Chemical and physical analyses

        Chemical analyses were carried out using standard
methods (AOAC, 1981). Endosperm degradation was measured by
a Hagberg semi-automatic falling number apparatus after
grinding on a falling number mill and expressed in seconds.
Apparent metabolisable energy (AME) was determined using
adult cockerels as described by Farrell (1978) with a 42h
collection period. Analyses of variance were carried out to
determine treatment differences as described by Steel and
Torrie (1960).

RESULTS AND DISCUSSION

        The falling number of wheats 1 and 2 was reduced by
controlled sprouting to levels which were similar to those
of field-sprouted wheats (4 and 5) (TABLE 1). There was a
slight increase in the gross energy (GE) and in the metab-
olisability of energy (AME/GE) and hence in the AME of
wheats 1B, 1C and 2B, 2C relative to their sound controls
(1A and 2A respectively). This effect on AME of sprouting
for wheat within the ranges of the present study has not
previously been reported, and was similar to results found
in micro-sprouting studies on these and other wheat samples
(R. J. Johnson, unpublished observations). Previous studies, in which comparisons were made between sound and
sprouted wheats of different origin, had concluded that the
availability of energy was slightly less for sprouted than
for sound grain (Taverner et al., 1975; King, 1976; Taverner and Karunajeewa, 1980). The AME of sound Australian
wheat for poultry ranged from 11.2 to 13.6 MJ/kg with a coefficient of variation (CV) at the mean of approximately 21%
(Farrell, 1981). Similarly, in a recent survey of 18 wheat
samples harvested during the 1984/85 season the AME ranged
from 12.5 to 14.4 MJ/kg (Johnson, R. J., unpublished
observations). Such differences could easily mask true differences between sound and sprouted wheat.
        Results on the effects of the different cereal grains
on pig and poultry performance are given in TABLES 3 and 4
respectively. Growth rate and feed conversion efficiency of
weaner pigs were significantly (P<0.05) influenced by cereal
grain. Inclusion of controlled-sprouted wheat into diets
increased (P<0.05) the growth rate for wheat 1, with a tendency for the same effect for wheat 2.
        The improved performance of weaner pigs fed on diets
which contained sprouted wheat observed in the present study

TABLE 3. The growth performance of young pigs fed diets containing sprouted and sound cereal grains from 10 to 20 kg liveweight

		Diet					
		1	2	3	4	5	
	WHEAT	1	1	2	2	3	SEM
Parameter		sound	CS[+]	sound	CS	sound	
Growth rate (g/d)		543[b]*	616[a]	546[b]	593[ab]	610[a]	34.7
Feed intake (kg/d)		0.95	1.05	1.04	0.97	0.98	0.044
Feed conversion (g feed/g gain)		1.77[ab]	1.72[a]	1.92[b]	1.65[a]	1.60[a]	0.123

[+]CS = controlled sprouted.
*Means with different superscripts within a row differ significantly. (P<0.05).

TABLE 4. The liveweight (W) at 42d of age and feed conversion (FCR, g feed/g gain) from 0-42d of age of male broiler chickens fed diets containing sprouted and sound cereal grains.

		Diet							
		1	2	3	4	5	6	7	SEM
	WHEAT	1	1	2	2	3	4	5	
Parameter		sound	CS[+]	sound	CS	sound	NS	NS	
W (g)		2014	1949	1941	1930	2004	1966	1979	24*
FCR		1.93	1.99	2.00	2.03	1.93	1.94	1.98	0.03NS

[+]Cs = controlled sprouted, NS = natural sprouted
*P<0.05, NS is not significant.

may have been caused by an increase in the digestible energy (DE) content due to sprouting, similar to that observed for the AME determined with poultry. The digestibility of energy and amino acids for these grains is currently being determined.

Liveweight and feed efficiency of male broiler chickens fed diets which were balanced for major nutrients was not influenced (P>0.05) by controlled - sprouting of wheat (TABLE 4). Similarly there were not significant differences between birds fed diets containing the sound wheat 3 and the natural-sprouted wheats (4 and 5). These results are similar to those reported for sorghum by Rowland et al., (1978)

who found no difference in the growth rates or feed conversion of broilers or turkey poults fed diets based on
sound or sprouted sorghum.

In conclusion the present studies have shown that there
are considerable changes in the physical and chemical composition of wheat due to sprouting. The AME of wheat was
not reduced due to severe sprouting, and there were no
detrimental effects of inclusion of sprouted wheat in diets
for pigs and poultry. Indeed growth of pigs was improved
when sprouted wheat was included in their diets.

REFERENCES

Association of Official Analytical Chemists (AOAC). 1981.
     Official methods of analysis of the Association of
     Official Analytical Chemists.
Agricultural Research Council (ARC). Nutrient requirements
     of pigs.
Farrell, D. J. 1978. Br. Poult. Sci. 19, p. 303.
Farrell, D. J. 1981. Wld's Poult. Sci. J. 37, p. 72.
King, R. H. 1976. Aust. J. Agric. Exptal. Husb. 16, p.
     668.
Rowland, L. O. and Oldfield, J. E. 1978. Poult. Sci. 57,
     p. 180.
Standing committee on Agriculture (SCA) 1983. Feeding
     standards for Australian Livestock-poultry. Farrell,
     D. J. (ed.).
Steel, R. G. D. and Torrie, J. H. 1960. Principles and pro-
     cedures of statistics. McGraw-Hill Book Company.
Taverner, M. R., Rayner, C. J. and Biden, R. S. 1975. Aust.
     J. Agric. Res. 26, p. 1109.
Taverner, M. R., Hume, I. D. and Farrell, D. J. 1981. Br.
     J. Nutr. 46, p. 159.
Taverner, M. R. and Karunajeewa, H. 1980. Chem. Aust. 47,
     p. 9.

Section IV

Grain Development

Recent Progress in the Physiology and Biochemistry of Immature Cereal Grains in Relation to Pre-Harvest Sprouting

C. M. Duffus, Department of Agricultural
Biochemistry, School of Agriculture,
West Mains Road, Edinburgh EH9 3JG,
Scotland.

SUMMARY

Some recent advances in our understanding of the
physiology and biochemistry of grain growth and maturation
are described with particular reference to the control mechanisms that might be involved in pre-harvest sprouting. The
factors controlling water supply and assimilate uptake are
considered together with the mechanism whereby, during the
later stages of grain filling, water loss from the interior
of the caryopsis can take place without interrupting the
supply of assimilate. Some of the morphological changes
accompanying embryogenesis are described in relation to the
surrounding tissues. The mechanisms involved in the supply
of nutrients and water to the developing embryo remain
largely unknown. Sucrose is the major sugar present in most
developing cereal endosperms and its concentration is a
function both of its rate of supply and of its rate of conversion to starch. It seems likely that, at least during
the period of grain-filling, physiological events in the
endosperm control sucrose levels and the rate of starch synthesis.
    A study of the relationship between sucrose supply,
grain water content and subsequent germination characteristics suggests that water potential of the grain may be a
key factor in the control of pre-harvest sprouting.

INTRODUCTION

During post-harvest germination in detached cereal
seeds, water may be taken up via the attachment point, the
micropyle and the outer layers. Water may then enter the

endosperm either through the gap in the testa at the micro-
plyar region or at the junction with the pericarp at the
crease. In germinating barley, most water is thought to
enter through the micropylar region. Increase in embryo
size is initially a function of water uptake and cell expansion. Growth involving cell division commences somewhat later. The substrates and energy required to drive the
early events of embryo growth are considered to be derived
from embryo reserves. Subsequently, degradation products
derived from endosperm reserves are translocated to the
growning embryo following absorption and metabolism by the
scutellum. It is not clear however, to what extent the
events concerned in pre-harvest sprouting resemble those
associated with post-harvest germination in detached seeds.
One obvious difference is that during pre-harvest sprouting
the grains are attached to the ear and therefore may be
under the influence of the intact plant via the vascular
tissue of the pericarp. Another difference is the physiological status of the embryo. For example, immature excised
embryos are capable of germination from as early as 10 to 12
days post-anthesis (in conditions where the number of days
between anthesis and harvest-ripeness is around 60). Normally however, this is not observed in vivo under field conditions. During the middle and later stages of grain
maturation, the outer layers dry out. It is during this
period that the embryo may sprout on the ear in response to
wet or humid environmental conditions. Superimposed on the
environmental effects is the fact that in most cereals, susceptibility to sprouting is influenced by genotype.

Little is known of the fundamental mechanisms which are
responsible for the onset of pre-harvest sprouting in cereals. The objective of this review then is to examine in
some detail recent advances in our knowledge of the physiology and biochemistry of grain growth and maturation. By
including work of direct relevance to embryo growth, maturation and germination it is hoped to be able to identify some
of the events concerned in the onset of pre-harvest sprouting.

WATER RELATIONS OF THE DEVELOPING CEREAL GRAIN

The factors controlling the water economy of the developing and maturing cereal grain are receiving increased
attention from plant physiologists. Water may enter the
caryopsis through the vascular bundles of the rachilla. It
may also enter through the epidermis of the pericarp. In

this case the site of entry may be via a small group of
cells next to the junction with the rachilla and where the
pericarp cuticular layer may be absent. Water may then
reach the embryo via the gap in the testa in the micropylar
region (Cochrane and Duffus, 1983). Water could also enter
the grain through a crack or tear in the pericarp brush
region. Such cracks appear to be formed during maturation
and their presence may be a function of genotype or of
environmental conditions (Woodbury and Wiebe, 1983). More
recent work has been concerned with the morphology of the
crease region in relation to assimilate uptake and water
loss during caryopsis development. Interest in this topic
stems from the observation that while both assimilates and
water accumulate in the grain in the early stages of growth
and development, during the mid and later stages dry matter
continues to accumulate but water content does not. Water
loss from the grain is initially from the outer layers and
in the final stages of maturation there is a steady loss of
water from the interior of the grain.

It has been shown by Jenner (1982a) that, during the
linear phase of dry matter increase, the wheat grain does
not accumulate as much water as would be expected if all
its assimilates were imported at the concentration thought
to be present in the phloem. From measurements of the net
flux of water through the stalk attaching the grain to the
rachis it was concluded that proportional unloading of
solute and solvent into the apoplast of the grain did not
occur. It is generally agreed that most of the assimilates
and mineral ions entering the endosperm move along the vascular strand of the pericarp and into the endosperm cavity
via the tissues of the crease (Frazier and Appalanaidu,
1965).

Cochrane (1983) has suggested that the xylem parenchyma cells may control the amount of water entering the
endosperm. These cells lie between the phloem and the
chalaza and thus a direct symplastic pathway for assimilates
passes through them. Excess water may be lost from the
pericarp through the cuticular layer and through the stomata
at the brush end of the caryopsis or exported via the xylem.
Basipetal movement of water in the pericarp xylem would
cease in the region of xylem discontinuity. Following
transfer to the xylem of the rachilla, the water could move
to the glumes and paleae and enter the atmosphere by transpiration as suggested for the rice caryopsis (Oparka and
Gates, 1981). It has been observed that grain water potential may be relatively unaffected by water stress (Barlow et
al., 1980; Brooks et al., 1982) although the extent of such

protective mechanisms clearly depends on the timing and duration of the period of water deficit (Nicholas et al., 1985). It may be that, in the region of xylem discontinuity, water exported from the caryopsis could transfer to the phloem of the pericarp vascular tissue and thus re-enter the caryopsis (Cochrane, 1983). This might provide a mechanism for conserving water and maintaining assimilate supply under drought conditions.

Of clear relevance to the problem of pre-harvest sprouting is the mechanism whereby, during the later stages of grain filling, water loss from the interior of the caryopsis can take place without interrupting the supply of assimilate. It has been suggested by Cochrane (1983) that this process is mediated by cell wall modifications in the band of cells occupying the gap between the ends of the testa in the crease region. Towards the end of the linear phase of grain filling these cells, referred to as the chalaza, develop a layer of suberin outside the membrane. This effectively isolates the symplast from the apoplast. At about the same time, the primary wall becomes lignified thus reducing its permeability to water. A similar observation has been made by Lingle and Chevalier (1985) and it was noted that this had no effect on the grain-filling rate. It may be then, that assimilates continue to enter the endosperm via the symplast of the chalaza while simultaneously water leaves the endosperm via the apoplast of the chalaza. It is possible that water leaves the endosperm with little accompanying movement of solutes since lignification of the chalazal cells may result in their becoming impermeable to solutes such as sucrose. During the later stages of caryopsis development when, under non-sprouting conditions, the endosperm is losing water rapidly, a core of cell wall material forms inside the nucellar projection and extends deep into the endosperm. Thus water may then pass from the endosperm in the apoplast to the chalaza. Since lignification ceases at the outer edge of the chalaza and the apoplast in continuous throughout the pericarp, water could be drawn out of the endosperm as a result of water loss from the pericarp. Passive loss of water could take place from the pericarp as described above. Water loss could also take place if some part of the apoplast contained a higher concentration of solutes than the apoplast of the nucellar projection. This would be an active process since energy would be required to maintain the concentration gradient. Passive and active components of water loss in maturing cereal grains have been described previously by Meredith and

Jenkins (1975). It has been suggested (Cochrane, 1983) that
the xylem parenchyma cells, since they are metabolically
active until at least 50 'days' post-anthesis, and hence may
be capable of raising their apoplastic sucrose levels above
those of the apoplast in the nucellar projection, are
responsible for active water loss from the pericarp. Water
drawn from other parts of the apoplast could then move from
the parenchma cell walls into the lignified walls of the
xylem. A water potential gradient caused by transpiration
from the lemma and palea would then be responsible for removing the water from the xylem. More recently Cochrane
(1985) has suggested that blockage of xylem elements by
pectinaceous material might, by controlling the supply of
water to the grain, play a key role in the regulation of the
final stages of grain dehydration.

It is not known however, to what extent modifications
to these mechanisms may be involved in the retention and
further uptake of water in pre-harvest sprouting. Certainly
it is known (Mares, 1983) that cultivars which differ in ear
and grain water uptake also differ in their degree of
sprouting. Furthermore it has been shown that 'wetting' of
ears is slower in awnless cultivars and that sprouting in
these cultivars is correspondingly less than in those with
awns (King and Richards, 1984). Interestingly the awns
themselves do not appear to be important as their removal
has no effect on ear wetting. Whether or not there are
structural modifications in the awned cultivars which lead
to increased water uptake is unknown.

DEVELOPMENTAL PHYSIOLOGY

The mechanisms involved in the control of embryogenesis
are of particular relevance in any study of pre-harvest
sprouting. However, little is known of the factors influencing embryo growth during caryopsis development. The source
of nutrients and water for the developing and sprouting embryo remain a subject for speculation. The transport pathway and sites of uptake are unknown.

The morphological changes accompanying embryogenesis in
barley and maize embryos have been described by Merry (1941)
and Randolph (1936) using light microscopy. More recently
however, Smart and O'Brien (1983) have described the overall
morphology of the developing wheat embryo in relation to its
surrounding tissues, using light and electron microscopy.
This work is of special interest since it includes some discussion of the pathways of nutrients destined for incorpora-

tion in the embryo.

Initially the embryo sits in a small pouch lined by nucellus and by 7 days post-anthesis (in conditions where the number of days from anthesis to 'maturity' was 80) is enveloped almost completely by modified endosperm cells. Unlike the typical starchy endosperm cell, these accumulate no starch and are densely cytoplasmic with very few vacuoles. Embryo growth is rapid and by 23 days, when the embryo was half its final size, the cells of the modifed endosperm have been totally digested along with the nucellar parenchyma cells. Slightly later, an extensive area of 'empty' endosperm cells appears which are adjacent to the scutellum. It was concluded from these observations that the developing embryo is a powerful sink for nutrients supplied to the caryopsis as a whole, since once development begins, the neighbouring cells are either hydrolysed or depleated of reserves. It was noted that the protein and starch reserves of the basal scutellum and coleorhiza accumulate earlier than those of the apical scutellum and coleoptile. It was suggested that this could be explained by the fact that the source of assimilate supply, that is, the vascular bundles of the pericarp, are nearer to the cells of the basal scutellum and coleoptile. As noted previously (Cochrane and Duffus, 1983) there was no evidence of plasmodesmata between the endosperm and embryo, and so all nutrients must enter the embryo by an apoplastic route.

However, the mechanisms involved in the differential loading of nutrients and water in developing embryos, the efflux of water from the embryo during maturation and the uptake of water by the embryo during pre-harvest sprouting remain unknown.

One of the first easily observable events following imbibition in germinating detached cereal grains is a rise in oxygen uptake (Chen and Varner, 1970). The source of the oxygen required for sprouting on the ear is unknown. It may come in, dissolved in water from the vascular bundles in the pericarp or from the atmosphere via the micropylar region (Cochrane and Duffus, 1983).

SUCROSE METABOLISM IN DEVELOPING CARYOPSES

The influence of nutrient supply on development and germination in immature cultured excised embryos has been described previously (Cameron-Mills and Duffus, 1980). It was suggested that pre-harvest sprouting may be a function of endosperm water potential. Water-soluble molecules

present in the endosperm which may control water potential
include sucrose and amino acids. Certainly increased sucrose concentrations in the range from 3 to 12 per cent in
the culture media cause a progressive decline in the germination growth rate.

In recent years there has been considerable interest in
the factors controlling the uptake, metabolism and concentration of sucrose in developing and maturing cereal
grains. Sucrose is the major sugar found in developing
wheat, barley and maize kernels (Cerning and Guilbot, 1973;
Chevalier and Lingle, 1983; Styer and Cantliffe, 1984). In
some sweet corn populations from a sugary enhancer breeding
programme the kernels accumulate relatively high levels of
maltose but these are rarely more than half those observed
for sucrose (Carey et al., 1984). In barley endosperm,
levels of free amino acids are similar to those observed for
sucrose over the developmental period (Cameron-Mills and
Duffus, 1980).

In wheat, and probably in barley also, most of the assimilates and mineral ions reaching the endosperm move along
the vascular strand of the pericarp and across the tissues
of the crease into the endosperm cavity (Frazier and Appalanaidu, 1965). Recently Donovan et al., (1983) have concluded that amino acids are transported along at least part of
the pathway, the longitudinal axis of the grain, by the same
mechanism as sucrose. Although there is evidence that the
transport of sucrose is subject to an upper limit to the in-
flux of sucrose into the grain (Jenner, 1976) it was suggested that the decrease in the accumulation of starch at
later stages of development in wheat is not due to a decrease in the sucrose supply (Jenner, 1982b). This has been
confirmed in barley (Cochrane, 1985) where it has been shown
that assimilate continues to reach the endosperm after the
end of the grain-filling period. Levels of soluble substrates in the endosperm must depend on differences in the
rate of their supply and the rate of their conversion to
insoluble reserve materials. In turn these are affected by
environmental conditions as well as the stage of grain
development.

The fate of sucrose in developing cereal endosperms
during the grain-filling period is not fully understood,
but it seems likely that most of the incoming sugar is converted to UDP-glucose and fructose in the starchy endosperm
in a reaction catalysed by UDP-dependent sucrose synthase.
Subsequent reactions may involve the conversion of UDP-
glucose to glucose-1-phosphate and then to starch via ADP-
glucose in a reaction involving both UDP- and ADP-pyrophos-

phorylase. There may however, be restrictions on the uptake
of intermediates by the amyloplast. Thus the first products
of sucrose metabolism may be degraded to dihydroxyacetone
phosphate in the cytoplasm of the endosperm cell. This
molecule may then cross the amyloplast inner membrane where
it may be converted to starch via the enzymes of gluconeo-
genesis, ADP-glucose pyrophosphorylase and ADP-glucose -
dependent starch synthase (Duffus and Cochrane, 1982).
Further work supporting the hypothesis (Jenner, 1982b) that
sucrose supply does not limit starch accumulation in the
later stages of caryopsis development comes from Lingle and
Chevalier (1984). For example, in 2h U-^{14}C-sucrose uptake
experiments via the awns of developing barley kernels it
was found that significant amounts of sucrose continued to
be taken up at the stage where dry matter accumulation had
apparently reached a plateau. Furthermore, the percentage
of non-leachable ^{14}C in the endosperm present as starch was
shown to decline markedly over the developmental period.
From this and other evidence (Duffus, 1982) it can be suggested that, at least during the period of grain-filling, it
is rate of starch biosynthesis which is important in the
control of sucrose uptake. It may be for example that, by
analogy with starch synthesis in leaves (Preiss and Levi,
1980), the enzyme ADP-glucose pyrophosphorylase is the major
regulatory enzyme on the pathway between sucrose and starch
synthesis.

SUCROSE - WATER RELATIONSHIPS

That there may be a correlation between assimilate
supply and grain water potential has been suggested by Craig
(1983). In her experiments, different concentrations of
sucrose were supplied to detached immature barley ears for a
10 day period at a number of stages of grain development.
As the sucrose concentration was increased, so the grain
water content, both as a total and a percentage, decreased.
For similar experiments with isolated barley embryos in culture it was shown (Dunwell, 1981) that increasing the sucrose concentration in the growth media resulted in a
decrease in the percentage water content after incubation
for 21 days.

These observations suggest that the concentration gradient set up by the increased sucrose supply may be responsible for drawing out water from the grain. Thus grain water
potential may be regulated by the concentration of assimilate supplied to the grain. The relationship between

assimilate supply and water loss during the final stages of
grain maturation has been investigated by Cochrane (1985).
Using detached ears of barley supplied with $U-^{14}C$ sucrose
for 4h, ^{14}C could be detected in the endosperms of grains
from ears harvested 50 'days' after anthesis. This was
10-15 'days' after the cell wall modifications, described
above, had taken place in the chalazal cells of the caryopsis. In 55-'day' ears supplied with $U-^{14}C$ sucrose for 20h
however, no ^{14}C could be detected in the upper part of the
rachis or in grains with a moisture content of less than
about 25 percent. Cytochemical reactions indicated that the
xylem elements in the upper part of the rachides of 55-'day'
barley ears were filled with pectic material. From these
observations it seems possible that progressive blockage of
the xylem elements could ultimately prevent the flow of
water and assimilates to the grain thus effectively cutting
off the grain from the influence of the rest of the plant.
At this stage therefore, any resistance to pre-harvest
sprouting must be associated with the grain itself. It has
already been suggested (Cameron-Mills and Duffus, 1980) that
such resistance may be related to low endosperm water potential. Certainly it is clear that this must decrease markedly in the later stages of maturation. Thus as we have
seen, sucrose and no doubt other soluble compounds, continue
to enter the grain in the period between the end of grain
filling and the final blockage of the xylem vessels. At the
same time water is being lost from the grain and the amount
of water entering is decreasing by the mechanisms described
above. Eventually the water potential will reach a level
where cellullar activity is inhibited and germination is
prevented. Work with high sugar mutants of maize (Styer
and Cantliffe, 1984) suggests a link between germination and
kernel sucrose content. Thus while the total measured sugar
concentrations, on a dry weight basis, in the mutant whole
seeds were twice those in the normal genotype, percentage
germination was greatest in the normal genotype. The
sucrose: starch ratio was highest in the embryos of the
genotype shrunken-1(sh2) which had the lowest percentage
germination.

Figures for water content in these genotypes were not
given but calculations based on data from Lee and Tsai
(1985) indicate that the percentage water content in kernels
of high sugar mutants was higher than in kernels from the
normal genotype over most of the developmental period. However, the osmotic potential in the mutant kernels at 28 days
post-pollination was shown to be significantly less than in
the normal kernels. This was attributed to the large con-

centrations of sucrose in the mutant kernels. It may be
that low osmotic potential favours movement into the kernels, thus accounting for the higher percentage water content in the high sucrose mutant at this stage of development. However, as maturation proceeds, the percentage water
content falls and the differential between osmotic poten-
tials of the genotypes is likely to be maintained. Any correlation suggested by these results between grain sucrose
content, osmotic potential or water potential, and germination is however at best circumstantial and factors other
than sucrose content may well be the cause of low seed vigor
in the sh2 genotype. For example, it may be that the low
starch content of the sh2 endosperm is unable to sustain
early growth of the germinating embryo. Furthermore, increasing the sucrose concentration of media in the range
from 1-10 percent had little effect on the percentage germination of the mutant and normal embryos in culture (Styer
and Cantliffe, 1984).

Water-soluble molecules other than sucrose which may
influence grain water potential are also present in the
immature cereal grains (Cerning and Guilbot, 1973: Singh
and Jenner, 1982; Shaw and Dickinson, 1984). Recent work
by Ho and Gifford (1984) has confirmed the presence of soluble fructan oligosaccharides in the endosperm cavity of
immature wheat grains during the linear phase of grain
growth. These were present at 9 times the concentration of
sucrose, and while it seemed unlikely that they were involved in sucrose metabolism, it may be that they too may
influence embryo development, maturation and germination
through an effect on water potential.

The mature seed has an extremely low water potential
(see Hadas, 1982). If water is available therefore, it will
start taking up water rapidly, thus increasing the water
potential and, in the absence of dormancy, germination will
take place.

The results reported here suggest that, in the period
between the end of grain filling and final blockage of the
xylem elements, the steadily decreasing water potential of
the grain is sustained by both active and passive processes
of water removal. If low water potential is indeed involved
in the regulation of pre-harvest sprouting then it can only
be effective once the grain has been sealed off from the
rest of the plant.

CONCLUSION

This discussion has examined a number of physiological and biochemical events accompanying grain development and maturation which are likely to be of relevance to embryo growth, maturation and germination. From a consideration of the mechanisms controlling grain water content, and sucrose supply and concentration it seems likely that, although direct evidence is not available, grain water potential may be a key factor in the control of pre-harvest sprouting. While oxygen may be required for sprouting, the mechanisms regulating its availability to the embryo are unknown. So far it has not been possible to use existing knowledge with any confidence to explain the primary causes of pre-harvest sprouting. Only when fundamental studies concerned with water relationships and embryo development are carried out under sprouting conditions with susceptible and non-susceptible genotypes can real progress be made.

REFERENCES

Barlow, E. W. R., Lee, J. W., Munns, R. and Smart, M. G. 1980. Water relations of the developing wheat grain. Aust. J. Plant Physiol. 7, pp. 519-525.

Brooks, A., Jenner, C. F. and Aspinall, D. 1982. Effects of water deficit on endosperm starch granules and on grain physiology of wheat and barley. Aust. J. Plant Physiol. 9, pp. 423-436.

Cameron-Mills, V. and Duffus, C. M. 1980. The influence of nutrition on embryo development and germination. Cereal Research Commun. 8, pp. 143-149.

Carey, E. E., Dickinson, D. B. and Rhodes, A. M. 1984. Sugar characteristics of sweet corn populations from a sugary enhancer breeding program. Euphytica 33, pp. 609-622.

Cerning, J. and Guilbot, A. 1973. Changes in the carbo-hydrate composition during development and maturation of the wheat and barley kernel. Cereal Chem. 50, pp. 220-232.

Chen, S. S. C. and Varner, J. E. 1970. Respiration and pro-tein synthesis in dormant and after-ripened seeds of Avena fatua. Plant Physiol. 46, pp. 108-112.

Chevalier, P. and Lingle, S. E. 1983. Sugar metabolism in developing wheat and barley kernels. Crop Sci. 22, pp. 272-277.

242

Cochrane, M. P. 1983. Morphology of the crease region in relation to assimilate uptake and water loss during caryopsis development in barley and wheat. Aust. J. Plant Physiol. 10, pp. 473-491

Cochrane, M. P. 1985. Assimilate uptake and water loss in maturing barley grains. J. exp. Bot. 36, pp. 770-782.

Cochrane, M. P. and Duffus, C. M. 1979. Morphology and ultrastructure of immature cereal grains in relation to transport. Ann. Bot. 44, pp. 67-72.

Cochrane, M. P. and Duffus, C. M. 1983. Observations on the development of the testa and pericarp in barley. In: Third Int. Symp. on Pre-harvest Sprouting in Cereals. Kruger J. E. and LaBerge, D. E. (eds.). Westview Press Boulder, Colorado. pp. 154-161.

Craig, L. M. 1983. Water relations in developing barley grains. M.Sc. Thesis, University of Edinburgh.

Donovan, G. R., Jenner C. F., Lee, J. W. and Martin, P. 1983. Longitudinal transport of sucrose and amino acids in the wheat grain. Aust. J. Plant Physiol. 10, pp. 31-41.

Duffus, C. M. and Cochrane, M. P. 1982. Carbohydrate metabolism during cereal grain development. In: The Physiology and Biochemistry of Seed Development, Dormancy and Germination. Khan, A. A. (ed.). Elsevier Biomedical Press, Amsterdam. pp. 43-66.

Dunwell, J. M. 1981. Influence of genotype and environment on growth of barley embryos in vitro. Ann. Bot. 48: pp. 535-542.

Frazier, J. C. and Appalanaidu, B. 1965. The wheat grain during development with reference to nature, location and role of its translocatory tissues. Am. J. Bot. 52, pp. 193-198.

Hadas, A. 1982. Seed-soil contact and germination. In: The Physiology and Biochemistry of Seed Development, Dormancy and Germination. Khan, A. A. (ed.). Elsevier Biomedical Press, Amsterdam: pp. 507-527.

Ho, L. C. and Gifford, R. M. 1984. Accumulation and conversion of sugars by developing wheat grains $\underline{V}$. The endosperm apoplast and apoplastic transport. J. exp. Bot. 35, pp. 58-73.

Jenner, C. F. 1976. Physiological investigations on restrictions to transport of sucrose in ears of wheat. Aust. J. Plant Physiol. 3, pp. 337-347.

Jenner, C. F. 1982a. Movement of water and mass transfer into developing grains of wheat. Aust. J. Plant Physiol. 9, pp. 69-82.

Jenner, C. F. 1982b. Storage of starch. In: Encyclopaedia

of Plant Physiology (Plant Carbohydrates I), Vol. 13A.
Loewus F. A. and Tanner, W. (eds.). Springer Verlag,
Berlin.

King, R. W. and Richards, R. A. 1984. Water uptake in
relation to pre-harvest sprouting damage in wheat:
ear characteristics. Aust. J. Agric. Res. 35, pp.
327-336.

Lee, L. and Tsai, C. Y. 1985. Effect of sucrose accumulation on zein synthesis in maize starch-deficient mutants. Phytochem. 24, pp. 225-229.

Lingle, S. E. and Chevalier, P. 1982. Movement and metabolism of sucrose in developing barley kernels. Crop
Sci. 24, pp. 315-319.

Lingle, S. E. and Chevalier, P. 1985. Development of the
vascular tissue of the wheat and barley caryopsis as
related to the rate and duration of grain-filling.
Crop Sci. 25, pp. 123-128.

Mares, D. J. 1983. Investigation of the pre-harvest sprouting damage resistance mechanisms in some Australian
white wheats. In : Third Int. Symp. on Pre-harvest
Sprouting in Cereals. Kruger J. E. and LaBerge, D. E.
(eds.). Westview Press, Boulder, Co. pp. 59-65.

Meredith, P. and Jenkins. L. D. 1975. Loss of moisture from
developing and ripening cereal grains. New Zealand J.
Sci. 18, pp. 501-509.

Merry J. 1941. Studies on the embryo of Hordeum sativum.
I. The development of the embryo. Bull. Torrey Bot.
Club. 68, pp. 585-598.

Nicholas, M. E., Gleadow, R. M. and Dalling, M.J. 1985.
Effect of post-anthesis drought on cell division and
starch accumulation in developing wheat grains. Ann.
Bot. 55, pp. 433-444.

Oparka, K. J. and Gates, P. 1981. Transport of assimilates
in the developing caryopsis of rice (Oryza sativa L.).
The pathways of water and assimilated carbon. Planta.
152, pp. 388-396.

Preiss, J. and Levi,C. 1980. Starch biosynthesis and
degradation. In: The Biochemistry of Plants. Vol. 3.
Carbohydrates: structure and function. Stumpf P. K.
and Conn, E. E. (eds.). Academic Press, New York.
pp. 371-423.

Randolph, L. F. 1936. Developmental morphology of the caryopsis in maize. J. Agric. Res. 53, pp. 881-916.

Shaw, J. R. and Dickinson, D. B. 1984. Studies of sugars
and sorbitol in developing corn kernels. Plant
Physiol. 75, pp. 207-211.

Singh, B. K. and Jenner, C. F. 1982. Association between concentrations of organic nutrients in the grain, endosperm cell number and grain dry weight within the ear of wheat. Aust. J. Plant Physiol. 9, pp. 83-95.

Smart, M. G. and O'Brien, T. P. 1983. The development of the wheat embryo in relation to the neighbouring tissues. Protoplasma. 114, pp. 1-13.

Styer, R. C. and Cantliffe, D. J. 1984. Dependence of seed vigor during germination on carbohydrate source in endosperm mutants of maize. Plant Physiol. 76, pp. 196-200.

Woodbury, W. and Wiebe, T. J. 1983. A possible role for the pericarp in control of germination and dormancy of wheat. In: Third Int. Symp. on Pre-harvest Sprouting in Cereals. Kruger, J. E. and LaBerge, D. E. (eds.). Westview Press, Boulder, Co. pp. 51-58.

The Effect of the Environment during Grain Growth in Barley on Gibberellin Physiology

P. B. Nicholls, The Department of Plant
Physiology, Waite Agricultural Research
Institute, University of Adelaide, Pte.
Mail Bag, Glen Osmond, S.A. 5064. Australia.

SUMMARY

Two physiological types of barley kernels have been identified. They differ in their requirement for gibberellic acid (GA_3) to induce alpha-amylase synthesis in de-embryonated kernels. Type A grains produce alpha-amylase only when germinated with an embryo or incubated in the presence of GA_3. On the other hand alpha-amylase is produced in Type B grains in the absence of both an embryo and GA_3.

Type B grains have been experimentally produced in a number of Australian barley cultivars and have been found in samples of Clipper from the 1983 South Australian crop. Inhibitors of gibberellin biosynthesis applied to Type B kernels in the absence of GA_3 have no effect on the production of alpha-amylase, indicating that an adequate amount of gibberellin may be present in the dry kernel. Although Type B kernels have been produced experimentally, the factors causing this development have not been fully delineated.

It is adduced that factors other than those examined may be of significance in the field situation and their possible interaction with the biosynthetic pathway of gibberellins is discussed.

INTRODUCTION

The appearance of pre-harvest sprouting in the field is an erratic phenomenon, it occurs only when the appropriate combination of factors are present (Stoy 1983). One of these is the potential of the ripening kernel to respond to the existing weather conditions. This potential is in part a function of the genetic constitution of the plant and in part a function of the interaction of the developing kernel with its surrounding environment. It should be noted that the environment during kernel development may be distinct from that precipitating sprouting of the nonharvested grain.

It was reported (Nicholls 1983) that two elements of the environment (temperature and light quality) affected the developing kernel to yield a mature aleurone layer which had the potential to produce alpha-amylase in the absence of both an embryo and added GA_3. Such kernels undergo endosperm modification (MacGregor et al., 1983) and produce the full complement of alpha-amylase isozymes (Nicholls et al., 1986). Implicitly, such kernels have the potential to respond to appropriate pre-harvest weather conditions by endosperm modification even though the embryo may be dormant. Further progress on this research will be reported in this paper.

MATERIALS AND METHODS

Kernels from various cultivars of Hordeum vulgare were grown in pots of recycled soil in the glasshouse until shortly before awn emergence and then transferred to a divided controlled environment room, one half of which was lit with metal halide lamps and the other with high pressure sodium lamps. The whole room was maintained at the given temperature on a 14h day/10h night cycle. These lighting systems gave an irradiance at ear height from 350 to 500 umole photon m^{-2} s^{-1} for metal halide lamps and from 700 to 900 umole photon m^{-2} s^{-1} for high pressure sodium lamps. The spectral distribution of the various lamps sources has been described by Cathey and Campbell (1980). At designated times after awn emergence the ears were harvested and dried in paper bags at 20°C.

All incubations were conducted as described by Nicholls (1982) and enzyme assays as described by Nicholls (1980). Gibberellin A_1 (GA_1) was a gift from Professor L. N. Mander (Australian National University), FR1587 [N,N,N,trimethyl-1-methyl-(2^r, 6',6'-trimethylcyclohex-2'-en-1'-yl)prop-2-enyl ammonium iodide] was a gift from Professor Tamura (The University of Tokyo), CCC [1-chloro-2-(N,N,N,-trimethyl)ethyl ammonium chloride) was obtained from Cyanamid Australia as cycocel 100A and +/- abscisic acid (ABA) was a gift from R. J. Reynolds Tobacco Company. In some instances incubations were carried out on samples of field grown barley supplied by the Australian Barley Board and by the S.A. Malting Co.

Response Index (Index) is calculated by expressing the alpha-amylase activity generated in the absence of GA as a percent of that in the presence of 10^{-6}M GA. An index of >1% was arbitrarily chosen to indicate Type B kernels.

RESULTS AND DISCUSSION

Sources of Variability in the Induction of Type B Grains

One source of variability in field crops is seasonal as illustrated in TABLE 1 where four samples of the 1983 crop of Clipper from different areas of South Australia yielded Type B kernels whereas the 1979 season sample quoted was typical of a large number of samples tested from the 1979, 1980 and 1981 crops in yielding Type A kernels.

TABLE 1. Production of alpha-amylase (EU) and Response Index of samples of field grown Clipper barley collected from silos in South Australia. The de-embryonated kernels were incubated for 66h at 15°C in 20mM $Ca(NO_3)_2$ solution with or without $GA_1(10^{-7}M)$.

| Season | 1979 | 1983 | 1983 | 1983 | 1983 |
Silo	Bordertown	Coonalpyn	Gladstone	Ardrossan	Pt. Lincoln
-GA	0.298	24.0	11.4	9.9	15.0
+GA	31.7	44.8	43.7	41.6	48.1
Index	0.94	53.6	26.1	23.8	31.2

Variability is also generated by the culture of growing ears in the laboratory at different temperatures and light conditions (Nicholls 1983). However, as summarized in TABLE 2, these procedures have not yielded consistent responses.

A third source of variability is related to the age of kernels at harvest (Nicholls 1983). Immature kernels (20 to 32 days from awn emergence) were markedly Type B whereas those collected at the harvest ripe stage were less so. However, as shown in TABLE 3, further work has again yielded inconsistent results.

The genetic constitution of the cultivar may be a fourth possible source of variation in the induction of Type B kernels (TABLE 4). Six of the seven cultivars used, behaved in a similar manner, all having Prior's Chevalier in their parentage, however, Shannon [Cl-3208/1 x Proctor (4x)] was quite different.

TABLE 2. Summary of three years experimentation with High
Pressure Sodium Lamps using Clipper barley.

Year	Growth temperature	Response index
1981	21/16, 27/22, 18/13,	94.2, 96.1, 90.7,
1981	24/19(SD), 24/19, 15/10	81.0, 62.7, 88.3
1983	20/16 (expt. 6.)	0.013
1984	20/16 (expt. 1.)	<1
1984	20/16 (expt. 2.)	24.5
1984	30/25 (expt. 3.)	62.5

1981 data are presented in chronological order.

TABLE 3. The effect of age at harvest (days after awn
emergence) on response index and the production of alpha-
amylase in de-embryonated kernels from three samples of
Clipper barley.

Planting	4/81		2/84		3/84	
Growth temperature	24/19		20/16		30/25	
Age at harvest	29	41	28	42	28	42
(-) GA	16.4	3.55	0.69	6.81	9.77	24.5
Index	81.0	24.3	2.8	24.0	36.4	62.5

TABLE 4. Production of alpha-amylase (EU) in the absence of
applied GA by de-embryonated kernels and response index for
a number of barley cultivars grown at 30/25°C and harvested
42 days after awn emergence.

	Weeah	Clipper	Schooner	Shannon	Stirling	Grimmett	Forest
Alpha Amy.	18.5	24.5	17.6	0.56	21.0	9.96	18.5
Index	73.4	62.5	59.1	1.5	67.3	39.6	61.9

In summary, none of the laboratory techniques used for
the induction of Type B kernels have been found to give consistent results. This variation in response does not appear
to be related to seasonal conditions which induce the
development of these grains at intermittent intervals in
field grown crops.

Partial Characterisation of Type B Kernels

It has been established that Type B kernels produce alpha-amylase isozyme 2 in the absence of an embryo and GA_3 (Nicholls et al., 1986), and similarly the starchy endosperm of such kernels undergoes modification (MacGregor et al., 1983). Neither CCC nor FR1587, both of which inhibit the synthesis of gibberellins at the cyclisation step (Hedden et al., 1977; Gilmour and Macmillan 1984), inhibit the production of alpha-amylase in Type B endosperms in the absence of GA_1 (TABLE 5.1). Incubation of Type B endosperms with 5×10^{-6}M ABA, which in most experiments inhibits the transcription of m-RNA coding for alpha-amylase, causes a decrease in the "acceleration rate" of the production of alpha-amylase and most of this is reversed by adding 10^{-6}M GA_3 (TABLE 5), however, no change in the lag time was found (data not shown).

TABLE 5. The effect of CCC and FR1587 and ABA on 1. the acceleration rate of alpha-amylase synthesis in Response B kernels (planting 3/82) in the absence of added GA and 2. the effect of ABA on different Response Types of Clipper barley.

1. Treatment	Fitted regression line	Acceleration Rate*
A. Control	$EU^{\frac{1}{2}} = 0.1155T - 1.534$	0.0267
$\quad 10^{-3}$M CCC	$EU^{\frac{1}{2}} = 0.1211T - 1.695$	0.0299
B. Control	$EU^{\frac{1}{2}} = 0.1233T - 1.544$	0.0304
$\quad 10^{-5}$M FR1587	$EU^{\frac{1}{2}} = 0.1336T - 1.804$	0.0357
C. Control	$EU^{\frac{1}{2}} = 0.0919T - 0.695$	0.0169
$\quad 5 \times 10^{-6}$M ABA	$EU^{\frac{1}{2}} = 0.0566T - 0.612$	0.0064
$\quad + 10^{-7}$M GA_3	$EU^{\frac{1}{2}} = 0.0903T - 1.636$	0.0163

2. Sample	Bordertown 79	Ardrossan 83	3/82
Index	0.42	20.5	70.1

Treatment	Acceleration Rate*		
Control	0.0001	0.0061	0.0225
5×10^{-6}M ABA	<0.0001	0.0016	0.0030
10^{-6}M GA_1	0.0238	0.0297	0.0321
both	0.0233	0.0190	0.0148

*Acceleration rate = $2 \times (slope)^2$

From these observations it would appear that there was an adequate concentration of endogenously, physiologically active gibberellin present in mature Type B kernels. Immunoassay analysis of extracts of two samples of Type B kernels indicated that there was 10 fold more GA_1 in these than in Type A Clipper kernels from the 1979 crop (Nicholls and Weiler, unpub.). If this is confirmed, then it appears that environmental conditions during grain growth significantly modify gibberellin metabolism. The naturally occuring, physiologically active gibberellin in developing barley kernels is GA_1, but in the only well-documented analysis of the gibberellins in developing barley kernels it was found to occur in low amounts relative to other more hydroxylated gibberellins (GA_{34} and GA_{48}) both of which contain a hydroxyl on the 2-beta position (Gaskin et al., 1983). Gibberellins containing a hydroxyl in this position are usually inactive (eg GA_8, 2-beta-OH derivative of GA_1) and the enzyme which catalyses this step has been recently well characterised (Smith and MacMillan, 1984). Ferrous ions, a source of reducing power, eg ascorbate, and 2-oxo-glutarate are essential components of this enzyme system. Manganous ions have been found to be essential in some in vitro enzyme extracts from developing seeds, which are capable of synthesising the first unique substance on the gibberellin biosynthetic pathway (GA_{12}ald) from the common starting material (mevalonic lactone) for a range of plant constituents eg. sterols, ABA, GAs, carotenes, and terpenes (Hedden and Graebe 1982). The demonstration of the involvement of these minor mineral nutrients in gibberellin metabolism introduces another dimension to studies on the interaction of environment and plant cell during growth and may suggest an explanation for induction of Type B kernels in both field and laboratory conditions. If environmentally induced changes in the $Fe^{2+}/[Fe^{2+}+Fe^{3+}]$ are an important component of gibberellin metabolism then a number of environmental factors may have effects through modulating this ratio eg. using High Pressure Sodium lamps as the major light source, growth temperature and variations in the nitrate/ammonium ratio in the supply of nitrogen (Bennett et al., 1984)

ACKNOWLEDGEMENTS

The author acknowledges the financial support provided by the Barley Industry Research Council of Australia and the technical assistance ably given by Mrs. Heather Paull.

Bennett, J. H., Krizek, D. T., Wergin, W. P., Flemming, A. L., Mirecki, R. M., and Wyse, R. E. 1984. Physiological and ultrastructural changes in the chloroplasts of snap been plants under LPS lamps during chlorosis and regreening. J. Plant Nutr. 7, pp. 819-832.

Cathey, H. M. and Campbell, L. E. 1980. Light and lighting systems for horticultural plants. Hort. Rev. 2, pp. 491-537.

Gaskin, P., Gilmour, S. J., Lenton, J. R., MacMillan, J., and Sponsel, V. M. 1983. Endogenous gibberellins and kaurenoids identified from developing and germinating barley grain. J. Plant Growth Regul. 2, pp. 229-242.

Gilmour, S. J. and MacMillan, J. 1984. Effects of inhibitors of gibberellin biosynthesis on the induction of alpha-amylase in embryoless caryopses of Hordeum vulgare cv. Himalaya. Planta 162, pp. 89-90.

Hedden, P. and Graebe, J. E. 1982. Co-factor requirement for the soluble oxidases in the metabolism of the C-20 gibberellins. J. Plant Growth Regul. 1, pp. 105-116.

Hedden, P., Phinney, B. O., MacMillan, J. and Sponsel, V. M. 1977. Metabolism of kaurenoids by Gibberella fujikuroi in the presence of the plant growth retardant, N,N,N,-trimethyl-1-methyl-(2',6',6',-trimethylcyclohex-2'-en-1'yl)prop-2-enyl ammonium iodide. Phytochem. 16, pp. 1913-1917.

MacGregor, A. W., Nicholls, P. B. and Dushnicky, L. 1983. Endosperm degradation in barley kernels that synthesize alpha-amylase in the absence of embryo and exogenous gibberellic acid. Food Microstructure. 2, pp. 13-22.

Nicholls, P. B. 1980. Development of responsiveness to gibberellic acid in the aleurone layer of immature cereal caryopses: effect of temperature. Aust. J. Plant Physiol. 7, pp. 645-653.

Nicholls, P. B. 1982. Influence of temperature during grain growth and ripening of barley on the subsequent response to exogenous gibberellic acid. Aust. J. Plant Physiol. 9, pp. 373-383.

Nicholls, P. B. 1983. Environment, developing barley grain, and subsequent production of alpha-amylase by the aleurone layer. In: Third Int. Symp. on Pre-harvest Sprouting in Cereals. Kruger J. E. and LaBerge D. E. (eds.). Westview Press, Boulder, Co. pp. 147-153.

Nicholls, P. B., MacGregor, A. W. and Marchylo, B. A. 1986. Production of alpha-amylase isozymes in barley caryopses in the absence of embryos and exogenous gibberellic acid. Aust. J. Plant Physiol. 13, in press.

Smith, V. A. and MacMillan, J. 1984. Purification and part-
    ial characterization of a gibberellin 2-beta hydroxyl-
    ase from Phaseolus vulgaris. J. Plant Growth Regul.
    2, pp. 251-264.
Stoy, V. 1983. Progress and prospect in sprouting research.
    In: Third Int. Symp. on Pre-harvest Sprouting in
    Cereals. Kruger, J. E. and LaBerge, D. E. (eds.).
    Westview Press, Boulder, Co. pp. 3-7.

Hormonal Changes during Cereal Grain Development

J. R. Lenton and M. D. Gale,* University
of Bristol, Department of Agricultural
Sciences, Long Ashton Research Station,
Bristol, England.
*Plant Breeding Institute, Cambridge, England.

INTRODUCTION

In general developing fruits and seeds are rich sources
of plant growth substances (hormones). Often there are
sequential, ontogenetic changes in the maximal concentrations of the different groups of hormones during seed
development (Goodwin, 1978) and this is also the case for
developing cereal grains (Wheeler, 1972). However there
is considerable controversy concerning the biological
significance of such fluctuations in growth substances.
Opinions vary from the assumption that hormone concentration controls development and determines 'sink activity', to
the belief that such changes reflect either 'luxury' uptake
from the parent plant or 'uncontrolled' synthesis in situ of
no physiological significance.

The relative importance of endogenous hormone concentration and the capacity of tissue to respond to it is a
matter of much lively debate (Trewavas, 1981; 1982). It is
well established that plant hormones elicit a wide range of
responses in a variety of cells, tissues and organs but the
molecular mechanisms involved in these responses are largely
unknown. Hormones induce rapid 'membrane' responses as well
as effects on gene expression. In addition, non-hormonal
secondary messengers, such as Ca-calmodulin (Hepler and
Wayne, 1985), should be considered as part of the signal
transduction system.

Despite recent advances in analytical techniques for
quantifying plant growth substances there are relatively
few laboratories equipped for such measurements. This is
particularly true for the cytokinins and gibberellins which
are groups of hormones containing large numbers of compounds
ranging widely in polarity. Highly sophisticated and expen-

sive combined gas chromatography-mass spectrometry (GC-MS) techniques are required for identification and quantification of these compounds. The situation is different for the auxins and the inhibitor, abscisic acid (ABA), as they can be measured reliably by other physical methods. More recently, immunoassays have been developed for plant hormone analysis but, in general, they have not been validated against physical methods. So-called quantitative estimates of hormone concentration from more classical bioassays performed on relatively crude plant extracts should be viewed with extreme caution.

It is against this background that the following discussion on hormonal changes in developing cereal grains should be considered.

Cytokinins

A transient peak of biologically-active cytokinins has been detected in developing wheat grains during the first six days after anthesis (FIGURE 1) (Wheeler, 1972; Jameson et al., 1982; Durley and Morris, 1983). Zeatin, zeatin riboside and their corresponding side-chain O-beta-D glucosides have been identified tentatively (Jameson et al., 1982; Durley and Morris, 1983). In addition, 1-hydroxy-methyl zeatin has been shown to reach a peak of activity at about 11 days after anthesis (Rademacher and Graebe, 1984). More recently a group of polar conjugates of zeatin have been detected in young wheat grains (Lenton and Appleford, in preparation). At present there are no reliable quantitative data on the cytokinin complex of young wheat grains.

It has been suggested that the source of cytokinins may be roots (Wheeler, 1972; 1976) but polar cytokinins are produced in situ in detached, cultured ears (Lenton and Appleford, in preparation). It has been argued that cytokinins are important in initiating reproductive 'sink activity' (Lenton, 1984) although the mechanism by which this may be achieved is not known. They might act indirectly either by stimulating endosperm nuclear and cell division or by inducing rapid expansion of the ovule and surrounding ovary tissues. Alternatively they may direct assimilate to young grains. Cytokinin content has been related to grain size in barley (Michael and Seiler-Kelbitsch, 1972) but the close relationship between grain size and cell number in wheat endosperm (Dunstone and Evans, 1974; Brocklehurst, 1977; Singh and Jenner, 1982) has not been examined in terms of differences in endogenous cytokinins.

FIGURE 1. Cytokinins, gibberellins, auxin and abscisic acid during wheat grain development.

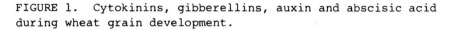

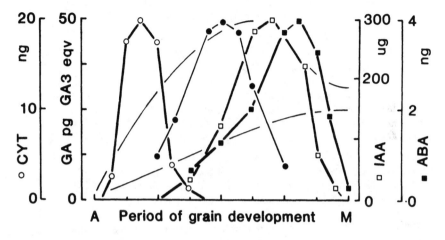

A Period of grain development M

Note. Hormone levels (per grain) and times of accumulation represent a general pattern. A = anthesis, M = maturity. Broken curves represent grain fresh (upper) and dry (lower) weights.

Auxins

Large amounts of free auxin, IAA, accumulate at the time of maximal grain fresh weight, then decline rapidly in both wheat (FIGURE 1) (Wheeler, 1972; Radley, 1976a; Mengel et al., 1985) and barley (Mounla et al., 1980). Most of the auxin is located in the endosperm but it is also present at high concentrations in young embryos (Rademacher and Graebe, 1984). Free IAA or its precursor, tryptophan, may be imported in the assimilate stream, possibly from senescing tissue. Little is known of the metabolic fate of IAA in wheat and barley although complex myo-inositol esters and oxindole metabolites occur in maize and rice (Bandurski, 1984).

Comparison of normal (Bomi) and high-lysine (Riso 1508) barleys showed large differences in IAA concentration but similar patterns of dry matter accumulation (Mounla et al., 1980). However, differences in IAA level were positively correlated with positional effects on grain size within the wheat cultivars, Solo and Kolibri, but not between cultivars or in different seasons (Bangerth et al., 1985). There is little positive information on the function of auxin in developing grain but it is unlikely to be rate-limiting for

cell expansion or auxin-directed assimilate transport
(Bauermeister et al., 1980).

Abscisic acid

As with auxin, the ABA content of developing cereal
grains follows the pattern of accumulation of dry matter
and a proportion is co-transported from leaves with assimilates. Again the majority of ABA is present in the endosperm although high concentrations occur in the embryo.
Grain ABA levels reach maximum after IAA and begin to
decline relatively slowly, prior to the net loss of water
from grains (FIGURE 1). Metabolites of ABA, such as phaseic
and dihydrophaseic acids, accumulate as ABA declines and
grains have the capacity to both synthesize and catabolise
ABA (see King, 1982 for review).

The proposed function of ABA in stimulating apoplastic
unloading of sucrose requires clarification as, logically,
it should inhibit active reloading of sucrose into endosperm
cells (see Lenton, 1984 for discussion). ABA is involved in
normal embryogenesis in barley (Umbeck and Norstog, 1979)
and soybean (Ackerson, 1984) and has been implicated in the
control of precocious germination in certain viviparous
mutants of maize (Brenner et al., 1977) and in an induced
ABA synthesis mutant of Arabidopsis (Karssen et al., 1983).
Some evidence for an involvement with dormancy in cereals
has been gained through correlations of extractable ABA with
germination in different varieties (King, 1976) and with
germination at times after maturity (Slominski et al.,
1979). Other studies, which have shown a lack of such correlations (Radley, 1979; King et al., 1979), indicate that
factors other than ABA may also be involved. Work on the
mechanism of action of ABA at the molecular level has shown
that it can induce the production of alpha-amylase inhibitor
(Mundy, 1984), and can also inhibit GA-induced alpha-amylase
formation (Chandler et al., 1984).

Gibberellins

The fluctuation in the amounts of extractable gibberellins (GAs) in developing cereal grains is more complex than
for the other groups of hormones (FIGURE 2).

An early peak of biologically-active GAs has been observed in whole ears of wheat during anthesis (Wheeler,
1972) (FIGURE 2) but it is not known if these are characteristic of 'seed' or 'shoot'. There is good evidence
from Phaseolus that GA_1 produced in the suspensor supports

FIGURE 2.   Gibberellin levels during grain development.

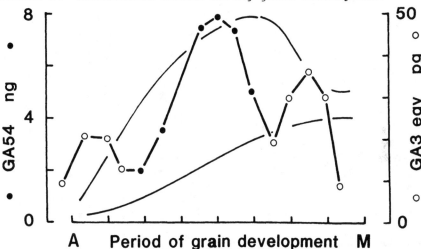

Note: Amount (per grain) and times of accumulation repre-
        sent a general pattern.   A = anthesis, M = maturity.
        Broken curves represent grain fresh (upper) and dry
        (lower) weights.

growth of heart-shaped embryos (Cionini et al., 1976; Alpi
et al., 1979).   By analogy, the early peak of GAs in cereals
might be equally important in embryogenesis and endosperm
formation.   The major peak of biological activity occurs
around the time of maximum grain volume (Wheeler, 1972;
Radley, 1976a;   Mounla, 1978;   Slominski et al., 1979) but
it is now clear that the GAs present in developing grains
are different from those found in vegetative tissues.

     There is good evidence from a range of single gene
deficient mutants in peas, maize and rice that GA_1 is the
main biologically-active GA responsible for stimulating
shoot growth in vegetative tissues (Phinney, 1985).   The
biosynthetic pathway to GA_1 involves both early 13- and late
3-hydroxylation steps (FIGURE 3).   In a comparative review
of 'shoot' and 'developing seed' GAs of six species, including wheat, barley and rice, Ingram et al., (1986) have concluded that the early 13-hydroxylation pathway for shoot GAs
is highly conserved whereas the GAs of developing seed have
different, and often species-characteristic, hydroxylation
patterns.   For example, in wheat a group of 1-hydroxylated
GAs (GA_{54}, GA_{55}, GA_{60}, GA_{61}, GA_{62}) have been identified in
endosperm at the time of maximum grain fresh weight (Gaskin
et al., 1980).   Both a non-13- and a 13-hydroxylation pathway may be present in wheat endosperm but 1-hydroxylation
presumably precedes 3-hydroxylation to account for the pro-

presumably precedes 3-hydroxylation to account for the production of GA_{54} and GA_{55} (Kirkwood and MacMillan, 1982)
rather than their non-1-hydroxylated counterparts, GA_4 and
GA_1, which are found in vegetative tissues (FIGURE 3). In
contrast, the GAs present in developing barley grain are
12-hydroxy and 18-hydroxy compounds (Gaskin et al., 1983).
For further discussion of the GAs in reproductive tissues,
including other cereals, see the detailed review by Pharis
and King (1985).

FIGURE 3.   Hypothetical scheme to account for the presence
of 1-hydroxylated gibberellins in developing wheat endosperm
(note the absence of 1-hydroxylated GAs from vegetative
tissues).

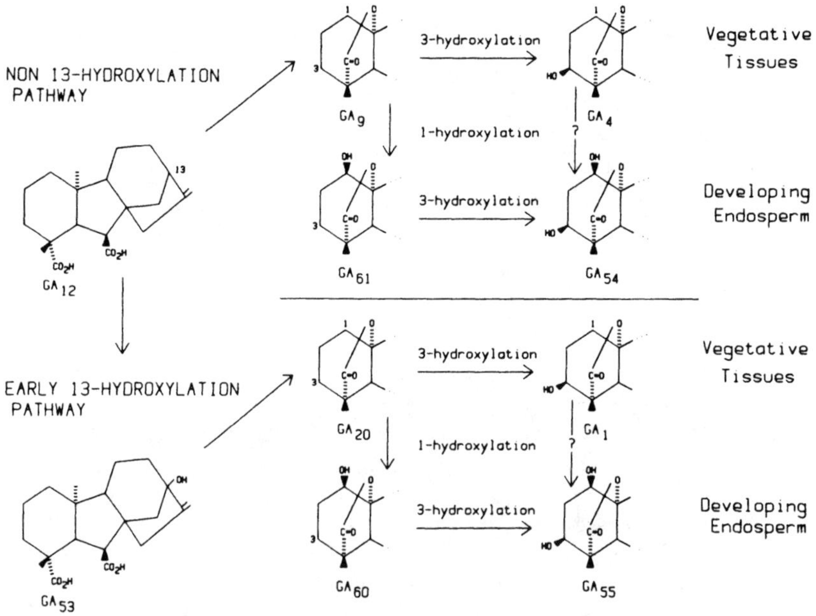

This wide range of structural types of seed GAs pre-
sents difficulties in preparing labelled compounds for
quantitative and metabolic studies. It is possible that
species-specific hydroxylated GAs are confined to triploid
endosperm tissue and that developing diploid embryos contain
the 'normal' shoot GA complement. There is some evidence
from inhibitor and substrate feeds to detached wheat ears
that they have the capacity for GA synthesis (Radley,
1976b). In addition, longer-term culture of detached wheat
ears has extablished that 1-hydroxylated GAs accumulate in
grains to the same extent as in ears of intact plants
(Lenton, unpublished). Although the concentration of GA_{54}
was high in grains of intact plants harvested 18 days after

was high in grains of intact plants harvested 18 days after anthesis it was accompanied by similar concentrations of ABA (Gale et al., this volume). More recently the GA_1 concentration of the same grains was shown to be 5000-fold less than that of GA_{54} (Lenton, unpublished). Although GA_{54} is 30 to 100-fold less active than GA_1 at inducing alpha-amylase production in mature, distal half-grains it was considered unlikely that it induces 'green' pericarp amylase production in developing grains (Gale et al., this volume). These alpha-amylases are produced by the same compound genes which produce the GA-induced 'green' enzyme during germination but comprise only a subset of the 17 alpha-AMY-2 isozymes (Gale and Ainsworth, 1984). A detailed analysis of the alpha-Amy-2 mRNAs has confirmed this finding (A. Hutley and D. C. Baulcombe, pers. comm.). Nevertheless there is an obvious requirement for accurate measurements of GAs, ABA and alpha-amylase isozymes in different tissues during development.

The detection, in some studies, of biologically-active GAs late in grain development (Rejowski, 1964; Radley, 1976a; Dathe and Sembdner, 1978) is of considerable interest to prematurity amylase production. A good example of this phenomenon was shown by the cultivar, Kleiber grown outdoors during the cool, wet season of 1974 and under low temperatures in controlled environments (Radley, 1976a). Such a late peak of GA activity could be the postulated embryo stimulus for germination alpha-amylase production observed late in grain development in tall genotypes (Gale et al., this volume). Although the circumstantial evidence is strong it remains to be determined if embryo produced GA_1, the normal germination GA, is the stimulus for prematurity alpha-amylase production. It is known that a change in sensitivity of aleurone cells to GAs during grain development is influenced by both temperature (Nicholls, 1980) and rate of dehydration (Nicholls, 1979; King and Gale, 1980; Armstrong et al., 1982). Therefore, at this stage, we are unable to agree with the conclusions of Pharis and King (1985) that "high levels of endogenous GAs in maturing grain has no relevance to seed development, at least in relation to alpha-amylase synthesis".

The dwarfing genes conferring GA-insensitivity in wheat, Rht1 and Rht3, cause a reduction in shoot extension and an accumulation of GA_1 in vegetative tissues, (Gale et al., this volume) but have no direct effect on ear and grain development (Youssefian and Gale, unpublished). This would suggest that GAs are not required during ear and grain development, or that the genes are not expressed in reproductive tissues following initiation of the collar, except in the case of the precociously germinating embryo. Alternatively it is possible that grain growth is mediated by l-hydroylated GAs, such as GA_{54}, but that the GA-sensitive dwarfing genes specifically block the utilization of GA_1. Changes in endogenous GAs during stem elongation and ear development in genotypes containing different combinations of dwarfing genes is currently under investigation. The genetic and physiological basis of the apparent insensitivity of developing cereal ears and grains to their endogenous GAs requires further elucidation at the molecular level, particularly as considerable progress has been made on the mechanism of GA-induced alpha-amylase formation during seed germination (Baulcombe et al., 1985).

In summary, gibberellins, like the other three main groups of hormones, appear in the developing ear and grains of wheat, barley and rye in a consistent, ordered manner; (i) an early peak, of unknown GAs, at and just after anthesis, associated in time with pericarp growth and the coenocytic phase of endosperm growth, (ii) a major peak of activity, consisting mainly of GA_{54}, associated in time with maximum grain growth, (iii) a late peak, of unknown GAs shortly before maturity, which may be dependent upon ripening conditions or variety.

Some obvious gaps in our knowledge may be filled by resolving such questions as (i) are the different GAs found in young grains restricted to grains or are they present throughout the tissues of the developing spike?, (ii) do the 'seed' GAs have a different role from 'shoot' GAs?, and (iii) does the late peak consist of the same 'seed' GAs, or could they be embryo derived 'shoot' GAs which are involved in prematurity alpha-amylase production and other preharvest sprouting phenomena?

CONCLUSIONS

In this brief and selective review we have considered only certain aspects of hormonal status and responsiveness of developing cereal grains. The relative importance of changes in hormone concentration and sensitivity can be investigated by the use of (1) seed developmental mutants, (2) hormone synthesis or sensitivity mutants, (3) hormone biosynthesis or action inhibitors. In addition, work on intact systems should be complemented with techniques of ear culture in vitro. More fundamental information is required both on the expression of Rht dwarfing genes during ear and grain development and on the molecular mode of action of GAs in relation to preharvest alpha-amylase production.

261

REFERENCES

Ackerson, R. C. 1984.   Regulation of soybean embryogenesis
    by abscisic acid.   J. exp. Bot. 35, pp.   430-413.
Alpi, A., Lorenze, R., Cionini, P. G., Bennici, A. and
    D'Amato, R. 1979.   Identification of gibberellin A_1
    in the embryo suspensor of Phaseolus coccineus.
    Planta 147, pp. 225-228.
Armstrong, C., Black, M., Chapman, J. M., Norman, H. A. and
    Angold, R. 1982.   The induction of sensitivity to gib-
    berellin in aleurone tissue of developing wheat grains.
    1.   The effect of dehydration.   Planta 154, pp.
    573-577.
Bandurski, R. S. 1984.   Metabolism of indole-3-acetic acid.
    In: The biosynthesis and metabolism of plant hormones.
    Crozier, A. and Hillman, J. R. (eds.).   Cambridge
    University Press, Cambridge.   pp. 183-200.
Bangerth, R., Aufhammer, F. and Baum, O. 1985.   IAA level
    and dry matter accumulation at different positions
    within a wheat ear.   Physiol. Plant. 63, pp. 121-125.
Bauermeister, A., Dale, J. E., Williams, E. J. and Scobie,
    J. 1980.   Movement of ^{14}C and ^{11}C labelled assimilate
    in wheat leaves : the effect of IAA.   J. exp. Bot. 31,
    pp. 1199-1209.
Baulcombe, D., Martienssen, R. and Lazarus, C. 1985.   Alpha-
    amylases and gibberellins: molecular analysis of co-
    regulated gene families in wheat.   In: Molecular form
    and function of the plant genome.   Vloten-Doting, L.,
    Groot, G. S. P. and Hall, T. C. (eds.).   Plenum, New
    York.   pp. 155-166.
Brenner, M. L., Burr, B. and Burr, F. 1977.   Correlation of
    genetic vivipary in corn with abscisic acid correla-
    tion.   Plant Physiol. 55S, pp. 85-86.
Brocklehurst, P. A. 1977.   Factors controlling grain weight
    in wheat.   Nature. 266, pp. 348-349.
Chandler, P. M., Zwar, J. A., Jacobsen, J. V., Higgins, T.
    J. V. and Inglis, A. S. 1984.   The effects of gib-
    berellic acid and abscisic acid on alpha-amylase mRNA
    levels in barley aleurone layers: studies using an
    alpha-amylase cDNA clone.   Plant Molecular Biology 3,
    pp. 407-418.
Cionini, P. G., Bennici, A., Alpi, A. and D'Amato, F. 1976.
    Suspensor, gibberellin and in vitro development of
    Phaseolus coccineus embryos.   Plant 131, pp. 115-117.
Dathe, W. and Sembdner, G. 1978.   Distribution of gibberel-
    lins and abscisic acid in different fruit parts of rye.
    Gibberellins LXVII Biochem. Physiol. Pflanzen. 173, pp.
    440-447.

Dunstone, R. L. and Evans, L. T. 1974. Role of change in
cell size in the evolution of wheat. Aust. J. Plant
Physiol. 1, pp. 157-165.

Durley, R. C. and Morris, R. O. 1983. Cytokinins in
developing wheat seeds. Plant Physiol. 72, (Suppl.)
p. 114.

Gale, M. D. and Ainsworth, C. C. 1984. The relationship
between alpha-amylase species found in developing and
germinating wheat grain. Biochem. Genet. 22, pp.
1031-1036.

Gale, M. D., Salter, A. M. and Lenton, J. R. 1986. The
induction of germination alpha-amylase during wheat
grain development in unfavourable weather conditions,
(this volume).

Gaskin, P., Gilmour, S. J., Lenton, J. R., MacMillan, J. and
Sponsel, V. M. 1983. Endogenous gibberellins and
kauranoids identified from developing and germinating
barley grain. J. Plant Growth Regul. 2, pp. 229-242.

Gaskin, P., Kirkwood, P. S., Lenton, J. R., MacMillan, J.
and Radley, M. E. 1980. Identification of gibberellins
in developing wheat grain. Agricultural and Biological
Chemistry. 44, pp. 1589-1593.

Goodwin, P. B. 1978. Phytohormones and fruit growth. In:
Phytohormones and Related Compounds: A comprehensive
treatise. Vol. II. Letham, D. S., Goodwin, P. B. and
Higgins, T. J. V. (eds.). Elsevier, Amsterdam. pp.
175-204.

Hepler, P. K. and Wayne, R. O. 1985. Calcium and plant
development. Ann. Rev. Plant Physiol. 36, pp. 397-
439.

Ingram, T. J., MacMillan, J. and Sponsel, V. 1986. Gibberellin distribution and metabolism: a comparison
between seeds, shoots and roots. Brno Proc. Symp.
Integrity of Whole Plants (in press).

Jameson, P. E., McWha, J. A. and Wright, G. J. 1982. Cytokinins and changes in their activity during the
development of grains of wheat (Triticum aestivum L.).
Z. Pflanzenphysiol. 106, pp. 27-36.

Karssen, C. M., Brinkhorst van der Swan, D. L. C., Breek-
land, A. E. and Koornneef, M. 1983. Induction of dormancy during seed development by endogenous abscisic
acid: studies on abscisic acid deficient genotypes of
Arabidopsis thaliana (L.) Heynh. Planta 157, pp. 158-
165.

King, R. W. 1976. Abscisic acid in developing wheat grains
and its relationship to grain growth and maturation.
Planta 132, pp. 43-51.

King, R. W. 1982. Abscisic acid in seed development. In: The physiology and biochemistry of seed development, dormancy and germination. Khan, A. A. (ed.). Elsevier Biomedical Press, Amsterdam, pp. 157-181.

King, R. W. and Gale, M. D. 1980. Pre-harvest assessment of potential alpha-amylase production. Proc. 2nd Int. Sprouting Symposium. Gale, M. D. and Stoy, V. (eds.). Cereal Research Communications. 8, pp. 157-166.

King, R. W., Salminen, S. O., Hill, R.D. and Higgins, T. J. V. 1979. Abscisic acid and gibberellin action in developing kernels of triticale (cv. GA 190). Planta 146, pp. 249-255.

Kirkwood, P. S. and MacMillan, J. 1982. Gibberellins A_{60}, A_{61} and A_{62}: partial synthesis and natural occurence. J. Chem. Soc. PT(1) pp. 689-697.

Lenton, J. R. 1984. Are plant growth substances involved in the partitioning of assimilate to developing reproductive sinks? Plant Growth Regulation 2, pp. 267-276.

Lenton, J. R. and Appleford, N. E. J. (in preparation). Cytokinins: Plant hormones in search of a role. British Plant Growth Regulator Group Monograph 14.

Mengel, K., Friedrich, B. and Judel, D. G. 1985. Effect of light intensity on the concentrations of phytohormones in developing wheat grain. J. Plant Physiol. 120, pp. 255-266.

Michael, G. and Seiler-Kelbitsch, H. 1972. Cytokinin content and kernel size of barley grain as affected by environmental and genetic factors. Crop Science 12, pp. 162-165.

Mounla, M. A. K. 1978. Gibberellin-like substances in parts of developing barley grain. Physiol. Plant. 44, pp. 268-272.

Mounla, M. A. K., Bangerth, F. and Stoy, V. 1980. Gibberellin-like substances and indole type auxins in developing grains of normal and high-lysine genotypes of barley. Physiol. Plant. 48, pp. 568-573.

Mundy, J. 1984. Hormonal regulation of alpha-amylase inhibitor synthesis in germinating barley. Carlsberg Res. Commun. 49, pp. 439-444.

Nicholls, P. B. 1979. Induction of sensitivity to gibberellic acid in developing wheat caryopsis: effect of rate of desiccation. Aust. J. Plant Physiol. 6, pp. 229-240.

Nicholls, P. B. 1980. Development of responsiveness to gibberellic acid in the aleurone layer of immature wheat and barley: effect of temperature. Aust. J. Plant Physiol. 7, pp. 645-653.

Pharis, R. P. and King, R. W. 1985. Gibberellins and
reproductive development in seed plants. Ann. Rev.
Plant Physiol. 36, pp. 517-568.

Phinney, B. O. 1985. Gibberellin A₁ dwarfism and shoot
elongation in higher plants. Biologia Plantarum.
27, pp. 172-179.

Rademacher, W. and Graebe, J. E. 1984. Hormonal changes in
developing kernels of two spring wheat varieties differing in storage capacity. Ber. Deutsch. Bot. Ges.
Bd. 97S, pp. 167-181.

Radley, M. 1976a. The development of wheat grain in relation to endogenous growth substances. J. Exp. Bot.
27, pp. 1009-1021.

Radley, M. 1976b. Effect of variation in ear temperature on
gibberellin content of wheat ears. Ann. appl. Biol.
82, pp. 335-340.

Radley, M. 1979. The role of gibberellin, abscisic acid and
auxin in the regulation of developing wheat grains. J.
exp. Bot. 30, pp. 381-389.

Rejowski, A. 1964. Gibberellins in maturing wheat seeds.
Bull. Acad. pol. Sci. Cl. U. Ser. Sci. Biol. 12, pp.
233-236.

Singh, B. K. and Jenner, C. F. 1982. Association between
concentrations of organic nutrients in the grain, endosperm cell number and grain dry weight within the ear
of wheat. Aust. J. Plant Physiol. 9, pp. 83-95.

Slominski, B., Rejowski, A. and Nowak, J. 1979. Abscisic
acid and gibberellin contents of ripening barley seeds.
Physiol. Plant. 45, pp. 167-169.

Trewavas, A. J. 1981. How do plant growth substances work?
Plant Cell Environ. 4, pp. 203-228.

Trewavas, A. J. 1982. Growth substance sensitivity: The
limiting factor in plant development. Physiol. Plant.
55, pp. 60-72.

Umbeck, P. F. and Norstog, K. 1979. Effects of ABA and
ammonium ion on morphogenesis in cultured barley
embryos. Bull. Torrey Bot. Club. 58, pp. 340-346.

Wheeler, A. W. 1972. Changes in growth substance contents
during growth of wheat grains. Ann. appl. Biol. 72,
pp. 327-334.

Wheeler, A. W. 1976. Some treatments affecting growth
substances in developing wheat ears. Ann. appl. Biol.
83, pp. 455-462.

Alpha-Amylase Activity and Falling Number in Spring Wheat, Rye and Triticale during Ripening

*J. Mjaerum, E. Mosleth and K. Ringlund,
Department of Crop Science, Agricultural
University of Norway, N-1432 As-NLH.*

INTRODUCTION

Observations have shown that triticale (x <u>Triticosecale</u> Wittmack) grown under Norwegian climatic conditions is damaged severely by sprouting even at very early stages of maturity. The purpose of these investigations was to study the development and breakdown of starch throughout the later stages of grain maturation for triticale, rye and wheat.

MATERIAL AND METHODS

Four Norwegian spring wheat lines, a North-European rye and three triticale lines of Mexican origin were grown under field conditions during the summer of 1985. The trial was laid out with three replications. Samples were harvested for seven weeks during the maturity period. The samples were carefully threshed and the moisture content determined. Two samples of grain from each plot were germinated in sand at 10°C and 20°C, respectively, to evaluate percent germination. A dormancy index was calculated according to Strand (1965) and the samples were dried to a moisture content of about 8 percent.

Portions of the samples were ground on a falling number mill, and falling numbers were determined. The alpha-amylase activity was determined by a colorimetric method (Olered 1967). Separation and identification of different alpha-amylase isoenzymes were done by isoelectric focusing on an Ampholine PAG plate, pH range 3.5-9.5 with 30 watt for 1.5h according to LKB recommendations. 1g falling number flour was extracted with 2ml, 10^{-3}M $CaCl_2$ overnight at 5-6°C, centrifuged at 36000 g at 4°C for 20 min and aliquots of the clear extract loaded onto the PAG plate. The alpha-

amylase activity in the bands was detected by Phadebas
sprayed on filterpaper as described by Lindblom (1980). To
illustrate the effects of sprouting damage, some of the
grains were examined by scanning electron microscopy on a
JSM using a 10KV accelerating voltage.

ANALYSIS

In TABLE 1 the moisture content of the grains is given
for the different harvest times. Moisture content decreased
with delayed harvest in all species, but was fastest in
wheat and slowest in triticale.

TABLE 1. Moisture content and days from heading (DAH) in
wheat, rye and triticale at the different harvest times.

Harvest	Wheat		Rye		Triticale	
	Moisture content (%)	DAH	Moisture content (%)	DAH	Moisture content (%)	DAH
1	49.0	(45)	58.3	(54)	60.5	(45)
2	40.3	(51)	52.1	(60)	55.6	(51)
3	27.8	(58)	43.3	(67)	50.5	(58)
4	27.8	(65)	39.3	(74)	50.4	(65)
5	19.3	(71)	17.2	(80)	39.5	(71)
6	19.2	(78)	20.6	(87)	33.1	(78)
7	15.0	(85)	18.8	(94)	24.6	(85)

The germination tests were affected by contamination with
Fusarium which reduced germination ability, especially in
wheat and triticale. The dormancy figures obtained, therefore, have high experimental errors. Dormancy was low in
triticale and higher in rye and wheat. This is in agreement
with observations from previous years. In wheat the dormancy increased towards the end of the ripening period.
This indicates that secondary dormancy was induced by cold
and moist weather. This phenomenon has been described previously by Belderok and Habekotte (1980) and Skinnes and
Buraas (1984). Dormancy indexes for the three species are
shown in FIGURE 1a.
    Spring wheat maintained a high falling number during
most of the ripening phase. A decrease in starch quality
was first visible after several periods of rainfall.

FIGURE 1. (a) Dormancy index (DI), (b) falling number and (c) alpha-amylase activity in wheat (———), rye (----) and triticale (-·-·-) at different moisture contents.

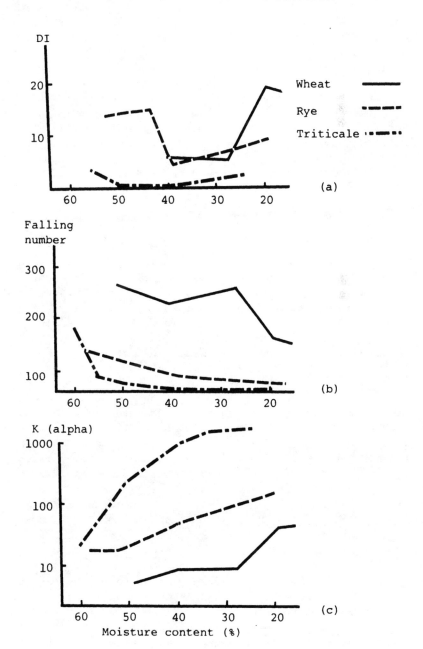

Results from the falling number analysis are shown in FIGURE 1b. Even at the last harvest stage all the wheat varieties had falling numbers that would satisfy Norwegian quality requirements for bread wheat. There was no lodging in the wheat plots. The falling number of the rye variety decreased during ripening, however, since quality requirements are lower in rye, the rye samples from the earliest harvest stages would also have been graded as food grain. In triticale the falling number was relatively high at the harvest, but thereafter decreased very rapidly. By the time the moisture content had reached approximately 50% the falling number had fallen to the minimum level. Alpha-amylase analyses showed that enzyme activity increased with decreasing falling number (FIGURE 1c). These results are in agreement with the findings of McEwan and Haslemore (1983).

ELECTRON MICROSCOPY

Scanning electron microscopy studies showed spherical, undamaged starch granules in all species at the earliest harvest stage. Starch degradation by alpha-amylase was observed in both triticale and rye at the last harvest date (FIGURE 2a, b) and was characterized by the presence of discrete holes in the starch granules. The degradation of starch granules commenced in the endosperm near the scutellum. By the time degradation was almost complete, the granules were hollow. This indicates that the surface of the starch granules is more resistant to alpha-amylase attack. Amylase action in wheat was not so severe, and gave the appearance of concentric craters on the granule surface (FIGURE 2c). Another interesting observation was that very few cell walls were visible in triticale kernels. According to MacGregor (1983) cell walls are degraded before starch hydrolysis. This is a further indication that enzymic degradation was more advanced in triticale than in the other species, and confirms the results from the other analyses.

ISO - ELECTRIC FOCUSING

Alpha-amylase isoenzymes were separated by iso-electric focusing into two groups of isoenzymes. These groups are, known as malt amylases and green amylases Daussant et al., (1980). Normally, wheat varieties have a high content of green alpha-amylase in the very early stages of maturity (Gale 1983), but these isoenzymes are inactivated during

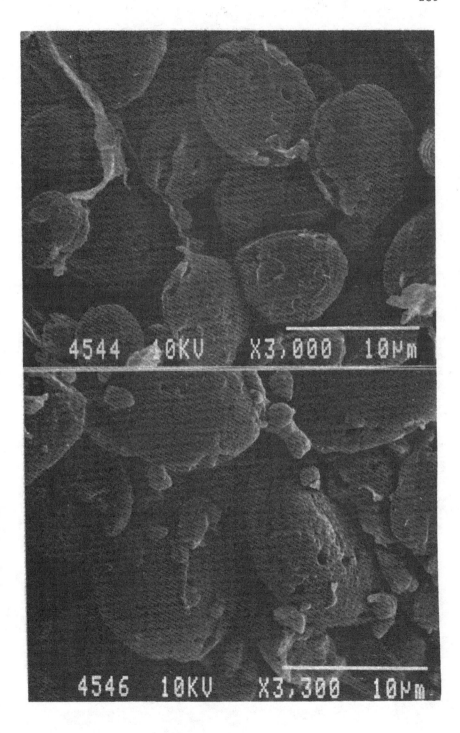

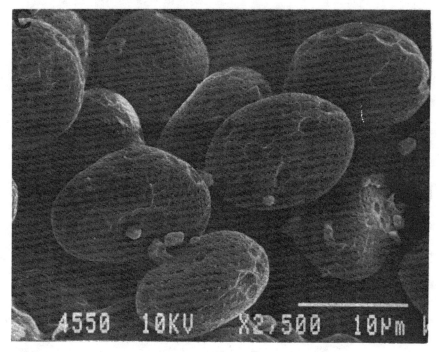

FIGURE 2. Degradation of starch granules by alpha-amylase
in (a) rye, (b) triticale and (c) wheat.
(Photos : AGREM, Agr. Univ. of Norway).

ripening. After a period of low enzyme activity, the activity of the malt amylases increases if sprouting occurs. In
this investigation, triticale differed from rye and wheat in
two ways.
1. Malt alpha-amylase was already present in triticale at
   a moisture level of 60%, and high concentrations were
   found at 50%.
2. Green alpha-amylase maintained its activity throughout
   the ripening period.

DISCUSSION

Triticale grown under dry conditions had high falling
numbers throughout ripening. High falling numbers and corresponding low alpha-amylase activities were also obtained
for triticale grown in a greenhouse. This suggested that a

humid environment and low temperatures during maturity may, therefore, be important external factors which favour increased enzyme activity. Olered and Jønsson (1970) found that green alpha-amylase activity was not reduced under conditions of delayed drying. Triticale lost water very slowly under our conditions. Kruger (1976) found that even dormant triticale had high enzyme activity. In our experiments we have also observed quite high dormancy in some of the triticale material with very high alpha-amylase activity.

Scanning electron microscopy showed a normal amount and shape of starch granules in all three species at the early stages of maturity. Alpha-amylase attack at later harvest stages were more severe in rye and triticale than in wheat.

The electrofocusing studies of triticale showed that malt amylases were already present at 60% moisture content. This is in agreement with gel electrophoretic surveys of Silvanovich and Hill (1979). An early mobilization of malt amylases may be caused by a lack of inhibitors in triticale. This theory needs further physiological investigation. Weselake et al., (1983) found endogenous inhibitor activity in both wheat, rye and triticale. These were only partially identical to the purified barley inhibitor, which suggests that each species may have a specific set of inhibitors. Since triticale is a genetic hybrid developed from two different generas, it is possible that there are errors in the enzymatic regulation processes.

REFERENCES

Belderok, B. and Habekotte, A. 1980. Induction of secondary dormancy in wheat crops by low temperatures and high relative humidities in the field. Cereal Res. Comm. 8, pp. 167-173.

Daussant, J., Mayer, C. and Renard, H. A. 1980. Immunochemistry of cereal alpha-amylases in studies related to seed maturation and germination. Cereal Res. Comm. 8, pp. 49-60.

Gale, M. D. 1983. Alpha-amylase genes in wheat. In: Proc. Third Int. Symp. on Pre-harvest Sprouting in Cereals. Kruger, J. E. and LaBerge, D. E. (eds.). Westview Press, Boulder, Co. USA. pp. 105-110.

Kruger, J. E. 1976. Biochemistry of pre-harvest sprouting in cereals and practical applications in plant breeding. Cereal Res. Comm. 4, pp. 187-194.

Lindblom, H. 1980. Changes in growth regulating substances in wheat kernels during the vegetative period. Cereal

Res. Comm. 8, pp. 139-142.

McEwan, J. M. and Haslemore, R. M. 1983. The response of Triticale and related cereals to conditions inducing pre-harvest sprouting. In: Proc. Third Int. Symp. on Pre-harvest Sprouting in Cereals. Kruger, J. E. and LaBerge, D. E. (eds.). Westview Press, Boulder Co. USA. pp. 279-286.

McGregor, A. W. 1983. Cereal endosperm degradation during initial stages of germination. In: Proc. Third Int. Symp. on Pre-harvest Sprouting in Cereals. Kruger, J. E. and LaBerge, D. E. (eds.). Westview Press, Boulder, Co. USA. pp. 162-168.

Olered, R. 1967. Development of alpha-amylase and falling number in wheat and rye during ripening. Plant Husbandry. 23, pp. 1-106.

Olered, R. and Jønsson, N. 1970. Electrophoretic studies of alpha-amylase in wheat II. J. Sci. Food Agric. 21, pp. 385-392.

Silvanovich, M. P. and Hill, R. D. 1977. Alpha-amylases from triticale 6A 190: Purification and characterization. Cereal Chemistry. 54, pp. 1270-1281.

Skinnes, H. and Buraas, T. 1984. Development of seed dormancy in barley, wheat and triticale under controlled conditions. Acta Agr. Scand. 35, p. 2.

Strand, E. 1965. Studies on seed dormancy in barley. Sci. Rep. Agr. College Norway. 44, p. 7.

Weselake, R. J., McGregor, A. W. and Hill, R. D. 1985. Endogenous alpha-amylase inhibitor in various cereals. Cereal Chemistry. 62, pp. 120-123.

The Induction of Germination Alpha-Amylase during Wheat Grain Development in Unfavourable Weather Conditions

M. D. Gale, A. M. Salter and J. R. Lenton,*
Plant Breeding Institute, Cambridge, England.
*University of Bristol, Department of Agricultural
Sciences, Long Ashton Research Station, Bristol,
England.

INTRODUCTION

At the last symposium we described an 'early sprouting damage risk period' during which some varieties of winter wheat, including cultivars released by the Plant Breeding Institute, were prone to produce germination alpha-amylases in immature grain before the onset of dormancy and without visible signs of germination (Gale, Flintham and Arthur, 1983). It was noted that the syndrome was exacerbated by cool, wet weather conditions that caused slow grain drying before harvest.

In the UK in 1985, conditions of precisely this sort were encountered. A very cool, wet July led to delayed maturity and harvest in August with grain containing elevated alpha-amylase levels and thus poor bread-making quality. A comparison of the Cereal Quality Survey data for 1985 and 1977, the last year in which the UK experienced similar weather damage (FIGURE 1), shows clearly that, while the mean Hagberg Falling Numbers were comparably low in the two seasons (167 and 127), the 1977 failure was associated with inadequate dormancy leading to visible sprouting but in 1985 the crop contained high levels of alpha-amylase in apparently sound grain.

In this paper we report results which, using new genetic stocks and material grown in 1985, (1) describe the location of germination alpha-amylases within the ear and their appearance within grains during maturation, (2) implicate gibberellins in the expression of the syndrome and (3) lead us to speculate that the gibberellins involved are not the 1-beta-hydroxylated compounds normally found in developing wheat grains but GA_1 released from the maturing embryo.

FIGURE 1. Visible sprouting in UK cereal quality survey
samples in 1977 and 1985.

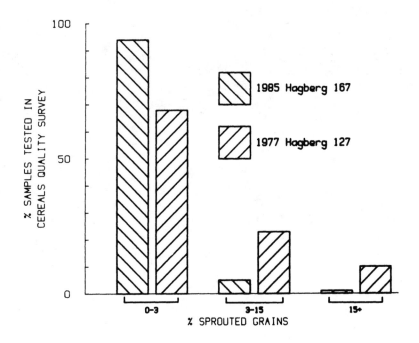

Sources. 'The quality of wheat and barley from the 1977
harvest'. Home Grown Cereals Authority, London, p. 16, and
'Cereals Quality Survey - Interim Results', HGCA Weekly
Digest 12 (18), pp. 1-2.

Genetic stocks and methods

All of the genotypes used are isogenic stocks based on
the varietal background of the winter wheat, Maris Huntsman.
These lines were produced by backcrossing six times to the
tall parent while keeping the GA-insensitive dwarfing genes
Rht1 and Rht3 heterozygous, with subsequently selfing and
selection of homozygous tall (rht) and dwarf (Rht) isogenic
pairs. Genetic duplicates, produced independently for each
pair have been used in some experiments.
Alpha-amylase extracts of individual grains were made
in 1 ml $CaCl_2$ (0.2% w/v) agitated for 2 h. Quantitative
assays were made using an autoanalyser method (Smith, 1970)
calibrated using the Phadebas (Pharmacia Diagnostics, AB

Uppsala) method, where 1 unit (U) of enzyme catalyses the breakdown of 1μ mole of glucosidic linkage per min. at 37°C. Analyses of the isozymes were performed by flat-bed iso-electric focussing (Gale et al., 1983).

For the quantitative GS-MS estimation of abscisic acid (ABA) and gibberellins (GA), ethyl acetate soluble acids were obtained from methanolic extracts of tissues and puri-fied on a QAE Sephadex A-25 anion exchange column and reverse-phase HPLC. Deuterated and tritiated internal standards were added to the methanolic extracts to correct for recovery losses and to locate the compounds resolved by HPLC. Samples were derivatised and analysed by capillary GC-MS. The relative intensities of equivalent ions of endo-genous compound and internal standard were measured using selected ion monitoring (Hedden, 1986).

RESULTS AND DISCUSSION

Sampled spikes, previously tagged for anthesis date, from 1985 drilled trials of a tall Huntsman line (Hrht1A), a semi-dwarf (HRht1A) and an extreme dwarf (HRht3A) were employed to provide information on the amount, distribution and developmental aspects of germinative alpha-amylase found in harvest ripe grains at 56 days post anthesis (dpa).

Alpha-amylase type

Electrophoresis of individual grain extracts showed that any grains with greater than 1 mU alpha-amylase at 42 dpa or later contained germination amylases consisting of both alpha-AMY-1 and alpha-AMY-2 isozymes.

Location of alpha-amylase within the spike

Inspection of the pattern of activity in spikes of the tall (rht) genotype in FIGURE 2a shows that (a) just over a half of the grains have been 'triggered' to develop germina-tion alpha-amylases, (b) these grains contain differing levels of the enzyme and (c) both the incidence of alpha-amylase and its level show a definite pattern within the ear. In general, grains in the lower central region of the ear, particularly in floret 2, contain most activity, while the extreme spikelets and distal floret positions within spikelets show least activity.

This pattern is consistent with the hypothesis that expression of the syndrome is related to drying rate and

FIGURE 2. Germination alpha-amylase (alpha-AMY-1 and alpha-AMY-2) location and levels in 1985 field grown (a) Maris Huntsman and (b) isogenic lines carrying Rht1 and (c) Rht3 alleles.

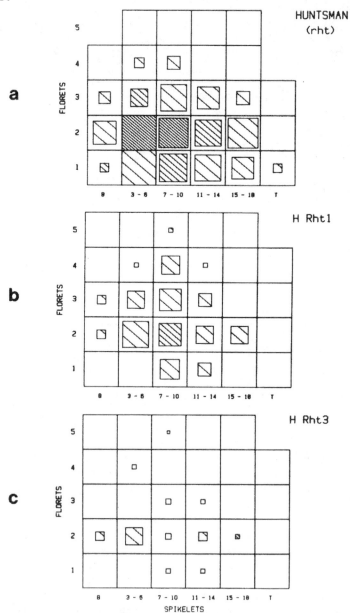

Note : Each diagram is a composite of individual grain
alpha-amylase assays of all grains from six spikes per
genotype. The hatched areas within spikelet region/floret
compartments represent the proportion of grains 'triggered'
to produce germination alpha-amylases. The density of
hatching indicates the mean quantity of enzyme : low < 5 mU
per grain, intermediate 5-10 mU, high >10 mU.

duration of development. Grains in lower florets of spikelets 3-10 will be the oldest grains and, because of their
position, would be least likely to dry rapidly after water
uptake from rain or dew. Conversely, the distal grains may
be up to a week younger and would be those likely to respond
most rapidly to drying effects of the sun or wind after
rain.

The probable participation of gibberellins in expression of the syndrome is shown by the effects of the GA-
insensitive dwarfing genes which reduce the numbers of
grains with germination alpha-amylase (Rht1, 30% and Rht3,
8% while the rht genotype showed 51%) and the relative
amounts of enzymes in these grains. The pattern of expression of the syndrome within the spike remains unaffected
(FIGURE 2b and c).

Appearance of alpha-amylase during grain ripening

The semi-dwarf genotype, HRht1, which showed an intermediate alpha-amylase phenotype was chosen for the developmental study. FIGURE 3 shows that the first appearance of
the germination enzymes coincides with latest observable
activity of the 'green' enzyme in the pericarp (42 dpa at
42% moisture content). The pattern is consistent with continued alpha-amylase production in these grains over the
next 10 days or so, after which the enzyme remains stable
until harvest in the absence of any conventional sprouting.

Source of the stimulus

Another developmental analysis with similar genotypes
is shown in FIGURE 4. In this experiment the separation of
grains taken from the lower florets of central spikelets
into proximal, embyro containing, and distal endosperm portions, confirms (a) the differential responses of rht, Rht1
and Rht3 gentoypes in producing germination alpha-amylase
(b) the timing of the first appearance of the germination
enzyme at about 42 dpa, and (c) the presence of the 'green'
enzyme mainly in the endosperm portion, which includes the
bulk of pericarp material, in all genotypes.

FIGURE 3. The presence and levels of 'green' pericarp
alpha-AMY-2 and germination alpha-amylases during grain
development of 1985 field grown Huntsman Rht1.

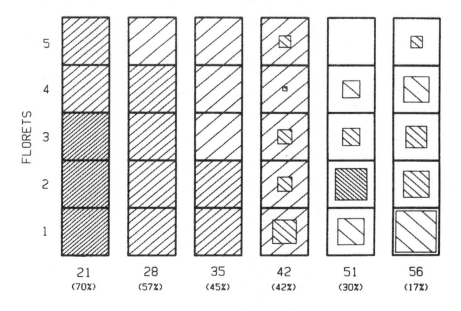

Note : Each sample time is a composite of individual grain
alpha-amylase assays from all grains in spikelets 7-10 from
6 spikes. Levels of 'green' enzyme /// and the proportion
and levels of 'germination'enzymes \\\ as for FIGURE 2.

　　　 Germination enzyme activity is seen first, and accumulates more rapidly, at the embryo end of the grain (rht
embryo 40 dpa; rht endosperm 48 dpa). This may indicate
that alpha-amylase is generated in the region of the embryo
from whence it diffuses into the endosperm or that a stimulus (presumably a gibberellin) is released from the embryo
which causes alpha-amylase production in scutellar or aleurone tissues near the embryo first. In either event the
embryo is implicated in a causal role in the syndrome.

GA-insensitivity genes and developing grain gibberellins

　　　 Rht1 and Rht3 (for review, see Gale and Youssefian,
1985), are alleles at the same locus and are characterised
by the relative insensitivity of plants carrying them to
applied GA. Their most obvious effects are on extension

FIGURE 4. Alpha-amylase in embryo and endosperm half grains in Maris Huntsman and Rht1 and Rht3 isogenic lines during grain development.

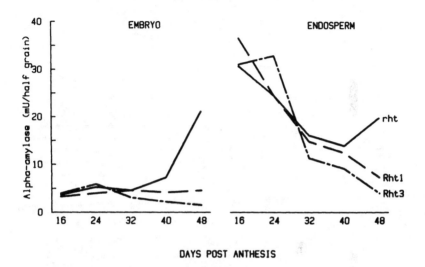

DAYS POST ANTHESIS

Note : Material grown in controlled environments, up to 22 dpa with 16 h day at 700 μ mol m^{-2} s^{-1} at 18°C and 8 h dark at 14°C, from 23 dpa to maturity the day/dark temperatures were reduced to 15°C/11°C.

growth of stems and leaves, probably by reducing cell elongation. However, they do not affect growth or development of the reproductive plant parts (S. Youssefian, pers. comm.). Thus they reduce growth only where the early-13-hydroxylation pathway GAs, particularly GA_1, predominate (Stoddart, 1984). At present there is no information of the GA complement of the developing spike but a different spectrum of GAs, including 1-beta-hydroxylated GAs, especially GA_{54}, predominate in developing grains (Gaskin et al., 1980).

The differential effects of the Rht alleles on the extractable amounts of GA_1 and GA_{54} in vegetative tissues and young grains respectively can be seen in TABLE 1. The Rht genes cause a 6-fold (Rht_1) and 25-fold (Rht_3) increase in the GA_1 content of vegetative seedling tissue but have no appreciable effects on levels of GA_{54} in the grain. The ABA results are included to demonstrate that the GA-insensitivity is not caused via effects on this inhibitor. As expected, the hormone levels are much greater in developing seeds than vegetative tissue.

TABLE 1. Gibberellins and abscisic acid in vegetative tissues and developing grains of Maris Huntsman and isogenic lines carrying Rht1 and Rht3.

Huntsman Genotype	7 day seedlings (ng. g^{-1} fwt)		18 dpa grains (ng. g^{-1} fwt)	
	GA_1	ABA	GA_{54}	ABA
rht	0.19	5.7	169	144
Rht1	1.16	6.7	196	225
Rht3	5.00	5.3	161	133

How then do the Rht genes affect the production of germinative alpha-amylase in the developing grain where they have no effects on spike growth or grain GAs? One possibility might be that, in genotypes prone to pre-maturity alpha-amylase production under slow drying conditions, GA_1, the germination gibberellin, is released from the maturing embryo. It is possible that the late second peak of biologically active GAs occasionally observed during wheat grain development, notably by Radley (1976), is GA_1, while the earlier peak of activity consists mainly of GA_{54} (see Lenton and Gale, 1986, for more discussion).

As yet there is no evidence that GA_1 is produced during late grain development and it was not detected in grains of tall lines harvested 43 days post anthesis in material used to obtain the information shown in FIGURE 4. However, it has been shown previously (Gaskin et al., 1980) that embryos of developing wheat grain do contain the immediate precursor to GA_1, namely GA_{20}. A detailed analysis of gibberellins in 1985 field-grown material is underway.

CONCLUSIONS

Early pre-maturity development of germination alpha-amylases may be more widespread, certainly in temperate wheat growing areas, than previously thought. Certainly, as experienced in the UK in 1985, it can have just as disastrous economic consequences as the more familiar problem of pre-harvest sprouting. The eventual elucidation of the underlying physiological mechanism may indicate directed genetic breeding strategies to reduce weather damage losses from this source. At the moment breeding screens are designed to identify lines with extended dormancy, rather than lines which are prone to prematurity alpha-amylase

production. Separate tests, possibly based on methods of
artificially slowing the ripening process but which stop
short of sprouting proper, may be necessary.

ACKNOWLEDGEMENTS

    The Plant Breeding Institute and Long Ashton Research
Station are financed through the Agricultural and Food
Research Council. We thank N.E.J. Appleford for skilled
technical assistance with the plant hormone analyses and Dr.
C. L. Willis and Professor Jake MacMillan, Chemistry Department, University of Bristol for generous gifts of deuterated
GA_1 and GA_{54}.

REFERENCES

Gale, M. D., Flintham, J. E. and Arthur, E. D. 1980. Alpha-
    amylase production in the late stages of grain develop-
    ment - an 'early sprouting damage' risk period? In:
    Proc. Third Inter. Symp. on Pre-harvest Sprouting in
    Cereals. Kruger, J. E. and LaBerge, D. E. (eds.).
    Westview Press, Boulder, Co. U.S.A. pp. 29-35.
Gale, M. D., Law, C. N., Chojecki, A. J. and Kempton, R. A.
    1983. Genetic control of alpha-amylase production in
    wheat. Theoretical and Applied Genetics. 64, pp.
    309-316.
Gale, M. D. and Youssefian, S. 1985. Dwarfing genes in
    wheat. In: Progress in Plant Breeding. Russell, G.
    E. (ed.). 1, pp. 1-35.
Gaskin, P., Kirkwood, P. S., Lenton, J. R., MacMillian, J.
    and Radley, M. E. 1980. Identification of gibberellins
    in developing wheat grain. Agricultural and Biological
    Chemistry. 44, pp. 1589-1593.
Hedden, P. 1986. The use of combined gas chromatography-
    mass spectrometry in the analysis of plant growth sub-
    stances. In: Modern Methods of Plant Analysis.
    Linskens, H. F. and Jackson, J. F. (eds.). Springer-
    Verlag, Heidelberg. 3, pp. 1-22. (in press)
Lenton, J. R. and Gale, M. D. 1986. Hormonal changes during
    cereal grain development. In: Proc. Fourth Inter.
    Symp. on Pre-harvest Sprouting in Cereals. Mares, D.
    J. (ed.). Westview Press, Boulder, Co. U.S.A. (in
    press).
Radley, M. 1976. The development of wheat grain in relation
    to endogenous growth substances. J. Exp. Bot. 27, pp.

1009-1021.

Smith, D. B. 1970. An automatic method for the determination of alpha-amylase activity in cereal extracts. In:
Automation in analytical chemistry. Technicon Symposium 1970. Basingstoke: Technicon Instrument Co.
Ltd.

Stoddart, J. L. 1984. Growth and gibberellin-A$_1$ metabolism
in normal and gibberellin insensitive (Rht3) wheat
(Triticum aestivum) seedlings. Planta 161, pp.
432-438.

Sensitivity of Developing Wheat Grains to Gibberellin and Production of Alpha-Amylase during Grain Development and Maturation

C. A. Cornford,* M. Black and J. Chapman,
Biology Department, King's College, University
of London, Campden Hill Road, London, W8 7 AH, U.K.
*Current address: Agronomy Group, Institute of
Agriculture, University of Western Australia,
Nedlands, WA 6009, Australia.

SUMMARY

Three examples of either natural or artificially induced alpha-amylase production by tissues of immature wheat grains are examined.

1. Aleurone tissue, predominantly that associated with the ventral portion of the grain appears to be responsible for the high levels of enzyme which can occur naturally in non-sprouted, pre-mature grains of the cv. Fenman.
2. Aleurone tissue of immature grains which is usually non-responsive to GA_3, due to a block on gene expression at the transcription level, can be made to respond to the growth substance by various treatments such as drying and preincubation.
3. Scutellar tissues can also be induced to produce alpha-amylase if the embryo is excised from the immature grain. Unlike aleurone-derived alpha-amylase, enzyme produced by the scutellum is predominantly alpha-AMY 2 (low PI) isoenzymes. Each example presented reflects a breakdown of the temporal controls governing the switch from developmental to germinative mode with respect to alpha-amylase.

INTRODUCTION

Gibberellin-induced alpha-amylase production is an example of a hormone-mediated germinative event which is normally suppressed in developing grains (Black et al., 1983). The mechanism(s) which serve to inhibit this and other germinative processes during the developmental mode remain unknown, although it has proved possible to induce the immature aleurone tissue to produce alpha-amylase (Evans et al., 1975; Nicholls, 1979; King and Gale, 1980; Armstrong et al., 1983; Cornford and Black, 1985).

This paper reports three examples of changes from 'developmental' to 'germinative' mode with respect to alpha-amylase production in immature wheat grains. Two examples are concerned with aleurone-derived enzyme production while the third is concerned with the behaviour of embryo/scutellar tissues, often neglected as a source of alpha-amylase in wheat.

METHODS AND MATERIALS

Wheat plants (Triticum aestivum L cvs Fenman and Sappo) were grown in controlled environment rooms operating a 16h 18°C day and an 8h 14°C night. (Under these conditions grains reached harvest ripeness at about 55-60 days after anthesis with a maximum dry weight at about 30-35 days after anthesis.)

The location of alpha-amylase activity within individual wheat grains was determined indirectly using a combined histochemical and blotting procedure modified by J. Daussant from that described by Okamato and Akazawa (1979). Hand cut sections of grain were placed on 1.2% agarose gel (pH 5.5) containing 2% beta-limit dextrin. After 5-10 minutes incubation at room temperature tissue was removed and the gel stained with KI solution.

Sterilization, dehydration and incubation conditions for de-embryonated grains plus alpha-amylase determinations were those described by Armstrong et al., (1982) as modified by Cornford and Black (1985). RNA extractions and hybridizations with cDNA probes were as carried out by Baulcombe and Buffard (1983).

Embryo/scutellar tissue was dissected intact from immature grains under sterile conditions and incubated on agar plates containing 0.5% starch and other supplements as required, at 25°C in darkness. Isoelectric focussing of tissue extracts was as described by Daussant and Renard (1983).

RESULTS

A.   Aleurone-derived alpha-amylase in non-sprouted pre-
mature wheat grains.

Grains of certain wheat cultivars e.g. Fenman can
contain high levels of alpha-amylase prior to harvest ripeness (FIGURE 1). This occurs in the absence of any morphological sign of germination (Cornford and Black, 1985) and
reflects a breakdown of control mechanism(s) which serve to
suppress such alpha-amylase production during grain development.   In agreement with other studies e.g. Gale and Ainsworth (1984), alpha-amylase present initially during grain
development consists entirely of alpha-AMY 2 (low PI) isoenzymes which are predominantly associated with the outer
pericarp tissues.   Activity is evenly distributed between
sampled grains and declines as the pericarp dies.   In contrast the enzyme responsible for the high levels in maturing
grains appears late during development (ca 40dpa) is aleurone derived, predominantly alpha-AMY 1 (high PI) (data not
presented) and is unevenly distributed within and between
ears.   Note also how variable grain water content is between
ears during this phase of development.   It is not clear however, if there is any relationship between water content and
alpha-amylase level since high enzyme activity is observed
in both high and low-water content ears.
      Within a high alpha-amylase grain the pattern of alpha-
amylase production is unlike that of a sprouted grain.
Enzyme activity is usually associated with aleurone tissue
in the   ventral portion of the grain (FIGURE 2a) whereas in
sprouted grains enzyme first appears adjacent to the scutellum and progresses from there throughout the grain (FIGURE
2b).
      Aleurone derived alpha-amylase is classically con-
sidered to be gibberellin regulated.   In order that the high
alpha-amylase production observed for Fenman can occur at
least two requirements must be fulfilled, viz (a), the aleurone tissue which is normally non-responsive to added GA_3 in
immature wheat grains must become capable of responding to
the control signal and (b), adequate levels of GA must be
present to initiate the response even though no germinative
growth is detected.   In the following section we address the
first of these requirements.

FIGURE 1. Alpha-amylase and % water contents of grain of individual ears of Fenman harvested at various points during grain development and maturation.
(A) 8dpa, (B) 33dpa, (C) 37dpa, (D) 41/42dpa, (E) 52dpa. From each ear a sample of 10 grains was assayed for alpha-amylase activity and the remainder used for water content measurements. % water content is given above each column.

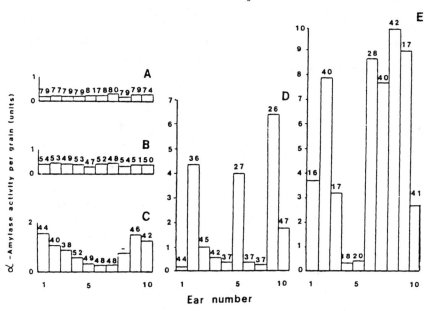

B.   Sensitivity of the aleurone tissue to GA₃

The aleurone tissue of immature wheat grains is usually considered to be non-responsive to gibberellin, at least with respect to alpha-amylase production. Responsiveness to the growth substance is attained as a consequence of grain maturation but can also be induced in immature grains by drying treatments (Nicholls, 1979; King and Gale 1980; Armstrong et al., 1982). A capacity to respond to the growth substance can also be induced in non-dehydrated immature tissue if half-grains are first incubated (usually for 72h) in the absence of GA₃. The longer the pre-incubution treatment, up to 96h, the greater the resultant GA responsiveness of the tissue (Cornford et al., in preparation).

Drying and pre-incubation treatments both appear to overcome a block on alpha-amylase production at the transcription level. Using cDNA probes for alpha-amylase, mRNA

FIGURE 2. Localisation of alpha-amylase in blots of sections of (A) high alpha-amylase non-sprouted and (B) 3 day germinated (sprouted) harvest-ripe grains of Fenman. Sections were made as follows (i) transverse sections along the length of the grain (ii) median longitudinal sections along the crease dividing the grain into two mirror images (iii) median longitudinal section dividing the ventral and dorsal portion of the grain.

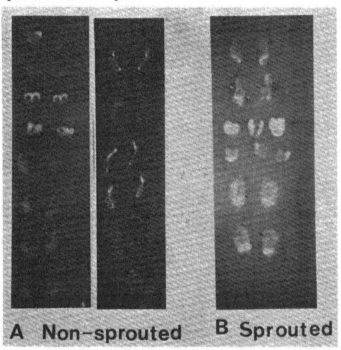

A Non-sprouted   B Sprouted

appears to be absent from extracts of immature tissue unless that tissue experiences either treatment prior to GA3 addition. (FIGURE 3). Detaching grains from the ear and maintaining them under conditions of high humidity for an equivalent time to the drying treatment occasionally results in the induction of a GA-responsive state (although this treatment is not as reproducible as either drying or pre-incubation as a sensitizing treatment).

Drying and pre-incubation may appear disparate treatments yet they both result in the induction of GA-responsivity and remove a block on the transcription of the alpha-amylase genes. How this is achieved is not yet known but the leakage and/or breakdown of compounds antagonistic to GA action, such as ABA, during pre-incubation and reimbibition

288

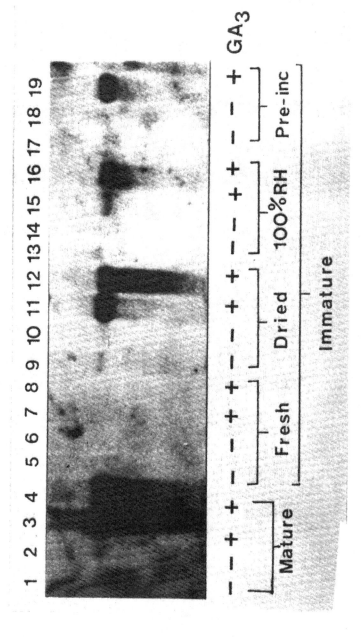

FIGURE 3. Hybridisation of an alpha-amylase cDNA probe to RNA isolated from aleurone tissue of mature (tracks 1-4) and immature (tracks 5-19) grains of Sappo following a 48h incubation either in the presence or absence of GA3. The cDNA probe used hybridises with both alpha-AMY 1 and alpha-AMY 2 mRNA.

of dried grains seems possible, on the basis of initial observations.

C.   Alpha-amylase production by embryo/scutellum

Although the aleurone tissue is recognised as a major source of alpha-amylase in germinating wheat grains and in non-sprouted high alpha-amylase varieties, the embryo/ scutellar tissues are often ignored as a potential source of enzyme. Embryo/scutellum tissue within a developing wheat grain contains little alpha-amylase but the enzyme is read- ily produced when the tissue is excised from the grain and incubated on agar (FIGURE 4). Enzyme production, after iso- lation is strongly influenced by embryo age, although germ- ination itself does not appear to be a pre-requisite for enzyme production. For example, 35dpa embryos which show a decrease in germinability exhibit an increased production of alpha-amylase relative to younger material.

Addition of GA_3 promotes both alpha-amylase production and germination. The increase in enzyme activity is due predominantly to an increase in alpha-AMY 2 (low PI) iso- enzymes as shown by isoelectric focusing (FIGURE 5) and direct tissue immunoelectrophoresis (data not shown). Abscisic acid inhibits GA-induced germination and greatly reduces alpha-amylase production. Alpha-AMY 1 (high PI) isoenzymic forms disappear whilst two others apparently increase in activity.

While scutellum cells appear capable of alpha-amylase production the maximum levels detected, under the experi- mental conditions used here, fall well below those demon- strated for aleurone tissue. Enzyme production is inhibited in the intact grain but is stimulated when embryos are removed from the grain. Such treatment may also release the embryo from constraints on germination as indicated pre- viously but a close relationship between germination and amylase production has not been observed.

CONCLUSIONS

These three examples illustrate the operation of either a natural or artificial switch from 'developmental' to 'germinative' mode with respect to alpha-amylase production in developing wheat grains. In each case a role for plant growth regulators is implicated but as yet not unequivocally proven. Equally unknown is the fate of developmental

FIGURE 4. Alpha-amylase production and % germination of embryos of various ages isolated from immature grains of Sappo following 48h incubation on starch/agar at 25°C. Germination is defined as the protrusion of either radicle or coleoptile.

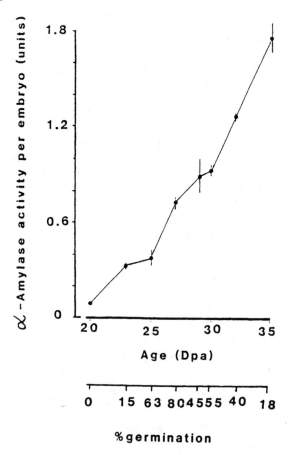

processes in such tissues following this switch in metabolism.

ACKNOWLEDGEMENTS

We wish to thank the Agriculture and Food Research Council for financial support; Dr D. Baulcombe (PBI, Cambridge) for the cDNA probes and Dr. J. Daussant and Miss C. Mayer (CNRS, Meudon) for the IEF and immunological studies. The latter was made possible by a Royal Society travel grant to one of us (CAC).

FIGURE 5. Isoelectric focusing of alpha-amylase from extracts of wheat embryos (cv Sappo) prepared immediately upon excision and following 72h incubation in the presence or absence of the plant growth regulators GA_3 (2×10^{-6}M) and/ or ABA (2×10^{-5}M).

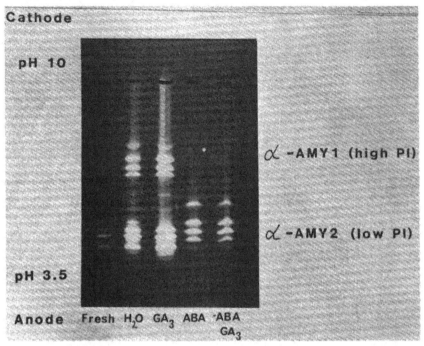

REFERENCES

Armstrong, C., Black, M., Chapman, J. M., Norman, H. A. and Angold, R. 1982. The induction of sensitivity to gibberellin in aleurone tissue of developing wheat grains. 1. The effect of dehydration. Planta 154, pp. 573–577.

Baulcombe, D. C. and Buffard, D. 1983. Gibberellic-acid-regulated expression of alpha-amylase and six other genes in wheat aleurone layers. Planta 157, pp. 493–501.

Black, M., Chapman, J. and Norman, H. 1983. The ability of wheat aleurone tissue to participate in endosperm mobilisation. In: Mobilisation of reserves in germination, Recent Advances in Phytochemistry Vol. 17. Nozzolillo, C., Lea, P. and Loewus, F. (eds.). Plenum Press, New York and London. pp. 193–211.

Cornford, C. A. and Black, M. 1985. Alpha-amylase content of pre-mature unsprouted wheat grains. J. Cereal Sci. 3, pp. 295-304.

Daussant, J. and Renard, H. A. 1983. Induction of alpha-amylase in immature wheat seeds. Proc. 7th World Cereal and Bread Congress, Prague, Dev. Food. Sci. 514, pp. 121-126.

Evans, M., Black, M. and Chapman, J. 1975. Induction of hormone sensitivity by dehydration is one positive role for drying in cereal seed. Nature. 258, pp. 144-145.

Gale, M. D. and Ainsworth, C. C. 1984. The relationship between alpha-amylase species found in developing and germinating wheat grain. Biochem. Genet. 22, pp. 1031-1036.

King, R. W. and Gale, M. D. 1980. Pre-harvest assessment of potential alpha-amylase production. Cereal Res. Comm. 8, pp. 157-165.

Nicholls, P. B. 1979. Induction of sensitivity to gibberellic acid in developing wheat caryopses: effect of rate of dessication. Aust. J. Plant Physiol. 6, pp. 229-240.

Okamoto, K. and Akazawa, T. 1979. Enzymic mechanisms of starch breakdown in germinating rice seeds. Plant Physiol. 63, pp. 336-340.

Section V

Physiology, Biochemistry and Molecular Biology of Germination

Molecular Biology of Expression of Alpha-Amylase and other Genes following Grain Germination

P. M. Chandler, Z. Ariffin, L. Huiet,
J. V. Jacobsen and J. Zwar. Division of
Plant Industry, CSIRO, Canberra, ACT 2601,
Australia.

In this paper we will discuss some of the events following germination of barley (Hordeum vulgare L.) grains
which occur in the aleurone tissue attached to the growing
seedling. Our particular interest lies in the regulation of
aleurone gene expression (and hence enzyme/protein production) by the plant hormones gibberellic acid (GA3) and
abscisic acid (ABA). Experiments utilizing isolated aleurone layers, or protoplasts prepared from such layers will
also be described since these have contributed greatly to
our understanding of aleurone response. Finally, we will
describe the response of seedlings to dehydration (again
from the perspective of possible hormone involvement and
changes in aleurone gene expression) since there may be
similarities in the response of mature germinated grain to
subsequent dehydration, and those mechanisms responsible for
preventing sprouting of developmentally mature grain while
still attached to the parent plant.

The extent to which the physiological and biochemical
changes associated with germination (and subsequent dehydration) of mature grain bears any actual relationship with
changes in a crop which is sprouting (or, equally significantly, not sprouting) is at present far from clear. Our
hope is that by understanding the former system, which is
much more convenient experimentally, key features can be
identified which can then be investigated in grain which
is prematurely germinating.

Germination, seedling development and alpha-amylase

After imbibing water the embryo of a cereal caryopsis
enters a growth stage which is nourished by sugars and amino

acids derived from the hydrolysis of endosperm reserves.
This hydrolysis is brought about in large part by enzymes
secreted from the aleurone layer. In the best studied
example, alpha-amylase, gibberellin produced by the embryo
stimulates aleurone cells to synthesize alpha-amylase which
is then secreted from the aleurone layer.

We now have a moderately detailed understanding of
events involved in the regulation of alpha-amylase synthesis
by GA$_3$ in aleurone cells. The accumulation of alpha-amylase
activity in grains of barley cv Himalaya following imbibition of water is shown in FIGURE 1. Most of the enzyme is
derived from the aleurone layer and only a small amount from
the scutellum (1) in contrast to the situation in, for
example, rice (2). The level of alpha-amylase mRNA in the
aleurone layers attached to such seedlings shows a similar
developmental profile (FIGURE 2) and is measured by hybridization of a radioactive alpha-amylase cDNA clone to equal
amounts of total RNA which has been size-fractionated on a
gel and blotted to a filter prior to hybridization. Highest levels of alpha-amylase mRNA are seen in the RNA from
seedlings 3-4d after imbibition.

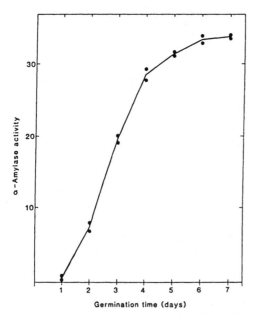

FIGURE 1. Accumulated
alpha-amylase activity in
the endosperm of barley
grains following germina-
tion. (Modified from
Ranki and Sopanen, 1984.
Plant Physiol. 75, pp.
710-715.)

Under the conditions
of hybridization used in
FIGURE 2 the radioactive
probe, which corresponds
to a mRNA for a high pI
isozyme of alpha-amylase
(3, 4) will be detecting
mRNAs encoding both the
low and high pI families
of alpha-amylase. The
levels of individual
mRNAs in both the low pI and high pI families have been
assessed in these RNA samples by primer extension analysis.
Each group of mRNAs (for either the low pI or high pI family
of isozymes) is apparently coordinately regulated, but the

FIGURE 2. Alpha-amylase mRNA levels in the aleurone of
barley seedlings.
Grains were planted in moist vermiculite and seedlings grown
near a laboratory window at approximately 23°C with daily
watering. At 2, 3, 4, 5 and 6 days after the commencement
of imbibition aleurone layers were excised and RNA isolated.
Samples of 10 µg of RNA from each stage were electrophoresed, blotted to a filter, hybridized with a ^{32}P-label-
led alpha-amylase cDNA clone, and autoradiographed as described in Chandler, et al., (1984).

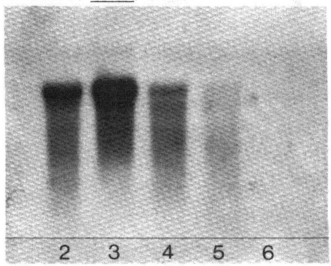

high pI group peaks at 3 days after imbibition, whereas the
low pI group peaks one day later (PMC unpublished data).

Control of alpha-amylase mRNA levels in aleurone by GA$_3$ and ABA

The changes in levels of alpha-amylase mRNA seen in the
aleurone of developing seedlings are also seen in isolated
aleurone layers treated with GA$_3$. An important advantage of
the isolated aleurone layer system is that responses to
known concentrations of GA$_3$ can be studied. In addition it
is possible to examine the influence of other hormones such
as ABA, on GA$_3$-induced expression of alpha-amylase. In
FIGURE 3 the relative levels of alpha-amylase mRNA are shown
in RNA isolated from layers incubated without hormone, with
GA$_3$, with GA$_3$ plus ABA or with ABA alone. There is a 48-
fold increase in levels of alpha-amylase mRNA from the GA$_3$-

FIGURE 3. Alpha-amylase mRNA levels in aleurone layers
incubated with or without hormones. RNA was isolated from
aleurone layers incubated for 24h without hormone (lane a)
or with GA_3 (lane b), GA_3 + ABA (lane c) or ABA (lane d)
alone. Samples were analysed as described in the legend to
FIGURE 2 (modified from Chandler et al., 1984. Plant Molecular Biology, 3, pp. 407-418).

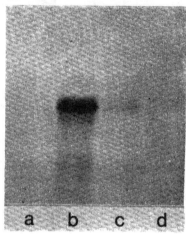

treated layers relative to the controls, but if ABA is also
present the GA_3-induction is only 3.4-fold.
        Changes in the levels of alpha-amylase mRNA seen in
isolated aleurone layers following hormone treatment may
reflect altered rates of transcription of alpha-amylase
genes, or changed stability of transcripts once formed, or
some combination of these two processes. Hybridization such
as those in FIGURE 2 and FIGURE 3 provide no information as
to the mechanism of change in levels of particular mRNAs,
since they measure the steady-state level of that mRNA in a
tissue. Investigation of the relative importance of transcriptional changes requires the isolation of nuclei from
suitably-treated tissue so that "run-on" transcription
experiments can be performed. In such experiments nuclei
are isolated and allowed to continue elongation of already-
initiated RNA transcripts in the presence of a ^{32}P-labelled
RNA precursor (e.g. UTP). The relative rate of transcription of a particular gene is measured by determining the
proportional representation of transcripts to immobilized
cloned DNA of the gene) in relation to total transcripts.
This type of experiment is difficult to perform in aleurone

layers as their thick cell walls and contaminating starch grains hinder attempts to prepare transcriptionally-active nuclei. However by modifying procedures (5) for the isolation of GA-responsive aleurone protoplasts from barley (6). Protoplasts are convenient starting material for nuclear isolations, and "run-on" transcription experiments have now been performed in aleurone protoplasts of both barley (7) and wild oat (8). The results for barley (FIGURE 4) demonstrate that GA_3 causes a reduction in total transcription in aleurone protoplasts (in part through a reduction in the rate of ribosomal RNA transcription) but causes a significant increase in the representation of alpha-amylase transcripts is seen. The results with wild oat (8) were very similar and both studies lead to the conclusion that GA_3 treatment of aleurone cells results in an increased transcription of alpha-amylase genes, and that this effect can be abolished by ABA.

It is likely, based on the protoplast result described above, that the increasing levels of alpha-amylase mRNA seen in the aleurone attached to seedlings 2-4 days post transcription of alpha-amylase genes in response to gibberellin from the embryo. At present we do not know if the increased rate of transcription can by itself account for the rise in mRNA levels, or whether mRNA stability changes are also occuring. The decline in mRNA levels from 4 to 6 days post imbibition may or may not involve transcriptional changes; this is difficult to assess as the aleurone cells are beginning to "self-destruct" and the quality of the RNA extracted from such layers declines rapidly with further developmental age.

Aleurone response to ABA

The observation that ABA can prevent the GA-induced synthesis of alpha-amylase indicates that aleurone is a tissue which is ABA-responsive, and raises the possibility that there may be a physiological role for its sensitivity to ABA. In addition to its affect on alpha-amylase production, ABA treatment of isolated aleurone layers stimulates synthesis of a new set of proteins (9), supporting the case for a physiological role. Although the functions of the ABA-induced proteins are mostly unknown, one of them has recently been identified as a bifunctional alpha-amylase/ subtilisin inhibitor (10). Since levels of ABA in mature grain are low (11), it is possible that the ABA-responsiveness of aleurone from mature grain is of no physiological significance; it could for instance be a relic of an

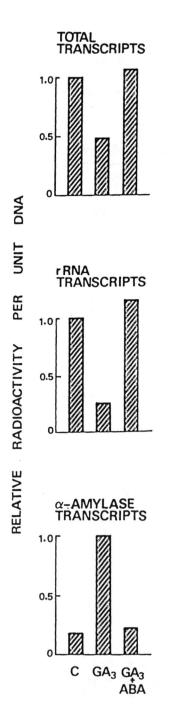

RELATIVE RADIOACTIVITY PER UNIT DNA

TOTAL TRANSCRIPTS

rRNA TRANSCRIPTS

α-AMYLASE TRANSCRIPTS

C   GA₃ GA₃ + ABA

FIGURE 4. Levels of transcripts made by nuclei isolated from aleurone protoplasts incubated without hormones (C) or in the presence of GA₃ or GA₃ + ABA for 24h (modified from Jacobsen and Beach, 1985). (Reprinted by permission from Nature, 316, pp. 275-277. Copyright (C) 1985 Macmillan Journals Ltd.)

unknown requirement for an ABA-responsive aleurone in mature grain is an adaptive property which allows young seedlings to cope with environmental stress e.g. dehydration. In view of the evidence which indicates that ABA is involved in preventing germination of developmentally mature seeds, the proteins induced by ABA in aleurone may function to prevent endosperm mobilization under stressful environmental conditions. The indentification of one of these proteins as an endogenous alpha-amylase inhibitor is consistent with this proposal.

In vivo expression of ABA-inducible mRNAs in aleurone seedlings

We have constructed cDNA clones from the RNA of aleurone layers which had been treated with ABA. From these clones several have been characterized which correspond to particular mRNAs which are inducible by ABA in isolated aleurone layers. Although we do not yet know which of the ABA-induced proteins particular cDNA clones correspond to, these clones are a useful assay for aleurone exposure to ABA.

During normal seedling development (i.e. with adequate water) mRNAs for these clones are present in aleurone RNA at very low levels, if at all. However when seedlings (2-3 days after imbibition, coleoptile/shoot 10-30 mm, roots approx 50mm) are exposed to dehydration large increases in the levels of the ABA-inducible mRNAs are seen in the aleurone (FIGURE 5). Control seedlings, which continue development with adequate water, show no sign of these mRNAs prior to degradation of their aleurone RNA which occurs 6-8 days after imbibtion.

The induction of these mRNAs in the aleurone of dehydrating seedlings occurs within 2 to 3 days of dehydration when the seedling is exposed to a relative humidity of 85-90%. Higher relative humidities are also effective, however the response is not seen if dehydration occurs too quickly. In all cases the seedlings can withstand the dehydration, and respond to reapplied water by formation of new roots and continued shoot development.

FIGURE 5. Appearance of pHVA39 mRNA in aleurone of dehydrating seedlings of barley.
Seedlings (2.5d after imbibition) were allowed to continue development on moist filter paper, or placed in a desiccator over 25% glycerol (relative humidity 85-90%). RNA was extracted from the endosperm of seedlings after two additional days development on moist filter paper (lane a), or after 1, 2, 3 or 4d dehydration (lanes b-e respectively) and processed as described in the legend to FIGURE 2. The hybridization probe was pHVA39, a cDNA plasmid corresponding to an ABA-inducible mRNA in mature aleurone (Ariffin, Z. and Chandler P. M. unpublished).

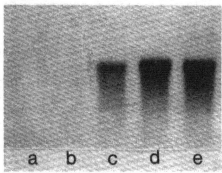

The observation that mRNAs which correspond to ABA-inducible mRNAs in isolated aleurone layers appear in the aleurone of dehydrating seedlings does not necessarily imply that ABA triggered their appearance in the seedling, but this is an obvious possibility to investigate. It is also possible that some factor other than ABA, but which is associated with dehydration has caused their appearance.

CONCLUSIONS

We have described some of the major hormonal controls affecting expression of alpha-amylase genes in aleurone of barley, discussed the cellular mechanisms by which they operate, and attempted to relate these to germination and early seedling development. Much of this work has been carried out on isolated aleurone layers because this system offers significant experimental advantages. There is a need however to return now to the whole grain/seedling system and investigate important components of regulation which have been insufficiently studied e.g. the regulation of gibberellin production by the embryo, and the significance of aleurone responsiveness to ABA.

From the perspective of pre-harvest sprouting these latter two areas are probably of most relevance. It is certainly conceivable that the tendency of a grain to sprout is related to gibberellin production by the embryo. It is certainly conceivable that the tendency of a grain to sprout is related to gibberellin production by the embryo. It is also possible that response to ABA may be an important component in "non-sprouting", since maize mutants which lack ABA germinate precociously ("vivipary").

Whether the accumulation of ABA-inducible mRNAs in the aleurone of dehydrating seedlings (FIGURE 5) occurs in response to ABA or to some other compound is not known. Similarly the function of these proteins is not known. Some of the mRNAs are relatively conserved however, since the barley clones detect a similar accumulation in the aleurone of dehydrating wheat seedlings. If these RNAs are an indicator of a seedling's tendency to resist further development under certain environmental conditions, and for it to return to a "quiescent" (albeit pregerminated) state, they could conceivably be useful markers for the tendency of grain to sprout. It would certainly be of interest to examine production of these mRNAs in dehydrating seedlings of wheat lines which differ in their tendency to sprout.

REFERENCES

Akazawa, T. and Miyata, S. 1982. Biosynthesis and secretion of alpha-amylase and other hydrolases in germinating cereal seeds. Essays in Biochemistry. 18, pp. 40-78.

Chandler, P. M., Zwar, J. A., Jacobsen, J. V., Higgins, T. J. V. and Inglis, A. S. 1984. The effects of gibberellic acid and abscisic acid on alpha-amylase mRNA levels in barley aleurone layers: studies using an alpha-amylase cDNA clone. Plant Molec. Biol. 3, pp. 407-418.

Hooley, R. 1982. Protoplasts isolated from aleurone layers of wild oat (Avena fatua L.) exhibit the classic response to gibberellic acid. Planta. 154, pp. 29-40.

Jacobsen, J. V. and Beach, L. 1985. Control of transcription of alpha-amylase and rRNA genes in barley aleurone protoplasts by gibberellin and abscisic acid. Nature. 316, pp. 275-277.

Jacobsen, J. V., Zwar, J. A. and Chandler, P. M. 1985. Gibberellic acid responsive protoplasts from mature aleurone of Himalaya barley. Planta. 163, pp. 430-438.

Ranki, J. and Sopanen, T. 1984. Secretion of alpha-amylase by the aleurone layer and the scutellum of germinating barley grain. Plant Physiol. 75, pp. 710-715.

Svensson, B., Mundy, J., Gibson, R. M. and Svendsen, I. 1985. Partial amino acid sequences of alpha-amylase isozymes from barley malt. Carlsberg Res. Commun. 50, pp. 15-22.

Zwar, J. A. and Hooley, R. 1986. Hormonal regulation of alpha-amylase gene transcription in wild oat (Avena fatua L.) aleurone protoplasts. Plant Physiol. (in press).

Molecular Biology of Barley (1-3, 1-4) -Beta-Glucanases

G. B. Fincher, Department of Biochemistry,
La Trobe University, Bundoora, Victoria,
3083, Australia.

INTRODUCTION

The two $(1\rightarrow3),(1\rightarrow4)$-β-glucanase isoenzymes which participate in cell wall degradation in germinating barley (Hordeum vulgare) represent the products of separate genes. This conclusion was reached by comparison of the sequences of their 40 NH_2-terminal amino acids; although a high degree of homology (approx. 90%) was found, differences which could be attributed to commonly observed, single nucleotide substitutions in the DNA sequences were also detected (Woodward et al., 1982). Since established barley varieties are essentially homozygous (Briggs, 1978), it is likely that the enzymes represent true genetic isoenzymes. The amino acid sequence homology at the NH_2-terminus, which has since been shown to extend along the proteins (unpublished data), suggests that the two genes encoding the barley $(1\rightarrow3),(1\rightarrow4)$-β-glucanase isoenzymes arose by duplication of a common ancestral gene (Woodward et al., 1982).

Expression of the $(1\rightarrow3),(1\rightarrow4)$-β-glucanase genes appears to be developmentally regulated, since the enzymes or their precursor forms cannot be detected in mature barley endosperm but are synthesized rapidly in the germinating grain (Stuart & Fincher, 1983). In studies with isolated tissues it has been shown that, in the presence of Ca^{2+}, the phytohormone gibberellic acid (GA_3) enhances $(1\rightarrow3),(1\rightarrow4)$-β-glucanase isoenzyme II secretion from aleurone layers and that the pattern of isoenzymes secreted from the aleurone layers differs markedly from that secreted by isolated scutella (Stuart et al., 1986). Although care must be exercised in the interpretation of results obtained with isolated tissues, the data suggest that the expression of $(1\rightarrow3),(1\rightarrow4)$-β-glucanase genes may also be under hormonal control in whole germinating barley and that genes encoding individual isoenzymes are expressed at different rates in different tissues.

The barley $(1\rightarrow3),(1\rightarrow4)$-β-glucanase isoenzymes therefore appear to represent a relatively simple, developmentally-regulated system for the study of tissue-specific regulation of gene expression and for the investigation of hormone action at a genetic level. Investigations of these aspects of $(1\rightarrow3),(1\rightarrow4)$-β-glucanase synthesis ultimately require detailed structural analyses of the genes encoding the isoenzymes. In the initial steps towards achieving this objective, mRNAs isolated from the scutellum and aleurone of germinating barley have been examined for the presence of $(1\rightarrow3),(1\rightarrow4)$-β-glucanase mRNA species (Mundy et al., 1985), the effects of phytohormones on levels of translatable mRNA encoding $(1\rightarrow3),(1\rightarrow4)$-β-glucanase have been investigated using isolated aleurone layers (Mundy & Fincher, 1986), and a cDNA encoding isoenzyme II has been identified and sequenced (Fincher et al., 1986).

mRNAs ENCODING $(1\rightarrow3),(1\rightarrow4)$-β-GLUCANASES

At 2, 5 and 8 days after the initiation of germination in whole barley grain, mRNA was isolated from excised aleurone layers and scutella. Roots and shoots were removed from the grain to expose the scutellum, which was subsequently prised out with a scalpel and freed from adhering pericarp-seed coat before homogenisation. Light microscopy showed that the scutella were essentially free of starchy endosperm, pericarp-seed coat and other embryonic tissues (Mundy et al., 1985). The starchy endosperm was subsequently squeezed out of the grain and the "aleurone layer", consisting of both the aleurone layer and the pericarp-seed coat, was homogenized for mRNA isolation. Extractions were performed with guanidinium chloride to minimise enzymic degradation of the RNA (Chirgwin et al., 1979) and polyadenylated mRNA was enriched by affinity chromatography on poly(U)-Sepharose 4B (Mundy et al., 1985).

The mRNA species encoding $(1\rightarrow3),(1\rightarrow4)$-β-glucanase were identified by immunoprecipitation (Jonassen et al., 1982) of total in vitro translation products of the mRNA preparations. Immunoprecipitation of the in vitro translation products of both scutellar and aleurone mRNA preparations with $(1\rightarrow3),(1\rightarrow4)$-β-glucanase antibodies revealed the presence of a single major polypeptide of M_r 33,000 (Mundy et al., 1985). This is higher than the M_r values of 28,000 (isoenzyme I) and 30,000 (isoenzyme II) obtained by gel electrophoresis for the mature enzymes (Woodward & Fincher, 1982), presumably due to the presence of signal peptides on the NH_2-termini of the unprocessed in vitro translation products. No difference could be observed in the size of the immunoprecipitated polypeptides from scutellar or aleurone mRNA preparations and it is not possible to determine whether the immunoprecipitated polypeptide represents a precursor form of isoenzymes I, II or III or a mixture of these (cf. Stuart et al., 1986). From the relative intensities of the immunoprecipitated polypeptide bands on autoradiograms, it appeared that the mRNA encoding (1→3),(1→4)-β-glucanase was approx. 20-fold more abundant in aleurone preparations than in scutellar mRNA preparations (Mundy et al., 1985).

The detection of mRNA encoding (1→3),(1→4)-β-glucanase in both aleurone and scutellar mRNA preparations is consistent with observations that isolated aleurone layers and scutella can synthesize and secrete specific (1→3),(1→4)-β-glucanase isoenzymes (Stuart et al., 1986) and confirms a role for both tissues in the secretion of (1→3),(1→4)-β-glucanase in germinating barley.

Effect of GA$_3$ and ABA

In isolated aleurone layers, GA$_3$ markedly stimulates the secretion of (1→3),(1→4)-β-glucanase isoenzyme II (Stuart et al., 1986). To assess the effects of GA$_3$ at the mRNA level, aleurone layers were incubated with and without GA$_3$ prior to extraction of polyadenylated mRNA and immunoprecipitation of (1→3),(1→4)-β-glucanase precursors from in vitro translation products of the mRNAs (Mundy & Fincher, 1986). With mRNA from untreated layers, a translation product of M_r 33,000 was immunoprecipitated with the (1→3),(1→4)-β-glucanase antibody, but the relative abundance of this polypeptide increased at least 10-fold amongst the in vitro translation products of mRNA from GA$_3$-treated aleurone layers (Mundy & Fincher, 1986). Simultaneous addition of abscisic acid decreased the relative abundance of mRNA for (1→3),(1→4)-β-glucanase precursor polypeptide.

Thus, the GA$_3$ enhancement of levels of translatable mRNA for (1→3),(1→4)-β-glucanase in aleurone layers (Mundy & Fincher, 1986) parallels its influence at the enzyme level (Stuart et al., 1986). It seems likely therefore that GA$_3$ exerts its effect by increasing the rate of transcription of the (1→3),(1→4)-β-glucanase gene(s), although it is formally possible that increases in levels of translatable (1→3),(1→4)-β-glucanase mRNA result from increased stability or more efficient processing of the mRNA.

The hormonal regulation of translatable mRNA encoding (1→3),(1→4)-β-glucanase appears to be similar to the regulation of mRNA for α-amylase in barley aleurone layers, for which GA$_3$-enhancement is also blocked by abscisic acid (Chandler et al., 1984). It has been suggested that GA$_3$ may decrease the overall number of proteins synthesised by barley aleurone layers while re-directing gene expression towards the rapid synthesis of a relatively small number of proteins (Mozer, 1980; Jacobsen & Beach, 1985). It now seems likely that, in addition to α-amylase, (1→3),(1→4)-β-glucanase isoenzyme II is a member of the GA$_3$-stimulated group of proteins in isolated barley aleurone layers (Mundy & Fincher, 1986).

IDENTIFICATION OF A cDNA ENCODING (1→3),(1→4)-β-GLUCANASE

cDNA synthesis

In view of the high relative abundance of mRNA encoding (1→3),(1→4)-β-glucanase in aleurone layers, mRNA from GA$_3$-treated aleurone layers was used as a template for cDNA synthesis. The mRNA preparation was fractionated by ultracentrifugation on a sucrose density gradient, where two major peaks of mRNA were resolved (Figure 1). The first peak contained mRNA species encoding polypeptides of M$_r$ 25,000 to 35,000, while the second included mRNAs encoding α-amylase.

Using mRNA in the first peak (Figure 1) as template, cDNA was prepared by the ribonuclease H method (Gubler & Hoffman, 1983). The procedure was modified to prevent massive over-synthesis of the second strand of the cDNA; the reaction temperature for second strand synthesis was maintained at 12°C and the amount of second strand monitored until it was equal to the amount of first strand template (Fincher et al., 1986). Further, it was necessary to include DNA ligase in the reaction mixture (cf. Gubler & Hoffman, 1983) to prevent tailing of single-stranded nicks during the subsequent terminal transferase reaction. The cDNA was inserted into Pst I-cut plasmid pUC9 (Viera & Messing, 1982) by conventional dG-dC tailing procedures and cloned into E. coli K-12 strain JM101 by the method of Hanahan (1983).

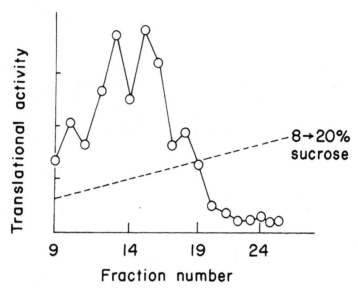

Figure 1.   Fractionation of aleurone poly(A)$^+$-mRNA by sucrose gradient ultracentrifugation.   Peak 1 (fraction 13) was enriched in mRNA encoding (1→3,1→4)-β-glucanase and was used for cDNA synthesis.   Peak 2 (fractions 15 and 16) was enriched in mRNA for α-amylase.

Screening the cDNA library

Purified (1→3),(1→4)-β-glucanase isoenzymes (Woodward & Fincher, 1982) were hydrolysed with trypsin, resultant peptides were fractionated by reversed phase HPLC, and their NH$_2$-terminal amino acid sequences were determined by automated Edman degradation (Fincher et al., 1986).   The amino acid sequence information was used to synthesize a mixed oligonucleotide probe (15-mer, 32-fold redundant), which was labelled with ^{32}P using T4 polynucleotide kinase. A (1→3),(1→4)-β-glucanase cDNA clone was selected by colony hybridization, plasmid dot blot and Southern transfer analysis with the labelled oligonucleotide probe (Fincher et al., 1986).

Properties of the cDNA encoding (1→3),(1→4)-β-glucanase

The cDNA clone, consisting of 874 nucleotide pairs (excluding the dG-dC tails), has been completely sequenced by the chain termination method of Sanger et al. (1977) and has one open reading frame on each strand. Comparison of the nucleotide sequence with NH_2-terminal and tryptic peptide amino acid sequences showed that one of the open reading frames encoded (1→3),(1→4)-β-glucanase isoenzyme II (Fincher et al., 1986), the major isoenzyme secreted from barley aleurone layers (Stuart et al., 1986). The NH_2-terminal amino acid of the mature enzyme is encoded by the seventh codon; the first six codons presumably encode part of a signal peptide. The 3' end of the cDNA extends into the COOH-terminal tryptic peptide, enabling the complete sequence of 306 amino acids to be deduced.

The number and composition of amino acids deduced from the nucleotide sequence are similar to those determined by amino acid analysis of the purified enzyme (Woodward & Fincher, 1982) and confirm the very basic nature of the protein (Table I).

Table I
Amino acid composition of barley (1→3),(1→4)-β-glucanase isoenzyme II

Number of Residues[a]

Amino acid	cDNA sequence	Enzyme composition	Amino acid	cDNA sequence	Enzyme composition
Lys	9	9	Pro	18	19
His	7	6	Gly	33	34
Arg	8	9	Ala	43	43
Trp	4	3	Cys	2	2
Asn	24]	32	Val	30	27
Asp	6		Met	11	11
Thr	15	16	Ile	11	11
Ser	23	24	Leu	17	18
Gln	10]	17	Tyr	15	14
Glu	6		Phe	14	14
			Total	306	309

[a]Comparing compositions determined by amino acid analysis of the purified enzyme (Woodward & Fincher, 1982) and that deduced from the nucleotide sequence of a cDNA encoding the enzyme (Fincher et al., 1986).

Preferential codon usage

The codon usage for amino acids 1 to 285 of the barley $(1\rightarrow3),(1\rightarrow4)$-β-glucanase isoenzyme II is shown in Table II. The overall (G+C) content in the region of the gene is 67% and this is due in part to a strong bias for G and C in the third position of degenerate codons. Of 285 codons, 267 have G or C in the wobble position. This preferential codon usage contrasts to hordein and other genes expressed during cereal endosperm development (Fincher et al., 1986), but is similar to the gene for α-amylase which is also expressed under the influence of GA$_3$ in barley aleurone layers (Rogers & Milliman, 1983; Chandler et al., 1984). It might be speculated that the conservation of this codon usage pattern could be important in the observed response of GA$_3$-treated aleurone layers in re-directing protein synthesis to a few major polypeptides.

Related to the preferential use of G and C in the third position of codons is the observation that the complementary strand of the cDNA also has an open reading frame which is in phase with the open reading frame of the coding strand. Whether or not the complementary strand of the gene is transcribed remains unknown.

Potential applications of the cDNA clone

The cDNA encoding barley $(1\rightarrow3),(1\rightarrow4)$-β-glucanase can now be used as a probe for the isolation of $(1\rightarrow3),(1\rightarrow4)$-β-glucanase genes for structural analysis. Comparisons of the $(1\rightarrow3),(1\rightarrow4)$-β-glucanase gene with those for α-amylase (Rogers & Milliman, 1983) might enable important promoter sequences to be identified, they might reveal possible sites of action for the enhancement of gene expression induced by phytohormones, and could lead to an understanding of the physiological basis for tissue-specific differences in the expression of $(1\rightarrow3),(1\rightarrow4)$-β-glucanase isoenzymes (cf. Stuart et al., 1986). Further, the cDNA could be applied to map the genes for the enzymes, since this information would be useful for barley breeders.

Of potential commercial importance could be the insertion of a cDNA encoding the complete barley $(1\rightarrow3),(1\rightarrow4)$-β-glucanase into a brewing yeast. High levels of soluble, cell wall $(1\rightarrow3),(1\rightarrow4)$-β-glucans in malt extracts can lower the rate of beer filtration and may contribute to the formation of precipitates in the final beer (Bamforth, 1982; Woodward & Fincher, 1983). If

Table II
Codon usage in the cDNA for (1→3),(1→4)-β-glucanase
isoenzyme II[a]

Leu	n	16	Pro	n	16	His	n	6
TTA	%	0	CCT	%	6	CAT	%	17
CTT		6	CCA		17	CAC		83
CTA		0	CCC		22			
TTG		0	CCG		56	Gln	n	8
CTC		50				CAA	%	0
CTG		44	Thr	n	15	CAG		100
			ACT	%	7			
Ser	n	23	ACA		0	Asn	n	21
TCT	%	0	ACC		63	AAT	%	0
TCA		0	ACG		30	AAC		100
AGT		0						
AGC		44	Ala	n	43	Lys	n	9
TCC		28	GCT	%	14	AAA	%	11
TCG		28	GCA		5	AAG		89
			GCC		51			
Arg	n	8	GCG		30	Asp	n	6
CGT	%	0				GAT	%	0
CGA		0	Gly	n	31	GAC		100
AGA		0	GGT	%	0			
AGG		24	GGA		5	Glu	n	5
CGC		38	GGC		68	GAA	%	0
CGG		38	GGG		29	GAg		100
Val	n	28	Tyr	n	13	Cys	n	2
GTT	%	4	TAT	%	0	TGT	%	0
GTA		0	TAC		100	TGC		100
GTC		48						
GTG		48						

[a]Expressed as percentages of possible codons for amino
acids, where n is the number of codons for each amino acid
in the cDNA (Fincher et al., 1986).

residual $(1\rightarrow3),(1\rightarrow4)-\beta$-glucan could be depolymerized during the fermentation stage, then these problems may be overcome. Accordingly, there has been considerable interest in the engineering of brewing yeasts. When the gene encoding the Bacillus subtilis $(1\rightarrow3),(1\rightarrow4)-$ β-glucanase was transferred into yeast, low levels of the active enzyme could be detected in cell homogenates, but the enzyme was not secreted (Hinchliffe, 1984; Hinchliffe & Box, 1984). This may be related to aberrant folding or incomplete processing of the prokaryotic enzyme in the rather different cellular environment of the eukaryotic yeast. It is noteworthy in this regard that DNA encoding proteins from other eukaryotes, namely the α-amylases from mouse salivary glands (Thomsen, 1983) and from germinating wheat (Rothstein et al., 1984), have been successfully transferred into yeast where the resultant polypeptides are correctly processed and secreted. Secretion of a bacterial cellulase by yeast has been achieved, but requires insertion of a DNA fragment encoding the signal peptide of a secreted yeast proteinimmediately upstream from the bacterial DNA sequence (Skipper et al., 1985). It might be anticipated from these results that the barley $(1\rightarrow3),(1\rightarrow4)-\beta$-glucanase could be synthesized by and secreted from yeast cells. Furthermore, the pH optimum of the barley enzyme is 4.7 (Woodward & Fincher, 1982) compared with values of 6.0 to 6.5 for the Bacillus enzyme (Hinchliffe & Box, 1984). Since the pH decreases from values of 5.2-5.7 to 4.0-4.5 during the fermentation phase of the brewing process (Hough et al., 1971), the barley enzyme might more efficiently depolymerise its substrate under the conditions found in a fermentation tank.

ACKNOWLEDGMENTS

This work was supported by grants from the Australian Research Grants Scheme. The important contributions of Dr. R.E.H. Wettenhall, Mr. P.A. Lock, Ms M.M. Morgan and Mr. J. Mundy are gratefully acknowledged. I also wish to thank Professor Diter von Wettstein, Department of Physiology, Carlsberg Laboratory, Copenhagen, Denmark for his support.

LITERATURE CITED

Bamforth, C.W. (1982) Brew. Dig. 57, 22-27, 35.
Briggs, D.E. (1978) in: Barley pp 419-480, Chapman and Hall, London.

Chandler, P.M., Zwar, J.A., Jacobsen, J.V., Higgins, T.J.V. and Inglis, A.S. (1984) Plant Molec. Biol. 3, 407-418.

Chirgwin, J.M., Przybyla, A.E., MacDonald, R.J. and Rutter, W.J. (1979) Biochemistry 18, 5294-5299.

Fincher, G.B., Lock, P.A., Morgan, M.M., Lingelbach, K., Wettenhall, R.E.H., Mercer, J.F.B., Brandt, A. and Thomsen, K-K. (1986) Proc. Natl. Acad. Sci. (USA), in the press.

Gubler, U. and Hoffman, B.J. (1983) Gene 25, 263-269.

Hanahan, D. (1983) J. Molec. Biol. 166, 557-580.

Hinchliffe, E. (1985) J. Inst. Brew. 91, 384-389.

Hinchliffe, E. and Box, W.G. (1984) Curr. Genet. 8, 471-475.

Hough, J.S., Briggs, D.E. and Stevens, R. (1971) in: Malting and Brewing Science, pp 512-518, Chapman and Hall Ltd., London.

Jacobsen, J.V. and Beach, L.R. (1985) Nature 316, 275-277.

Jonassen, I., Ingversen, J. and Brandt, A. (1981) Carls. Res. Commun. 46, 175-181.

Mozer, T.J. (1980) Cell 20, 479-485.

Mundy, J. and Fincher,G.B. (1986) FEBS Lett., in the press.

Mundy, J., Brandt, A. and Fincher, G.B. (1985) Plant Physiol. 79, 867-871.

Rogers, J.C. and Milliman, C. (1983) J. Biol. Chem. 258, 8169-8174.

Rothstein, S.J., Lazarus, C.M., Smith, W.E., Baulcombe, D.C. and Gatenby, A.A. (1984) Nature 308, 662-665.

Sanger, F., Nicklen, S. and Coulson, A.K. (1977) Proc. Natl. Acad. Sci. (USA) 74, 5463-5467.

Skipper, N., Sutherland, M., Davies, R.W., Kilburn, D., Miller, R.C., Warren, A. and Song, R. (1985) Science 230, 958-960.

Stuart, I.M., and Fincher, G.B. (1983) FEBS Lett. 155, 201-204.

Stuart, I.M., Loi, L. and Fincher, G.B. (1986) Plant Physiol., 80, 310-314.

Thomsen, K.K. (1983) Carls. Res. Commun. 48, 545-555.

Viera, J. and Messing, J. (1982) Gene 19, 259-268.

Woodward, J.R. and Fincher, G.B. (1982) Eur. J. Biochem. 121, 663-669.

Woodward, J.R. and Fincher, G.B. (1983) Brew. Dig. 58, 28-32.

Woodward, J.R., Morgan, F.J. and Fincher, G.B. (1982) FEBS Lett. 138, 198-200.

The Vascular System of the Wheat Spikelet

T. P. O'Brien, M. E. Sammut[1], J. W. Lee[2], and
M. G. Smart[3], Botany Department, Monash
Univesity, Clayton, Victoria, 3168,
Australia.

SUMMARY

The attachment region of a mid-spike spikelet was sectioned serially. These sections were used to construct an
accurate 3-dimensional model of the course of the vascular
system that supplies the organs of the a and b florets, and
the rachilla of the c and d florets. All organs are interconnected by vascular tissue, but certain parts of the system are phloem-only. In particular, the supply to the
groove bundle of the pericarp, widely held to be the most
important pathway to the grain, is made via an annulus of
phloem to which lemma, palea and lodicules have phloem-only
connections. The vascular system is sufficiently different
from the pattern encountered in vegetative nodes to warrant
treatment sui generis.
    The relationships between different cell types need
greater histological study, especially in the complex composite bundles. This analysis shows that bundle shape in
cross-section and the arrangement of zylem and phloem vary
sharply over very short distances (100μm). The distribution
of xylem and phloem transfer cells agrees with the proposal
that significant solute relocation takes place in the
regions where the vascular supplies to different organs
meet. The area in the ovary neck that encompasses the
fusion zone of the supplies to lemma, palea and pericarp
emerges as a zone in need of detailed study, both in spikelet positions within a cultivar of known, but different,
grain performance, and as a region to analyse for inter-
cultivar comparisons.

This chapter is a revised version of an article that originally appeared in Aust. J. Plant Physiol. 12, 487-511, 1985.

INTRODUCTION

The vascular anatomy of vegetative nodes of wheat and a
selection of other grasses has received some attention (e.g.
see Hitch and Sharman 1971; O'Brien and Zee 1971; Patrick
1972a, 1972b; Bell 1976; Busby and O'Brien 1979, and
references cited therein). Very much less is known about
the vascular supply to the grass spikelet. Zee and O'Brien
(1971a) give an outline of vascular supply in this region,
but the work was focused on the distribution of vascular
transfer cells, not on the details of the vascular system of
the spikelet and its organs. Most publications deal with
the vascular system in the pericarp, or just below it (e.g.
Frazier and Appalanaidu 1965; Zee and O'Brien 1970a, 1970b;
Sakri and Shannon 1975 in wheat; Cochrane and Duffus 1980
in barley; Oparka and Gates 1981 in rice; Cook and Oparka
1983 in wheat and barley; and Felker and Shannon 1980 in
maize). Hanif and Langer (1972) traced the vascular system
of the spikelet of wheat but give very little detail about
the behaviour of the tissues, e.g. phloem, largely ignore
the supply to the glumes and lemmas, and concentrate chiefly
on the supply to the rachis segments of the spikelet. A
more detailed study of the development of the rachis and
rachilla system in barley is given by Kirby and Rymer (1974)
and a moderately detailed study of vascularization of the
barley spikelet, but again with little emphasis on tissues,
is given by Kirby and Rymer (1975). Whingwiri et al. (1981)
give a detailed analysis of vascular behaviour in the wheat
rachis. The only detailed analysis of the vascular supply
in any grass spikelet appears to be that by Kawahara et al.,
(1977a, 1977b) in rice. Unfortunately, the text of these
papers is in Japanese, but the figures show a number of features relevant to this study.

Between 50 and 60% of the grains's eventual mass
accumulates without further increase in its water content
(Sofield et al., 1977). To obtain a fuller interpretation
of these and other studies (e.g. Bremner and Rawson 1978;
Jenner 1982) an accurate picture of spikelet and floret
vascular supply must be available.

MATERIALS AND METHODS

Wheat, Triticum aestivum cv. Sun 9E, was grown either
in a field plot or in growth cabinets at the CSIRO Wheat
Research Unit in North Ryde. In the latter case, the plants
were grown in 20-cm pots in a 1 : 1 : 1 mixture of peat

moss, perlite and vermiculite and irrigated daily with water and twice weekly with Aquasol (Hortico Ltd, Melbourne). A photoperiod of 16h day/8h night and a temperature of 21°C day/16°C night were maintained. Individual ears were tagged on the day of anthesis. Mid-spike samples (usually spikelet No. 11 from the base) were taken at 12 days post-anthesis after preliminary observations had shown that all vascular tissue in the spikelet was mature by that stage. Two spike-lets were analysed fully by serial sectioning, the results agreeing closely with one another. The major features were checked in hand sections of numerous samples of fresh mater-ial over a period of 3 years. The region studied in de-tail lies between levels 'b' and 'c' in FIGURE 2, a dist-ance of c. 1mm.

All tissues were fixed with aspiration in 4% glut-araldehyde (made from 8% solution of spectroscopically pure glutaraldehyde) in 0.025 M phosphate buffer at pH 6.8 over-night then post-fixed in 1% OsO_4 in the same buffer for 2h. Tissues were dehydrated in a graded acetone series, infil-trated gradually in Spurr's resin, with a total of nine daily changes (see O'Brien and McCully 1981). After embed-ding, serial sections were cut 2μm thick, flattened on drops of water in a xylene atmosphere, and dried on a hot-plate at 45°C. Sections were stained in 0.1% toluidine blue O at pH 9.6. Every eighth section was photographed and mon-tages were prepared. The course of all bundles was deter-mined and a Perspex model prepared by transferring the out-lines of all vascular tissues to Perspex sheets; one sheet corresponded to a vertical height of 16μm. Vascular sup-plies to various organs were colour-coded, the sheets were drilled and bolted together, and the sides were polished. This 3-dimensional model, studied either as a whole, or in groups of parts, was an invaluable aid in 3-D reconstruc-tion.

Having established the course of the vascular system, certain levels, were selected for more detailed study. Using a projection microscope at c. 313X total magnification (objective x25; eyepiece x12.5), large (c. 1.0 x 0.7m) transparent overlays were prepared on which the areas occu-pied by xylem, phloem, xylem transfer cells, the xylem discontinuity in the ovary neck, and bundle sheath cells were recorded. These overlays were made into Letraset diagrams and photographically reduced.

OBSERVATIONS AND DISCUSSION

FIGURE 1 shows the attachment region of an intact spikelet, while FIGURE 2 illustrates a sagittal section of a comparable spikelet. Three partly filled grains are shown; the regions of attachment of the sterile glumes, lemmas and paleas of the a and b florets are evident, as in the rachilla that supplies the c floret. FIGURE 3 shows the arrangement of vascular tissues in the rachis segment at level 'a' in FIGURE 1, while FIGURE 4 shows the pattern seen at level 'b' in FIGURES 1 and 2. The 3-D model (not shown) shows that the peripheral bundles (pb) seen in FIGURES 3 and 4 form an interconnected net in the zone 200μm above level 'b'. Some parts of this network make contact with the larger bundles that supply the sterile glumes and which arise from the central core (ib) of FIGURE 4. Other parts give rise directly to some of the numerous tiny strands seen in profile in the sterile glumes while the bundles in the wings of FIGURE 3 (wb) have direct connections to the bundles seen in the wings of the rachis segment above this spikelet. The bundles of the central core (ib) give off large traces which turn outwards and run obliquely through the attachment region of the sterile glumes. Some fine branches from these larger traces contribute to some of the finer veins seen in the sterile glumes.

The full details of the vascular system of the wheat spikelet have been published recently in the Australian Journal of Plant Physiology 12, 487-511, 1985, using colour-coded diagrams that cannot be reproduced here. What follows is a summary of the main points of that paper.

With the exception of the peripheral bundles from the wings mentioned above, the bundles that supply the rachis of the spikelet above this one are derived from the central core. Seven strands leave the central core, the upper five departing at a lower level in the node. Eventually, all seven strands turn over sharply and assume an oblique course through the node as they move out into the upper rachis segment.

400μm from level 'b' (see FIGURES 1 and 2) all bundles that will supply the sterile glumes and those that are going to the rachis segment above this spikelet have departed from their connections with the central core. The central core has undergone significant alteration in shape over this interval and consists now of a series of discrete strands. These strands supply the vasculature of all other organs of the spikelet.

FIGURE 1. Photomicrograph showing a segment of the rachis (r) and the region of the spikelet attachment. The two sterile glumes (sg) and the paleas of the two lower florets (a and b) are evident. Note the slight difference in level of attachment of the two sterile glumes. a, level of FIGURE 3;   b, level of FIGURE 4.

FIGURE 2. Approximately sagittal view of a spikelet similar to that shown in FIGURE 1.   The a, b and c grains, the attachments of the sterile glumes (sg), the rachilla (ra) to grain c and the lemma (l) and palea (pa) of floret a are shown. The rachis (r) is sectioned obliquely. b, level of FIGURE 4. c, 1040μm further up.

FIGURE 3. Hand-cut cross-section of the rachis below the spikelet at level 'a' on FIGURE 1. At this level the rachis has an inner group of 12 bundles (ib; the two at 12 o'clock are engaged in fusion), and an outer ring of peripheral bundles (pb). Some of these peripheral bundles enter the sterile glumes directly while the ones in the wings (wb) fuse with the sterile glume traces and then enter the wings of the rachis segment above this spikelet. Toluidine blue stained.

FIGURE 4. Montage of epoxy sections at level 'b' on FIGURES 1 and 2. This central core of vascular tissue (ib) splits off all bundles that supply the inner bundles of the rachis segment above this spikelet and also gives rise to the majority of the vascular tissues that supply this spikelet (the peripheral bundles are connected to the sterile glume traces and the inner bundles and the 3-dimensional model shows that even those peripheral bundles that enter the sterile glume directly still have connections to the central bundles).

FIGURES 1-4 reproduced with permission from Aust. J. Plant Physiol. 12, pp. 487-511, 1985.

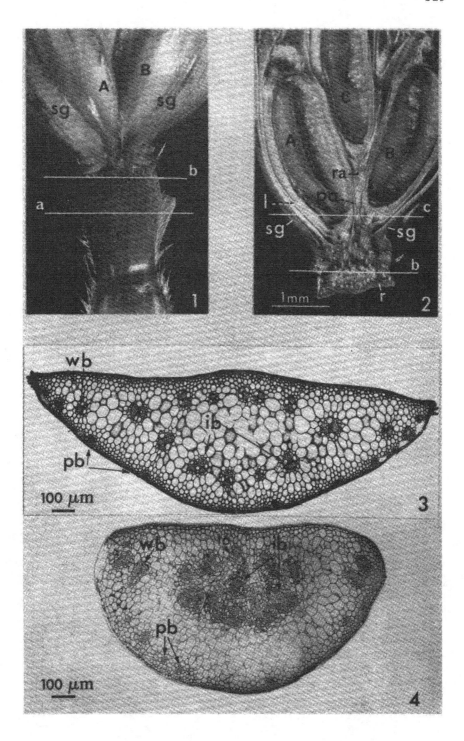

LEMMA AND PALEA

The origins of several stands can be traced from this region. The bundle that becomes the mid-rib of the lemma subtending the a floret begins to depart from the central core about 384μm from level 'b' (see FIGURES 1 and 2). Regions that will give rise to the larger laterals of the lemma of floret a can be traced readily through the next 600μm, as can the bundles that will become the lemma traces of floret b. The bundles of the palea of floret a and those of the palea of floret b also originate close to this level and may be traced readily upwards. This first bundle that can be identified reliably as a contributor to the vascular supply of the rachilla above the b floret also appears at this level. Other such bundles appear in quick succession until all are clearly evident at level 'c' (see FIGURE 2).

THE PERICARP SUPPLY

The origin of the vascular system that supplies the pericarp bundles is fascinating and complex. It is easier to deal with floret a. Up to the level 400μm from level 'b' in FIGURES 1 and 2, analysis of the 3-D model shows that there are always regions known as composite bundles that can be traced directly to similar regions in the central core of FIGURE 4. However, from this level upwards, it is essentially impossible to be certain that this is still the case, since the supply from the more median bundles of the lemma all impinge of this region.

Part of this set of tissues has a very direct connection to the annulus of phloem that surrounds the blocked xylem supply to the pericarp. The palea and lemma are both connected by phloem-only strands to this annulus. Phloem connections, unaccompanied by xylem, were detected in the region beneath the collapsed lodicules. No attempt was made to track the regions of attachment of the anthers as these had almost totally disappeared by the stage these spikelets were selected. Unblocked tracheary elements embedded within the phloem supplying the pericarp bundle in the crease of the a grain reappear about level 'c' in FIGURE 2.

Analysis of serial sections in this region of the xylem block (see Zee and O'Brien 1970a) shows very rapid change in the vascular pattern over a distance of 64μm. This underlines the fact that closely adjacent sections must be studied sometimes if features of importance are to be documented adequately.

THE b FLORET

Analysis of slides and micrographs showed that the
vascular connections beneath the b floret were essentially
identical to those of the a floret.

THE INTERCONNECTEDNESS OF THE VASCULAR SYSTEM

The peripheral bundles of the rachis segment seen in
FIGURE 3, in addition to becoming the numerous tiny strands
of the sterile glumes, are themselves intimately cross-
connected by horizontal bundles in the first 300μm above
level 'b'. The wing regions of the peripheral bundles are
also connected directly to the wing regions of peripheral
bundles of the rachis segment above this spikelet. Similarly, both the peripheral and central bundles of the rachis
segment are extensively connected to the core regions of
FIGURE 4 (ib), from which all other parts of the vascular
system of the spikelet are derived. This same core region
provides the inner bundles of the rachis segment above this
one and in view of the work of Whingwiri et al., (1981), it
seems likely that this pattern would be repeated at each
spikelet throughout the head. Thus, all organs of any one
spikelet appear to have some vascular connections with all
the other organs. In the majority of cases, these connections are probably both via the xylem and phloem, with two
exceptions. As shown earlier by Zee and O'Brien (1971b),
lodicules in wheat have only a phloem supply; secondly,
no tracheary elements with unblocked luminal pathways cross
the region in the base of the ovary neck, so that the ultimate supply to the pericarp bundles is also a phloem supply.
We suspect that the anther supply may also be phloem only
although, like the groove pericarp bundle, there are tracheary elements above the regions of attachment of the anther
filament (unpublished observations).
The unravelling of the exact histological nature of all
of the connections between the organs of a spikelet is a
formidable task that lies ahead: there is rapid alteration
in the fine-scale pattern of tissue distribution in many
regions of this spikelets's vasculature. Nonetheless, there
are large areas in the composite bundles where apparent
zones of contact between xylem and phloem are evident. It
seems likely that, at the level of tissue systems, there is
continuity in the network. However, we are equally certain
that some of the connections consist solely of phloem. A
careful study of the palea bundles shows that they have

phloem-only connections to the annulus of phloem that supplies the grain. If one traces this region downwards, no
xylem connection can be found. The palea connections for
the b grain show the same phenomenon.

CONCLUSIONS

    A clear exposition of the behaviour of bundles in vege-
tative nodes of wheat is given in Busby and O'Brien (1979).
We had expected a similar behaviour in these nodes at the
base of the spikelet, but the structure found here departs
from that of the vegetative nodes in a number of ways.
First, despite the richness of the vascular supply to the
sterile glumes and the lemmas, each glume makes only three
or four connections to the rachis bundles while the lemmas
have five. As the 3-D model shows, this is because most of
the lemma bundles fuse into just five strands before entering the central core. With the glumes, there is also a lot
of fusion prior to entry to the central core but, in addition, some of the peripheral bundles of the rachis simply
pass straight into the sterile glumes, contracting the central core only indirectly by connections to other bundles
with direct connection to the core.
    Each lemma has a group of large bundles in the swollen
margins of the organ. The strands from this region of both
lemmas share a common origin but do not appear to interact
with any other bundles than those of the central core. However, the 3-D model does show that the other bundles of the
lemma are superimposed above the major bundles of the sterile glume, in what looks like an obstructed/obstructing bundle relationship. However, we cannot find any convincing
evidence that the lemma traces 'fork' around the obstructing
traces of the sterile glumes, either beneath the a or b
floret. The palea attachments likewise show no signs of
forking but perhaps this is to be expected since the palea
bundles might behave as if they were alternating bundles
to the lemma traces. This would depend on the morphological
interpretation of the anthers, lodicules, paleas and ovary,
a matter beyond the scope of this paper.
    We do not feel that it is particularly helpful to
approach the structure of this spikelet network from the
point of view used by Busby and O'Brien (1979) for tiller
nodes. The highly condensed nature of the attachment
region, and the very unusual vascular system found at the
base of the ovary are better viewed, in our opinion, as a
unique structure.

The other major conclusions are that: (1) all organs share vascular connections to the system in the rachis segments above and below a spikelet attachment (see also Kirby and Rymer 1974 for barley; Whingwiri et al., 1981) for wheat); (2) xylem and phloem connections predominate, except that the lodicules and ovary have a phloem-only connection; and (3) a very strong phloem connection links lemma and palea traces to the phloem annulus that supplies the pericarp vasculature (see also Oparka and Gates 1981; Kawahara et al., 1977a, 1977b for rice). The histological nature of phloem connections in the more complex bundles needs further study, as does the fine structure of the cell types found throughout the ovary attachment. The distribution of xylem transfer cells is consistent with the hypothesis that they abstract solutes from the free space and deliver them to the phloem for ultimate passage to the grain, an hypothesis much in need of direct experimental support (see discussion in Busby and O'Brien 1979). The unexpectedly rich vascularization of the sterile glumes suggests that these organs may be playing an important role in nutrient cycling in the spikelet, a matter which also needs further testing (see also Lee 1978).

An interesting minor point concerns the number of inner bundles in the rachis segment above and below this spikelet (No. 11 in the head from the base). Beneath the spikelet, the number is about 12 (see FIGURE 3), in reasonable agreement with mid-spike expectations from Whingwiri et al., (1981) (see their figure 7 which predicts 10-11 such bundles for the spikelet position we have analysed). However, only seven inner bundles leave this spikelet for the next rachis segment. The expectation from Whingwiri et al.'s figure 7 would be at least nine such inner bundles. This raises the possibility that some inner bundles branch after they enter the next rachis segment. This needs to be checked by surveying bundle structure throughout the length of rachis segments.

The zone in the ovary neck where the grain draws upon its connections to palea, lemma, and lodicules stands out as a zone worthy of intensive study in between-cultivar comparisons and, within a cultivar, of between-spikelet comparisons where these have known but different grain performances. We are attempting to quantitate tissue distribution in this region to see if one can produce a model that is predictive of grain performance.

324

ACKNOWLEDGMENTS

We wish to thank Mrs. Ann Dalton for her expert
assistance in the early stages of this project. The work,
although carried out at CSIRO Wheat Research Unit in Sydney,
was funded by an external grant from CSIRO to one of us
(TPO'B) and the assistance of Dr. David Simmonds in initiating that liaison and funding, and of Dr. Colin Wrigley
in the later stages of the project, is gratefully acknowledged. Thanks are due also to Ms. Jan Palanci for typing
and to Mr. John Nailon for help with the references. One of
us (MGS) acknowledges his debt to the kidney donor.

REFERENCES

Bell, A. D. 1976. The vascular pattern of Italian ryegrass
    (Lolium multiflorum Lam.). 3. The leaf trace system
    and tiller insertion, in the adult. Ann. Bot. (London)
    40, pp. 241-250.
Bremner, P. M. and Rawson, H. M. 1978. The weights of in-
    dividual grains of the wheat ear in relation to their
    growth potential, the supply of assimilate and inter-
    action between grains. Aust. J. Plant Physiol. 5, pp.
    61-72.
Busby, C. H. and O'Brien, T. P. 1979. Aspects of vascular
    anatomy and differentiation of vascular tissue and
    transfer cells in vegetative nodes of wheat. Aust. J.
    Bot. 27, pp. 703-711.
Cochrane, M. P. and Duffus, C. M. 1980. Nucellar projection
    and modified aleurone in the crease region of develop-
    ing caryopsis of barley (Hordeum vulgare L. var.
    distichum). Protoplasma 103, pp. 361-375.
Cook, H. and Oparka, K. J. 1983. Movement of fluorescein
    into isolated caryopses of wheat and barley. Plant
    Cell Environ. 6, pp. 239-242.
Craig, S. and O'Brien, T. P. 1975. The lodicules of wheat :
    pre- and post-anthesis. Aust. J. Bot. 23, pp. 451-458.
Felker, F. C. and Shannon, J. C. 1980. Movement of ^{14}C-
    labeled assimilates into kernels of Zea mays L. Plant
    Physiol. 65, pp. 864-870.
Frazier, J. C. and Appalanaidu, B. 1965. The wheat grain
    during development with reference to nature, location,
    and role of its translocatory tissues. Am. J. Bot.
    52, pp. 193-198.
Hanif, M. and Langer, R. H. M. 1972. The vascular system of
    the spikelet in wheat. Ann. Bot (London) 36, pp. 721-
    727.

Hitch, P. A. and Sharman, B. C. 1971. The vascular pattern
    of festucoid grass axes, with particular reference to
    nodal plexi. Bot. Gaz. 132, pp. 38-56.
Jenner, C. F. 1982. Movement of water and mass transfer into
    developing grains of wheat. Aust. J. Plant Physiol.
    9, pp. 69-82.
Kawahara, H., Matsuda, T. and Chonan, N. 1977a. Studies on
    morphogenesis in rice plant. IX. On the structure of
    vascular bundles and phloem transport in the spikelet.
    Proc.Crop Sci. Coc. Jpn. 46, pp. 82-90.
Kawahara, H., Matsuda, T. and Chonan, N. 1977b. Studies on
    morphogenesis in rice plant. X. Ultrastructure of the
    dorsal vascular bundle in the ovary and a transport
    mechansism. Proc. Crop Sci. Jpn 46, pp. 91-96.
Kirby, E. J. M. and Rymer, J. L. 1974. Development of the
    vascular system in the ear of barley. Ann. Bot.
    (London) 38, pp. 565-573.
Kirby, E. J. M. and Rymer, J. L. 1975. The vascular anatomy
    of the barley spikelet. Ann. Bot. (London) 39, pp. 205-
    211.
Lee, J. W. 1978. Influence of nitrogen source on nitrogen
    metabolism in detached wheat heads. Aust. J. Plant
    Physiol. 5, pp. 779-785.
O'Brien, T. P. and McCully, M. E. 1981. 'The Study of Plant
    Structure : Principles and Selected Methods.'
    (Termarcarphi: Melbourne.)
O'Brien, T. P. and Zee, S-Y. 1971. Vascular transfer cells
    in the vegetative nodes of wheat. Aust. J. Biol. Sci.
    24, pp. 207-217.
Oparka, K. J. and Gates, P. 1981. Transport of assimilates
    in the developing caryopsis of rice (Oryza sativa L).
    Planta 152, pp. 388-396.
Patrick, J. W. 1972a. Vascular system of the stem of the
    wheat plant. I. Mature state. Aust. J. Bot. 20, pp.
    49-63.
Patrick, J. W. 1972b. Vascular system of the stem of the
    wheat plant. II. Development. Aust. J. Bot. 20, pp.
    65-78.
Sakri, R. A. K. and Shannon, J. C. 1975. Movement of [14]C-
    labeled sugars into kernels of wheat. Plant Physiol.
    55, pp. 881-889.
Simmonds, D. H. and O'Brien, T. P. 1981. Morphological and
    biochemical development of the wheat endosperm. In :
    Advances in Cereal Science and Technology. Vol. 4.
    Pomeranz, Y. (ed.). Am. Assoc. Cereal Chem. St Paul,
    Minnesota.

Smart, M. B. and O'Brien, T. P. 1983. The development of the wheat embryo in relation to the neighbouring tissues. Protoplasma 114, pp. 1-13.

Sofield, I., Wardlaw, I. F., Evans, L. T. and Zee, S-Y. 1977 Nitrogen, phosphorus and water contents during grain development and maturation in wheat. Aust. J. Plant Physiol. 4, pp. 799-810.

Swift, J. G. and O'Brien, T. P. 1970. Vascularization of the scutellum of wheat. Aust. J. Bot. 18, pp. 45-53.

Whingwiri, E. E., Kuo, J. and Stern, W. R. 1981. The vascular system in the rachis of a wheat ear. Ann. Bot. (London). 201, pp. 189-201.

Zee, S-Y. and O'Brien, T. P. 1970a. A special type of tracheary element associated with 'xylem discontinuity' in the floral axis of wheat. Aust. J. Biol. Sci. 23, pp. 783-791.

Zee, S-Y. and O'Brien, T. P. 1970b. Studies on the ontogeny of the pigment strand in the caryopsis of wheat. Aust. J. Biol. Sci. 23, pp. 1153-1171.

Zee, S-Y. and O'Brien, T. P. 1971a. Vascular transfer cells in the wheat spikelets. Aust. J. Biol. Sci. 24, pp. 35-49.

Zee, S-Y. and O'Brien, T. P. 1971b. The vascular tissue of the lodicules of wheat. Aust. J. Biol. Sci. 24, pp. 805-809.

[1]Wheat Research Unit, CSIRO, Private Bag, North Ryde, NSW, 2113. Present address, 30 Bentrield Gardens, Stantsted, Essex CM24 8HG, U.K.

[2]Wheat Research Unit, CSIRO, Private Bag, North Ryde, NSW, 2113, Australia.

[3]Wheat Research Unit, CSIRO, Private Bag, North Ryde, NSW, 2113. Present address, Botany School, University of Melbourne, Parkville, Victoria, 3052, Australia.

Ear and Grain Wetting and Pre-Harvest Sprouting

R. W. King, CSIRO, Division of Plant
Industry, Canberra, Australia.

SUMMARY

Some of the factors which may control ear and grain
wetting are discussed. These include rainfall characteristics, ear wetting and drying rates, ear and grain structure and anatomy and the pathways of water penetration.
Simply inherited characters which give some control of wetting include ear nodding angle the use of awnless lines and
normal, non-compact, ears. Effects of glaucousness and
glabrousness require further examination. Simple correlations of grain characters with grain water uptake are yet to
be found.

INTRODUCTION

In his opening address to the first of these confer-
ences MacKey (1976) considered and discarded the possibility
that the rate of wetting of the cereal ear and grain could
be important in the control of sprouting. However, the
amount and rate of supply of water to the grain must influence its germination and, as discussed below, there are now
reasonable grounds for believing that wetting of ear and
grain can be manipulated to give some control of sprouting.
Furthermore, this solution is free of the penalties and
difficulties of breeding for greater grain dormancy.

MEASUREMENT TECHNIQUES

A simple measurement of weight change suffices for following ear and grain wetting by rain. However, to express

uptake as change over initial weight emphasizes differences such as ear size which may be of little direct relevance. By contrast, calculations based on final weights are also of limited value because they give a truncated or asymptotic approach to 100%. However, when water uptake ranges up to about 50% then this non-linerarity is of little importance. This latter calculation is preferred, therefore, as it places less emphasis on varietal differences in ear weights (King and Richards 1984).

The simulation of rainfall is a much greater problem. To properly simulate rainfall, drop size should increase and number decrease as intensity increases (Hardy and Dingle, 1960; Mueller and Jones, 1960). Current "rain simulators", however, provide droplets of a fairly uniform size, either as a fine mist (King and Richards, 1984) or in larger drops (see McMaster and Derera, 1976). Intensity receives little attention and even so intensity is changed by altering the duration of exposure with no change in drop size. Nevertheless, for wheat ears King and Richards (1984) did find ear wetting increased with increased intensity, to give saturation at about 8 to 10mm h^{-1}. Furthermore, when ears of the same variety were exposed to natural rainfall for an hour their wetting was comparable to that expected for a similar intensity of simulated rain.

Possibly the most serious shortcoming of rainfall simulators is their inability to deliver rain particles with the same kinetic energy and detachment power as occurs at terminal velocity of natural rain. Even so, for natural rainfall it has proven difficult to precisely describe the randomness of drop sizes particularly at low rainfall intensities <20-25mm h^{-1} (see Kinnell, 1973, 1981) and, hence, it becomes almost impossible to simulate all or any geographical locations. Nevertheless, provided intensities are below those giving a saturation of ear wetting rate then some effort should be made to examine effects of drop size and, clearly, a properly designed rainfall tower should be employed so that terminal drop velocities are reached.

CULTIVAR COMPARISONS

Studies of King and Richards, (1984) with fifty cultivars demonstrated some correlation between ear-wetting and in-ear sprouting. The difficulty, however, is to distinguish differences due to water uptake by the ear (which is easily measured) and the separate effects on both water penetration to the grain and on grain water uptake. For example, King (1984) reported large differences between cultivars for water uptake by isolated grain and there was a reasonable correlation to germination (r^2 = 0.37). Clearly, therefore, when seeking correlations to in-ear sprouting it is necessary to separately summate grain and ear water up-take as attempted by King (1984). Assessment of penetration differences has yet to be attempted but could be quite sig-nificant. Mares (1983) for instance, found that when ear wetting was no different and after grain wetting differences were accounted for, there were still differences in grain water uptake and in time to reach an adequate water content for sprouting. Such varietal differences are seen in his comparison between cv's Shortim and Songlen (TABLE 1). Clearly this evidence points to differences in transport pathways to the grain.

TABLE 1. The rate of ear uptake and of movement of water into grain in the ear or after its isolation. Potential to sprout was based on the grain reaching a 40% water content (adapted from Mares, 1983).

Cultivar	Saturation for ear wetting (hours)	Isolated grain wet to 50% (hours)	Hours to 40% threshold water for sprouting
Songlen	3	11	24
Shortim	3	10	40
Kite	3	14	43
Sun 44E	3	16	50

As an aside, a problem associated with dormancy arises if ear wetting/sprouting studies are begun at harvest. In the experiments of King and Richards, (1984) post harvest dormancy was avoided by holding ears dry for six months prior to wetting and assessment of actual sprouting. How-ever, the approach adopted by Mares (1983) with fresh ears would appear to be equally valid. Studies with wheat, for example Owen (1952), Lush, Groves and Kaye, (1981), Mares (unpublished), indicate a threshold water requirement for germination of about 40% water content on a dry weight basis. Thus, irrespective of any dormancy differences measurements of the time to reach a "germination" water con-tent will adequately define sprouting limitations imposed by water uptake processes. Care should be exercised, nevertheless, in harvesting grain near to its maturity. If at all immature then wetting the ear or isolated grain leads to no water uptake (FIGURE 1). It may take up to 4 weeks after first dehydration of the grain before grain and ear water uptake reach their maximum values. Furthermore, it cannot be assumed that there is either uniform drying or of wetting of the once dried grain. For barley, at least, it is the surface layers including husk, pericarp and embryo which first reach full hydration. The starchy endosperm lags considerably (Reynolds and MacWilliam, 1966).

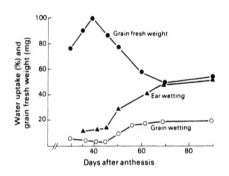

FIGURE 1. Effect of stage of ear maturity on water uptake by the ear and of isolated grain. Natural drying of the grain also shown for the wheat cultivar, Gabo (King, unpublished).

EAR AND GRAIN CHARACTERISTICS AND WETTING

Several workers (Pool and Patterson, 1957, Clarke, 1982) have studied the effect of awns on ear drying rates in the field. Waxiness may have an effect (Pool and Patterson, 1958) and ear nodding angle is of significance for wetting damage (Brinkman and Luk, 1979). King and Richards (1984) took these studies a step further in showing that awned ears took up water more rapidly and there was enhanced in-ear sprouting. This relationship was evident for a number of isogenic awned/awnless lines and for a comparison of fifty cultivars. The sequence of events for the isogenic lines is quite predictable; more rapid ear wetting leads to more rapid grain wetting and to an eventual triggering of germination some 10 to 15 hours earlier in the awned line (FIGURE 2).

In these studies the presence of awns accounted for some of the cultivar differences (r^2 = 17%). Of other contributing factors the presence of hairs (pubescence) or wax made little difference to ear wettability although further examination of these characters is required (see below). The club head character does, however, lead to enhanced ear water uptake. Nodding angle should be examined further although it was unimportant in these studies as ears were

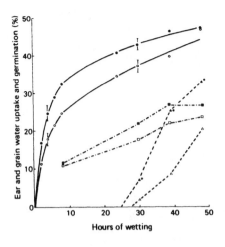

FIGURE 2. Time course of
ear and grain water uptake
and of sprouting in the ear
of awned/awnless lines.
(From King and Richards,
1984).

held upright.

By contrast with these positive results for whole ears,
for grain germinating on filter paper none of the characteristics tested to date can account for the up to 2-fold differences that exist between cultivars in their rate of water
uptake and hence in their sprouting. In earlier studies
suggestions had been made about effects of seed coat colour
(Wellington, 1956) or grain hardness (Butcher and Stenvert,
1973; Moss, 1973) but grain of lines isogenic for these
characters showed no differences in water uptake or germination once dormancy differences had passed (King, 1984).
Comparison between cultivars for grain protein content and
testa thickness also failed to show correlations to grain
wetting rate or to germination (King 1984).

Given the present impasse, any number of other factors
could be considered in searching for control of grain water
uptake. Grain surface to volume ratios could be significant
and differences in this ratio between cultivars have been
documented by Marshall et al., (1984). Surface rupturing
and the formation of fissures could differ between varieties. As yet, there is no evidence of such cultivar differences only that water uptake might be through fissures
(Woodbury and Wiebe, 1983). Physico-chemical properties of
the grain could differ but it seems unlikely that the embryo
would be affected. Furthermore, water imbibition in cereals
shows an equivalence between volume and weight changes
(Chung et al., 1961; Leopold, 1983) so that there is none
of the massive swelling of seed polymers that drives water
uptake in some seeds.

Another possible restriction to seed water uptake suggested by some workers relates to the structure of the seed

coat and, particularly, to lignification of the testa. In
barley for instance the testa blocks transverse movement into the endosperm of dyes (Briggs and MacDonald, 1983) or of
an aqueous solution of GA_3 (Freeman and Palmer, 1984).
Similarly, in wheat Hinton (1955) reported that the testa/
pericarp was the major site of restriction of direct water
entry. The testa/pericarp was 3 times more resistive than
the hyaline layer, the aleurone or the starchy endosperm.
Such a restriction to water passage may reflect lignin
deposition in the grain as this occurs for wheat just at
maturity (Zee and O'Brien, 1970) but it is not known whether
this occurs in the testa or just in the xylem discontinuity
zone above the rachilla. Such a control over permeability
gains some credence from the reports for legumes that their
seed coat impermeability first becomes detectable at seed
dessication at which time there is a highly active polyphenol oxidase in the seed coat and, simultaneously, there
is a deposition of lignin (Egley et al., 1985). However, any
restrictions imposed by the testa may be insignificant as
Briggs and MacDonald (1983) showed with barley that there is
unrestricted lateral flow of water from the distally wet
half of a seed up to the embryo. Apparently the pericarp is
an effective wick provided the non-wet proximal half of the
seed is held in high humidity to prevent lateral evaporation
of water. The endosperm itself took up little water directly in these experiments but germination began. Clearly,
restriction of water uptake by the testa is of little significance to germination. It remains as reported by Woodbury and Weibe (1983) that passage of water along the pericarp wick from a distally wet seed can take some hours (up
to 40) and so be important in slowing sprouting particularly
for grain in the ear. There is, however, no information on
any varietal differences in pericarp water conductivity.

GENERAL CONSIDERATIONS

     It is now clear that altering ear and grain water
absorption characteristics offers scope for combatting pre-
harvest sprouting in a way that will be additive to any
changes that can be made by selecting for dormancy. For
water uptake by ear and grain the identification of simply
inherited characteristics of use to the breeder is, however,
incomplete. Furthermore, as evident in the report by Mares
(1983), penetration of water to the grain may be of equal
importance.
     One suite of ear and grain characters for which effects

on wetting could be expected are the surface characters including hairiness and waxiness. Furthermore, these are
known to be simply inherited and hence of immediate use to
breeders. Literature on foliar uptake of pesticides by
plant surfaces provides comprehensive details of the slowing
of water absorption in the presence of wax or hairs. For
wheat leaves, for example, removal of surface wax enhances
wetting (Troughton and Hall, 1967), in fact, in their early
studies of ear drying after wetting Pool and Patterson
(1958) reported slower water loss from ears of waxy vs non-
waxy isogenic lines. However, it is now known that these
early selections of glaucous material may have produced
material with a changed pattern of wax deposition rather
than in amount deposited (Troughton and Hall, 1967). More
recently developed isogenic lines do differ in wax deposited
but for one set of these there was no detectable reduction
in ear wetting in the waxy genotype (King and Richards,
1984). Ear wetting was similarly no different for hairy vs
hairless lines (King and Richards 1984). There may be a
counterbalancing effect in the 30-fold more rapid evaporation rate at least of waxy lines (Brain and Butler, 1985).
Further testing is warranted, however, particularly of water
penetration to the grain for both waxy and hairy lines.

One last area requiring more attention is the actual
simulation of rainfall. The awned/awnless lines which
showed different ear water uptake when examined by King and
Richards (1984) failed to show differences (King, unpublished) in the rain simulator used by Mares (1983) and
adapted by others. Both rainfall intensity and droplet size
would have differed in these trials. Clearly, therefore, it
must be a priority to establish the adequacy of the various
approaches to simulation of rainfall.

REFERENCES

Brain, P. and Butler, D. R., 1985. A model of drop size
distribution for a system with evaporation. Plant Cell
and Environ. 8, pp. 247-253.
Briggs, D. E. and MacDonald, J. 1983. The permeability of
the surface layers of cereal grains and implications
for tests of abrasion in barley. J. Inst. Brewing.
89, pp. 324-332.
Brinkman, M. A. and Luk, R. M. 1979. Relationship of spike
nodding angle and kernel brightness under simulated
rainfall in barley. Can. J. Plant Sci. 59, pp. 481-
485.

Butcher, J. and Stenvert, N.L. 1973. Conditioning studies on Australian wheat. III. The role of the rate of water penetration into the wheat grain. J. Sci. Food Agric. 24, pp. 1077-1084.

Chung, D. S., Fan, L. T. and Shellenberger, J. A. 1961. Volume increase of wheat kernels accompanying absorption of liquid water. J. Biochem. and Microbiol. Tech. and Eng. 3, pp. 377-393.

Clarke, J. M. 1982. Effect of awns on drying rate of wind-rowed and standing wheat. Can. J. Plant Sci. 62, 1-4.

Egley, G. H., Paul, R. N., Duke, S. O. and Vaughin, K. C. 1985. Peroxidase involvement in lignification in water-impermeable seed coats of weedy leguminous and malvaceous species. Plant Cell and Environ. 8, pp. 253-260.

Freeman, P. L. and Palmer, G. H. 1984. The influence of the pericarp and testa on the direct passage of gibberellic acid into the aleurone of normal and abraded barley. J. Inst. Brewing. 90, pp. 95-104.

Hardy, K. R. and Dingle, A. N. 1960. Raindrop size distributions in a cold frontal shower. Proc. 8th Weather Radar Conference pp. 179-186. Cambridge, Mass., USA.

Hinton, J. J. C. 1955. Resistance of the testa to entry of water into the wheat kernel. Cereal Chem. 32, pp. 296-306.

King, R. W. 1984. Water uptake and pre-harvest sprouting damage in wheat: grain characteristics. Aust. J. Agric. Res. 35, pp. 337-345.

King, R. W. and Richards, R. A. 1984. Water uptake and pre-harvest sprouting damage in wheat: ear characteristics. Aust. J. Agric. Res. 35, pp. 327-336.

Kinnell, P. I. A. 1973. The problem of assessing the erosive power of rainfall from meterological observations. Soil Sci. Soc. Am. Proc. 37, pp. 617-621.

Kinnell, P. I. A. 1981. Rainfall intensity-kinetic energy relationships for soil loss prediction. Soil Sci. Soc. Am. Proc. 45, pp. 153-155.

Leopold, A. C. 1983. Volumetric components of seed imbibition. Plant Physiol. 73, pp. 677-680.

Lush, W. M., Groves, R. H. and Kaye, P. E. 1981. Presowing hydration-dehydration treatments in relation to seed germination and early seedling growth of wheat and ryegrass. Aust. J. Plant Physiol. 8, pp. 409-425.

MacKey, J. 1976. Seed dormancy in nature and agriculture. Cereal Res. Commun. 4, pp. 83-91.

Mares, D. J. 1983. Investigation of the pre-harvest sprouting damage resistance mechanisms in some Australian

white wheats. In: Proc. Third Int. Symp. on Pre-
harvest Sprouting in Cereals. Kruger, J. E. and
LaBerge, D. E. (eds.). pp. 59-65. Westview Press,
Boulder, Co., USA.

Marshall, D. R., Ellison, F. W. and Mares, D. J. 1984.
Effects of grain shape and size on milling yields in
wheat. I. Theoretical analysis based on simple geo-
metric models. Aust. J. Agric. Res. 35, pp. 619-630.

McMaster, G. J. and Derera, N. F. 1976. Methodology and
sample preparation when screening for sprouting damage
in cereals. Cereal Res. Commun. 4, pp. 251-254.

Moss, R. 1973. Conditioning studies on Australian wheat.
II. Morphology of wheat and its relationship to conditioning. J. Sci. Food Agric. 24, pp. 1067-1076.

Mueller, E. A. and Jones, D. M. A. 1960. Drop-size distributions in Florida. Proc. 8th Weather Radar Conference. pp. 299-305. Cambridge, Mass., USA.

Owen, P. C. 1952. The relation of germination of wheat to
water potential. J. Expt. Bot. 3, pp. 188-203.

Pool, M. and Patterson, F. L. 1958. Moisture relations in
soft red winter wheats. II. Awned versus awnless and
waxy versus non-waxy glumes. Agron. J. 50, pp. 158-
160.

Reynolds, T. and MacWilliam, I. C. 1966. Water uptake and
enzymic activity during steeping of barley. J. Inst.
Brewing. 72, pp. 166-170.

Troughton, J.H. and Hall, D. M. 1967. Extracuticular wax
and contact angle measurements on wheat (Triticum
vulgare L.). Aust. J. Biol. Sci. 20, pp. 509-525.

Wellington, P. S. 1956. Studies on the germination of
cereals. 2. Factors determining the germination
behaviour of wheat grains during maturation. Ann.
Bot. 20, pp. 481-500.

Woodbury, W. and Wiebe, T. J. 1983. A possible role of the
pericarp in control of germination and dormancy of
wheat. In: Proc. Third Int. Symp. on Pre-harvest
Sprouting in Cereals. Kruger, J. E. and LaBerge, D. E.
(eds.). pp. 51-58. Westview Press, Boulder, Co., USA.

Zee, S. Y. and O'Brien, T. P. 1970. A special type of
tracheary element associated with "xylem discontinuity"
in the floral axis of wheat. Aust. J. Biol. Sci. 23,
pp. 783-791.

Hormone Responses in the Cereal Aleurone Tissue

C. A. Brearley, D. N. Vakharia and D. L. Laidman,
Department of Biochemistry and Soil Science,
University College of North Wales, Bangor,
Gwynedd LL57 2UW, Wales, United Kingdom.

INTRODUCTION

One of the manifest features of germination in cereal grains is the complete hydrolysis of the starchy endosperm to produce soluble nutrients for the sustenance of the growing seedling. Since the starchy endosperm is a dead tissue, the hydrolysis is catalysed by hydrolases secreted from the neighbouring scutellum and aleurone tissues. The relative contributions of these tissues have been a matter of conjecture for a century and they remain so today. Nevertheless, it is clear that the aleurone tissue plays an important part in the process and that it is an important site of hydrolase synthesis and secretion during germination. Moreover, the synthesis and secretion of the hydrolases is controlled by the embryo via hormonal mechanisms. The discovery by Paleg and Yomo independently in 1960, that exogenously applied gibberellin A_3 can initiate the synthesis and secretion of alpha-amylase from the aleurone tissue of embryoectomised grains, implicated the gibberellins as the responsible hormone in vivo. It was soon found that the synthesis and secretion of other hydrolases, including proteinase, were also a part of the response. The hormone abscisic acid was found to inhibit the gibberellin response. As a consequence of these observations, many workers have studied the hormonal response of the aleurone tissue and a considerable body of data has become available.

In addition to its role in producing and exporting hydrolases which degrade the starchy endosperm reserves, the aleurone tissue also contains protein and glycoprotein reserves in its own right, as well as appreciable reserves of lipid and most of the seed's reserves of minerals in the

336

form of phytin. The mobilisation of these reserves during germination is also controlled by hormonal factors some of which emanate from the embryo. In some cases gibberellin appears to be the responsible factor, but in other cases the responsible factors have not yet been characterised. Thus, several factors are apparently involved and the control of aleurone tissue function by the embryo is many-faceted. Several recent reviews have been published covering various aspects of the subject (Akazawa and Hara-Nishimura, 1985; Bewley and Black, 1978; Laidman 1982) and the reader is referred to them for their detailed accounts and bibliography. Within the space of this paper we will restrict ourselves to an overview and attempt to delineate the present state of knowledge on the subject.

HYDROLASE SYNTHESIS AND SECRETION

One important aspect of the response of the aleurone tissue to gibberellin undoubtedly lies in a redirection of genetic expression, and a large amount of experimental data is available to describe the system. Briefly, the induced transcriptional activity is manifest as the synthesis of a set of mRNAs including those for alpha-amylase and protease. The new mRNAs, isolated as an enriched Poly(A) RNA fraction, have been shown to function in cell-free translation systems where they direct the synthesis of alpha-amylase and protease. In the case of alpha-amylase, the mRNA has been reverse-transcribed and the product DNA was shown to be complementary to the alpha-amylase RNA by hybridisation. Induction of the synthesis of other species of RNA is apparently not involved in the gibberellin response. Inhibitors of RNA and protein synthesis prevent the gibberellin response in vivo. Also the cereal alpha-amylase gene has been cloned into yeast where it is actively expressed.
Convincing evidence that the newly-formed hydrolases are synthesised de novo from amino acids in the aleurone tissue has been obtained from labelling experiments using radioactive and heavy isotope-labelled precursors. The newly synthesised enzymes are rapidly secreted from the aleurone cell and accumulate in the starchy endosperm. Available evidence from inhibitor studies suggests that the secretion is also under gibberellin control but independent of the synthesis, inferring that gibberellin action involves the induced synthesis of additional functional proteins which are involved in the secretory process. This view is supported by the existence of some enzymes (eg. beta-1,3-

glucanase), the secretion but not the synthesis of which is induced by gibberellin.

Topographically, in the aleurone cell, the alpha-amylase is synthesised at the rough endoplasmic reticulum where it is vectorially transported into the cisternal lumen. Following proteolysis of its leader sequence it is transported via transport vesicles and possibly the Golgi complex to the plasmamembrane where it is expelled from the cell. Unlike the alpha-amylase of the scutellum, the aleurone tissue enzyme is probably not glycosylated, and the involvement of the Golgi complex in its processing is therefore speculative. Relatively little information is available regarding the intracellular processing and transport of the other hydrolytic enzymes the synthesis and secretion of which is controlled by gibberellin. As stated above, the exocytotic mechanism for the hydrolases appears to be under individual control by gibberellin.

Thus, gibberellin appears to control amylolysis and proteolysis in the starchy endosperm by directing transcription and translation in the aleurone tissue leading to the synthesis and secretion of the appropriate hydrolases.

The hormone abscisic acid inhibits the gibberellin-induced synthesis of alpha-amylase and it may be important in maintaining dormancy in cereals. For several years it was believed that abscisic acid acted by directly preventing either the transcriptional or the translational step of alpha-amylase induction. More recently, however, it has been found that abscisic acid induces the synthesis of new protein in aleurone tissue, and it has therefore been proposed that these proteins are responsible for the inhibition of the gibberellin response or of alpha-amylase activity itself. The mode of action of abscisic acid thus remains an open question.

MOBILISATION OF PROTEIN AND PHYTIN RESERVES IN THE ALEURONE TISSUE

The aleurone tissue contains appreciable reserves of protein and glycoprotein and almost all of the seed's reserves of phytin which are deposited within aleurone bodies. Mechanisms for the control of the hydrolysis of these reserves have not been extensively studied and they are only vaguely understood. Both protease and phytase are present in the quiescent tissue and they become active following the imbibition of water early during germination. Marked increases in the levels of hydrolysis products (amino acids, mineral ions) are observed following imbibition. Moreover, this increase occurs in embryoectomised as well as in whole seeds. It therefore appears that the hydrolysis of these reserves in the aleurone tissue is not initiated by gibberellin or by any other hormone emanating from the embryo. There is, in fact, a two-fold increase in phytase activity in the aleurone tissue of wheat during germination and this increase is dependent upon the presence of the embryo. The active principal is not known however (it apparently is not gibberellin), and the additional phytase activity appears to be superfluous to requirement.

If the control of the hydrolysis of reserves in the aleurone tissue is still poorly appreciated, regulation of the release of their hydrolysis products is more clearly understood. The hydrolysis products of phytin (phosphate, potassium, magnesium, calcium) are released rapidly from aleurone tissue isolated from quiescent wheat seeds when the tissue is incubated in an aqueous bathing medium. By the end of the first day of germination, the rate of release decreases to less than half of the initial value. After the first day of germination, the rate of release increases again reaching the quiescent seed value by the fourth day of germination. Removal of the embryo does not prevent the decrease in release capacity observed during the first 24 hours, but it abolishes the subsequent increase. This suggests that the increase might be mediated by a factor emanating from the embryo. (Embryo diffusates when added to incubating embryoectomised seeds can induce a similar increase in the rate of ion release). The decrease in ion release capacity during the first 24 hours is abolished if the starchy endosperm is removed from the aleurone tissue, suggesting that a regulatory factor from the former tissue is involved in the response.

The factor emanating from the embryo can be replaced by exogenous gibberellin. Auxin enhances and absicisic acid inhibits the action of the gibberellin. In wheat at least, studies using inhibitors of RNA and protein synthesis indicate that transcriptional and translational control are not an essential part of this gibberellin response, in contrast to the gibberellin induction of hydrolase synthesis and secretion. The factor from the starchy endosperm, which decreases the release of mineral ions during the first 24 hours of germination, has been studied. The requirement for the starchy endosperm can be replaced by extracts of the starchy endosperm and partially by the application of cytokinins. Inhibitor studies have indicated that the cytokinin-induced response is dependent upon protein synthesis, but transcriptional activity may not be necessary. Although a cytokinin has been reported to be present in the starchy endosperm of barley, it remains to be proven that this is the active hormone controlling ion release from the aleurone tissue in vivo.

The control of amino acid release from the aleurone tissue is generally similar to that described above for the mineral ions. A detailed study of the efflux patterns of the amino acids from isolated aleurone tissue into aqueous incubation media has revealed some additional features however. There are at least two pools of these compounds within the aleurone cell, and the control of each pool size appears to be somewhat different. One pool has an amino acid composition similar to that of the storage protein in the aleurone grain and it is presumably derived from that storage body. This pool, which increases several-fold in size during germination, is released from the cell when the aleurone tissue is isolated and incubated in an aqueous medium. It presumably represents the hydrolysis products of the storage protein en route to secretion from the cell. The increase in the size of the pool during germination is dependent upon presence of the embryo, and exogenously applied gibberellin is able to substitute for the embryo in embryoectomised seeds. A second pool of amino acids is retained within the aleurone tissue when it is excised from the germinating seed and incubated in an aqueous medium. This pool has a significantly different composition characterised by high levels of tryptophan and glycine betaine. The size of the pool increases about three-fold during germination, but this increase is independent of the presence of the embryo. Unexpectedly, incubation of embryoectomised seeds with gibberellin lead to a sharp decline in the pool size. A situation thus exists in the control of amino acid pool sizes in which gibberellin does not accurately mimic the embryo in the responses which they elicit in the aleurone tissue. It has been proposed that factors other than gibberellin, and which emanate from the embryo, are involved in this regulatory function. The possible factor(s) have not been identified. Also, the likely interaction of abscisic acid with these gibberellin responses has not been studied.

MEMBRANE MORPHOGENESIS AND PHOSPHOLIPID METABOLISM

The hormonal control of membrane morphogenesis and metabolism in the cereal aleurone tissue has been studied by several groups of workers and a fairly clear if incomplete

several groups of workers and a fairly clear if incomplete
picture is emerging. During the first day of germination,
in wheat at least, there is a large net synthesis of phospholipid and formation of new membrane. Initiation of both
activities is independent of hormonal control from the
embryo, depending only upon the hydration of the tissue.
The synthesis of the new phospholipid occurs in the aleurone
grain-oleosome complex and possibly also on free oleosomes.
It has been proposed that the triacylglycerol reserves in
these structures provide the substrate, thus avoiding the
need for the de novo synthesis of phospholipid. The newly-
formed membrane is observed as prolamellar bodies which are
later converted to endoplasmic reticulum. Although, the
available evidence is incomplete, it appears that the conversion is also independent of the presence of the embryo.
Gibberellin action may, however, lead to an acceleration of
the process. Recent data from our own laboratory shows an
unusually rapid turnover of phosphatidyl choline during this
early net synthesis of phospholipid, and the turnover can be
regulated by addition of exogenous gibberellin (see pp.
370-376).

After 24 hours of germination, when hydrolase synthesis
and secretion commences in the wheat aleurone tissue, the
endoplasmic reticulum cisternae become distended, fenestrat-
ed and form transport vesicles which have their function in
the intracellular transport and secretion of the newly synthesised hydrolases. As described above, this endomembrane
process is under control from the embryo, and gibberellin
can substitute for the embryo in incubating embryoectomised
seeds. The action of the gibberellin also leads to decreased phospholipid levels occuring at the same time as the
hormone-induced vesiculation of the endoplasmic reticulum.
Because this gibberellin response includes the stimulation
of phosphatidyl inositol turnover, and because gibberellin
also induces changes in membrane permeability to calcium
while calcium is necessary for the induction of alpha-
amylase synthesis/secretion in the tissue, it has been proposed that phosphatidyl inositol turnover linked to calcium
gating may constitute a hormone signal-transducing mechanism
in cereal aleurone tissue (cf. agonist-induced secretary
activities in animal tissues). The response in the aleurone
tissue is very slow compared with that in animal tissues,
however, and further work is needed to confirm or reject the
proposal. Another note of caution is also necessary. The
decrease in phospholipid levels, which is induced in embryoectomised seeds by gibberellin, does not occur in the aleurone tissue of normal germinating seeds. Thus, the response

elicited by the embryo and that induced by gibberellin are
not exactly the same in this respect. This has led to the
proposal that factors other than gibberellin, or in addition
to gibberellin, are involved in the control of aleurone tissue metabolism by the embryo during germination.

The possible role of absicisic acid in controlling membrane metabolism in the aleurone tissue has not been studied
to any meaningful extent.

MOBILISATION OF LIPID RESERVES

The control of lipolysis and fatty acid catabolism in
germinating cereals has not been extensively studied and it
remains poorly understood. Most studies have been carried
out on wheat, but the available evidence from barley suggests that it is similar.

In the aleurone tissue, mobilisation of the triacylglycerol reserves begins within the first day of germination
and continues for several days. It correlates with a
decrease in the number of oleosomes. Removal of the embryo
has profound effects on this process. Thus, in the aleurone
tissue of incubating embryoectomised seeds about 20% of the
total reserve is broken down within 24 hours. Then no further breakdown is observed. Removal of the starchy endosperm from the aleurone tissue abolishes the mobilisation of
triacylglycerols entirely. This suggests that two pools of
triacylglycerol are present within the aleurone tissue. One
is mobilised in response to a factor emanating from the
starchy endosperm, while the other is mobilised in response
to a factor from the embryo. The endosperm factor can be
replaced by exogenously applied cytokinins (cf. the control
of mineral ion efflux), but the pathways by which this triacylglycerol mobilisation proceeds has not been characterized. The embryo-controlled mobilisation, on the other
hand, corresponds to the appearance of neutral lipase activity in the tissue. In contrast to its ability to induce
alpha-amylase activity for carbohydrate mobilisation, gibberellin is not able to replace the embryo in inducing the
neutral lipase or its associated triacylglycerol mobilisation. Both lipase activity and the lipolysis can, however
be induced in embryoectomised seeds by certain nitrogenous
compounds, such as glutamine, in the presence of indole
acetic acid. Experiments using metabolic inhibitors have
suggested that the induction of lipase activity by the
glutamine and indole acetic acid is dependent upon both RNA
and protein synthesis. Although these in vitro effects of

glutamine and indole acetic acid are very interesting, there is no evidence available at present to support the idea that they might be the active agents in vivo.

The appearance of the embryo-induced lipase activity in the aleurone tissue coincides with the appearance of enzymes for fatty acid metabolism in the form of the beta-oxidation pathway and the glyoxylate cycle. Both of these metabolic activities are confined to glyoxysomes in the aleurone tissue, and ultrastructural studies have indicated that the limiting membrane of the glyoxysome is formed by differentiation from the endoplasmic reticulum during germination. Furthermore, both the formation of the glyoxysomes themselves are controlled by the embryo. In contrast to the situation with lipase, gibberellin can replace the embryo in this role. Thus, a fascinating situation is presented in which gibberellin integrates the synthesis and secretion of new protein (alpha-amylase, protease etc.) with production of the necessary energy from fatty acid oxidation.

RECENT DOUBTS ABOUT THE ROLE OF GIBBERELLINS

In the above account we have described gibberellin-induced responses in the aleurone tissue of embryoectomised wheat seeds which are not the same as the responses elicited by the embryo during germination. These findings from our own laboratory have lead us to suggest that hormones other than, or in addition to, gibberellin might be involved in these responses. More recently, Akazawa and Weiler and their colleaques have presented direct challenges to the traditional view about the role of gibberellin in germinating cereals. In particular, Weiler and colleagues have presented experimental evidence which indicates that alpha-amylase synthesis in the aleurone tissue might be induced by gibberellin produced in situ rather than by gibberellin from the embryo. If this is so, the question must be asked, "How does the embryo induce the many-faceted response in the aleurone tissue, and what are the chemical factors involved?" If the proposals of Weiler and Akazawa are wrong, then one must ask, " Is the gibberellin produced in the aleurone tissue comparted in such a way that it is not biologically active?" These questions, together with the fact that a gibberellin receptor has not yet been identified in the aleurone tissue, make it clear that the study of hormone responses in germinating cereals is far from exhausted.

344

ACKNOWLEDGEMENTS:

We are indebted to the Science Research Council for the award of a research studentship to C.A.B. and to The Association of Commonwealth Universities for a studentship to D.N.V.

REFERENCES

Akazawa, T. and Hara-Nishimura, L. 1985. Ann. Rev. Plant Physiol. 36, pp. 441-472.
Bewley, J. D. and Black, M. 1978. In: Physiology and Biochemistry of Seeds in Relation to Germination, Vol. 1. Springer-Verlag, Berlin-Heidelberg-New York.
Laidman, D. L. 1982. In: The Physiology, and Biochemistry of Seed Development, Dormancy and germination. Khan, A. A. (ed.). Elsevier Biomedical Press, Amsterdam.

Transfer of Information within the Germinating Grain

T. Obata, Division of Bio-function Research,
Biomedical Research Laboratories, The Jikei
University School of Medicine, Tokyo 105,
Japan.

The seed germination processes of cereal grains have
clearly divided functional differentiation. In particular,
the system that supplies nutrition for germination is
anatomically divided from the embryo, even though its func-
tioning is controlled by information from the embryo (Armstrong and Jones, 1973). Within this "nutrition supply
system," the mechanism that regulates the process by which
gibberellic acid induces the secretion of hydrolase within
aleurone cells is still not well understood. This problem
will be the subject of my presentation.

It is well known that the function of gibberellic acid
is to aid the supply of nutritional resources, such as mono-
saccharide, amino acids, inorganic phosphate, or metallic
ions, to the germinating embryo by means of inducing hydrolases in the aleurone cell that will catalyze the stock
resources found within the aleurone cells and starchy endosperm of the grain. This effects short and easily detected
processes. Therefore, it is relatively easy to study the
mode of action of this hormone in living cells.

In the past 26 years since Yomo and Paleg each independently reported on alpha-amylase induction by gibberellic
acid in cereal grain (Yomo, 1960; Paleg, 1960), many
scientists have studied and clarified the mode of action of
gibberellic acid in the aleurone cell. In short, gibberellic acid elicits the translation of alpha-amylase mRNA from
the DNA (Muthukrishnan et al., 1979), and induces the synthesis of alpha-amylase protein. Although gibberellic acid
not only affects but is also necessary for the enzyme secretion processes in the aleurone cell (Chrispeels and Varner,
1967), the study of the enzyme secretion process and its
regulating systems has not progressed.

I think there are two main reasons for this. One is

the difficulty of obtaining the secretory organelles from an aleurone cell into which gibberellic acid has been induced (Gibson and Paleg, 1972; Jones, 1972; Jones, 1980). The other is the difficulty of distinguishing between the enzyme synthesis process and the secretion process in such a way that specific secretion processes can be identified apart from and even after the beginning of enzyme synthesis.

My colleagues and I have studied the gibberellic acid induced-enzyme sucretion process in the barley aleurone cell and have identified a controlled condition by which the enzyme secretion phase is separated from the enzyme synthesis phase of the gibberellic acid-induction of hydrolase (Obata and Suzuki (1976).

It was well known that in the barley aleurone cell, gibberellic acid serves to induce alpha-amylase production and secretion. However, without the presence of gibberellic acid these cells could neither produce nor secrete alpha-amylase (FIGURE 1A). A dose of 1ppm gibberellic acid was

FIGURE 1. Imbibition time course of changes in the intra- (o) and extracellular (•) enzyme activities of barley aleurone layers in the presence (——) or absence (---) of 1ppm gibberellic acid (GA).

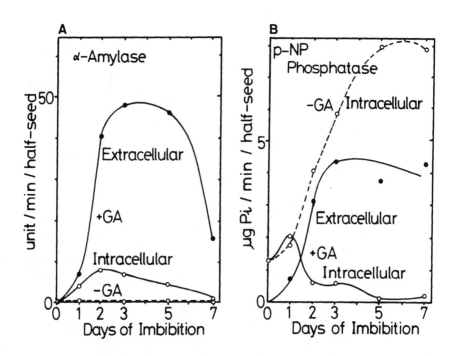

sufficient to induce alpha-amylase production and secretion
into the culture medium. But without any addition of gibberellic acid, the aleurone cells could not produce alpha-
amylase until after 7 days of imbibition.

During this imbibition, although non-specific acid
phosphatase activity increased within the cells even without
any addition of gibberellic acid, the phosphatase was not
secreted into the medium. Moreover, the phosphatase was
secreted into the culture medium only after an addition of
gibberellic acid (FIGURE 1B). (Note, for this non-specific
acid phosphatase, we used p-nitrophenyl phosphate for substrate at pH 5.0, which we term p-NP phosphatase.) We also
tried using other acid phosphatases under the same conditions. For example, we tried using phytic acid, ATP, alpha-
glycerophosphate, alpha-napthylphosphate as substrates,
all at pH 5.0. These other acid phosphatases all gave the
same results as p-NP phosphatase.

This showed us that in barley aleurone cells, non-
specific acid phosphatase activity can increase even without gibberellic acid. Only its secretion is dependent upon
the effect of gibberellic acid.

Next we studied the effect of gibberellic acid on
phosphatase secretion under various time durations. For
example, various imbibition durations result in different
amounts of enzyme accumulation in the cell. We used aleurone cells that had been imbibed for periods of 2, 4, and
7 days.

In these experiments, the amount of time required for
gibberellic-acid-induced p-NP phosphatase secretion to occur
increased in direct proportion to the degree by which the
imbibition period was extended (FIGURE 2A).

Although alpha-amylase production began simultaneously
in all these cells, the start of the gibberellic-acid-
induced alpha-amylase secretion was delayed just as in the
case of the p-Np phosphatase secretion mentioned above.
Particularly significant was the fact that although the
cells that were incubated for 7 days contained sufficient
amounts of enzyme activity, the alpha-amylase secretion did
not occur until almost 24 hours after the induction of gibberellic acid (FIGURE 2B).

The longer the period of imbibition, the later the
hydrolase secretion started. This seems to be a common
phenomenon of gibberellic-acid-induced enzyme secretion from
the aleurone cell. We considered the possibility that this
retardation or delay might be the result of an accumulation
of the enzyme in the periplasmic space on the aleurone
cell's surface. We tested this hypothesis by using an acid-

FIGURE 2. Incubation time course of changes in the intra-
(o) and extracellular (●) enzyme activities of barley aleurone layers preincubated in water for 2, 4, and 7 days in
the presence of 1ppm GA.

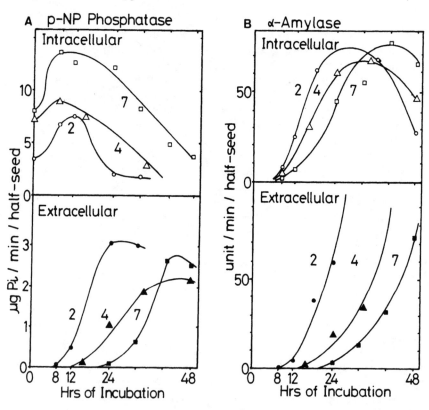

wash treatment, and calculating the amount of enzyme activity within the periplasmic space, cytoplasm, and culture
medium.

The amount of enzyme activity within each of these
three areas is revealed in FIGURE 3. It clearly shows that
periplasmic space has no relationship to the time span
required before the onset of enzyme secretion.

In order to find a possible structural mechanism for
this enzyme secretion delay, we made ultrastructural observations of the aleurone cell (Obata, 1979). Electron-
microscope-observations studies have been done since 1964
(Paleg and Hyde), and it is well known that endoplasmic
reticulum appears during the lag phase of gibberellic acid
induction.

Using aleurone cells that had been imbibed for periods

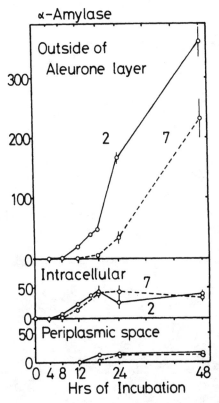

α-Amylase

Outside of Aleurone layer

300

200

100

0

2   7

Intracellular   7

50

0   2

50 Periplasmic space

0
0 4 8 12   24   48

Hrs of Incubation

FIGURE 3. Incubation time course of changes in alpha-amylase activity of barley aleurone layers preincubated in water for 2, 7 days. The activity in the periplasmic space was estimated the deduction with the acid-washed treatment (0.01N HCl, 5', 30°C) from the activity of whole and non-treated aleurone layer extract.

of 2 and 7 days, we made continous observations of the gibberellic-acid-induction culture for 60 hours.

In the aleurone cells that had been imbibed for 2 days, large size, condensed aleurone grains were observed. Moreover, these grains were surrounded by many spherosomes (PLATE 1A).

After 8 hours of gibberellic acid induction, some rough endoplasmic reticulum began to appear, and spherosomes were dispersed into the cytoplasm, so that the aleurone grains took on the appearance of an electron transparency.

After 24 hours of gibberellic acid induction, large, stacked, rough endoplasmic reticulum was observed. This indicated the occurence of active protein synthesis in the cell. In addition, the following phenomena were also observed. The cell nuclei became amoeboid in shape, the number of mitochondria increased, and the vacuole became swollen (PLATE 1B).

After 48hrs, large central vacuoles occupied a large amount of cell volume. Also some stacked, rough endoplasmic reticulum could still be observed.

In contrast to these, in the aleurone cells that had been imbibed for 7 days, the gibberellic acid induced secretion was delayed 24hrs.

We noted that the development of rough endoplasmic reticulum also was delayed. In particular, stacked, rough, endoplasmic reticulum were not observed until 24hrs after

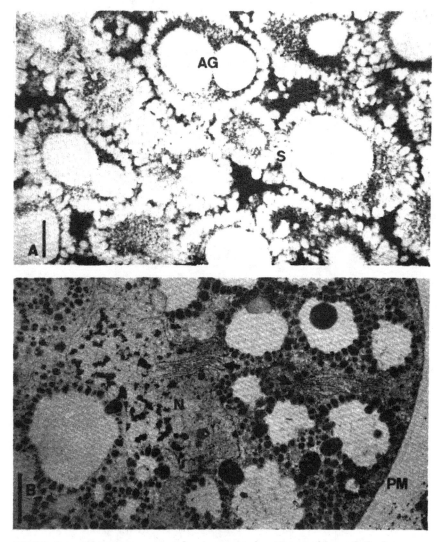

PLATE 1. Ultrastructural changes in the cells of 2 days pre-cultured half-seeds during treatment with gibberellic acid (GA). Scale = 1μ.

  A. The cell before GA-treatment. Aleurone grain (AG) sur-
      rounded by spherosome (S).
  B. After 24hr GA-treatment. Note the dynamic membrane-
      flow of stacked rER (arrow) from amoeboid-shape nuc-
      leolus (N) to plasmamembrane (PM).
All samples were fixed by OsO4 after glutaraldehyde pre-
fixation.

PLATE 2. Ultrastructural changes in the cells of 2 days pre-cultured half-seeds during treatment with gibberellic acid. Scale = 1μ.

A. After 24hr GA-treatment. Nucleolus became an amoeboid shape and rER became stacked.
B. After 36hr GA-treatment. Rough ER wound round the mitochondria (M) and sometimes beside the nucleolus.
C. After 24hr GA-treatment. Rough ER became stacked.
D. After 24hr GA-treatment. Mitochondria are clustered beside the stacked rER stream and this stacked rER stream seems to be directed towards the plasmamembrane.

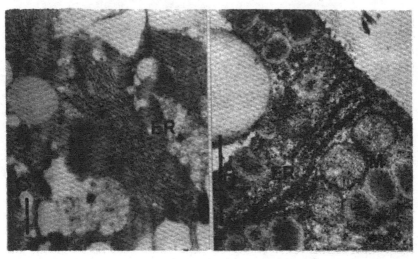

the induction of gibberellic acid.

48hrs after the induction, the gibberellic-acid-induced enzyme secretion, itself, started, and we could easily identify the stacked, rough endoplasmic reticulum.

These observations indicate that a characteristic structure of the enzyme-secretion phase is the formation of and stacking of rough endoplasmic reticulum. And they also indicate that these subcellular organelle structures are related to the enzyme secretion process in the aleurone cell.

During the active enzyme-secretion phase of an aleurone cell, the nucleolus is transformed from its usual round-shape into an amoeboid-like structure of unstable shape (PLATE 2A).

During this phase, usually the nucleolus is surrounded by the stacked rER and the mitochondria. This suggests to us that this transformation is closely related to the formation or function of the stacked rER. Moreover, the number of mitochondria increased. And after a short time, they either surround the nucleolus and stacked rER or they are clustered beside the stacked rER stream (PLATE 2B). This suggests that the stacked rER structures need something that is supplied by the mitochondria.

The stacked rER shown in this PLATE 2C are one of the most characteristic subcellular structures seen during the active enzyme-secretion phase of the aleurone cell. This stacked rER structure always binds ribosomes on its surface. Furthermore, this stacked rER structure seems to have the function of transporting enzymes from the nucleolus directly to the plasmamembrane even while the enzyme proteins themselves are being synthesized (PLATE 2D). This is to say that the enzyme proteins do not appear to be first synthesized and then vesiculated before being transmitted to the plasmamembrane as is normally the case in other plant cells.

Attaining these results, we next tried to speculate on what could be the regulation mechanism of this enzyme secretion process. We hypothesized that the formation of this highly developed rER network may be controlled by a cytoskeleton. In other plants, it is known that it is the intercisternal element network that regulates both the development of the Golgi vesicles (Mollenhauer and Morré, 1975) and the direction of the secretory-vesicles's movement (Mollenhauer and Morré, 1976). Therefore in the aleurone cell, as well, it might be a cytoskeleton system which regulates the formation of stacked rER. If so, then Ca-binding protein, in turn, might regulate this system.

To test these hypotheses, we tried affinity-column

chromatography, using a preparation of the Ca-binding protein, calmodulin (Charbonneau and Cormier, 1979).

A psychoactive phenothiazine drug, fluphenazine, was
coupled to the Sepharose. On this affinity gel, the calmodulin will bind to the Sepharose in the presence of calcium ions, but will release from the Sepharose upon the
removal of calcium ions. The tissue (30 grams of barley
aleurone layers that were collected from 2000 half-seeds)
was homogenized with 1mM EGTA and 10mM PIPES buffer, at pH
7.0. Then the homogenate was centrifuged, and the supernatant fluid was adjusted to a concentration of 2mM by the
addition of $CaCl_2$, and applied to the fluphenazine-Sepharose
column. Next the column was washed with PIPES buffer containing 1mM $CaCl_2$. Also it was eluted in two steps, first
with 10mM PIPES containing 1mM $CaCl_2$ and 0.5M NaCl and then
with 50mN Tris-HCl, at pH 8.0, containing 0.5M NaCl and 10mM
EGTA.

The resulting protein elution profile shows three peaks
(FIGURE 4). The first peak includes a non bound protein
fraction, the second peak consists of three components.

FIGURE 4. Elution profile of calmodulin-affinity chromato-
graphy. Details of conditions are described in the test.
[PIPES : Piperazine-N,N'-bis 2-ethanesulfonic acid, Tris :
Tris (hydroxymethyl) aminomethane]

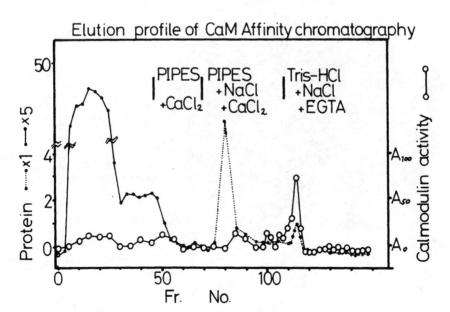

354

However, the binding of these three proteins was not affected by the presence of calcium ions. These proteins might
constitute what could be termed a group of "natural cal-
modulin-binding proteins". Finally, the third peak consists
of a rather small protein fraction, but this is the only
peak that shows any calmodulin activity.

The proteins in this fraction, the ones from the third
peak, were separated and tested for calmodulin by means of
alkaline gel-electrophoresis.

In this test, the control sample is purified calmodulin
prepared from rat's brain according to Yazawa's method
(Yazawa et al., 1980). Calmodulin activity was assayed by
using the activation rate of calmodulin-dependent Ca-Mg
ATPase in erythrocyte membrane (Gopinath and Vincentzi,
1977). One of the criteria for testing calmodulin is the
speed of the alkaline-gel electrophoresis in the presence
or absence of Ca ions. Normally, the rat calmodulin tests
faster in the absence of Ca ions. The calmodulin from the
aleurone cells also shows differences of electrophoretic
distance depending upon the presence or absence of Ca ions,
as does the rat calmodulin (FIGURE 5).

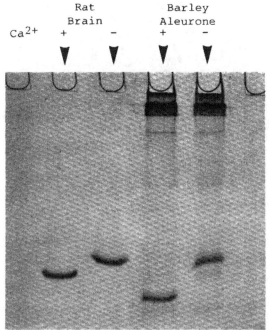

FIGURE 5. Electro-
phoresis of various
calmodulins on 15%
polyacrylamide gel at
pH 8.3 in the presence
(5mN $CaCl_2$) or absence
(10mM EGTA) of calcium
ions.

But the aleurone calmodulin differs from the rat calmodulin is that, when there are no Ca ions present, it shows greater differences in distance than does the rat calmodulin. This may be due to some difference between the protein structures or the amino acid components of the two types of calmodulin.

As part of our research, we also estimated the molecular weight of aleurone camodulin by SDS-polyacrylamide gel electrophoresis. It is in the range of 13,500 to 16,000. Therefore, aleurone calmodulin is smaller than rat calmodulin.

Finally, to further elucidate the role of calmodulin in the aleurone cell, we studied the effect of calmodulin antagonists on the gibberellic acid-induced enzyme secretion processes (Obata, Tamiguchi and Maruyama, 1983) (FIGURE 6).

FIGURE 6. Pulse inhibition by chlorpromazine (CPZ) on GA induced p-NP phosphatase and alpha-amylase secretion. After 6hr of GA incubation, this induction medium was discarded and replaced by a fresh induction medium, CPZ being introduced into the medium either before or after 6hr of GA incubation. After 24hr of GA incubation, intra- and extracellular enzymes were assayed.

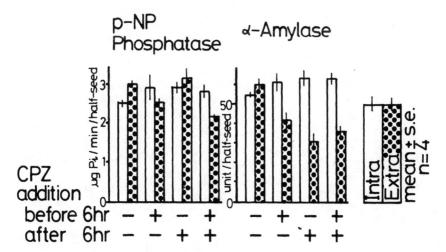

This figure shows the results of a culture-medium-exchange experiment after 6hrs of gibberellic acid induction. This evidence indicates that p-NP phosphatase secretion was inhibited only when the calmodulin antagonist, chlorpromazine, was added before the 6hr induction period.

356

There seems to have been no effect if it was added after the 6hr induction period. In contrast to the p-NP phosphatase, the alpha-amylase secretion was inhibited by the chlorpromazine. In particular, it was more effective if added after the 6hr induction period than before the 6hr period.

We concluded that these two enzymes, the p-NP phosphatase and the alpha-amylase, were secreted by different processes and/or under different conditons. This may indicate the existance of multiple enzyme secretion processes. However, there still remains the possibility that the alpha-amylase secretion process might be regulated by calmodulin action.

To briefly summerize the following main points.

(1) In studying the gibberellic-acid-induced enzyme secretion process and its regulation in the barley aleurone cell, prolonged imbibition revealed that the process of enzyme secretion exists during a separate phase, distinct from that of enzyme synthesis.

(2) An ultrastructural-observation study of these two phases revealed that the secretion-specific structure is most likely the formation of stacked rER.

(3) Calmodulin prepared from barley aleurone cells activates calmodulin dependent Ca-Mg ATPase in rat erythrocytes and also has the same general affects on gel electrophoresis as does vertebrate calmodulin. However, aleurone calmodulin has a lower molecular weight and might also have a different amino-acid composition.

(4) Calmodulin antagonist does inhibit the gibberellic-acid-induced enzyme secretion process. In this inhibition action, it has different effects on the process of alpha-amylase-secretion and on that of p-Np phosphatase-secretion. Of these two, the secretion of alpha-amylase might, at least to some extent, be regulated by the calmodulin action.

ACKNOWLEDGEMENTS

The author wishes to thank Professor Y. Maruyama, The University of Tokyo, Department of Agricultural Chemistry, for his encouragement and helpful guidance during the course of this work. Also the author is grateful to Mr. William Bodiford for his invaluable advice in preparing the manuscript.

REFERENCES

Armstrong, J. E. and Jones, R. L. 1973. J. Cell Biol. 59, pp. 444-455.

Charbonneau, H. and Cormier, M. J. 1979. Biochem. Biophys. Res. Comm. 90, pp. 1039-1047.

Chrispeels, M. J. and Varner, J. E. 1967. Plant Physiol. 42, pp. 1008-1016.

Gibson, R. A. and Paleg, L. G. 1972. Biochem. J. 128, pp. 367-375.

Jones, R. L. 1972. Planta. 103, pp. 95-109.

Jones, R. L. 1980. ibid. 150, pp. 58-87.

Gopinath, R. and Vincentzi, F. F. 1977. Biochem. Biophys. Res. Comm. 77, pp. 1203-1209.

Mollenhauer, H. H. and Morré, D. J. 1975. J. Cell Sci. 19, 231-237.

Mollenhauer, H. H. and Morré, D. J. 1976. Protoplasma. 87, 39-48.

Muthukrishnan, S., Chandra, G. R. and Maxwell, E. S. 1979. Proc. Natl. Acad. Sci. USA. 76, pp. 6181-6185.

Obata, T. 1979. Ann. Bot. 44, pp. 333-337.

Obata, T. and Suzuki, H. 1976. Plant and Cell Physiol. 17, pp. 63-71.

Obata, T., Taniguchi, H. and Naruyama, Y. 1983. Ann. Bot. 52, pp. 877-883.

Paleg, L. G. 1960. Plant Physiol. 35, pp. 293-299.

Paleg, L. G. and Hyde, B. 1964. Plant Physiol. 39, pp. 673-680.

Yazawa, M., Sakuma, M. and Yagi, K. 1980. J. Biochem. 87, pp. 1313-1320.

Yomo, H. 1960. Hakko Kyokai-shi. 18, pp. 600-602.

Mobilization of Endosperm Reserves during Germination of Wheat

*J. E. Kruger, Grain Research Laboratory,
Canadian Grain Commission, 1404-303 Main
Street, Winnipeg, Manitoba R3C 3G8, Canada.*

An understanding of the mobilization of endosperm reserves during germination is essential to understand how quality is affected by sprouting. In simplistic terms, there are three periods in the germination process which influence the quality of cereals. During the very early onset of germination, there is the breakdown of the biochemical components required for the synthesis of enzymes participating in the later mobilization stages. This undoubtedly occurs in the aleurone layer and it would appear also in the embryo. From a quality point of view, such events must be important in exerting, if not control, at least some influence, on the rate and overall amounts of hydrolytic enzymes that are formed.

The second period of germination which is considered to have a major influence on quality is that in which enzyme levels have increased substantially but no substantial breakdown of reserves has occured. Under field conditions, this effect could come from slight weathering of the overall crop, from a few highly sprouted kernels, or from a combination of both. Probably there is no typical condition and the cultivars grown, the weathering conditions experienced and the grading system will all influence what is actually experienced. In any event, the major quality affecting enzyme is alpha-amylase which, in excess, can lead to a wet sticky loaf of bread during a breadmaking process. Less influential but nevertheless important in certain wheat end-product processes would be proteolytic enzymes.

The third stage in the germination process of major

1 Paper No. 577 of the Canadian Grain Commission, Grain Research Laboratory, 1404-303 Main Street, Winnipeg, Manitoba. R3C 3G8.

importance is the breakdown of the storage reserves which
affect quality. In particular, wheat proteins have unique
visco-elastic properties and their degradation must, of
necessity, cause a deterioration of quality. The relative
importance of direct enzyme effects versus the actual breakdown of reserves will certainly be dependent on the end-
product process. For example, in most western-style breads,
it can be expected that the amount of hydrolytic enzymes
which form during early germination will be sufficient to
cause disastrous effects on quality before any substantial
degradation of reserves has occurred. On the other hand,
there are other wheat end-products, i.e. pasta products such
as spaghetti, where limited moisture would inhibit interaction of enzyme with substrate. In such cases, greater
amount of sprouting may occur before quality is affected and
some deterioration of endosperm reserves could occur.
Finally, there is a whole range of other products such as
flat breads, chapatties, etc., where our knowledge of the
effects of sprouting are incomplete and where quality
deterioration other than oven-spring or stickiness are
important. For many such products, a darkening of the crumb
is experienced. Presumably, this arises due to the
increased formation of oxidases, i.e. polyphenol oxidase,
peroxidase, etc.

Onset of germination

There is a paucity of information and new research on
the control exerted on the synthesis of enzymes required for
early mobilization of reserves. A few preliminary research
findings have been reported at plant physiology meetings.
For example, there has been confirmation of earlier research
that alpha-amino acids for GA_3 induced synthesis of wheat
alpha-amylase are formed from degradation of pre-existing
aleurone proteins (Jofre-Gaifias et al., 1984). The protease responsible for this is unaffected by gibberellic acid
and therefore unlike proteases that are formed later and
participate in germination and in the subsequent degradation
of storage reserves. A similar GA_3-independant endoprotease
was reported present in barley (Rostogi and Oaks, 1984) and
involved in the initial stages of protein hydrolysis.
Further research is essential to fully understand their role
in the synthesis of other hydrolytic enzymes.

Early germination and formation of enzymes
_____

     At present, a number of the major enzymes that form
during germination have been isolated and rigorously characterized. On the other hand, there is less information on
the manner in which they degrade their natural substrates in
vivo and in vitro as well as during typical end-product processing. I would like to describe some of our in vitro
studies on the breakdown of starch granule and their soluble
components, amylose and amylopectin, by the alpha-amylases
present in wheat. We have isolated the group I (green, low
pI) and group III (germinated, high pI) isoenzymes and
studied the breakdown pattern of the two groups by high-
performance gel permeation chromatography (Kruger and Marchylo, 1982). To purify the two groups, we used affinity
chromatography on a column of cyclohepta-amylose-epoxy-
sepharose 6B followed by chromatofocusing. The activities
of the two alpha-amylase groups were balanced and tested
initially for their relative abilities to break down beta-
limit dextrin, amylopectin and amylose. The results indicated that no appreciable differences were apparent in the
action pattern of the two groups upon the three substrates.
If we plotted the distribution of a particular molecular
weight fraction versus reaction time for the two groups,
there was a difference in the mode of breakdown of amylopectin in comparison to beta-limit dextrin. The GI and GIII
groups degraded amylopectin to intermediate and low molecular weight dextrins of less than 6000 while still maintaining a substantial proportion (14-20%) of amylopectin
with a molecular weight greater than the exclusion limits of
the column. By contrast, beta-limit dextrin contained a
much smaller amount of high molecular weight material relative to intermediate and low molecular weight products.
Amylose breakdown again was dissimilar from that of amylopectin and beta-limit dextrin. Whereas the latter substrates formed intermediate molecular products simultaneously with production of low molecular products, amylose
hydrolysis was characterized by the progressive degradation
of a more uniform and symmetrical product that decreased
continually in molecular weight. This suggests that initial
substrate and its earlier degradation products was favored
over cleavage near the ends of the substrate chains.
     The effects of equivalent amounts of the two types of
alpha-amylases also was tested against isolated large and
small starch granules. At 37°C, only the GIII group absorbed onto large and small starch granules. Absorption was
not a prerequisite for starch granule breakdown as indicated

by liberation of reducing sugars by both groups. The GIII group seemed to be more effective than the GI group. Thus, after 6 hours, 61% of the small starch granules were solubilized by the GIII components, compared to 27.9% by the GI components. Absorption of the granules was very temperature dependant as at 2°C both groups were absorbed.

The above results differ from those of Sargeant and Walker (1978) who felt that the GIII group was necessary to solubilize starch granules whereas the GI group broke down the fragments released by the GIII components.

It should be stressed that the above are in vitro experiments and that caution should be used in extrapolating the findings to processes that actually occur in vivo during cereal germination. Different methods of isolation could impart different binding characteristics to the isoenzymes. Furthermore several other factors could cause discrepancies, including the method in which starch granules are prepared, temperature, pH, ratio of enzyme to substrate, etc.

The effects of the two alpha-amylase groups on the amylograph also were determined (Kruger and Marchylo, 1982). It was found that identical decreases occurred with increasing alpha-amylase extracts (FIGURE 1). This was somewhat suprising in that the green alpha-amylase is more thermolabile than the germinated group and should be inactivated earlier in the amylograph cycle thereby being less effective in starch degradation. Evidently, stability is conferred by the high starch to liquid ratio and points out that caution should be used in extrapolating results with purified enzymes in test tube experiments to actual conditions existing in an end-product process. We also found that the two groups had identical effects in increasing the gassing power of a flour (FIGURE 2), confirming again the similarity of their action patterns.

Breakdown of storage reserves

Our recent research in this area has concentrated on the mobilization of storage proteins. Degradation of wheat proteins was followed over a five-day period using the technique of HPLC in the reversed-phase and gel permeation modes. In the reversed-phase mode proteins are separated on a silica based column to which are attached hydrocarbon chains of varying length, typically C-3, C-8 or C-18. The silica spheres are normally 5 to 10μ with large pore size, i.e. 300A° in order to accommodate high MW proteins. The proteins are absorbed onto the column and eluted with an

362

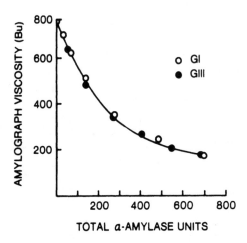

FIGURE 1. Decrease in the amylograph viscosity of wheat flour with increasing addition of GI and GIII alpha-amylases from germinated wheat. One unit of alpha-amylase denotes the amount that hydrolyzes 1mg of maltose/min x 10^{-3} at 37°C from reduced starch (Kruger and Tipples 1981).

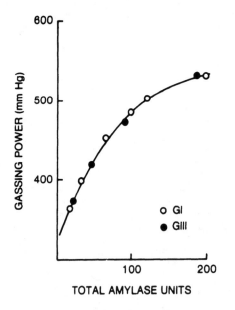

FIGURE 2. Increase in the gassing power of a wheat flour with increasing addition of GI and GIII alpha-amylases from germinated wheat. One unit of alpha-amylase denotes the amount that hydrolyzes 1mg of maltose/min x 10^{-3} at 37°C from reduced starch (Kruger and Tipples 1981).

increasing gradient of acetonitrile. This elutes the proteins in order of increasing surface hydrophobicity. We initially removed the albumins and globulins with salt and then extracted the remaining storage proteins with 50% 1-propanol, 1% acetic acid, 4% dithiothreitol (DTT). Five cultivars were examined (Kruger and Marchylo, 1985). It was found, as expected, that there was considerable intervarietal variation. Comparison of the ungerminated and germinated extracts for a particular variety, however, indicated that the patterns were almost identical. Typical chromatograms of ungerminated and 4-day germinated Neepawa and Glenlea are shown in FIGURE 3. Other stages of germination also

FIGURE 3. RP-HPLC separation of proteins from (A) and (B) 96-hour germinated Neepawa and Glenlea wheat. Run time 105 min. A buffer : acetonitrile:water (15:85) containing 0.1% trifluoracetic acid. B buffer : acetonitrile:water (80:20) containing 0.1° trifluoracetic acid. Gradient 20-58% B. 25ul injection.

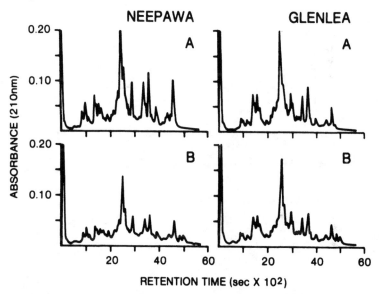

were analyzed and the same phenomena was observed. Of course, relative amounts of the proteins decreased with time. In addition, if we analyzed the water-soluble proteins, there was an increase in amounts with time. What was surprising, of course, is that there was such little variation in the relative proportions of the storage proteins during germination and that no new partially degraded components were evident.

Now we also examined the molecular weight changes in the storage proteins using high-performance gel permeation chromatography (HPGPC). Typical elution profiles for Neepawa and Glenlea are shown in FIGURE 4. In this case, salt-soluble proteins were not removed. In general there was no major observable shift in the molecular weights of the major components with time except for the progressive formation of low molecular weight species. This is more easily observed if we divide the molecular weight profile into six groups: (1) 1 million (excluded); (2) 173-1000 x 10^3; (3) 43-173 x 10^3; (4) 8-43 x 10^3; (5 and 6) 8 x 10^3 (below the lower exclusion limit of the column). Group 6, in fact, is material absorbed to the column. Plotting the areas of

FIGURE 4.   Changes in HPGPC protein elution profiles of
Neepawa and Glenlea wheats upon germination. Column: 30 x
0.75cm Sherogel TSK 3000 (Altex Scientific, Berkley, CA).
Buffer 0.1 M sodium phosphate, PH 7.0 containing 2% sodium
dodecyl sulfates.

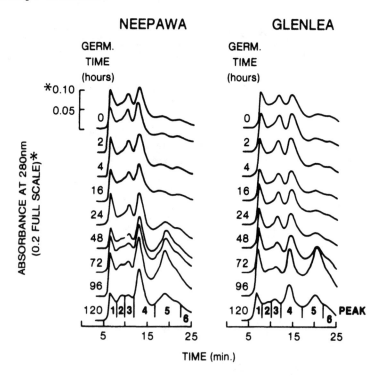

these components versus germination time, as illustrated for
Neepawa and Glenlea wheat in FIGURE 5, revealed very slight
changes in components 1-4 whereas component 5 progressively
increased as one might expect.

How do we account for this behaviour where there is
very little evidence of transient polypeptides of intermediate molecular weight? Well, there are a few likely explana-
tions. First of all, as germination proceeds, there appears
to be a sink adjacent to the scutellum which gradually
increases with germination time and in which metabolic
events such as protein and starch hydrolysis rapidly occur.
Only a small part of the storage proteins are degraded at
any one time as the area of this sink increases. The bulk
of the storage proteins presumably remain untouched and
would give the same pattern as found in the mature seed by
RP-HPLC. Until the wheat is extensively germinated, there
would always be a preponderance of undegraded reserve pro-

FIGURE 5. Changes in component (%) and peak area upon germination for molecular weight components shown for Neepawa and Glenlea wheat in FIGURE 4.

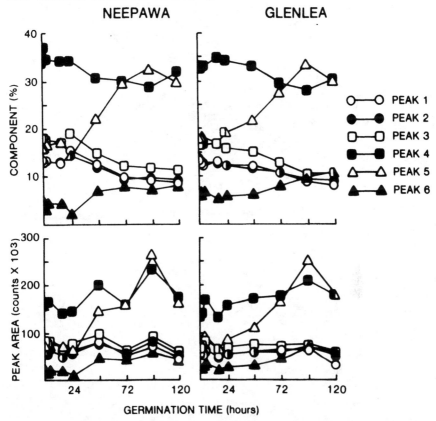

teins relative to transient polypeptides, such that the latter would only be detected with difficulty.

A second reason is suggested by our previous studies on wheat which indicate that hydrolytic products resulting from enzymic degradation of proteins were predominantly amino acids with peptides remaining at the same level (Preston et al., 1978). These studies indicated that the carboxypeptidases in germinated wheat very rapidly hydrolyze wheat endosperm proteins once limited endoproteolysis has taken place (Preston and Kruger, 1976a, b; Kruger and Preston, 1977). This was confirmed in a study in which HPGPC was used to follow the increases in buffer-soluble proteins with time (Kruger, 1984). It was found that the major change in the molecular weight distribution of the buffer-soluble proteins was the formation of low molecular weight species. Results for Glenlea and Wascana wheat are

FIGURE 6. Molecular distribution of buffer-soluble proteins in Glenlea and Wascan ungerminated and germinated wheat. Injection volumes were 100μl for wheat germinated for zero and four days and 50μl for wheat germinated for five and six days. Column: as for FIGURE 4. Buffer 0.05 M sodium phosphate, pH 7.0 containing 0.5 M sodium chloride.

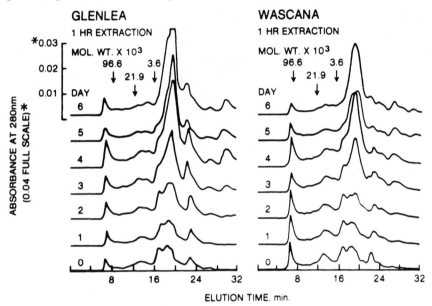

shown in FIGURE 6. Very little differences were found in the amount of intermediate or high molecular weight buffer-soluble material. It is interesting to note that there also were two strongly absorbed components which increased with germination time. Unfortunately, at least three days of germination were required for their production so they would not be suitable markers of early germination. The amount of the low molecular weight products could be used, however, as a very rough guide of the extent of germination as shown in FIGURE 7 for four wheats.

It is obvious that some drastic changes do occur in the quality determinants of wheat during germination as evidenced by pronounced changes in rheological properties. Recently, we have carried out a study on changes in the relative proportion of groups of barley proteins fractionated by different solvents during germination which indicate that subtle differences do occur (Marchylo et al., 1986). In the barley study, we were actually comparing two varieties of different malting quality by analysis of their protein patterns using RP-HPLC.

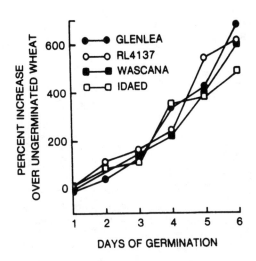

FIGURE 7. Percent increase in the amount of protein eluting between 19 and 22 minutes upon germination.

The proteins were extracted sequentially with 0.5 M salt, 50% 1-propanol, 50% 1-propanol containing dithiothreidol (DTT) and finally urea and DTT. The elution profiles of the propanol-soluble proteins did not change drastically over five days of germination, although cultivar differences in protein patterns were found. They consisted of the C or sulfur poor hordeins. Both the propanol-DTT agent and the urea-DTT solvents extracted D and B hordeins. Quantitative analyses of chromatograms indicated that during germination the hordeins were degraded significantly faster than either the B or C hordeins.

In wheat, it has been suggested that the high molecular weight glutenin proteins (comparable to D hordeins) are responsible for breadmaking quality (Payne et al., 1979). Could it be that, as in barley, these molecular weight components are preferentially degraded and that there may be varietal differences in the rate of breakdown? We are presently investigating this by the analysis of different cultivars using RP-HPLC.

SUMMARY

We are beginning to fit in the jigsaw pieces related to the events occuring during the mobilization of the endosperm reserves but we have a long way to go before the complete picture is known. Some of the problem areas that we should continue to concentrate on are:

(1) What is the role of proteases in the mobilization of proteins required for early synthesis of the enzymes responsible for degradation of the endosperm reserves?
(2) Numerous other enzymes such as disulfide reductases, polyphenol oxidases, etc., increase during germination. What is their role in end-product processing?

(3) For the assorted myriad of products for which wheat is suited, what is the relative importance of direct enzyme effects compared to effects attributable to storage reserve degradation?
(4) What are the subtle effects of endosperm protein modification which affect quality?
(5) What is the best way to ameliorate the disasterous effects of enzymes on quality?

REFERENCES

Jofre-Garfias, A., Garcia-Maya, M. and Hamabata, A. 1984. On the amino acid source of alpha-amylase synthesis induced by GA₃ in wheat aleurone layers. Plant Physiology Supplement. 75, No. 1, abst. 1052.

Kruger, J. E. 1984. Rapid analysis of changes in the molecular weight distribution of buffer soluble proteins during germination of wheat. Cereal Chem. 61, p. 205.

Kruger, J. E. and Marchylo, B. A. 1985. A comparison of the catalysis of starch components by isoenzyme from the two major groups of germinated wheat alpha-amylase. Cereal Chem. 62, p. 11.

Kruger, J. E. and Marchylo, B. A. 1985. Examination of the mobilization of storage proteins of wheat kernels during germination by high-performance reversed-phase gel permeation chromatography. Cereal Chem. 62,

Marchylo, B. A., Kruger, J. E. and Hatcher, D. 1986. Reversed-phase high-performance liquid chromatographic analysis of hordein during malting of two barley varieties of contrasting malting quality. Cereal Chem. 63, (in press).

Payne, P. I., Carfield, K. G. and Blackman, J. A. 1979. Identification of a high-molecular-weight subunit of glutenin whose presence correlates with bread-making quality in wheats of related pedigree. Theor. Appl. Genet. 55, p. 153.

Preston, K. R., Dexter, J. E. and Kruger, J. E. 1978. Relationship of exo- and endoproteolytic activity to storage protein hydrolysis in germinating durum and hard red spring wheat. Cereal Chem. 55, p. 877.

Preston, K. R. and Kruger, J. E. 1976a. Location and activity of proteolytic enzymes in developing wheat kernels. Can. J. Plant Sci. 56, p. 217.

Preston, K. R. and Kruger, J. E. 1976b. Purification and properties of two proteolytic enzymes with carboxypeptidase activity in germinated wheat. Plant Physiol. 58, p. 516.

Preston, K. R. and Kruger, J. E. 1977. Specificity of two
    isolated wheat carboxypeptidases. Phytochem. 16, p.
    525.

Rostogi, V. and Oaks, A. 1981. Hydrolysis of storage pro-
    tein in barley endosperms. Plant Physiology Supplement
    75, No. 1., abst. 174.

Sargeant, G. and Walker, T. S. 1978. Absorption of wheat
    alpha-amylase isoenzymes to wheat starch. Stäerke.
    30, p. 160.

Phospholipid Metabolism and Membrane Assembly in Wheat Aleurone Tissue

C. A. Brearley, D. N. Vakharia, M. C. Wilkinson
and D. L. Laidman, Department of Biochemistry
and Soil Science, University College of North
Wales, Bangor, Gwynedd LL57 2UW, Wales, U.K.

INTRODUCTION

An early event in the tissues of germinating seeds is
the proliferation of intracellular membranes (Mollenhauer et
al., 1978). In wheat aleurone tissue this process, which
occurs during the first 24 hours, is not reflected in an
increase in the activity of the endoplasmic reticulum (ER)
marker enzyme NADH-cytochrome C reductase until after 24
hours (Laidman, 1982) but it is reflected in a 50% increase
in phospholipid levels. In complete contrast to events
which occur after 24 hours, the earlier events do not require the presence of the embryo and they are independent of
gibberellin. Thus, early membrane synthesis in wheat aleurone tissue is not necessarily associated with a functional
ER. The well-documented lag phase leading to alpha-amylase
synthesis and secretion, represents transcriptional and
translational events. But, since secretion is mediated via
the endomembrane system, secretion must await the development of a functional membrane system.

Ultrastructural studies have implicated prolamellar
bodies as intermediates in the biogenesis of ER. These
bodies appear briefly in the tissue associated with protein
bodies and oleosomes early during germination. A direct
physical continuity between the prolamellar bodies and ER
has been observed in other seeds (Mollenhauer et al., 1978).
It has therefore been suggested that oleosomes might be the
source of lipid for membrane biogenesis. In an earlier paper
from this laboratory (Wilkinson et al., 1984), evidence was
presented for a phospholipid-synthesizing site associated
with free oleosomes and the oleosomes of the aleurone grain-
oleosome complex. Further evidence is presented here to
support this interpretation and to illustrate a possible
regulatory role for gibberellin.

370

EXPERIMENTAL

Embryoectomised seeds were incubated for up to 96h at 25° in either distilled water (control) or 4μM gibberellic acid. The aleurone tissue was then isolated, homogenised in ethylene glycol : water (80:20,v/v) and fractionated by differential centrifugation to yield three particulate fractions. These were a 1000xg pellet which contained mostly the aleurone grain-oleosome complexes, a 100,000xg pellet containing most of the lighter structures (glyoxysomes, mitochondria, microsomes), and a fatty surface pad which collected at the top of the 100,000xg supernatant and which consisted of free oleosomes. Marker enzyme studies showed only a small contamination (<10%) of the 1000xg pellet or the surface pad fraction with mitochondrial or microsomal material from the 100,000xg pellet. There was no contamination of the 100,000xg pellet with aleurone grains from the 1000xg pellet. Presumably, many of the free oleosomes had been displaced from the aleurone grain-oleosome complexes during homogenisation and centrifugation, and they should rightly belong to the 1000xg pellet. Total lipid extracts were prepared from the sub-cellular fractions and the extracts were fractionated and assayed for their phospholipid contents. The inclusion of phospholipase inhibitors in the fractionation medium ensured high recoveries of phospholipid from this procedure.

Phospholipid levels in the sub-cellular fractions are presented in FIGURE 1. The surface pad accounted for more than 60% of the total phospholipid recovered. During the first 48 hours of incubation there was a large increase in the phospholipid level in this fraction. Although the level in the 1000kg pellet was lower, it doubled during the first 24 hours. The increase in the 100,000kg pellet was proportionately less. Incubation of the embryoectomised seeds with gibberellin did not have a statistically significant effect on the increases during the first 24 hours. After 24 hours in the control (-GA) tissue, the phospholipid levels in the surface pad and the 100,000xg pellet continued to increase, but that in the 1000xg pellet decreased. Incubation with gibberellin arrested the increase in this fraction. Gibberellin action also led to a faster and more extensive decline in phospholipid level in the 1000xg pellet and a slower increase in the 100,000xg pellet.

In another experiment, the isolated aleurone tissue was radiolabelled for 30min. with 1μCi 10mM [Me^{14}C] choline (saturating uptake concentration). The lipid fraction of

FIGURE 1.   Phospholipid levels in sub-cellular fractions of aleurone tissue.

the tissue was then extracted and its radioactivity was determined.   All of the radioactivity was located in phosphatidyl choline, and the pattern of incorporation of radioactivity into phosphatidyl choline is presented in FIGURE 2.   In the control tissue, incorporation decreased progressively during incubation of the embryoectomised seed up to 96 hours.   Thus, the greatest incorporation occured during and shortly after imbibition.   The gibberellin-treated

FIGURE 2.   Incorporation of [Me -^{14}C] choline into phosphatidyl choline.

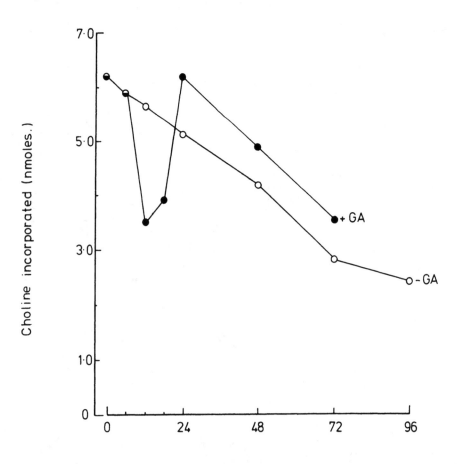

Incubation time (hours.)

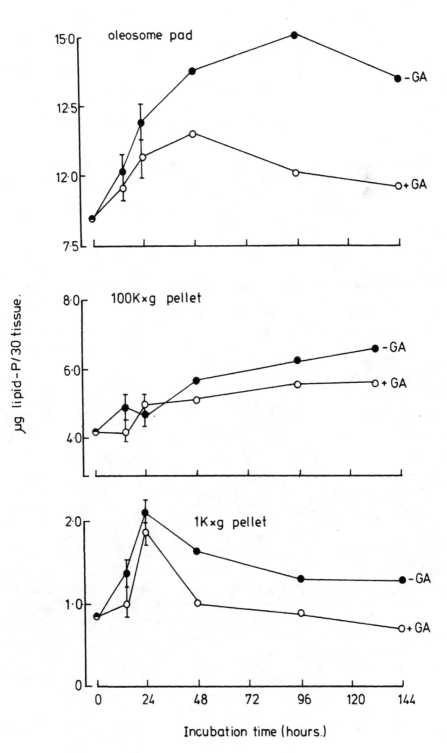

tissue initially incorporated the same amount of choline as the control tissue had done. By 14 hours however, the incorporation was reduced to 3.4 nmoles of choline compared with 5.6 nmoles in the control tissue. By 24 hours the value for the gibberellin-treated tissue had increased again to 6.2 nmoles, compared with 5.1 nmoles for the control, and it remained above the control at least up to 72 hours.

Pulse-chase experiments were carried out to determine the turnover rate of the [Me-^{14}C] choline-labelled phosphatidyl choline. The aleurone tissue for these experiments was taken from embryoectomised seeds which had been incubated for 14 hours either with or without gibberellin. This was the time at which the effect of gibberellin observed in FIGURE 2 was maximal. The isolated tissue was pulse-labelled for 30 minutes with 1μCi [Me-^{14}] choline and then chased for up to 48 hours in a medium containing unlabelled 1mM choline. In the control tissue, the turnover was very fast with a half-life value (T$\frac{1}{2}$) of 7.5 hours, while the gibberellin-treated tissue showed a much reduced turnover rate (T$\frac{1}{2}$ = 30h) (TABLE 1).

TABLE 1. Turnover of [Methyl - ^{14}C] choline labelled phosphatidyl choline

	Turnover half-life value (hr)	
	− GA	+ GA
Total tissue	7.5	30
1000 xg pellet	22	22
100,000 xg pellet	14	23
Oleosome pad	3.5	5.5

The pulse-chase experiments were extended to determine turnover rates in the subcellular fractions. In these experiments, the tissue was pulse-labelled and chased for various periods of time as described above. The tissue was then homogenised and fractionated as described above, except that a fully aqueous medium was used in place of the ethylene glycol : water. Phosphatidyl choline in the surface pad fraction (free oleosomes) showed the fastest turnover (TABLE 1); T$\frac{1}{2}$ for the fraction from the control tissue was 3.5 hours, while that for the fraction from the gibberellin-treated tissue was 5.5 hours. In the 100,000xg pellet

(glyoxysomes, mitochondria, microsomes) $T\frac{1}{2}$ was 14 hours
for the control tissue and 23 hours for the gibberellin-
treated tissue. (Centrifugation of this fraction in a
sucrose gradient has shown that most of the radioactivity
was associated with microsomes). In the 1000xg pellet
(aleurone grain-oleosome complex), $T\frac{1}{2}$ was 22 hours and it
was unaffected by gibberellin action.

Key experiments from the radiotracer experiments described above were repeated using choline labelled in the
methylene carbons, i.e. $[1,2-^{14}C]$ choline. It was found
that phosphatidyl choline labelled with this precursor did
not show the rapid turnover that had been observed using
$[Me-^{14}C]$ choline. Thus, $T\frac{1}{2}$ for phosphatidyl choline labelled with $[1,2-^{14}C]$ choline in the control tissue was greater
than 80 hours. The label from this precursor remained in
phosphatidyl choline throughout the chase period, and it did
not transfer to any other phosphatide such as phosphatidyl
ethanolamine.

DISCUSSION

The data showing phospholipid levels in sub-cellular
fractions confirm our earlier reports (Laidman, 1982;
Wilkinson et al., 1984) that the most rapid net synthesis of
these membrane lipids takes place during the first 24 hours
following the start of imbibition, and that the synthesis
occurs independently of control from the embryo. The synthesis takes place predominantly in the fractions containing
oleosomes and aleurone grain-oleosome complexes where new
membrane is believed to be formed as prolamellar bodies
attached to the oleosomes (Mollenhauer et al., 1978;
Buckhout and Morre, 1982). Our previously reported data
also showed a greatly reduced net synthesis of phospholipids
after 24 hours in aleurone tissue incubated in the absence
of gibberellin (control tissue), while tissue incubated with
gibberellin showed a net loss of phospholipid. The present
data show that this response to gibberellin is reflected at
the sub-cellular level especially in the oleosome fraction.
The data can also be interpreted as illustrating a transfer
of phospholipid from the oleosome - containing fractions
into the fraction containing ER as well as suggesting an
acceleration of the transfer process by gibberellin. Such a
transfer would support the findings of Buckhout and Morre
(1982); their ultrastructural studies indicated a transfer
of membrane material from prolamellar bodies into ER and the
transfer was accelerated following gibberellin application

376

to the tissue. Our data connot be conclusive in this respect however, because they are an admixture of synthetic,
breakdown and transfer processes, so that stoichometric
relationships between fractions cannot be demonstrated.

Earlier radiotracer experiments (Wilkinson et al.,
1984) have demonstrated the rapid labelling from [Me-^{14}C]
choline of the choline moiety of phosphatidyl choline in the
oleosomes during early germination. The present data show
that the labelled moiety is also turned over rapidly.
Gibberellin attenuates the turnover. Moreover, parallel
experiments to compare Me-^{14}C with 1,2-^{14}C-labelled choline
show that the turnover is specifically concerned with the
N-methyl groups of choline. This gibberellin-controlled
turnover appears to be concerned with the process by which
new membrane is formed in the prolamellar bodies attached to
the oleosomes. Since the process occurs during the lag
phase of gibberellin action, before the appearance of
induced alpha-amylase synthesis in the aleurone tissue, it
is tempting to speculate that attenuation of the N-methyl
group exchange is part of the signal-transduction mechanism
in the gibberellin response. Significantly, similar proposals have been put forward for agonist-induced responses
in animal tissues (see e.g. Mato and Alemany, 1983).

ACKNOWLEDGEMENTS

We are indebted to the Science Research Council for the
award of research studentships to C.A.B. and M.C.W., and to
the Association of Commonwealth Universities for a studentship to D.N.V.

REFERENCES

Buckhout, T. J. and Morre, D. J. 1982. Bot. Gaz. 143, pp.
    156-163.
Laidman, D. L. 1982. In: The Physiology and Biochemistry
    of Seed Development, Dormancy and Germination. Khan,
    A. A. (ed.) Elsevier, Amsterdam.
Mato, J. M. and Alemany, S. 1983. Biochem. J. 213, pp.
    1-10.
Mollenhauer, H. H., Morre, D. J. and Jelsema, C. L. 1978.
    Bot. Gaz. 139, 1-10.
Wilkinson, M. C., Laidman, D. L. and Galliard,T. 1984.
    Plant Sci. Lett. 35, pp. 195-199.

Section VI

Grain Dormancy

Control and Development of Dormancy in Cereals

M. Black, J. Butler and M. Hughes,
Department of Biology, Kings College London,
University of London, Campden Hill Road,
Kensington, London W8 7AH, England.

INTRODUCTION

Pre-harvest sprouting in its strict sense (germinative growth on the ear) obviously depends on the germinability of the grain. Several factors influence germinability but one of the most important is the condition of the grain itself with respect to dormancy. Clearly, the severity of sprouting is inversely related to the level of dormancy and if all grains were dormant during their development and maturation the sprouting problem (sensu stricto) would not exist. A full appreciation of the involvement of dormancy requires consideration of the following questions : (a) What is the nature of seed dormancy in cereals? (b) What factors cause dormancy - breakage and how do they operate? (c) What is the pattern of dormancy in developing and maturing grains? These questions will be addressed in this paper, drawing to some extent on our own studies of wheat and barley.

Dormancy

Dormancy is the failure to germinate because of some internal block that prevents the completion of the germination processes ie. the conversion of an embryo into a seedling. For completeness we should add that dormant seeds cannot germinate in the same conditions (eg. water, air, temperature) under which non-dormant seeds do so. Unlike seeds of many species cereal seeds or grains express their dormancy only when the temperature is relatively high (Bewley and Black, 1982; Gosling et al., 1981; Mares, 1984). Seeds of many grasses behave similarly, though a few cases are known, such as the very deeply dormant 'Montana'

strain of <u>Avena fatua</u> (selected in Naylor's laboratory in
Saskatchewan) in which dormancy occurs over a wide temperature range (eg. Sawhney and Naylor, 1979). The temperature
dependency of dormancy shown for wheat and barley in FIGURE
1 also illustrates the plasticity of the phenomenon, in that
grain dormancy often varies with provenance, sometimes in a
more striking fashion than is evident here. It is worth
emphasising that the depressed germination that occurs as
temperatures exceed a certain value (ca 14° in wheat cv.
Sappo) is truly an expression of dormancy and is not an
inevitable effect of temperature on germination, for it does
not take place in grains which have been after-ripened or
prechilled (Mares, 1984; see FIGURE 2).

FIGURE 1. Effect of temperature on germination of wheat and
barley.

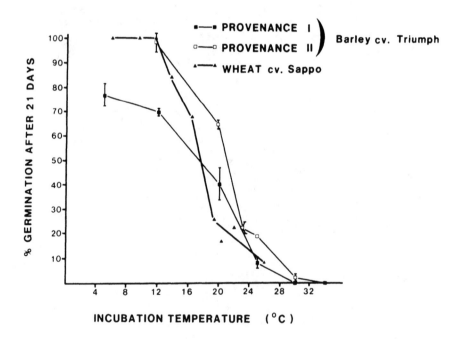

INCUBATION TEMPERATURE (°C)

It is well known that coat-imposed dormancy occurs in
cereals. This is evident because embryos isolated from
dormant grains generally can germinate. Dormancy is also
reduced or abolished if enclosing tissues (eg. glumes) are
removed or if the pericarp and testa are scratched or
pierced (eg. Bewley and Black, 1982). There are some
extreme cases, however, in which these surgical treatments
have no effect because the embryos themselves are dormant

(eg. in the Montana strain of <u>Avena fatua</u>; also see FIGURE 6). The fact that an isolated embryo can germinate does not, however, mean that all constraints on germination have been removed by excision. When rates and not capacities are determined it is seen that embryos taken from dormant grains germinate markedly slower than those from non-dormant grains (FIGURE 2). These results also support the point made above that dormancy is expressed at temperatures which do not necessarily prevent germination. A conclusion which arises out of such studies is that the coat alone is not respons- ible for dormancy. In dormant grains the embryos themselves suffer an internal constraint or inadequacy, which when exacerbated by the presence of the coat makes them unable to germinate.

FIGURE 2. Germination rates at different temperatures of embryos isolated from dormant and non-dormant barley (cv. Doublet). Dormancy was removed by 72h chilling at 5°C.

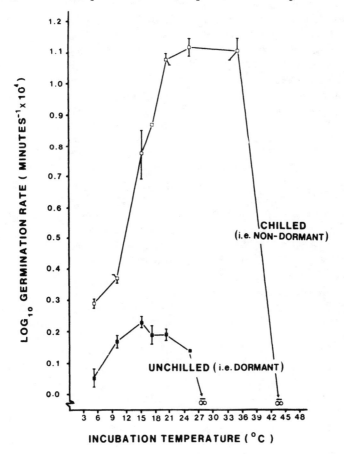

Maintenance of dormancy

Following from the conclusion in the last section it is clear that two sites must be considered when analysing the causes of dormancy, viz. the embryo and the enclosing tissues. We have to identify the physiological and biochemical basis of the inadequacy or debility of the embryo which renders it incapable of overcoming the constraints imposed by the coat. Features which have been investigated are: (a) hormonal effects (b) respiratory metabolism (c) membrane state (d) macromolecular synthesis. There is no evidence for a hormonal basis to the maintenance of dormancy, although as we shall see later, hormones may be involved in dormancy induction. Dormancy does not result from the presence of an inhibitor such as abscisic acid nor from the lack of a promoter such as gibberellin. For example, in our laboratory we have found no significant differences in embryo and endosperm ABA content between dormant and non-dormant mature wheat grains. Similarly, there is no convincing evidence that dormant grains contain less GA than their non-dormant counterparts. One of the most recent detailed investigations on GA status in relation to dormancy, by Metzger (1983) on Avena fatua, found nothing to suggest that non-dormant and dormant grains differ in the levels of GA biosynthesis in the imbibitional and post-imbibitional phases. Respiratory metabolism became implicated in the dormancy mechanism largely as a result of the observations that cyanide and other respiratory inhibitors, high oxygen levels and several electron acceptors such as nitrate and nitrite all break dormancy (see Roberts and Smith, 1977 for detailed information). A long-standing theory that owes its inception to Roberts (1969) is that dormancy is characterised by a relatively low activity of the pentose phosphate pathway (PPP) and that when dormancy is broken by inhibitors of the TCA cycle or terminal oxidation and by electron acceptors there is an accompanying elevated operation of the pathway. As far as cereals are concerned early work was concentrated on barley (Roberts and Smith, 1977) and brought some evidence in support of the theory but this has been criticised on several grounds (see Bewley and Black, 1982 for a full discussion). In Avena fatua, it seems clear that the levels of activity of key enzymes of the PPP, glucose-6-phosphate dehydrogenase and 6-phosphogluconate dehydrogenase are very similar in dormant and non-dormant caryopses (Uphadhaya et al., 1981; Gahan et al., 1986). To date, therefore, there is more evidence against the PPP theory of dormancy than in its favour (Fuerst et al., 1983). The

alternative respiratory pathway (ie. CN-resistant respiration) has recently received much attention, prompted to a
large extent by the studies of Esashi and his colleagues
(Esashi et al., 1979, 1981a). The suggestion here is that
in order to germinate the seed must have the correct balance
of CN-resistant and CN-sensitive respiration and that dormant seeds are deficient in this respect. The concept has
been intensively investigated using cocklebur (Esashi et al.
1981b; Esashi et al., 1982) but there is no evidence to
support its application to cereals. On the contrary, it is
concluded that CN-resistant respiration plays no special
role in the dormancy and germination mechanisms of A. fatua
(Adkins et al., 1984b): and in barley our own work shows
that dormancy-breakage proceeds even when alternative respiration is blocked by inhibitors (data not shown). The
special participation of the TCA cycle in dormancy and germination has been invoked in the case of A. fatua (Adkins,
et al., 1984) but more work is needed to assess its role in
cereals.

Several kinds of evidence have been presented by Hendricks and Taylorson to suggest that the state of cell membranes determines whether a seed is dormant or not (Hendricks and Taylorson, 1979; Bewley and Black, 1982). Much
of the evidence concerns the effects of temperature on dormancy exists only when seeds experience temperatures above
or below a critical value. This is thought to be due to
temperature - dependent membrane transitions. It is striking that dormancy in cereals seems to display some of these
features for, as we can see in FIGURE 1, dormancy begins to
be expressed once a certain temperature is exceeded (approx.
12°C in the examples shown here but sometimes slightly
higher, eg. 16°C - Roberts and Smith, 1977). Another argu-
ment in favour of the membrane theory of dormancy is based
on the action of chemicals (anaesthetics) which can release
seeds from dormancy. Seeds of several grasses respond to
these chemicals (Taylorson, 1980/81) but it has not yet been
shown satisfactorily that their effect is really on membranes. Ethanol, an anaesthetic which breaks dormancy of
Avena fatua is considered to act via effects on respiration
(Adkins et al., 1984c).

Dormant grains may lack the ability to synthesise key
germination enzymes, ie. they may be unable to express certain genes. It is now becoming clear that novel proteins
appear in seeds at about the same time as radicle elongation
commences (eg. Misra and Bewley, 1985) and it may be that if
their synthesis is suppressed in dormant embryos germination
cannot be completed. The methodology is now available to

pursue this rigorously and possibly to characterize genes whose expression is linked with the dormant state.

The likely effects of the tissues enclosing the embryo are to impose a physical restraint and/or to limit gaseous exchange by the embryo. The former is not easy to investigate since in most cereals it is difficult to remove the tissues from around the coleorhiza/radicle while preserving them in an intact state. Hence, direct measurements of resistance cannot readily be carried out. Interference with gas exchange by the embryo is thought by many workers to account for the inhibitory action of the glumes, pericarp and testa. One way in which this might occur is by the enzymic consumption of oxygen by these tissues thus depriving the embryo itself (Côme et al., 1984). Enzymes thought to be involved are polyphenol oxidases, which have been found in cereal caryopses (Kruger, 1976; Gordon, 1980), which act upon various grasses (Lenoir et al., 1983; Gubler and Ashford, 1985; Jayachandran-Nair and Sridhar, 1975; Renard and Capelle, 1976). The temperature dependency of dormancy is thought to be due to the elevated oxygen uptake by the coats, such as the glumes of barley (Lenoir et al., 1983; Côme et al., 1984). It is important to recall, however, that the temperatures which intensify the expression of dormancy in intact grains also depress the rate of germination of embryos isolated from them (FIGURE 2) and so the relatively high temperature cannot act only through effects on oxygen uptake by the enclosing tissues.

A feature worth noting is the variation in level of dormancy among grains, either those from different mother plants or those on the same ear. Such polymorphism or heteroblasty has been observed in several gramineous species including wheat (Datta et al., 1972; Gosling et al., 1981). It is obviously of importance to pre-harvest sprouting but its physiological and biochemical basis is unknown. Properties of the embryo and/or enclosing tissues may be responsible: clarification of the phenomenon may provide important clues as the mechanism of dormancy.

We have so far considered dormancy exclusively in the contest of germination, ie. the ability of the embryo to carry out processes culminating in growth and emergence of the embryonic axis from the grain. There may, however, be other aspects of grain physiology also affected by dormancy, such as the ability of the aleurone layer to produce alpha-amylase in response to gibberellin. This seems to be the case in Avena fatua where aleurone tissue from high-dormant lines are poorly responsive to GA, suggesting that aleurone responsivity has co-evolved with grain dormancy (Upadhyaya, et al., 1982). Whether such a correlation exists in culti-

vated cereals remains to be seen but treatments which produce low-dormant wheat grains (eg. chilling) in our hands
have no effect on the responsiveness of the aleurone cells
to GA (Black and Butler, unpublished). Interestingly, low
temperature alters responsivity of aleurone tissue from
certain dwarf wheats (Singh and Paleg, 1984) but it is
doubtful that this effect is related to dormancy.

Dormancy Breakage

In the context of pre-harvest sprouting the release of
grains from dormancy is obviously important since it may
occur on the ear and lead to germination. For the purposes
of this discussion we will exclude breakage brought about
by various chemicals (although an understanding of how these
operate may give important clues to the germination mechanism) and concentrate on afterripening and chilling. The
former is a time - and temperature - dependent process that
takes place in 'dry' grains (eg. at 15% water content) while
the latter occurs only in grains at above 25% water content
(eg. Mares, 1984). Since dormancy itself depends on the
condition of the embryo and the enclosing tissues the dormancy-breaking changes might occur in either or both of
these sites.

Little is known about the mechanism of afterripening.
Afterripened caryopses of Avena fatua show altered responsiveness to added nitrogenous compounds but the reasons for
this are not yet known (Adkins et al., 1984a). It has been
suggested that in barley the important changes taking place
in afterripening lie in the glumes and other tissues
covering the embryo. Glumes from afterripened grains have a
lower oxygen uptake than those from dormant grains, possibly
reflecting altered activities of polyphenol oxidases or
lowered levels of phenolics (Lenoir et al., 1983). But
afterripening also brings about changes in the embryo which
enhance its germination ability. This can be seen by comparing germination rates of embryos excised from dormant and
afterripened grains, the latter showing much more rapid
initiation of radicle growth (Black and Hughes, unpublished). Nothing is known about the changes that might take
place in the embryo during afterripening: not surprisingly,
GA levels do not alter (Metzger, 1983).

Exposure of imbibed grains to relatively low temperatures ('chilling' or stratification) prior to transfer to
higher germination temperatures (eg. 25°C) terminates dormancy. The requisite durations of chilling for breaking

dormancy of intact grains extends over many hours or a few days. In wheat, cv. Sappo, 18 hours at 5°C are fully effective (Black and Butler, unpublished) whereas in the cultivars Kenya 321 and RL 4137 50 hours at 5°C are needed for full effect (Mares, 1984). The Doublet cultivar of barley requires even longer exposure times (FIGURE 3). It is clear, however, that the presence of the coat increases the required chilling period for if embryos are isolated from grains having had as little as 20 hours at 5°C they show maximally enhanced rates of germination (Black and Hughes, unpublished). Isolated embryos themselves need surprisingly short exposure times, eg. as little as 2 hours at 5°C can be detected while 4 hours cold treatment is saturating (Black and Hughes, unpublished). It is clear, therefore, that the action of chilling is in the embryo and apparently not on the enclosing tissues although the presence of the latter does impose a requirement for longer periods of cold. An important point is that the temperatures effective in so-called chilling can be as high as 17°C in barley (FIGURE 4) and 15°C in wheat (Mares, 1984). The mechanism of action of chilling in dormancy-breakage is not known. Alterations in the balance of CN-resistant and CN-sensitive respiration, as suggested for cocklebur by Esashi et al., (1982) do not seem to be involved, at least in barley (Black and Hughes,

FIGURE 3. Dormancy-breakage by chilling in barley (cv. Doublet).

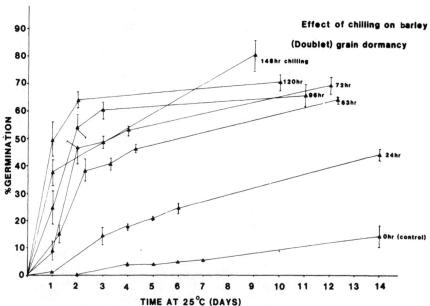

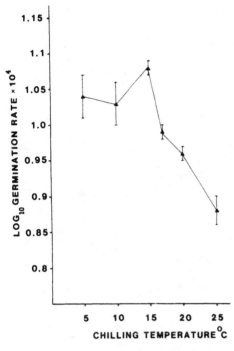

FIGURE 4. Temperature requirement for dormancy removal by "chilling". Isolated embryos of barley (cv. Halycon) were chilled for 4 hours and then transferred to 25°C when germination rates were determined $[\frac{1}{time\ to\ 50\%}]$

unpublished). We have found, however, that effective chilling requires the presence of oxygen. This implies that some oxidative processes are necessary and would seem to argue against a simple low - temperature inducted membrane transition, which has been suggested for lettuce seeds (Van de Woude and Toole, 1980).

Development of dormancy

With respect to pre-harvest sprouting it is the dormancy of developing and maturing grains which is important. Dormancy develops early during grains development but may disappear later during maturation (eg. King, 1976). The pattern which occurs before harvest-ripeness depends on the temperature conditions. At a relatively high day temperature, developing grains of wheat cv. Sappo pass through a period of high germinability at 20-26 days post anthesis (dpa) after which they enter a period of dormancy (FIGURE 5a). When the day temperature is lower, the period of high germinabiltiy is missing and the grains thus have an extended period of dormancy until about 45 dpa (FIGURE 5b). The important effect of temperature during development is illustrated strikingly by an extreme case, when grains are held at 10°C following anthesis. Such grains are deeply dormant throughout their development and into maturity at 65 dpa (FIGURE 6). Other studies produce results in general agreement with this : low temperatures during development and maturation encourage the production of highly dormant grains

FIGURE 5. Dormancy and germinability patterns in developing and maturing wheat (cv. Sappo) (a) Ears were held at 25°/20° after anthesis (b) Ears at 20°/20°C after anthesis. Germination was monitored at 5°C and 20°C. Dormancy is indicated by Δ germination at the two temperatures.

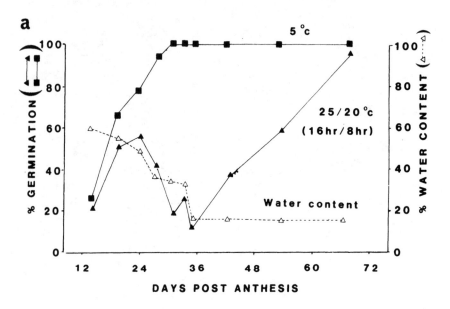

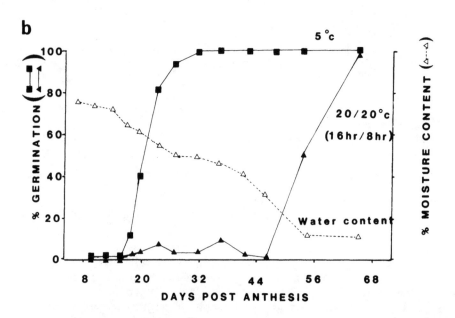

FIGURE 6.   Effect of temperature on development of dormancy. Ears were held at 10°, 15° and 20°C after anthesis.   Germination was tested at 36 dpa and 65 dpa at 2°, 5°, 10° and 20°, of whole grains or isolated embryos.   Note the whole grains developed at 10° are very dormant at all germination temperatures.

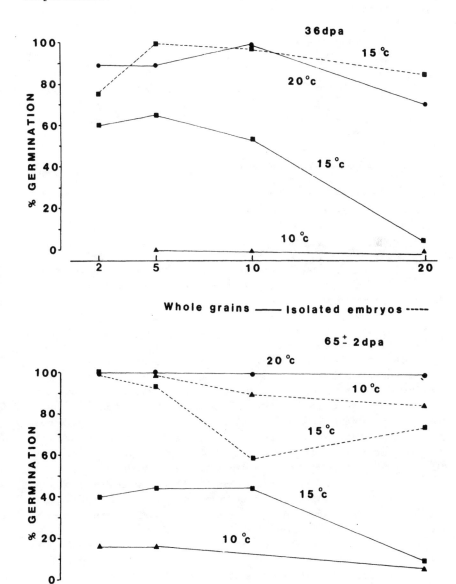

(eg. Sawhney and Naylor, 1979; Peters, 1982; Rauber, 1984).

It is important to note that the temperature during development and maturation can have at least two effects, viz (a) on the onset of dormancy, and (b) on the termination of dormancy. Our experience with wheat (cv. Sappo) shows that low temperatures early in development lead to the production of deeply dormant grains. But low temperatures at a later stage can affect the termination of dormancy by retarding afterripening. The increase in germinability that occurs once grains have reached ca. 20% water content (FIGURE 5) reflects the loss of dormancy by afterripening. Transfer experiments show that this is affected by the temperature (Black and Butler, unpublished). Dormancy might also be expected to be broken, on the ear, by chilling (see above) provided the water content of the grains is above about 25%. In our experience, however, chilling has only a marginal effect on dormant grains at 33-36 dpa (approx. 38% water). But when such grains are dried, reimbibed and then chilled dormancy is almost completely broken. This suggests that the low-temperature termination of dormancy would occur only when more mature (ie. dried) grains become re-wetted.

How is dormancy initiated? ABA may be implicated in the process as suggested by King (1976) who found a peak concentration of this hormone as grains enter dormancy. The strongest support so far available for the participation of ABA in dormancy inception comes from the work of Karssen et al., (1983) on mutants of Arabidopsis thaliana. Here, ABA - deficient developing seeds do not become dormant. This needs to be explored in other species and if it turns out to be the case in cereals it has important implications for our understanding of dormancy and pre-harvest sprouting.

It is appropriate that we now take an overview of dormancy and germination in the context of pre-harvest sprouting. Grains which develop at a relatively low temperature (a possible event in Northern Europe) have a fairly deep dormancy throughout and into maturity. After drying and rewetting, the dormancy of such grains may be removed by chilling; hence, sprouting would occur rather late in maturation. Given fairly high temperatures during their development and maturation (eg. 25°/20°C grains would pass through a period of high germinability (ca 30 dpa) when they would be susceptible to sprouting if wetted. If they avoid this, however, they then enter a period of dormancy, when sprouting is again possible but only at low temperatures (ie. when dormancy is not expressed). If this does not occur, grains then begin to afterripen, at rates dependent upon the temperature. The degree of afterripening, ie. loss of dormancy, then determines if the grains germinate when wetted at relatively low or high temperatures.

The important place of dormancy in this scenario is clear. While we are learning more about the general physiology of the phenomenon unfortunately we still lack understanding of the biochemical basis of dormancy inception, maintenance and loss. It is essential that these gaps in our knowledge are filled in order to develop a sound strategy for the control of pre-harvest sprouting.

REFERENCES

Adkins, S. W., Simpson, G. M. and Naylor, J. M. 1984a. Physiol. Plant. 60, pp. 227-233.
Adkins, S. W., Simpson, G. M. and Naylor, J. M. 1984b. Physiol. Plant. 60, pp. 234-238.
Adkins, S. W., Naylor, J. M. and Simpson, G. M. 1984c. Physiol. Plant. 62, pp. 18-24.
Bewley, J. D. and Black, M. Physiology and biochemistry of seeds. Vol. 2. Springer-Verlag, Berlin, Heidelberg, 1982.
Côme, D. and Corbineau, F. 1984. C. R. Acad. Agri. de France. 70, pp. 709-715.
Côme, D., Lenoir, C. and Corbineau, F. 1984. Seed Sci. and Technol. 12, pp. 629-640.
Corbineau, F., Lenoir, C., Lecat, S. and Côme, D. 1984. C. R. Acad. Agri. de France. 70, pp. 717-724.
Datta, S. G., Gutterman, Y. and Evenari, M. 1972. Planta. 105, pp. 155-164.
Esashi, Y., Ishihara, N., Saijoh, K. and Satoh, M. 1982. Plant Cell and Environ. 6, pp. 47-54.
Esashi, Y., Kusyuama, K., Tazaki, S. and Ishihara, N. 1981a. Plant Cell Physiol. 22, pp. 65-71.
Esashi, Y., Sakai, Y. and Ushizawa, R. 1981b. Plant Physiol. 67, pp. 503-508.
Esashi, Y., Wakabayashi, S., Tsukada, Y. and Satoh, S. 1979. Plant Physiol. 63, pp. 1039-1043.
Fuerst, E. P., Upadhyaya, M. K., Simpson, G. M., Naylor, J. M. and Adkins, S. W. 1983. Can. J. Bot. 61, pp. 667-670.
Gahan, P. B., Dawson, A. L., Chapman, J. M. and Black, M. 1986. Ann. Bot. (in press).
Gordon, I. L. 1980. Cereal Res. Comm. 8, pp. 115-129.
Gosling, P. G., Butler, R. A., Black, M. and Chapman, J. 1981. J. Exp. Bot. 32, pp. 621-627.

392

Gubler, F., and Ashford, A. E. 1985. Aust. J. Plant
   Physiol. 12, pp. 297-305.
Hendricks, S. B. and Taylorson, R. B. 1979. Proc. Natl.
   Acad. Sci. (USA). 76, pp. 778-781.
Jayachandran - Nair, K. and Sridhar, R. 1975. Biol. Plant.
   17, pp. 318-319.
Karssen, C. M., Brinkhorst-van der Swan, D. L. C., Breek-
   land, A. E. and Koornneef, M. 1983. Planta, 157, pp.
   158-165.
King, R. D. 1976. Planta. 132, pp. 43-51.
Kruger, J. E. 1976. Cereal Chem. 53, pp. 201-203.
Lenoir, C., Corbineau, F. and Côme, D. 1983. Physiol. Veg.
   21, pp. 633-643.
Mares, D. J. 1984. Aust. J. Agric. Res. 35, pp. 115-128.
Metzger, J. D. 1983. Plant Physiol. 73, pp. 791-795.
Misra, S. and Bewley, J. D. 1985. Pl. Physiol. 78, pp.
   876-882.
Peters, N. C. B. 1982. Weed Research. 22, pp. 205-212.
Rauber, R. 1984. Landwirtsch. Forschung. 37, pp. 102-110.
Renard, C. 1976. Bull. Soc. Roy. Bot. Belg. 109, pp. 227-
   230.
Renard, C. and Capelle, P. 1976. Aust. J. Bot. 24, pp.
   437-446.
Roberts, E. H. 1969. In: Dormancy and survival, Woolhouse,
   H. (ed.). SEB Symposium. 23, pp. 161-192.
Roberts, E. H. and Smith, R. D. 1977. In: The physiology
   and biochemistry of seed dormancy and germination.
   Khan, A. A. (ed.). Elsevier/North Holland. pp. 385-
   411.
Sawhney, R. and Naylor, J. M. 1979. Can. J. Bot. 57, pp.
   59-63.
Singh, S. P. and Paleg, L. G. 1984. Plant Physiol. 76, pp.
   139-142.
Taylorson, R. B. and Hendricks, S. B. 1980/81. Israel J.
   Bot. 29, pp. 273-280.
Upadhyaya, M. K., Simpson, G. M. and Naylor, J. M. 1981.
   Can. J. Bot. 59, pp. 1640-1646.
Upadhyaya, M. K., Naylor, J. M. and Simpson, G. M. 1982.
   Can. J. Bot. 60, pp. 1142-1147.
Van der Woude, W. J. and Toole, V. K. 1980. Plant Physiol.
   66, pp. 220-224.

Breaking of Seed Dormancy by Treatment with Ammonia

A. L. P. Cairns and O. T. de Villiers,
Department of Agronomy and Pastures,
University of Stellenbosch, Stellenbosch
7600, Republic of South Africa.

SUMMARY

Breaking of seed dormancy of wild oat (Avena fatua L.) with ammonia did not influence the activity of the pentose phosphate pathway enzymes glucose-6-phosphate dehydrogenase (G6PDH) and 6-phosphogluconate dehydrogenase (6PGDH). Alpha-amylase synthesis in de-embryonated endosperm halves which had been treated with ammonia and subsequently washed under tap water for 48h, increased five-fold. Washing did not influence alpha-amylase production of untreated endosperm halves.

Ammonia treatment led to an 81% increase in respiration in incubating seeds. This increase in respiration was partially inhibited by salicylic hydroxamic acid (SHAM). It is postulated that loss of dormancy induced by treatment with ammonia is associated with the induction of increased respiratory activity. There is, however, no conclusive evidence that this increased respiratory activity is via the alternative respiratory pathway.

INTRODUCTION

In a review article on wild oat (Avena fatua L.) dormancy, Simpson (1984) compared the relative efficacy of various dormancy-breaking compounds. These compounds were listed in descending order of efficacy viz. gibberellic acid, ethanol, respiratory inhibitors such as cyanide and azide and nitrogenous compounds such as nitrates and nitrites. In deeply dormant seed such as in the Montana 73 line, dormancy could be broken only by GA_3. The other dormancy-breaking compounds could break dormancy in seed from less dormant lines or seed from dormant lines which had been after-ripened.

The dormancy of deeply dormant wild oat seed of the Montana 73 line was broken by treating the air-dry seed with gaseous ammonia. The dormancy of several other grass weed species was also broken by ammonia but the gas had no effect on the germination of several dormant dicotyledonous weed species (Cairns and de Villiers, 1986). This is the first report of the dormancy-breaking properties of the gas and it was thus deemed necessary to examine the physiological basis of this finding.

MATERIALS AND METHODS

Seed from the dormant Montana 73 line was used in all experiments unless otherwise stated. The mature seed was harvested from plants grown under glasshouse conditions and stored for two weeks at 4°C. Thereafter the seed was trans-fered to sub-zero temperatures to maintain dormancy.

Air-dry seed was treated with ammonia in test tubes as previously described (Cairns and de Villiers, 1986). The seed was exposed to the air for 48h before incubation.

The activity of the pentose phosphate pathway enzymes, glucose-6-phosphate degydrogenase (G6PDH) and 6-phospho-gluconate dehydrogenase (6PGDH) was determined according to the methods of Upadhyaya et al., (1981).

Alpha-amylase was extracted according to the method of Nichols (1979) and the enzyme activity was assayed using a method based on that of Barnes and Blakeney (1974). Ten endosperm halves were used per determination and the endo-sperm halves were incubated with 5×10^{-6} mM GA_3.

The effect of ammonica, sodium azide and salicylic hydroxamic acid (SHAM) on the respiration of wild oat seed was investigated by incubating ammonia-treated seed and seed imbibed in 3 mM NaN_3 with and without SHAM at 3 mM. Prior to incubation, seed was surface sterilized with 1% NaOCl. After 48h incubation, respiration was determined by meas-uring O_2 uptake with an oxygen electrode Yellow Springs Instruments.

RESULTS AND DISCUSSION

The activity of 6PGDH in control and NH_3-treated seed during a 144h incubation period is illustrated in FIGURE 1.

FIGURE 1. Effect of NH3 on the activity of glucose-6-phos-
phate dehydrogenase (G6PDH) and 6-phosphogluconate dehydrogenase (6PGDH).

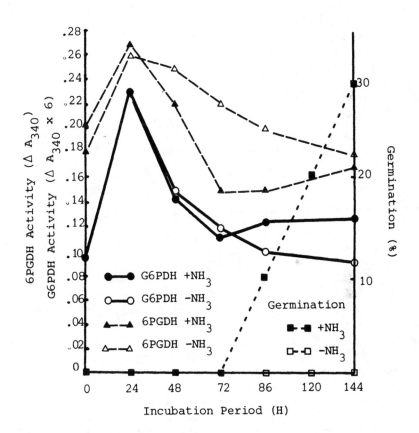

No difference in the activity of G6PDH could be observed
between the two treatments until after the initiation of
germination of the NH3-treated seeds. After the initiation
of germination the NH3-treated seed showed an increase in
G6PDH activity whereas the activity in the control seeds
declined further. The activity of 6PGDH in the control
seeds was slightly lower than in the NH3-treated seed during
the initial 24h incubation period. The enzyme activity of
both control and NH3-treated seeds declined during the following 48h. After the initiation of germination at 96h the
6PGDH activity in the NH3-treated seeds showed a gradual
increase whereas enzyme activity continued to decline in the
control seeds.

The above results indicate that there is no increase in
the above-mentioned pentose phosphate pathway enzymes prior

to the initiation of germination and points to the non-involvement of this pathway in the triggering of germination. The same conclusion was arrived at by Upadhyaya et al., (1981). Work by Simmonds and Simpson (1971) who found a correlation between increased activity of the pentose phosphate pathway and loss of dormancy in A. fatua can be discounted on the grounds that the increase of activity of the pentose phosphate pathway they observed was a consequence and not a cause of germination.

The effect of NH_3 on GA_3-mediated alpha-amylase synthesis in washed and unwashed endosperm halves of A. fatua.

Alpha-amylase synthesis in unwashed de-embryonated endosperm halves which had been pre-treated with NH_3 did not differ significantly from that of the control endosperm halves. However, in endosperm halves which had been washed under tap water for 48h prior to incubation, pre-treatment with NH_3 led to a massive increase in alpha-amylase synthesis (TABLE 1). The same result was observed in several similar experiments. Washing did not influence alpha-amylase synthesis in control endosperm halves.

The mechanism involved in the stimulation of alpha-amylase synthesis is not known at this stage but several possibilities may be considered. Ammonia may release bound inhibitors of alpha-amylase synthesis which are subsequently leached out with washing. An examination of the leachates of washed NH_3-treated and control seeds will have to be done to determine if this is the case.

TABLE 1. Effect of NH_3-treatment and washing the seed under running tap water for 48h prior to incubation on GA_3-stimulated synthesis of alpha-amylase in de-embryonated endosperm halves of A. fatua seed.

| | Alpha-amylase activity (mEU*/g endosperm half) | |
	Unwashed	Washed
$-NH_3$	2.55[a]	2.13[a]
$+NH_3$	0.71[a]	14.50[b]

* Milli enzyme units. An enzyme unit is defined as the amount of enzyme catalyzing the hydrolysis of 1 μm glucosidic linkage at 50°C.

Ammonia may also cause a structural change in cell walls and membranes which is exacerbated by hydrating the endosperm prior to incubation with GA_3. Electron micrographs of the treated endosperm show extensive damage to cell walls and membranes (results not shown). It must therefore be concluded that damage to cell walls exacerbated by pre-incubation hydration led to greater penetration of GA_3 to the site of action. The possible role of release and subsequent leaching out of bound inhibitors still has to be investigated.

Effect of NH_3, NaN_3 and SHAM on the respiratory activity of incubating A. fatua seed.

Ammonia and NaN_3 both stimulated respiration of incubated A. fatua seed (TABLE 2). Azide however, stimulated respiration by 25% whereas NH_3 stimulated respiration by 81%. SHAM at 3 mM had no significant effect on the respiratory activity of control seeds but inhibited respiration of the azide- and NH_3-treated seed by 12% and 17% respectively.

Upadhyaya et al., (1983) also found SHAM to have little effect on the respiration of control seeds but inhibited azide-stimulated germination by more than 90%. Upadhyaya et al., (1982) had previously demonstrated that SHAM completely inhibited azide-stimulated germination in several semi-dormant lines of A. fatua but that SHAM had no effect on the germination of non-dormant seed or on after-ripened seed of dormant lines. Upadhyaya et al., (1983) used SHAM at 10 mM whereas only 3 mM was used in the present study. This difference in concentration could explain the moderate inhibition of azide-stimulated respiration observed in the present study. However, SHAM is not a specific inhibitor of the alternative respiratory pathway (Rich et al., 1978) and results from experiments using relatively high concentrations of SHAM must be open to question. The fact that SHAM inhibited azide and ammonia-stimulated respiration can thus not be used as conclusive evidence of increased activity of the alternative respiratory pathway.

The dormancy-breaking effect of nitrite, nitrate and azide on dormant and after-ripened A. fatua seed from several lines was examined by Adkins et al., (1984a). Dormancy in seed from three dormant lines which had been after-ripened for 6 weeks could be broken by nitrate, nitrite and azide. None of these compounds broke dormancy in 6-week-old Montana 73 seed. This ecotype had to be after-ripened for

TABLE 2. Effect of azide and ammonia on the respiratory
rate of A. fatua as affected by the alternative respiratory
pathway inhibitor SHAM.

| | O_2 Uptake (nl caryopsis^{-1}min^{-1}) | |
	- SHAM	+ SHAM
Control	31.3[a]	31.6[a]
NaN$_3$ (1mM)	37.1[a]	33.6[a]
NH$_3$ (12ml/30ml)	54.9[c]	46.0[b]

6 months before becoming responsive to these dormancy breaking compounds. Cairns and de Villiers (1986) showed that
dormancy in a wide range of wild oat ecotypes, including
Montana 73, could be broken by ammonia. In the present
study it has been shown that NH$_3$ is only slightly inferior
to GA$_3$ in relieving dormancy of Montana 73. Adkins et al.,
(1984b) ascribe the failure of ammonium salts to promote
germination in dormant A. fatua seed to the fact that they
cannot act as electron acceptors. Compounds of higher nitrogen oxidation level such as nitrate and nitrite break
dormancy by their ability to act as electron acceptors. The
results obtained in the present study with ammonia are difficult to reconcile with this hypothesis. Ammonia cannot
act as an electron acceptor and thus another explanation
must be sought to elucidate the increased respiration and
dormancy-breaking effect of ammonia and remains the subject
of further investigations.

REFERENCES

Adkins, S. W., Simpson, G. M. and Naylor, J. M. 1984a.
    The physiological basis of seed dormancy in Avena fatua
    III. Action of nitrogenous compounds. Physiol. Plant.
    60, pp. 227-233.
Adkins, S. W., Simpson, G. M. and Naylor, J. M. 1984b. The
    physiological basis of seed dormancy in Avena fatua IV.
    Alternative respiration and nitrogenous compounds.
    Physiol. Plant. 60, pp. 234-238.
Barnes, W. C. and Blakeney, A. B. 1974. Determination of
    cereal alpha-amylase using a commercially available
    dye-labelled substrate. Die Stärke. 26, pp. 193-197.
Cairns, A. L. P. and de Villiers, O. T. 1986. Breaking dor-
    mancy of Avena fatua L. seed by treatment with ammonia.
    Weed Research. (In press).

Nicholls, P. B. 1979. Induction of sensitivity to gibberellic acid in developing wheat caryopses: Effect of rate of desiccation. Aust. J. Plant Physiol. 6, pp. 229-240.

Rich, P. R., Wiegand, N. K., Blum, H., Moore, A. L. and Bonner, W. D. 1978. Studies on the mechanism of inhibition of redox enzymes by substituted acids. Biochem. Biophys. Acta. 525, pp. 325-337.

Simmonds, J. A. and Simpson, G. M. 1971. Increased participation of the pentose phosphate pathway in response to after-ripening and gibberellic acid treatment of caryopses of Avena fatua. Can. J. Bot. 49, pp. 1833-1840.

Simpson, G. M. 1984. A review of dormancy in wild oats and a lesson it contains for today. Canadian Plains Proceedings 12. Wild Oat Symposium Vol 2. pp. 3-20. University of Regina, Regina, Canada.

Upadhyaya, M. K., Naylor, J. M. and Simpson, G. M. 1982. The physiological basis of seed dormancy in Avena fatua L. I. Action of respiratory inhibitors sodium azide and salicylic hydroxamic acid. Physiol. Plant. 54, pp. 419-454.

Upadhyaya, M. K., Naylor, J. M. and Simpson, G. M. 1983. The physiological basis of seed dormancy in Avena fatua L. II. On the involvement of alternative respiration in the stimulation of germination by sodium azide. Physiol. Plant. 58, pp. 119-123.

Upadhyaya, M. K., Simpson, G. M. and Naylor, J. M. 1981. Levels of glucose-6-phosphate and 6-phosphogluconate dehydrogenase in the embryos and endosperms of some dormant and non-dormant lines of Avena fatua during germination. Can. J. Bot. 59, pp. 1640-1646.

Selection for Seed Dormancy by using Germination Tests

S. Larsson, Research and Developmental Department, Svalöf AB, S-268 00, Svalöv, Sweden.

SUMMARY

Progenies from a cross between two spring wheat lines were selected for seed dormancy by using a simple germination test in petri dishes. Hand-threshed grains were germinated at room temperature for three days on filter paper wetted with distilled water. The germination test was combined with selection for other important agronomical traits, both from field observations and from chemical analysis of grain quality.

Improvements were found in seed dormancy and in alpha-amylase activity already after one selection step. Most of the selected progeny lines in F_4, which all exhibited acceptable plant height, earliness and grain quality, had a deeper dormancy than any of their parents. This indicates the possibility to achieve transgressive combinations with respect to this characteristic. Selection by using germination tests can be useful in handling a large amount of material in the early generations of the breeding process, as a supplement to other methods to improve sprouting and malting resistance in spring wheat.

INTRODUCTION

Different methods have long been discussed and more recently used for selection of pre-harvest sprouting and malting resistance, such as determination of alpha-amylase activity (Bingham and Whitmore, 1966; Derera and Moss, 1967; Moss et al., 1972; Persson, 1975; McMaster and Derera, 1976; Johansson, 1977), and falling number (Svensson and Lagerström, 1966; Moss et al., 1972;

Svensson, 1976; Weilenmann, 1976), and by using different
types of germination tests (Harrington and Knowles, 1940;
Harrington, 1949; Hutchinson et al., 1948; Everson and
Hart, 1961; Strand, 1965; Johansson, 1977; Reitan, 1983).
All these tests of enzymatic and germinative events have
their advantages and disadvantages in selectivity and
screening capacity (cf. Johansson, 1977; Hagemann and Ciha,
1984). In general germination tests are better in predicting sprouting susceptability, whereas enzymatic tests are
better in quantifying actual sprouting damage (Hagemann and
Ciha, 1984).

A close relationship between seed dormancy and alpha-
amylase activity has been reported by Ringlund and Strand
(1970) and by Johansson (1977), but the developmental patterns of the germinability and alpha-amylase activity do not
need to be parallel (Gordon, 1979). A deep dormancy, however, prevents the formation or activation of alpha-amylase
(Ringlund and Strand, 1970). In that way selection for seed
dormancy also effects the enzymatic pattern of the grains
during the after-ripening period. Combinations with both
seed dormancy and low alpha-amylase activity would naturally
be preferred (Bingham and Whitmore, 1966).

As the control of seed dormancy seems to be of multigenic nature (Derera et al., 1977) it would be valuable to
combine components from different genoypes in order to provide, if possible, transgressive progenies. The purpose of
this study was first of all an attempt to combine genotypes
with different dormancy components to get such a result. It
was also to find a quick, simple and reproducable test useful for further breeding of sprouting and malting resistance
in wheat. Such a laboratory test has to be included in the
ordinary breeding program. As seed germination already has
been suggested as useful in selection of a large amount of
material in early generations this technique was used here.

MATERIAL AND METHODS

Thirty-six crosses were made in 1980 between cultivars
or lines known to be more or less resistant to pre-harvest
sprouting. Some of these combinations became more valuable
than others due to the choice of parents. A cross between a
breeding line from Svalöf 'SvU 67653' and a German line
'LP.2659.74' gave many progenies with fairly good seed dormancy. 'SvU 67653' acquired its sprouting resistance from
the German line 'Strube 4926'. The results from selection
in this combination is used as an example in this study.

Ten ears per genotype were pollinated in both directions. The F_1 kernels were sown in separate rows for each
pollinated ear. From these rows ten vigorous plants were
selected. Kernels from these plants were sown in separate
plant rows as F_2.

Hand-threshed grains from F_2 and F_3 were then tested
and selected for seed dormancy in 1982 and 1983. Spikes
were randomly picked from each pedigree line and the threshed grains mixed before the test, since all selections both
from germination tests and field observations were made on a
pedigree line basis.

Germination test

Fifty grains per sample were placed on wetted filter
paper in petri dishes (diam. 9cm with 7ml distilled water).
The dishes were placed at room temperature (20°C) and the
germinated grains were counted after three days. To avoid
harsh treatment during threshing all material was threshed
by hand.

Selection in F_2

Of 200 pedigree lines grown in 1982 fifty-eight were
discarded due to late harvest ripening, poor vitality and
poor straw stiffness. Five spikes were randomly taken from
each of the remaining lines to be tested for seed dormancy.

1982 was a relatively wet year and the harvest was delayed. Due to this and to the weather during grain filling
the dormancy was relatively weak in the material. The germination tests had to be done immediately after harvest.
Despite this the material could be selected for seed dormancy, but not under optimal conditions. Samples with germination percentage above 95% were discarded, together with
some samples with poor grain quality (FIGURE 1). Grain
quality was measured by the NIR-technique.

Selection in F_3

420 new pedigree lines (84 x 5) were sown as F_3 in
double spaced plots. About 50% of these lines were discarded in the field because of their late harvest ripening, poor
vitality, weak straw and susceptability to yellow rust and
mildew. From each of the remaining 200 lines 20 spikes were

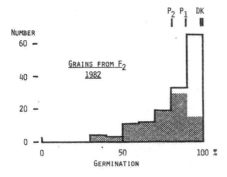

FIGURE 1. Germination test of grains from F_2-pedigree lines. The shaded area represents that part of the population which was selected for the next generation.
P_1 = SvU 67653
P_2 = LP.2659.74
D = WW Drabant
K = WW Kadett

randomly taken: 6 for sowing next year, 5 for germination test, 5 for analysis of grain quality by the NIR-technique and the last 4 for a laboratory determination of the alpha-amylase activity after exposure of the spikes to malting conditions in a moist chamber.

The weather this year was stable and the samples could be collected at the normal time of harvest, at harvest ripeness. Dormancy was deep in all material at harvest. The grains were therefore stored at room temperature for after-ripening about one month before the germination test. The germination percentages of the tested samples were this time more evenly distributed over the whole range of 0 to 100% (FIGURE 2). The selection was done not only according to seed dormancy but also to other aspects of grain quality, such as high protein content. Of the 200 tested lines 148 were stored for sowing in F_4.

FIGURE 2. Germination test of grains from F_3-pedigree lines. The shaded area represents that part of the population which was selected for the next generation.

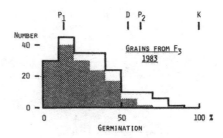

Germination test in F_4

In this generation the germination test was performed to get the combined results of the two preceding selection steps.

RESULTS

In relation to the parents the offspring received an apparently deeper seed dormancy already after one selection step, made on grains from F_2-plants. This selection resulted simultaneously in a lower alpha-amylase activity (FIGURE 3). Most of the offspring with low germination percentage exhibited lower alpha-amylase activity than any of their parents. It should be noted that the alpha-amylase activity in this case was determined on material in the after-ripening period, which was still more or less dormant.

FIGURE 3. Relationship between percent germination and alpha-amylase activity in the 200 tested samples from F_3-pedigree lines.

ALPHA-AMYLASE ACTIVITY

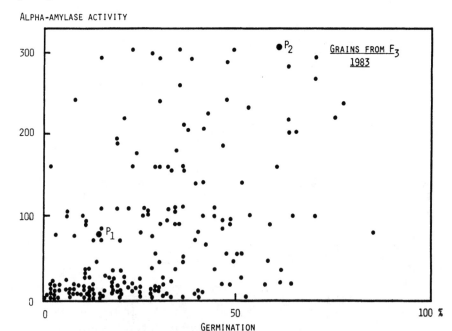

After selection in two generations for seed dormancy, the offspring exhibited a better dormancy than their parents and much better than the two cultivars on the market in Sweden today, 'WW Drabant' and WW Kadett' (FIGURE 4). Most of these breeding lines possessed a suitable earliness to harvest ripening, suitable plant height, suitable protein content and a suitable volume weight. Unfortunately the resistance to mildew was too bad in this combination, which made the material unsuitable for further breeding. This

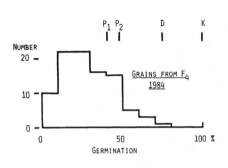

FIGURE 4. Germination test of grains from F_4- pedigree lines.

was, however, obvious already in F_1 and F_2. A number of new crosses were for that reason made in F_3 between the best lines and other breeding lines with better resistance to mildew.

DISCUSSION

The combination used did not result in any useful breeding lines, due mainly to their poor mildew resistance, but the use of the germination test was encouraging. Not only this combination but also several of the other tested combinations gave transgressive segregations both of seed dormancy and alpha-amylase activity. The germination test has a high screening capacity. It took two people two weeks to accurately test 3000 lines approximately. The most time-consuming step is threshing, which has to be done by hand to avoid undesirable stimulation of germination.

The most suitable time for testing varies from year to year, depending on the climatic conditions before and close to harvest-ripening. The germination test has to be done sometime during the period of after-ripening to get accurate separation between genotypes. Some years it is necessary to make the test immediately after harvest, other years it is better to wait several months. One way to lengthen the period for after-ripening, which enable greater quantities to be screened, is to store the material at low temperature. This can be done without changing the relative relationships of seed dormancy between genotypes (Mares, 1983), and makes the test procedure more flexible.

Interpretations only in terms of dormancy, however, could be misleading if no attention is given to differences in time of ripening between genotypes. Germination tests or alpha-amylase assays on grain samples during the after-ripening measure differences in maturity as well as differences in dormancy (Gordon et al., 1979). It is important therefore to include field observations and selections for earliness in the laboratory test. Otherwise there is a risk that much of the selected material becomes late ripening.

ACKNOWLEDGEMENTS

The author wishes to thankfully acknowledge the financial support provided by the Swedish Plant Breeding Board.
I am also very greatful to Dr. V. Stoy and Dr. D. O'Connor
for critical reading of the manuscript.

REFERENCES

Bingham, J. and Whitmore, E. T. 1966. Varietal differences
    in wheat in resistance to germination in the ear and
    alpha-amylase content of the grain. J. Agric. Sci.
    66, pp. 197-201.
Derera, N. F., Bhatt, G. M. and McMaster, G. J. 1977. On
    the problem of pre-harvest sprouting in wheat.
    Euphytica 26, pp. 299-308.
Derera, N. F. and Moss, H. J. 1967. Weather damage-grain
    dormancy-grain quality relationships. Nth. West Res.
    Inst. Rep., pp. 25-27.
Everson, E. H. and Hart, R. B. 1961. Varietal variation for
    dormancy in mature wheat. Q. Bull. Mich. Agric. Exp.
    Stn. 43, pp. 820-829.
Gordon, I. L. 1979. Selection against sprouting damage in
    wheat. III. Dormancy, germinative alpha-amylase, grain
    redness and flavanols. Aust. J. Agric. Res. 30, pp.
    387-402.
Gordon, I. L., Balaam, L. N. and Derera, N. F. 1979. Selec-
    tion against sprouting damage in wheat. II. Harvest
    ripeness, grain maturity and germinability. Aust. J.
    Agric. Res. 30, pp. 1-17.
Hagemann, M. G. and Ciha, A. J. 1984. Evaluation of methods
    used in testing winter wheat susceptibility to harvest
    sprouting. Crop Science 24, pp. 249-254.
Harrington, J. B. 1949. Testing cereal varieties for dor-
    mancy. Sci. Agric. 29, pp. 538-550.
Harrington, J. B. and Knowles, P. F. 1940. The breeding
    significance of after-harvest sprouting in wheat. Sci.
    Agric. 20, pp. 402-413.
Hutchinson, J. B., Greer, E. N. and Brett, C. C. 1948.
    Resistance of wheat to sprouting in the ear: prelimin-
    ary investigations. Emp. J. Exp. Agric. 16, pp.
    23-32.
Johansson, N. E. 1977. Studies of sprouting resistance in
    wheat-methods of determination, differences between
    varieties and pattern of inheritance. Diss. s. No.
    25137, Inst. for genetic and plant breeding.

Lantbrukshögskolan, Uppsala, Sweden.

Mares, D. J. 1983. Preservation of dormancy in freshly harvested wheat grain. Aust. J. Agric. Res. 34, pp. 33-38.

McMaster, G. J. and Derera, N. F. 1976. Methodology and sample preparation when screening for sprouting damage in cereals. Cereal Res. Comm. 4, pp. 251-254.

Moss, H.J., Derera, N. F. and Balaam, L. N. 1972. Effect of pre-harvest rain in germination in the ear and alpha-amylase activity of Australian wheat. Aust. J. Agric. Res. 23, pp. 769-777.

Persson, E. 1975. Otello - a result of amylase selection for sprouting resistance. Cereal Res. Comm. 4, pp. 101-106.

Reitan, L. 1983. Early generation mass selection for seed dormancy in barley. In: Third Int. Symp. on Pre-harvest Sprouting in Cereals. Kruger, J. E. and La-Berge, D. E. (eds.). Westview Press, Boulder, Co. USA. pp. 244-250.

Ringlund, K. and Strand, E. A. 1970. Seed dormancy and alpha-amylase activity in wheat Triticum aestivum L. Inst. Plantekultur, Norges Landbrukshøgsk., Melding. 49, (14).

Strand, E. A. 1965. Studies on seed dormancy in barley. Inst. Plantekultur, Norges Landbrukshøgsk., Melding. 44, (7).

Svensson, G. 1976. Screening methods for sprouting resistance in wheat. Cereal Res. Comm. 4, pp. 263-266.

Svensson, G. and Lagerström, G. 1966. Bestimmung der Auswuchsresistenz bei Weizen, Geste und Hafer. Agr. Hort. Genet. 24, pp. 11-47.

Weilenmann, F. 1976. A selection method to test the sprouting resistance in wheat. Cereal Res. Comm. 4, pp. 267-273.

Studies on Dormancy and Pre-Harvest Sprouting in Chinese Wheat Cultivars

Z. Wu, Nanjing Agricultural University,
Peoples's Republic of China.

SUMMARY

Seed germination tests on 374 wheat varieties from various regions of China indicated that dormancy characteristics among varieties differed markedly, varying from non-
dormancy to a dormancy period of 7 or more weeks. Seed
dormancy was generally associated with red seed coat. However, a few stable dormant types with white seed coats were
found. Dormancy and red seed coat were ecological characters of those indigenous varieties which were adapted to
the southern winter wheat regions and north-eastern spring
wheat region with heavy precipitation during ripening stage.
Reciprocal differences in four sets of crosses between
two dormant parents and two non-dormant parents showed the
presence of maternal effect. The observed F_2 segregation
data from these crosses showed that dormancy was not simply
inherited. The paucity of dormant types with white seed
coat confirmed the conclusion that there were several other
genes for dormancy not associated with red seed coat. Among
various mechanical treatments applied to the dormant seeds,
tearing off the seed coat over the embryo was the most
effective in breaking dormancy. This implied that the seed
coat is a source of germination inhibitor(s) and/or a
barrier to the leaching of inhibitor(s).
Investigations on germinability during grain development in 31 cultivars and lines indicated that germination
in the spike was significantly correlated with isolated
grain at each stage. There were significant varietal differences in germinability at various grain development
stages, but the heritabilities were relatively low.
A preliminary study involving 15 cultivars indicated
that varietal differences in endogenous ABA level and

falling number were significant, yet their correlations with
germinability were insignificant.

INTRODUCTION

China is the world's second largest wheat producer.
In 1984 wheat was grown on 35 million hectares with the
total production of 87 million tonnes giving an average
yield of 2.5 t/ha. Wheat is the second largest food crop in
China, exceeded only by rice. In the last 35 years, genetic
improvements in wheat as well as improved cultural practices
and irrigation, have resulted in an increase of one-third in
area, a three-fold increase in yield, and about a five-fold
increase in total production. In the major wheat producing
regions, a shift to new, improved cultivars has occurred at
7- to 8-year intervals. New cultivars have the following
characteristics: higher resistance to disease, especially
to rusts, wider ecogeographic adaptation, higher yield capacity, and shorter growth period for multiple cropping. However, breeding for resistance to weather damage and for
quality has not received adequate attention. Following the
release of white-seed cultivars, pre-harvest sprouting has
shown up as a serious problem especially in the southern
winter wheat regions and the north-eastern spring wheat
region where the wheat crop matures during a rainy period.
Breeding for sprouting resistant white-seeded wheat is now
in progress. The objective of this paper is to give a brief
account of some studies on dormancy in relation to pre-
harvest sprouting in Chinese wheat cultivars conducted at
Nanjing Agricultural University.

VARIATION IN DORMANCY AND OTHER TRAITS IN CHINESE WHEAT
CULTIVARS

The dormancy status of 374 wheat cultivars from various
regions in China was determined by a series of germination
tests. The cultivars tested were grouped into three categories according to their degree of dormancy. Those which
attained full germination within 1-2 weeks, 3-4 weeks and 5
or more weeks were considered to possess a low, medium and
high degree of dormancy, respectively.
Among 374 cultivars tested, 125 were classified to have
a low degree of dormancy, 208 to have a medium degree of
dormancy, and 41 to have a high degree of dormancy. The
degree of dormancy found in these varieties was found to be

associated with seed coat colour, and not with growth habit or early maturity. In addition, two red-seeded lines selected from natural variants of two white-seeded cultivars with non-dormancy appeared to have a medium level of dormancy. These results support the general association of red seed coat and dormancy (Gfeller and Svedja, 1960; Khan and Strand, 1977). The great majority of the indigenous varieties from the southern winter wheat regions and the northeastern spring wheat region were red-seeded wheats with a high level of dormancy. More than two-third of the indigenous varieties from the northern winter wheat regions were white-seeded wheats which showed a low degree of dormancy. Winter wheats in South China mature in April or May at which time they experience 200-300mm of rainfall from 10-15 rainy days each month. In North China winter wheat region wheat matures in June with only 50-75mm of rainfall from 5 rainy days. It is suggested that longer dormancy period assoiated with red-seeded coat of the indigenous cultivars from those southern regions and northeastern region resulted from natural selection in which heavier precipitation during maturation period had played an important role, and the dormancy period associated with seed coat colour should be considered as in important ecological character which reflects both natural and agricultural conditions (Wu and Bo, 1962). The observations on a few white-seeded cultivars with stable dormancy period, e.g., Xuzhou 438, provided evidence that there are exceptions to the general relationship of seed coat colour and dormancy and it is possible to combine dormancy and white seed coat colour (DePauw and McCaiq, 1983).

INHERITANCE AND MECHANISM OF DORMANCY

In order to study the inheritance of dormancy, all possible reciprocal crosses between four wheat cultivars, involving two white-seeded non-dormant cultivars, Nanda 2419 and Biyumai, and two red-seeded dormant cultivars, Jinda 2905 and Jiangdonman, were made. The parents, F_1's (reciprocal included) and F_2's were tested for dormancy.

There were significant recriprocal differences in dormancy of the F_1's in the four sets of crosses. When the dormant cultivars were used as female parents the degree of dormancy was high. This suggests that the effect of seed coat on the dormancy status is dependent on the maternal tissues. Neither reciprocal difference nor transgressive segregation for dormancy was found in F_2 seeds in the four sets of crosses. The observed F_2 segregation data for seed germination did not fit one or two factor ratios, and the parents were assumed to differ by more than two factors in the control of dormancy. These conclusions were somewhat different from those drawn by Bhatt et al., (1983), probably due to differences in specific parental combinations and in techniques. The paucity of dormant types of white seeds observed in F_2's confirmed the hypothesis that several mechanisms may be associated with dormancy besides seed coat colour (Bhatt and Derera, 1979;   De Pauw and McCaig, 1983).

For investigating the mechanism of dormancy, various mechanical treatments were given to the dormant seeds of Jinda 2905 and Jiangdonman, involving piercing the seed coat over the embryo, tearing off the pericarp and seed coat over the embryo, and excised embryos with or without pericarp and seed coat.

All mechanical treatments were effective in promoting germination. Tearing off the pericarp and seed coat over the embryo as well as excised embryo without the pericarp and seed coat were the most effective treatments in inducing germination, followed by excised embryo with the pericarp and seed coat, and piercing the seed coat over the embryo. The results supported the suggestions that dormancy is coupled to some factor situated in the seed coat (Olssen, 1975) and that the seed coat is a source of germination inhibitor(s) and/or a barrier to the leaching of inhibitors(s) (King, 1983).

SPROUTING RESISTANCE AND RELATED TRAITS

Fifteen commercial wheat cultivars grown in the Lower Yangtze region and 16 breeding lines were included in a field experiment. The experiment was a randomized complete block design, consisting of three replications. Samples of fifteen spikes were taken 25, 30, 35 and 40 days after anthesis from each plot and examined for germinability. Germination tests were carried out on intact spikes as well as on the isolated seeds for each sample, and falling number was determined by the standard method using 7 grams of ground wheat at 15% moisture and 25ml of water. In addition, samples of spikes were taken at harvest ripeness stage and 20 days after ripeness from each plot of the 15 cultivars. Two hundred seeds per sample were tested for germinability, 7 grams of seed of the same sample were used to determine falling number, and another 100 grams of seed were ground and evaluated for endogenous ABA content by radioimmunoassay method (Weiler, 1979).

412

Investigations on germinability during grain development in 31 cultivars and breeding lines indicated that the correlation between rate of seed germinability in the spike and that of isolated seed on filter paper at the same stage of development was highly significant. There were significant varietal differences in germinability at various grain development stages, however the heritability of this trait was relatively low. Germination rate of the seeds in the spike was relatively lower than that of the isolated seeds. Low germination rate in the spike was associated with white glume colour but not with tightness of the glume. A preliminary study of 15 cultivars indicated that varietal differences in endogenous ABA level and falling number were significant, yet their correlations with germinability were non-significant. Futher verification of these findings is in progress.

REFERENCES

Bhatt, G. M. and Derera, N.F. 1979. Potential use of Kenya 321-type dormancy in a wheat breeding programme aimed at evolving varieties tolerant to pre-harvest sprouting. Cereal Res. Commun. 8, pp. 291-295.

Bhatt, G. M., Ellison, F. W. and Mares, D. J. 1983. Inheritance studies in dormancy in three wheat crosses. In: Third Int. Symp. on Pre-harvest Sprouting in Cereals. Kruger, J. E. and LaBerge, D. E. (eds.). Westview Press, Boulder, Co. U.S.A. pp. 274-278.

De Pauw, R. M. and McCaig, T. N. 1983. Recombining dormancy from RL4137 with white seed colour. In: Third Int. Symp. on Pre-harvest Sprouting in Cereals. Kruger, J. E. and LaBerge, D. E. (eds.). Westview Press, Boulder, Co. U.S.A. pp. 251-259.

Gfeller, F. and Svedja, F. 1960. Inheritance of post-harvest seed dormancy and kernel colour in spring wheat lines. Can. J. Plant Sci. 40, pp. 1-6.

Khan, F. N. and Strand, E. A. 1977. Investigations into the genetics of kernel colour and dormancy in wheat (Triticum aestivum L.). Meldinger Fra Norges Landbrukahogskole. 56, pp. 1-12.

King, R. W. 1983. The physiology of pre-harvest sprouting - a review. In: Third Int. Symp. on Pre-harvest Sprouting in Cereals. Kruger, J. E. and LaBerge, D. E. (eds.). Westview Press, Boulder, Co. U.S.A. pp. 11-12.

Olssen, G. 1975. Breeding for sprouting resistance in wheat.
Proc. 2nd Int. Winter Wheat Conf. pp. 108-113.

Weiler, E. W. 1979. Radioimmunoassay for the determination
of free and conjugated abscisic acid. Planta. 144,
pp. 255-263.

Wu, Zhaosu and Bo, Yuanjia. 1962. Studies on the ecological
classification and zoning system of the wheat of China.
Scientia Agricultura Sinica. 11, pp. 12-20.

Dormancy in the Wild and Weedy Relatives of Modern Cereals

J. Mac Key, Department of Plant Breeding,
Swedish University of Agricultural Sciences,
Uppsala, Sweden.

The wild progenitors of modern cereals are specialized
to occupy habitats characterized by a shift from one season
of the year when plants can grow to one when they cannot
grow. They are adapted to avoid the adverse period as dry,
dormant seeds buried and stored in the surface soil or lit-
ter. This strategy necessitates a short life cycle ending
at the right time. An efficient dispersal unit adapted for
such a long period of inactivity is also important. Since
annual growth habit requires high efficiency every year at
re-establishment, the seeds have large endosperms. Such a
tactic gives immediate and plentiful access to energy at the
onset of growth.

The success of such a strategy depends largely on the
ability to synchronize cycles of growth, reproduction and
rest with changing season. Signals from the environment
must be interpreted and used to control metabolism and
development. Even the apparently inactive, dormant seed
must be able to react to such outside signals. An emphasis
on re-establishment also includes the potential for spreading. Accordingly, the wild cereals show their strength by
invading disturbed habitats such as those caused by weathering, erosion, fire, animal grazing, etc.

A STRATEGY OF PHENOTYPIC AND GENOTYPIC FLEXIBILITY

Their success is not all a matter of precision. Owing
to fluctuations over years in environmental signals, both
the active and the passive phase must be timed with a considerable buffering. This flexibility is achieved by a
strategy based on opportunism, partly solved by an ontogenetic variation in dormancy. Seeds from different parts of

the ear/panicle or even at different positions in a spikelet show different magnitudes of reaction to outside signals (Mac Key 1976).

Wild representatives like Hordeum spontaneum with one seed per spikelet have to rely primarily on variation along the ear which falls apart in triplets. Avena species with more than one seed per spikelet and where the whole spikelet acts as a dispersal unit shows a more gradual increase in dormancy from first basal grain inside each unit and up-wards. In the thaoudar variant of wild einkorn wheat or in Aegilops kotschyi, the system is even more elaborate but the other way round. Two seeds are normally formed per spike-let. The basal seed is smaller, richer in protein and darker and has a lower level of endogenous gibberellin and a hull offering a greater barrier, both physically and bio-chemically, than that of the second, larger seed. The lat-ter germinates in the thaoudar wheat in autumn, the small seed first in the spring and guarantees survival even after severe winters. Ae. kotschyi uses the same design to over-come the risk of failure in a desert climate. Less elabo-rate systems, found in Hordeum and Avena, are also able to show the same span in ontogenetic variation, which even ena-bles seeds to wait for a following favourable season.

In fact, dormancy appears never to be absolutely syn-chronized within genotype. Even in modern, highly homo-zygous cultivars, germinability revealed by pre-harvest sprouting follows a positional sequence.   In wheat and barley for example, it starts at the tip, proceeds at the base and ends in the middle of the ear. It is also well proven that temperature experienced by the maternal plant during seed development, and by seeds following maturation, has an important, phenotypic effect on depth and length of dormancy (Sexsmith 1969; Sawhney and Naylor 1979, 1982; Reddy et al., 1985).

Heterogeneity at the individual level is generally strengthened by heterogeneity at the population level (cf. Sexsmith 1967; Naylor 1983). Together they provide an insurance.

This does not imply that the environment dependent selection pressure is ineffective. With a mild type of autogamy, most of the wild cereals can alternate between genotypic fixation and recombination in a very efficient way for evolutionary progress.

AN EVOLUTIONARY DIVERSIFICATION PROCESS

The domestication process can be looked upon as a more radical example of such an evolutionary renewal. In this case, man guarantees preservation and sowing. As a consequence, nonshattering biotypes without deep seed dormancy become more functional.

Unconsciously, man opened up another evolutionary pathway by breaking up and preparing land. In parallel with the domestication process followed an adaptation to another existence on man-made land. The weedy representatives of the cereal species developed. Rye and oats had even started to adapt in cultivated fields as weeds, before they were themselves taken up for cultivation (Vavilov 1926; Coffman 1961).

During the long period of more or less unconscious selection by man to form his cultivars, the evolution towards weeds might have been just as successful. The opportunistic strategy of the annual wild cereals made them well fitted for both pathways. They were pre-adapted to man-made environments. Accordingly, we have weed wheats as well as weedy relatives within **Aegilops**, weed barleys, weed rices, weed maizes, weed sorghums, etc. (Harlan 1965).

The truly wild progenitors are specilaized to their endemic habitats. This does not imply that they were highly restricted in their intra-specific differentiation in ecotypes. Rather, their particular strategy for survival over years forced them to adapt to the edaphic and average climatic situation in each niche occupied. Rice with its adaptation to deeply flooded, swampy and upland habitats is perhaps the best example of adaptive breadth where water was a factor. Barley and wheat illustrate the capacity to adjust to a wide range in temperatures; having been able to climb from lowland to high altitudes. Wild oats have demonstrated their ability to spread widely into different climatic zones and daylength regimes.

It is reasonable to suggest that the wide spectrum in ecological adaptation is an accumulated result of the climatic shift in the region concerned (Mac Key 1976). When the latest glacial period ended about 20,000 years ago, a cool and humid climate was gradually exaggerated with respect to seasonal extremes. Climatic types such as the Mediterranean developed over much wider areas. Demands arose for adaptation to dry, hot summers rather than cold winters. Differences in altitude or other topographic variations preserved old achievements and allowed them to accumulate along with the new ones.

PARTIAL COEVOLUTION OF WEEDY AND CULTIVATED FORMS

A characteristic of weedy and cultivated derivatives is that man has encouraged them to adapt to even wider climatic regimes. By preparing land, man offered the disturbed habitats, into which the wild cereals, already so well prepared, could invade.

At the onset of agriculture, the capacity for a rapid shift in evolution was apparently already available. In the beginning, the distinction between truly wild, weedy and cultivated biotypes must have been diffuse. Gradually, they have become more divergent although still with patterns in common. Many taxonomists erroneously look upon this process as a completed speciation. The intermating is still a potent evolutionary factor.

The adaptation towards weediness made it necessary to retain the basic principle of seed shattering and dormancy. Adapted to more fertile and continuously disturbed land, the weedy forms have diverged from their ancestors in a number of other morphological and physiological characteristics. Where the two types coexist, the integrity of the wild type becomes more and more threatened owing to gene migration both from the weedy and the cultivated relatives. When found on slopes, road banks, etc. in the vicinity of agriculture, the genuine status of the wild type can always be questioned. Barley may serve as an illustrative example of such interwoven processes. The brittle six-rowed Hordeum agriocrithon has been taken both as merely a secondary hybrid derivative with weedy characters predominating (Zohary 1960) and as an alternative prototype (lit. cit. Hammer 1984).

In features where wild and weedy relatives have gradually diverged, the weedy and cultivated forms show coevolution. The weeds are not always as successful as when they are directly promoted by man, but the same incentives are there. Since weeds and cultivars inhabit the same type of land, occur intermixed and can intermate, the weedy types are, however, able to hitchhike directly and by hybrid introgression. They are thus able to tap efforts laid down by man. Even if such a process has definitely promoted the general adaptation to new agroecological niches, the capacity to change the weedy syndrome must mainly be an evolutionary process of its own.

Before present seed control systems, the gene flow in the direction of the crop plant was by no means completely blocked. The aim of the present symposium is i.a. to learn the best way to take up this possibility again in order,

atavistically, to improve resistance to pre-harvest sprouting.

The potential for hybrid introgression will greatly depend on the isolating mechanisms. For safe survival, the
weedy populations generally flower earlier than the surrounding cultivar resulting in a kind of isolation in time.
It is difficult to find genetic systems which have directly
evolved as a true speciation process. A rare case is the K_1
and K_2 genes in wheat prohibiting outcrossing with rye
(Riley and Chapman 1967). Genic barries such as complementary hybrid necrosis or chlorosis in wheat (Tsunewaki 1968)
have not been found to show a distinction between weed and
cultivar. Barriers have evolved as for example in rice (Oka
1974), but for other reasons. They are often dependent on
differences in ploidy level as in wheat and oats, but this
has nothing to do with the distinction between weed and cultivar. It might be observed that cross-pollinated species
like rye have difficulty in maintaining the two separate
types. For that reason, rye as weed is mostly found in
wheat fields.

When comparing different capacities of wild cereals to
develop weediness, there are apparently major differences in
time. A case study from Great Britain may illustrate this
(Thurston 1954). In this instance Avena fatua is an old
weed, whilst A. sterilis ssp. ludoviciana is a more recent
introduction. Both are originally adapted to mature and
develop seed dormancy in warm, dry summer weather with
germination occuring mainly in the autumn. A. sterilis is
still limited to the southeastern, warmest part of England
and to heavy soils where winter cereals are grown. A. fatua
has a much wider edaphic and climatic adaptation, going far
to the north and is particularly associated with spring
cereals. This noxious weed has also adapted therefore to
cool and rainy summers, hibernation and a germination in
spring late enough to escape being wiped out during seed-bed
preparation. In the north, A. fatua is adapted to regulate
its dormancy-germinability process by different signals from
those originally developed in the Mediterranean type of climate. Indeed it is more noxious than A. strigosa which was
already adapted as a wild plant to Northwestern Europe.

Other environmental differences will also enforce new
adaptation patterns. Survival on untouched as against
annually cultivated land emphasizes different capabilities.
The risk that animals will eat up the seeds is larger in
nature than if the plough buries the seed deep in the soil.
Observations tell us that up to two third of the seeds of
wild oats or rice may be lost by predation (Marshall and

Jain 1970;   Oka 1976;   cf. also Cavers 1983).   An efficient
self-burial offers an escape mechanism as well as less
strain during an adverse season and will thus give selective
advantage (Somody et al., 1985).

Weeds on ploughed land have other damands for survival.
Their seeds must be able to stay alive preferably over several years and maintain dormancy even in wet soils if buried
too deep.   But they must nevertheless be ready to germinate
when again brought up to the surface.   A. sterilis is not
able to cope with this situation but special A. fatua biotypes are (Whittington et al., 1970).   They may stay alive
for at least seven years (Banting 1965) and are able to
react by germination when first exposed to light and
increased temperature at an appropriate moisture regime.

The remarkable adaptability of A. fatua populations to
respond to selection pressures even from such factors as
different land management regimes has been demonstrated by
Naylor (1983).   The selective advantage for types with
longer dormancy could already be observed in some fatua
populations after only one summer fallow.   The ability to
stay alive over more than one year is more important in this
situation than under continuous cropping.

WILD AND WEEDY FORMS AS SOURCES FOR SPROUTING RESISTANCE

From what has been said, it should be quite clear that
wild and weedy types should not be considered alike nor
handled as a homogeneous group just because they all show
seed dormancy.   The weedy types in particular may represent
a wide spectrum of patterns adapted to different signals and
with different lengths and types of dormancy.   These differences must be borne in mind during attempts to tap weedy
types for genes in order to improve resistance to pre-
harvest sprouting among their cultivated relatives.   As a
general rule weedy representatives from the same area for
which the plant breeding programme is intended, should be
preferred.   Their pattern ought to be better tuned to the
relevant situation than that of others.

When dealing with the transfer of desirable genes for
seed dormancy from wild or weedy relatives, the literature
on genetic regulation gives little guidance.   In fact, the
few breeding experiments, almost entirely concentrated on A.
fatua, have raised more questions than they have answered.
With high phenotypic plasticity caused by a considerable
maternal influence, it is difficult to have to rely on such
an arbitrary measure as rate of germination.   Tests under

standard conditions may not recognize certain relevant genes
which are phenotypically silent and thus not recorded. Jana
and Naylor (1980) estimate, however, that half of the variation in their fatua populations could be ascribed to environmental and half to genetic effects. The dissemination
device is mostly regulated by simple inheritance or tight
linkage of major genes (Mac Key 1959, 1966; Takahashi and
Hayashi 1964). A similar selective advantage could also be
expected for the wild types for seed dormancy but appears
less likely although not proved. Rather the higher demand
on flexibility among the weeds would suggest a more poly-
genic regulation.

It is not understood how adaptation has evolved from
short- to long-term dormancy or how low instead of high
temperature becomes critical at maturation when the reverse
applies to germination. It appears as if the plant is working with different genetic blocks with distinct physiological functions triggered by such outside factors as temperature, oxygen, light, etc. (Sawhney and Naylor 1979; Jana
et al., 1979; Naylor 1983). It is quite possible that the
regulation system thus observed is superimposed on some
respiratory deficiency or block interconnected by something
like the citric acid cycle or the pentose phosphate pathway
(cf. Bewley and Black 1982).

Whatever it may be, seed dormancy could almost be compared to a set of fortifications one outside the other. For
obvious reasons, this system is much more elaborate in the
wild and weedy types. Many components may not even be of
interest in connection with pre-harvest sprouting resistance
in cultivars. The ability to stay dormant in deep soils for
example has little value.

Conflicts with other agronomic evaluations and specific
quality demands may hamper a transfer programme. The influence of ear characteristics on water uptake and evaporation
(Varis and Manneri 1963; King and Richards 1984) must be
matched with the knowledge that one third of all assimilates
going into the grain may come from the ear (Mac Key 1984).
Differences in the evaporation of ear water are illustrated
for awned and awnless isogenic lines of spring wheat in
FIGURE 1.

Glume or hull thickness and attachment to the caryopsis
are traits where quality considerations intervene. The reason why some spelt types of wheat are still grown in mount-
ainous regions is undoubtedly connected with their better
protection against rain showers.

A thick and tight seed coat and testa are also undesirable features characteristic of wild and weedy types. Not

FIGURE 1. Time course under field conditions of ear drying after two separate rain showers for one awned and one awnless isogenic line of spring wheat.

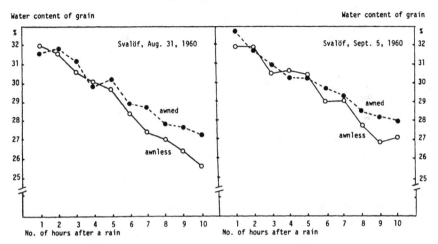

only physical but also biochemical protection mechanisms have to be weighed against nutritional quality or yield and whiteness of flour. Similar industrial constraints may be laid upon endosperm structure and metabolic status partly governed by signals from the embryo itself. It should, however, be noted that wild and weedy relatives are generally characterized by a higher protein concentration in the grain/groat (Youngs and Peterson 1973; Millet et al., 1984) which might have some positive connection with dormancy.

Exploiting the deeper seed dormancy of the wild and weedy relatives of our modern cereals is clearly not without problems. It would appear advisable to set up an ecoideotype for pre-harvest sprouting resistance, taking account of conflicts with quality demands and to use a more analytical approach than that generally applied in breeding for yield.

REFERENCES

Banting, J. D. 1965. Studies on the persistence of Avena fatua. Can. J. Plant Sci. 46, pp. 129-140.
Bewley, J. D. and Black, M. 1982. Physiology and biochemistry of seeds in relation to germination. 2. Viability, dormancy and environmental control. Springer-Verlag, Heidelberg. p. 375.

Cavers, P. B. 1983. Seed demography. Can. J. Bot. 61, pp. 3578-3590.

Coffman, F. A.(ed.) 1961. Oats and oat improvement. Amer. Soc. Agron., Madison. p. 650.

Hammer, K. 1984. Bestäubungsökologische Merkmale und Phylogenie von Hordeum L. subgen. Hordeum (Pollination ecological traits and phylogeny of Hordeum L. subgen. Hordeum). Flora. 175, pp. 339-344.

Harlan, J. R. 1965. The possible role of weed races in the evolution of cultivated plants. Euphytica. 14, pp. 173-176.

Jana, S., Acharya, S. N. and Naylor, J. M. 1979. Dormancy studies in seed of Avena fatua. 10. On the inheritance of germination behaviour. Can. J. Bot. 57, pp. 1663-1667.

Jana, S. and Naylor, J. M. 1980. Dormancy studies in seed of Avena fatua. Heritability for seed dormancy. Can. J. Bot. 58, pp. 91-93.

King, R. W. and Richards, R. A. 1984. Water uptake in relation to pre-harvest sprouting damage in wheat: ear characteristics. Aust. J. Agric. Res. 35, pp. 327-336.

Mac Key, J. 1959. Morphology and genetics of oats. Handb. Pfl. zücht. Kappert, H. and Rudolf, W. (eds.) Verlag P. Parey, Berlin, 2nd ed. 2, pp. 467-531.

Mac Key, J. 1966. Species relationship in Triticum. Proc. 2nd Inter. Wheat Genet. Symp., Lund 1963. Hereditas Suppl. 2, pp. 237-276.

Mac Key, J. 1976. Seed dormancy in nature and agriculture. Cereal Res. Comm. 4, pp. 83-91.

Mac Key, J. 1984. Assimilation and yield structure - a plant breeder's conclusion. Sveriges Utsädesf. tidskr. 94, pp. 135-144.

Marshall, D. R. and Jain, S. K. 1970. Seed predation and dormancy in the population dynamics of Avena fatua and A. barbata. Ecology. 51, pp. 886-891.

Millet, E., Levy, A. A., Avivi, L., Zamir, R. and Feldman, M. 1984. Evidence for maternal effect in the inheritance of grain protein in crosses between cultivated and wild tetraploid wheats. Theor. Appl. Genet. 67, pp. 521-524.

Naylor, J. M. 1983. Studies on the genetic control of some physiological processes in seeds. Can. J. Bot. 61, pp. 3561-3567.

Oka, H. 1974. Experimental studies on the origin of cultivated rice. Proc. 13th Inter. Congr. Genet., Berkely 1973. Genetics. 78, pp. 475-486.

Oka, H.-I. 1976. Mortality and adaptive mechanisms of Oryza perennis strains. Evolution. 30, pp. 380-392.

Reddy, L. V., Metzger, R. J. and Ching, R. M. 1985. Effect of temperature on seed dormancy of wheat. Crop Sci. 25, pp. 455-458.

Riley, R. and Chapman, V. 1967. The inheritance in wheat of crossability with rye. Genet. Res. Cambridge. 9, pp. 259-267.

Sawhney, R. and Naylor, J. M. 1979. Dormancy studies in seed of Avena fatua. 9. Demonstration of genetic variability affecting the response to temperature during seed development. Can. J. Bot. 57, pp. 59-63.

Sawhney, R. and Naylor,J. M. 1982. Dormancy studies in seed of Avena fatua. 13. Influence of drought stress during seed development on duration of seed dormancy. Can. J. Bot. 60, pp. 1016-1020.

Sexsmith, J. J. 1967. Variental differences in seed dormancy of wild oats. Weeds. 15, pp. 252-255.

Sexsmith, J. J. 1969. Dormancy of wild oat seeds produced under various temperature and moisture conditions. Weed Sci. 17, pp. 405-407.

Somody, C. N., Nalewaja, F. D. and Miller, S. D. 1985. Self-bureal of wild oat florets. Agron. J. 77, pp. 359-362.

Takahashi, R. and Hayashi, J. 1964. Linkage study of two complementary genes for brittle rachis in barley. Ber. Ohara Inst. Landw. Biol. 12, pp. 99-105.

Thurston, J. M. 1954. A survey of wild oats (A. Fatua and A. ludoviciana) in England and Wales in 1951. Ann. Appl. Biol. 41, pp. 619-636.

Tsunewaki, K. 1968. Origin and phylogentic differentiation of common wheat revealed by comparative gene analysis. Proc. 3rd Inter. Wheat Genet. Symp., Canberra 1968, Aust. Acad. Sci., Griffin Press, Netley. pp. 71-85.

Varis, E. and Manneri, M. 1963. The effect of the morphological properties of the ear on the susceptibility of winter wheat to sprouting in the ear. Maataloust. Aikak. 35, pp. 27-35.

Vavilov, N. I. 1926. Studies on the origin of cultivated plants. Bull. Appl. Bot. Pl. Breed. 16, (2) pp. 1-248.

Whittington, W. J., Hillman, J., Gatenby, S. M., Hooper, B. E. and White, J. C. 1970. Light and temperature effects on the germination of wild oats. Heredity 25, pp. 641-550.

Youngs, V. L. and Peterson, D. M. 1973. Protein distribution in the oat (Avena sterilis L.) kernel. Crop Sci. 13,

424

pp. 365-367.
Zohary, D. 1960. Studies on the origin of cultivated
    barley. Bull. Res. Council Israel. 9D, pp. 21-42.

Mechanisms of Dormancy in Wild Oats (Avena fatua)

A. I. Hsiao, Agriculture Canada, Research
Station, Box 440, Regina, Saskatchewan,
Canada, S4P 3A2.

SUMMARY

Duration of primary and secondary seed dormancy in
field populations of wild oats (Avena fatua L.) is
controlled by several genes which are expressed as different
physiological and biochemical traits. Mechanisms producing
variation in levels of primary and secondary dormancy
include hormonal and environmental control of biochemical
pathways which can interact with components of seed structure. Water is undoubtedly the one environmental factor
with greatest capacity for germination control, and it has
been suggested that seed dormancy in A. fatua is due to
factors that prevent the uptake of imbibed water by the
embryo in the amount required for the induction of germination. Light inhibits germination in seeds of genetically
nondormant populations. This light-induced germination
inhibition is intensity-dependent: the higher the light
intensity, the greater the inhibition. Germination inhibition by light is accentuated by higher incubation temperatures. Temperature can not only administer its effect
directly on seed metabolism, but can also mediate its effect
either through maternal tissues or on the immature ovule.
Thus temperature can regulate germination behaviour by its
effect on three stages of the life cycle of A. fatua plants,
i.e., vegetative growth, seed maturation, and the incubation
period following seed maturation. The total effect of
genetic and environmental factors at these stages of the
life cycle may provide flexible strategies for their survival. However, in spite of numerous studies on factors that
influence the degree of dormancy of seeds, there is still no
general agreement about the nature of the control mech-
anism(s) involved.

426

INTRODUCTION

Seed dormancy is the main survival mechanism in wild
oats (Avena fatua L.) (Sawhney et al., 1984, 1985). A.
fatua is a highly inbreeding species with the observed
frequencies of outcrossing being less than 1% (Imam and
Allard, 1965). This feature has facilitated both the demonstration of polymorphism controlling the germination
behaviour in individual populations and the isolation of
pure lines which differ in their requirements for germination. The differences in germination behavious have a
genetic basis showing the level of heritability of seed
dormancy to be about 0.5 under field conditions (Jana and
Naylor, 1980). Breeding experiments showed that parental
lines differ in at least 3 genes controlling rate of after-
ripening (Jana et al., 1979). The aspect of genetic control
of seed dormancy in A. fatua was reviewed by Naylor (1983).
However, after reviewing several hundred papers, Simpson
(1978) concluded that much of the information published,
especially before 1960, is of little value because the
genetic basis of seed dormancy was not appreciated. It is
desirable, in principle, to work with uniform, inbred populations when investigating the mechanism of seed dormancy.
The aspects of seed dormancy in A. fatua have been
reviewed extensively (Chancellor, 1976; Hsiao and Quick,
1983; Naylor, 1983; McIntyre and Hsiao, 1983; Quick and
Hsiao, 1983; Simpson, 1978, 1984). The objective of this
paper is to review mechanisms of dormancy in A. fatua,
providing mainly the new information available since the two
extensive reviews by Simpson (1978, 1984).

MECHANISMS OF DORMANCY IN A. FATUA

Role of seed structure

Little new information has been published on the role
of seed structure since Simpson's review in 1978 except the
studies carried out by Hsiao and his associates on the
effects of piercing and cutting on the germination of dormant caryopses.
Puncturing the seed coat increased germination (Hay,
1962; Hsiao, 1979; Hsiao and Simpson, 1971; Hsiao et al.,
1983; McIntyre and Hsiao, 1985; Raju et al., 1986) and
oxygen uptake (Hay, 1962). Hsiao et al. (1983) also
observed that (i) the rate of increase in germination by
piercing both the pericarp and aleurone layer of the imbibed

caryopses was inversely related to the distance of the hole
from the embryo, and (ii) the rate of germination induced by
piercing or cutting the imbibed caryopsis was significantly
increased by the application of water to the site of injury.
The induction of germination by mechanical injury might
result from a localized increase in water potential at the
site of injury. The water-potential gradient thus established could promote the flow of water to the embryo, raising its water potential to the critical level required to
trigger germination. Raju et al. (1986) demonstrated that
germination induction by piercing imbibed dormant caryopses
also promoted rapid elongation of the scutellum. This
response was associated with the development of scutellar
papillae and the induction of xylem differentiation and cell
division in the distal region of the scutellum. These
effects preceded protrusion of the radicle by about 24h.
This was consistent with previous evidence that embryo water
content is a limiting factor (McIntyre and Hsiao, 1983,
1985) and that seed dormancy in A. fatua is caused by factors that prevent the uptake of water by the embryo in the
amount required for germination (Hsiao et al., 1983;
McIntyre and Hsiao, 1985).

Role of environment

After-ripening. Seed dormancy in A. fatua is lost when
seeds are stored, generally under dry conditions. However,
no new information has appeared since the publication of
previous reviews (Chancellor, 1976; Simpson, 1978).
Water. Hsiao and Simpson (1971) demonstrated that any
desired percentage of germination could be obtained from a
population of nondormant seeds either by choosing a specific wavelength of light and varying the volume of water or
vice versa. The inhibition of germination by light is
mediated by a reduction in water potential. However, if
nondormant seeds were completely immersed in water they
became secondarily dormant (Hay, 1962; Hsiao and Quick,
1983). The induction of secondary dormancy by water immer-
sion was affected by genotype, age of seeds, and various
wetting and drying cycles (Hsiao and Quick, 1983). Variability in germination of nondormant seeds associated with
variation in available water is the result of an interaction
of water with hulls and, possibly, the pericarp and testa,
which in turn modifies the effects of light and movement of
gases, metabolites and growth regulators essential to germination.
There are generally two phases of water uptake by

seeds. In A. fatua water is readily absorbed to about 30%
dry weight within several hours (Atwood, 1914). This is
followed by a gradual increase taking 40 to 48h to reach 40%
moisture content (Hsiao et al., 1983) and about 7 days to
reach the maximum water content of about 75% (Atwood, 1914;
Hsiao et al., 1983). McIntyre and Hsiao (1985) suggested
that water uptake in dormant seeds is a limiting factor for
germination. They found that dormant caryopses with 75%
moisture content after 10 days of imbibition actually had
about 121 and 60% water (DW basis) in embryo and enclosing
tissues, respectively. Dormant embryos excised from fully
imbibed caryopses and placed on wet filter paper, showed
water content increases of about 10% within 30 min. This
response was followed, after a further 18h, by germination.
A similar increase in embryo water content occurred 12-18h
after piercing of the imbibed caryopsis and about 72h before
the resumption of growth and elongation of the radicle.
Embryo water content is thought to be a limiting factor and
seed dormancy in A. fatua is presumed to be caused by factors that prevent the uptake of imbibed water by the embryo
in the amounts required for germination.

Gases. Germination of A. fatua is increased by the
presence of added atmospheric oxygen and CO_2 (Hart and
Berrie, 1966; Hay, 1962; Simmonds and Simpson, 1971). Both
gases are essential for germination and have important
effects on germination through their effects on metabolism,
particularly in the embryo. Simpson's review (1978) still
provides a curent summary of this subject.

Light. Conflicting report exist concerning the effect
of light on germination of A. fatua. Atwood (1914) reported
no effect of light. Several researchers found no effect of
light on nondormant or dormant seeds but Hart and Berrie
(1966) noted light inhibition of partially dormant seeds
while Hilton (1984, 1985) and Hilton and Bitterli (1983)
reported the opposite. Hsiao and Simpson (1971) observed
inhibition or promotion of after-ripened seeds of a dormant
line by white light depending upon water availability.
Sawhney et al. (1986) explained these discrepancies on the
basis of diferences in genetic make-up of the seeds, their
innate dormancy status, after-ripening conditions and duration, place of ecological origin, light, temperature and
moisture conditons used, and differences in parental
environment.

Sawhney et al., (1986) found that light inhibited germination in freshly harvested seeds of several nondormant
lines. This inhibition was accentuated by the higher temperatures, indicating an interaction between these two

factors on the induction of secondary dormancy in nondormant
lines. This interaction may be of adaptive significance to
the survival of nondormant populations.

Temperature. Naylor and Fedec (1978) demonstrated the
interaction of temperature with genotype in controlling germination in A. fatua. The temperature to which plants are
exposed is known to have an important effect on the expression of genes conferring dormancy in A. fatua (Peters, 1982;
Sawhney and Naylor, 1979, 1980, 1982; Sawhney et al., 1984,
1985, 1986). The ambient temperature is able not only to
administer its effect directly on seed metabolism, but can
also mediate its effect either through maternal tissues or
on the immature ovule. Thus temperature can regulate germination behaviour by its effect on three stages of the life
cycle of A. fatua plants, i.e., vegetative growth (Sawhney
et al., 1985), seed maturation (Sawhney and Naylor, 1979,
1980, 1982), and the incubation period following seed maturation (Sawhney and Naylor, 1980, 1982; Sawhney et al.,
1984, 1986). The imposition of water stress on plants
during seed development also strongly reduces the level of
seed dormancy in dormant-type lines. Moreover, there are
significant differences among these families in the level of
response to both drought and temperature. Sawhney et al.
(1984, 1985, 1986) have suggested that this kind of phenotypic plasticity has adaptive value as a survival mechanism
for A. fatua.

Growth regulators. Simpson (1978) suggested that
growth regulators may have a role in the initiation, maintenance and termination of dormancy in A. fatua, however,
most evidence has been obtained from the effects of the substance applied exogenously. The most documented regulatory
compound is gibberellin A_3 (GA_3) by Naylor and Simpson
(1961b), i.e., excised embryos of A. fatua showed two responses to GA_3 with different concentrations. The higher
dose (ca. 50 ppm) was associated with conversion of starch
to sugars and the lower (ca. 0.1-1 ppm) with sugar utiliz-
tion. Comparing dormant and nondormant lines in terms of an
endogenous diffusible GA-like substance proved that synthesis of the GA-like substance(s) in the embryos was
blocked in mature dormant seeds (Simpson, 1965). Simpson
(1978, 1984) has provided detailed accounts of findings on
the role of GA_3 in the termination of seed dormancy in A.
fatua. Other chemicals with promotive effects on germination of dormant seeds are KNO_3 (Adkins et al., 1984b, c;
Hilton, 1984, 1985), H_2O_2 (Hay, 1962; Hsiao and Quick,
1984), sodium hypochlorite (NaOCl)(Hsiao, 1979; Hsiao and
Hanes, 1981; Hsiao and Quick, 1984, 1985; Hsiao et al.,

1979), ethylene (Adkins and Ross, 1981) substituted phthalimides (Metzger, 1983; Tilsner and Upadhyaya, 1985: Upadhyaya et al., 1986), aluminium phosphide (Cairns and De
Villiers, 1980), azide (Adkins et al., 1984a-d; Upadhyaya
et al., 1982b, 1983), ethanol (Adkins et al., 1984c,d), and
certain organic acids (Adkins et al., 1985).

A germination inhibitor found in both primary and secondary seeds was identified as abscisic acid (Berrie et al.,
1975). It could prevent the GA-promoted synthesis of alpha-
amylase. Naylor and Simpson (1961a) suggested that dormancy
is controlled by a GA/inhibitor antagonism and that the control of germination during after-ripening was due to changes
in the level of inhibitor and not of GA. Simpson (1966)
further suggested that after-ripening may be the gradual
loss of a natural growth retardant that prevents the synthesis of GA in dormant embryos.

Metabolic system

Endosperm. Only new information in the areas of enzymes and phosphorus is provided here.
1. Enzymes. Regulation of synthesis of a hydrolase in
the endosperm, alpha-amylase, showed that excised endosperm
segments of the dormant-type lines examined do not synthesize an appreciable amount of alpha-amylase, though they do
so readily when supplied with GA_3 (Upadhyaya et al., 1982a).
In contrast, nondormant lines differ widely in their capacity for this 'autonomous' (embryo-independent) production
of the enzyme in their aleurone tissue. The degree of endosperm autonomy increased with age in both dormant and non-
dormant lines of A. fatua. The expression of this autonomy
involves the translation and the transcription of alpha-
amylase. Embryo dormancy and endosperm integrity are governed by a multiple gene system within which the genes controlling the traits are linked (Jong, 1984). This indicates
that in nature there is strong selection pressure favouring
the coupling of long-term dormancy and a rigorous hormonal
control by the embryo of the production of at least one of
the hydrolases. However, this selection pressure is relatively relaxed in nondormant families.

Hooley (1984a,b) demonstrated that the aleurone layer
in A. fatua produces acid phosphatase and GA_3 stimulates
this production, its secretion across the plasma membrane
and release through and from the aleurone cell wall. Hooley
(1984b) showed that GA_3 also controls secretion of other
specific molecular forms of phosphatase that probably assist
in endosperm hydrolysis. Zwar and Hooley (1986) reported

that the alpha-amylase is accompanied by a rise in the level
of alpha-amylase mRNA in aleurone protoplasts of A.fatua.
They suggested that hormone-mediated changes in alpha-amylase mRNA levels can be largely accounted for by changes in
the level of transcription of alpha-amylase genes.

    2. Phosphorus. The level of endogenous inorganic
phosphate (Pi) available to the embryo of A. fatua is related to the ability of the seed to germinate (Hsiao et al.,
1984; Jain et al., 1982, 1983; Kovacs and Simpson, 1976;
Quick and Hsiao, 1983, 1984; Simpson, 1965). The Pi in
dry, mature, primary and secondary caryopses of dormant
lines increased with the duration of after-ripening. Very
dormant lines were found to possess lower levels of Pi than
less dormant lines grown under equivalent conditions (Jain
et al., 1982). Seeds produced by phosphorus-stressed plants
contained less Pi and were more dormant than seeds from
unstressed plants of the same line grown in the same controlled environment chamber (Quick et al., 1982/83). Secondary seeds from a given panicle contained less Pi and were
more dormant than primary seeds from the same panicle (Quick
and Hsiao, 1983); while immature seeds of dormant or non-
dormant lines alike were rich in Pi and germinated readily
before desiccation and reduction in Pi level that accompanied seed maturation (Quick and Hsiao, 1983).

    Embryo and scutellum. Simpson (1978, 1984) has given
a detailed account of the involvement of oxidative metabolism including the pentose phosphate (PP) pathway and Krebs
cycle. However, Fuerst et al., (1983) have demonstrated
that the PP pathway is not involved in the mechanism of
natural termination of dormancy in A. fatua. Studies on
the role of various respiratory pathways in the control of
seed dormancy in A. fatua suggested that an increase in the
activity of the Krebs cycle and/or glycolysis may play an
important part in the breaking of dormancy of partly after-
ripened caryopsis (Adkins et al., 1984a-d; Upadhyaya et al.
1982b, 1983).

    Many inhibitors of respiration, especially those of
mitochondrial electron transport, that block several pathways leading to acceptance of electrons by cytochromes, are
known to overcome dormancy in A. fatua (Adkins et al.,
1984a-d, 1985; Upadhyaya et al., 1982b, 1983). Azide was
effective in breaking dormancy in Anderson pure lines (AN
51, 86, 127, and 265) but failed to stimulate germination in
MON 73 (Upadhyaya et al., 1982b). SHAM (salicylhydroxamic
acid), an inhibitor of alternative respiration, prevented
this azide-induced stimulation while SHAM by itself had no
effect. This indicates that alternative respiration is

necessary for the stimulation of germination in the presence
of azide but it is not necessary for the germination of
genetically nondormant, GA3-treated dormant or after-ripened
seeds. Nitrate, organic acids and ethanol also break dormancy in A. fatua. CCC, (2-chloroethyl)trimethylammonium
chloride, an inhibitor of GA biosynthesis, inhibited the
germination stimulation induced by azide, nitrate and ethanol indicating that germination stimulation by these three
promoters require GA biosynthesis (Adkins et al., 1984a-c).
The stimulation of oxygen uptake and germination by nitrate,
ethanol and citric acid were not greatly inhibited by SHAM
indicating that the operation of alternative respiration was
not required for these three promoters to have an effect
(Adkins et al., 1984a-c). The nitrate- and ethanol-induced
stimulation of germination and oxygen uptake are not inhibited by azide, but are inhibited when both azide and SHAM
are applied together. This indicates that stimulation of
oxygen uptake and germination by nitrate and ethanol require
only the operation on one of the pathways of respiration.

The stimulation of germination by ethanol was moderate
on freshly harvested seeds and increased steadily during 6
months of after-ripening (Adkins et al., 1984c). Sensitivity to ethanol was detected much earlier during after-
ripening than to nitrate, organic acids and azide indicating
the mode of action of ethanol differs from that of nitrate,
organic acids and azide (Adkins et al., 1984d). Adkins et
al., (1984c) suggested that ethanol may have two roles in
breaking dormancy: as a metabolic substrate and by having
an anaesthetic effect on the membrane. Adkins et al.,
(1985) reported that organic acids such as citric, succinic,
fumaric, malic, pyruvic and lactic acids break dormancy in
two genetically dormant lines of A. fatua. They suggested
that the mode of action of organic acids in overcoming dormancy in partially after-ripened caryopses is by lowering
pH.

A system concept

Several concepts for explaining the mechanism of dormancy in A. fatua were proposed. Simpson (1965) was first
to use a system concept to explain the mechanism of dormancy. The role of GA3 in mobilization and utilization of
endosperm reserves by the embryo was linked with the observation that dormant embryos failed to secrete GA-like
substances. Dormancy was regarded as a single block in the
production of GA3 within the embryo. This in turn imposed
two other blocks due to the requirement for GA as a trigger
for synthesis of enzymes necessary (i) to mobilize endosperm

reserves, and (ii) for the embryo to utilize these reserves.

Hsiao and Simpson (1971) modified the scheme to account for demonstrated or assumed effects of both seed structure and environment on the release or formation of the GA-like substance in the embryo. The effects of gases, promoter/inhibitor ratio, light (presumedly via the phytochrome system), and variations in water status were all incorporated to explain the loss of dormancy through an after-ripening period.

Simmonds and Simpson (1972) presented a model indicating how dormancy control may depend upon Krebs cycle activity and suggested that one aspect of after-ripening may be the development of a mechanism which inhibits this activity, thus releasing oxygen for the alternative PP pathway which is essential to germination. Now it is believed that the stimulation of the PP pathway is not a prerequisite for the onset of germination in A. fatua (Adkins and Ross, 1981; Fuerst et al., 1983; Upadhyaya et al., 1981).

Studying the role of various respiratory pathways in the control of seed dormancy in A. fatua, Adkins et al., (1984c) demonstrated that the operation of the alternative pathway is not required for germination unless the cytochrome pathway is blocked. They consequently proposed that in A. fatua promoters of germination such as azide, nitrate, ethanol, and organic acids act by increasing the activity of glycolysis and/ or the Krebs cycle. This elevated activity may directly or indirectly overcome dormancy by increasing the supply of substrates for synthesis of numerous essential cellular components (Adkins et al., 1984a-d, 1985; Upadhyaya et al., 1982b, 1983). These compounds overcome dormancy by similar but not necessarily identical mechanisms (Adkins et al., 1985). This hypothesis is attractive as it accounts for the dormancy-breaking activity of all categories of compounds on partly after-ripened A. fatua seeds.

Hsiao and Quick (1984) suggested H_2O_2 and NaOCl may act as agents modifying the properties of cellular or organelle membranes, as with ethanol (Adkins et al., 1984a, d). Studies on the role of NaOCl and H_2O_2 all showed that both agents made the seeds more sensitive to exogenous GA_3 treatment. Hsiao and Quick (1984) suggested two roles for NaOCl and H_2O_2 in the termination of dormancy in A. fatua; a modification of the properties of the hull and seed coat membranes, and in provision of additional oxygen to the seed. The increased response to exogenous GA_3 treatment may have arisen from other effects of both agents such as a a reduction in seed coat restriction or a loss of some inhibitor(s) of GA_3 action.

Dormancy in A. fatua developed rapidly between 18 and 22 days from anthesis (McIntyre and Hsiao, 1983, 1985). At the beginning of this period the embryo water content had declined to 119% (DW) which was about the same embryo water content as mature caryopses after imbibition of water for 10 days. It appears that the tissues of intact caryopses restricted the uptake of water by the embryo, maintaining the embryo water content at a value (ca. 120%) that was below the critical value required for induction of germination. This value, at which water still appeared to be the limiting factor, was similar to that at which dormancy started to develop during seed maturation. McIntyre and Hsiao (1983, 1985) found that germination was preceded by a significant increase in embryo water content and suggested that this apparent lag of 12-18h between the piercing treatment and the increase in the embryo water content may reflect the time required for the transport of water to the embryo from the injured tissues. It was postulated that (i) piercing or cutting the imbibed caryopses caused an increase in water potential at the site of injury and that this would promote the flow of water to the embryo, triggering germination, and (ii) dormancy in A. fatua may be caused by factors that prevent the uptake of water by the embryo in the amount required for germination (Hsiao et al., 1983; McIntyre and Hsiao, 1983, 1985).

Simpson (1984) pointed out that most of our understanding of the control of dormancy in A. fatua has concentrated on the embryo. There is no doubt that all other parts of the seed/environment system contribute directly or indirectly, to the induction, maintenance or release from seed dormancy of A. fatua. Furthermore, there is genetic diversity in each of these subsystems. Although seed dormancy in A. fatua is an intensively studied systems, there is still no general agreement about the nature of the control mechanism(s) involved.

CONCLUDING REMARKS

The mechanism of dormancy in A. fatua is incompletely known. It does appear ultimately, to rest upon some trigger in the embryo which is set off, possibly by environmental conditions, and generates GA-like factors which in turn initiate growth of the embryo and release food from the endosperm for its growth. It would appear that the prevention of the breaking of seed dormancy is the best aspect of seed behaviour to exploit for control of this weed. A search should therefore be made for suitable treatments which could be applied to growing plants, maturing panicles or seed after it has been shed. Our aim is to find the physiological and genetic bases of seed dormancy so that a way to exploit weak points in the cycle of induction, maintenance and removal of dormancy of A. fatua can be discovered.

REFERENCES

Adkins, S. W. and Ross, J. D. 1981. Studies in wild oat seed dormancy. I. The role of ethylene in dormancy breakage and germination of wild oat seeds. (Avena fatua L.). Plant Physiol. 67, 358-362.

Adkins, S. W., Naylor, J. M. and Simpson, G. M. 1984a. The physiological basis of seed dormancy in Avena fatua. V. Action of ethanol and other organic compounds. Physiol. Plant. 62, 18-24.

Adkins, S. W. Simpson, G. M. and Naylor, J. M. 1984b. The physiological basis of seed dormancy in Avena fatua. III. Actions of nitrogenous compounds. Physiol. Plant. 60, 227-233.

Adkins, S. W. Simpson, G. M. and Naylor, J. M. 1984c. The physiological basis of seed dormancy in Avena fatua. IV. Alternative respiration and nitrogenous compounds. Physiol. Plant. 60, 234-238.

Adkins, S. W., Simpson, G. M. and Naylor, J. M. 1984d. The physiological basis of seed dormancy in Avena fatua. VI. Respiration and the stimulation of germination by ethanol. Physiol. Plant. 62, 148-152.

Adkins, S. W., Simpson, G. M. and Naylor, J. M. 1985. The phsyiological basis of seed dormancy in Avena fatua. VII. Action of organic acids and pH. Physiol. Plant. 65, 310-316

Atwood, W. M. 1914. A physiological study of the germination of Avena fatua. Bot. Gaz. 57, 386-414.

Berrie, A. M. M., Don, R., Buller, D., Alam, M. and Parker, W. 1975. The occurrence and function of short chain length fatty acids in plants. Plant Sci. Lett. 6, 163-173.

Cairns, A. L. P., and de Villiers, O. T. 1980. Effect of aluminium phosphide fumigation on the dormancy and viability of Avena fatua seed. S. African J. Sci. 76, 323.

Chancellor, R. J. 1976. Seed behaviour. Wild Oats in World Agriculture. Jones, D. P., (ed.). Agricultural Research Council, London. pp. 65-87.

Fuerst, E. P., Upadhyaya, M. K., Simpson, G. M., Naylor, J.
M. and Adkins, S. W. 1983. A study of the relationship
between seed dormancy and pentose phosphate pathway
activity in Avena fatua. Can. J. Bot. 61, 667-670.

Hart, J.W., and Berrie, A. M. M. 1966. The germination of
Avena fatua under different gaseous environments.
Physiol. Plant. 19, 1020-1025.

Hay, J. R. 1962. Experiments on the mechanism of induced
dormancy in wild oats (Avena fatua L.). Can. J. Bot.
40, 191-202.

Hilton, J.R. 1984. The influence of light and potassium
nitrate on the dormancy and germination of Avena fatua
L. (wild oat) seed and its ecological significance.
New Phytol. 96, 31-34.

Hilton, J. R. 1985. The influence of light and potassium
nitrate on the dormancy and germination of Avena fatua
L. (wild oat) seed stored buried under natural conditions. J. Exp. Bot. 36, 974-979.

Hilton, J. R., and Bitterli, C. J. 1983. The influence of
light on the germination of Avena fatua L. (wild oats)
seeds and its ecological significance. New Phytol.
95, 325-333.

Hooley, R. 1984a. Gibberellic acid controls the secretion
of acid phosphatase in aleurone layers and isolated
aleurone protoplasts of Avena fatua. J. Exp. Bot. 35,
822-828.

Hooley, R. 1984b. Gibberellic acid controls specific acid-
phosphatase isozymes in aleurone cells and protoplasts
of Avena fatua L. Planta. 161, 355-360.

Hsiao, A. I. 1979. The effect of sodium hypochlorite and
gibberellic acid on seed dormancy and germination of
wild oats (Avena fatua). Can. J. Bot. 57, 1729-1734.

Hsiao, A. I. and Hanes, J. A. 1981. Application of the
sodium hypochlorite seed viability test to wild oat
populations with different dormancy characteristics.
Can. J. Plant Sci. 61, 115-122.

Hsiao, A. I. and Quick, W. A. 1983. The induction and
breakage of dormancy in wild oats In: Wild Oat Symp.
Proc. No. 12. Smith, A. E. (ed.). Can Plains Res.
Center, Regina, Sask., Canada. pp. 173-185.

Hsiao, A. I. and Quick, W. A. 1984. Actions of sodium hypochlorite and hydrogen peroxide on seed dormancy and
germination of wild oats (Avena fatua). Weed Res. 24,
411-419.

Hsiao, A. I. and Quick, W. A. 1985. Wild oats (Avena fatua
L.) seed dormancy as influenced by sodium hypochlorite,
moist storage and gibberellin A_3. Weed Res. 25, 281-
288.

Hsiao, A. I. and Simpson, G. M. 1971. Dormancy studies in seed of Avena fatua. 7. The effects of light and variation in water regime on germination. Can. J. Bot. 49, 1347-1357.

Hsiao, A. I., MacGregor, M. E., and Banting, J. D. 1979. The use of sodium hypochlorite in testing the seed viability of wild oats. Can. J. Plant Sci. 59, 1047-1052.

Hsiao, A. I., McIntyre, G. I. and Hanes, J. A. 1983. Seed dormancy in Avena fatua. I. Induction of germination by mechanical injury. Bot Gaz. 144, 217-222.

Hsiao, A. I., Quick, W. A., and Jain, J. C. 1984. Phosphorus-containing compounds at comparable germination stages of caryopses of Avena species. J. Exp. Bot. 35, 617-625.

Imam, A. G., Allard, R. W. 1965. Population studies in predominantly self-pollinated species. VI. Genetic variability between and within natural populations of wild oats in differing habitats in California. Genetics. 51, 49-62.

Jain, J. C., Quick, W. A. and Hsiao, A. I. 1982. Studies of acid-soluble phosphorus compounds in genetically pure lines of Avena fatua with different dormancy characteristics. Can. J. Bot. 60, 2099-2104.

Jain, J. C., Quick, W. A. and Hsiao, A. I. 1983. ATP synthesis during water imbibition in caryopses of genetically dormant and non-dormant lines of wild oat (Avena fatua L.). J. Exp. Bot. 34, 381-387.

Jana, S., Acharya, S. N. and Naylor, J. M. 1979. Dormancy studies in seed of Avena fatua. 10. On the inheritance of germination behaviour. Can. J. Bot. 57, 1663-1667.

Jana, S. and Naylor, J.M. 1980. Dormancy studies in seed of Avena fatua. 11. Heritability for seed dormancy. Can. J. Bot. 58, 91-93.

Jong, F. S. 1984. The regulation of endosperm hydrolysis in Avena fatua. M. Sc. Thesis, Univ. of Saskatchewan, Saskatoon, Sask. pp. 76.

Kovacs, M. I. P. and Simpson, G. M. 1976. Dormancy and enzyme levels in seeds of wild oats. Phytochemistry. 15, 455-458.

McIntyre, G. I. and Hsiao, A. I. 1983. The role of water in the mechanism of seed dormancy in wild oats (Avena fatua L.). Wild Oat symp. Proc. No. 12. Smith, A. E. (ed.). Can. Plains Res. Center, Regina, Sask., Canada. pp. 187-200.

McIntyre, G. I. and Hsiao, A.I. 1985. Seed dormancy in

438

  Avena fatua. II. Evidence of embryo water content as a limiting factor. Bot. Gaz. 146, 347-352.

Metzger, J. D. 1983. Promotion of germination of dormant seeds by substituted phthalimides and gibberellic acid. Weed Sci. 31, 285-289.

Naylor, J. M. 1983. Studies on the genetic control of some physiological processes in seeds. Can. J. Bot. 61, 3561-3567.

Naylor, J. M. and Fedec, P. 1978. Dormancy studies in seed of *Avena fatua*. 8. Genetic diversity affecting response to temperature. Can. J. Bot. 56, 2224-2229.

Naylor, J. M. and Simpson, G. M. 1961a. Dormancy studies in seed of *Avena fatua*. 2. A gibberellin-sensitive inhibitory mechanism in the embryo. Can. J. Bot. 39, 281-295.

Naylor, J. M. and Simpson, G. M. 1961b. Bioassay of gibberellic acid using excised embryos of *Avena fatua* L. Nature. 192, 679-680.

Peters, N. C. B. 1982. The dormancy of wild oat seed (*Avena fatua* L.) from plants grown under various temperature and soil moisture conditions. Weed Res. 22, 205-212.

Quick, W. A. and Hsiao, A. I. 1983. The role of phosphorus in wild oat seed dormancy. In: Wild Oat Symp. Smith, A. E. (ed.). Proc. No. 12. Can. Plains Res. Center, Regina, Sask., Canada. pp. 161-172.

Quick, W. A. and Hsiao, A. I. 1984. Changes in inorganic phosphate and seed germinability during after ripening of wild oats. Can. J. Bot. 62, 2469-2472.

Quick, W. A., Hsiao, A. I. and Jain, J. C. 1982/83. Endogenous inorganic phosphate in relation to seed dormancy and germination of wild oats. Plant Sci. Lett. 28, 129-135.

Raju, M. V. S., Hsiao, A.I. and McIntyre, G. I. 1986. Seed dormancy in *Avena fatua*. III. The effect of mechanical injury on the growth and development of the root and scutellum. Bot. Gaz. (in press).

Sawhney, R. and Naylor, J.M. 1979. Dormancy studies in seed of *Avena fatua*. 9. Demonstration of genetic variability affecting the response to temperature during seed development. Can. J. Bot. 57, 59-63.

Sawhney, R. and Naylor, J.M. 1980. Dormancy studies in seed of *Avena fatua*. 12. Influence of temperature on germination behaviour of nondormant families. Can. J. Bot. 58, 578-581.

Sawhney, R. and Naylor, J. M. 1982. Dormancy studies in seed of *Avena fatua*. 13. Influence of drought stress during seed development on duration of seed dormancy.

Can. J. Bot. 60, 1016-1020.

Sawhney, R., Hsiao, A. I. and Quick, W. A. 1984. Temperature control of germination and its possible role in
the survival of a non-dormant population of Avena
Fatua. Physiol. Plant. 61, 331-336.

Sawhney, R., Quick, W. A. and Hsiao, A. I. 1985. The effect
of temperature during parental vegetative growth on
seed germination of wild oats (Avena fatua L.). Ann.
Bot. 55, 25-28.

Sawhney, R., Hsiao, A. I. and Quick, W. A. 1986. The influence of diffused light and temperature on seed germination of three genetically non-dormant populations of
wild oats (Avena fatua) and its adaptive significance.
Can. J. Bot. (in press).

Simmonds, J. A. and Simpson, G. M. 1971. Increased participation of the pentose phosphate pathway in response to
after-ripening and gibberellic acid treatment in caryopses of Avena fatua. Can. J. Bot. 49, 1833-1840.

Simmonds, J. A. and Simpson, G. M. 1972. Regulation of the
Krebs cycle and pentose phosphate pathway activities in
the control of dormancy of Avena fatua. Can. J. Bot.
50, 1041-1048.

Simpson, G. M. 1965. Dormancy studies in seed of Avena
fatua. 4. The role of gibberellin in embryo dormancy.
Can. J. Bot. 43, 792-816.

Simpson, G. M. 1966. The suppression by (2-chloroethyl)
trimethyl-ammonium chloride of synthesis of a gibberellin-like substance by embryos of Avena fatua. Can J.
Bot. 44, 115-116.

Simpson, G. M. 1978. Metabolic regulation of dormancy in
seeds - a case history of the wild oat (Avena fatua).
In: Dormancy and Development Arrest. Clutter, M.
(ed.). Academic Press, New York. pp. 167-220.

Simpson, G. M. 1984. A review of dormancy in wild oats and
the lesson it contains for today. In: Wild Oat, Vol.
2. Use and Mode of Wild Oat Herbicides. Can. Plains
Proc. 12. Smith, A. E. and Hsiao, A. I. (eds.). Can.
Plains Res. Center, Regina,Sask.,Canada. pp. 3-20.

Tilsner, H. R. and Upadhyaya, M. K. 1985. Induction and
release of secondary seed dormancy in genetically pure
lines of Avena fatua. Physiol. Plant. 64, 377-382.

Upadhyaya, M. K., Hsiao, A. I. and Bonsor, M. E. 1986.
Differential response of pure lines of Avena fatua L.
to substituted phthalimides: Germination and endo-
sperm-mobilization studies. Ann. Bot. (in press).

Upadhyaya, M. K., Simpson, G. M. and Naylor, J. M. 1981.
Levels of glucose-6-phosphate and 6-phosphogluconate

dehydrogenases in the embryos and endosperms of some dormant and nondormant lines of Avena fatua during germination. Can. J. Bot. 59, 1640-1646.

Upadhyaya, M. K., Naylor, J. M. and Simpson, G. M. 1982a. Co-adaptation of seed dormancy and hormonal dependence of alpha-amylase production in endosperm segments of Avena fatua. Can. J. Bot. 60, 1142-1147.

Upadhyaya, M. K., Simpson, G. M. and Naylor, J. M. 1982b. The physiological basis of seed dormancy in Avena fatua L. I. Action of the respiratory inhibitors sodium azide and salicylhydroxamic acid. Physiol. Plant. 54, 419-424.

Upadhyaya, M. K., Naylor, J. M. and Simpson, G.M. 1983. The physiological basis of seed dormancy in Avena fatua L. II. On the involvement of alternative respiration in the stimulation of germination by sodium azide. Physiol. Plant. 58, 119-123.

Zwar, J. A. and Hooley, R. 1986. Hormonal regulation of alpha-amylase gene transcription in wild oat (Avena Fatua L.) aleurone protoplasts. Plant Physiol. 80, 459-463.

Section VII

Enzymes in Germinating Grains

Biosynthesis and Secretion of Alpha-Amylase in Germinating Rice Seedlings

T. Akazawa and T. Mitsui, Research Institute
for Biochemical Regulation, School of Agriculture,
Nagoya University, Chikusa, Nagoya 464, Japan.

Historically, studies on alpha-amylase have been made in close relation to the brewery industry. Since germination of seeds constitutes the initial step in the whole life of a plant, it is one of the classical themes of plant physiology and has received the attention of investigators from various angles (Thimann, 1977; Bewley and Black, 1978). However, the biochemistry of alpha-amylase in germinating (sprouting) cereal seeds has had a relatively recent history. Morphologically, starchy cereal seeds such as rice, barley, wheat and maize are composed of two principal parts, viz., the embryo and the endosperm. All the genetic information required for seedling growth and development after germination is stored in the embryo from which both roots and coleoptiles originate. On the other hand, during the initial step of germination, reserve polysaccharide, i.e. starch, stored in the endosperm is degraded and supplies the energy and carbon skeletons necessary to sustain the growth of these young tissues. Several enzymes, of which alpha-amylase has the most crucial role, are involved in the hydrolytic breakdown of starch molecules in the endosperm. From the metabolic standpoint the plants become autotrophic upon emergence of green tissues after germination, due to the exhaustion of reserve materials in storage organ.

Contemporary studies on alpha-amylase biosynthesis in cereal seeds are based on the initial observation that a dilute solution of gibberellic acid enhances alpha-amylase activities in embryo-less half seeds of barley (Yomo, 1960). Subsequently, extensive research by Varner and his associates (Varner and Ho, 1976) established an hypothesis which

has prevailed in 20th century plant physiology. 'During the
initial step of cereal seed germination, gibberellic acid
synthesized in the embryo is transported to aleurone layers,
where the plant hormone triggers the inducible production of
alpha-amylase. Newly synthesized alpha-amylase is then
secreted to the endosperm....' (Thimann, 1977; Salisbury
and Ross, 1985). Currently, due to the development of
recombinant DNA techniques, the mechanism of gibberellic
acid-dependent inducible formation of alpha-amylase in barley seeds is receiving the vigorous attention of many
investigators at the molecular level of gene expression.

Histochemical studies by Okamoto and Akazawa (1979) and
Gibbons (1979) posed the simple question. 'What tissues in
the cereal seeds are initially important in the synthesis of
alpha-amylase molecules?' Indeed, this has long been an
unsolved problem of plant physiology research, and arguments
have persisted for many years. By means of the simple
starch-film technique, which was originally developed by
Daoust, we reported that during the initial stage of germination, the site of alpha-amylase formation is the upper-
most layer of scutellum referred to as the epithelium, which
separates the embryo from endosperm (Okamoto, Kitano and
Akazawa, 1980; Akazawa and Miyata, 1982). The enzyme molecules are then secreted across the cell wall to the endosperm tissues, followed by degradation of the reserve
starch. However, in some cereals such as barley, it
appears quite evident that in the later stages of germination, endosperm starch is broken down by alpha-amylase
produced and secreted from the surrounding aleurone layers.
Although it is beyond doubt that rice seed alpha-amylase is
synthesized in the scutellum in the initial stage of germination, some investigators contend that in other cereal
seeds such as barley the scutellar tissues play a minor role
in enzyme production, whereas the major part of enzymes
synthesis is in the aleurone layers (MacGregor et al., 1984;
Ranki and Sopanen, 1984; Mundy, Brandt and Fincher, 1985;
and Akazawa and Hara-Nishimura, 1985 for review). It is
most likely that alpha-amylase formation in the aleurone
layers is inseparable from the enhancing effect of gibberellic acid, but so far no firm experimental evidence is
available indicating the mode or mechanism of transport of
gibberellic acid from embryo to aleurone in any cereal
seeds. Alternatively one could postulate that the crucial
role of endogenous gibberellic acid present in the scutellum
is to support the biosynthesis of alpha-amylase in situ.
The exact role of gibberellic acid remains to be clarified.

In addition to the histochemical observations described

above, decisive experimental evidence concerning the formation of rice seed alpha-amylase in the scutellar tissues has been obtained from experiments dealing with the incorporation of [35S]-methionine into alpha-amylase both in vivo and in vitro (Akazawa and Miyata, 1982).

Rice seed alpha-amylase has been shown to be a typical secretory glycoprotein, and experimental results with this enzyme have increased our knowledge concerning the pathway and mechanism of secretory glycoprotein synthesis in higher plants. It should be noted that our current views on the pathway and intracellular transport of glycoproteins have been mostly based on studies using animal tissues (Palade, 1975; Lennarz, 1981). Central to this topic are the formation of precursor polypeptides, synthesized in rER (signal hypothesis) (Blobel, 1977) and subsequent processing leading to the formation of the mature form which involves secondary glycosylation reactions which occur in the Golgi complex (Lennarz, 1981). Among the numerous secretory proteins there are several non-glycoprotein-type single proteins such as serum albumin, insulin, and growth hormone, but the majority belong to the class of carbohydrate-bearing Asn-N-linked glycoproteins. Akazawa and his colleagues (Akazawa and Miyata, 1982; Akazawa and Hara-Nishimura, 1985) demonstrated that alpha-amylase secreted from the scutellar epithelium is a typical Asn-N-linked glycoprotein, based on experiments dealing with the inhibitory effect of tunicamycin, which blocks the initial step of the dolichol pathway. By comparison with animal systems, the functional role of the Golgi system in the modification of the oligosaccharide chain of the alpha-amylase molecule has been difficult to establish. Evidently, the small size of rice seedlings is a disadvantage when it comes to isolating the Golgi preparation for biochemical and/or enzymatic studies. However, monensin, a Na+/K+ ionophore, has been effectively used to clarify the role of Golgi in the sugar modification reaction in glycoprotein biosynthesis (Tartakoff, 1983). We have tested the effect of monensin on rice seed alpha-amylase biosynthesis and found that 10^{-7}M monensin induces a prominent morphological alteration (dilation) of the Golgi cisternae which accompanies the inhibition of the terminal sugar conjugation of the alpha-amylase molecule (Mitsui et al., 1985).

We have also found that rice seeds contain two types of secretory alpha-amylase molecules, which can be distinguished by their digestibility by endoglycosidase (Endo-beta-H) (Mitsui et al., 1985). We have designated them as Endo-beta-H resistant (R-form) and susceptible (S-form)

molecules. Several lines of evidence suggest that the two
molecular species correspond to the glycoprotein (alpha-
amylase) bearing complex-type and high mannose-type oligosaccharide chains (Lennarz, 1981; Mitsui et al., 1984).
The crucial subcellular site involved in the transglyco-
sylation reactions leading to the formation of the alpha-
amylase isoforms is the Golgi complex, but it is not known
whether the S-type alpha-amylase is produced and intracel-
lularly transported to the cell surface bypassing the Golgi
system (Akazawa and Hara-Nishimura, 1985).

An important element involved in the multi-pathways
forming R- and S-type alpha-amylase molecules in the rice
scutellum was found to be the intracellular level of Ca^{2+}
(Mitsui and Akazawa, 1986). By elevating the Ca^{2+} concentration, the synthesis and secretion of R-type alpha-
amylase were stimulated, whereas the S-type enzyme was
unaffected. It remains open for future rigorous investigations to determine whether the Ca^{2+} effect is mediated by
calmodulin, although we have found inhibition of alpha-
amylase secretion by the calmodulin antagonist, W-7 (Mitsui
et al., 1984).

Another important environmental factor that altered the
two molecular forms of alpha-amylase was found to be
temperature. By raising the incubation temperature, predominantly R-type alpha-amylase molecules were secreted from
rice scutellar tissues, whereas at low temperatures, the
S-type molecules were secreted (Mitsui and Akazawa, 1984).

Overall the results indicate that the terminal glycosylation reaction involved in the maturation of alpha-
amylase proceeds in the Golgi complex and the subsequent
intracellular transport of the enzyme molecules to the cell
surface are subject to control by at least two environmental
factors i.e. Ca^{2+} and temperature. Steps proceeding in the
Golgi complex are clearly distinguishable from events occurring in the rER, i.e. polypeptide elongation and the primary
glycosylation reaction (co-translational glycosylation) via
the dolichol pathway.

The hypothetical mechanism concerning the biosynthesis
and intracellular transport of alpha-amylase molecules in
the rice scutellar epithelium based on the experimental
results we have obtained is presented in FIGURE 1. As
illustrated in the figure, four isoforms of alpha-amylase,
i.e. A-S, B-S, A-R and B-R, are secreted by the rice
scutellum and we can postulate that at least two distinct
intracellular pathways may operate: the secretion of R-type
alpha-amylase (A-R and B-R) is under the control of Ca^{2+} and
high temperature (regulated pathway), whereas that of S-type

enzyme (A–S and B–S) is not affected by Ca^{2+} and temperature (constitutive pathway). Elucidation of the detailed sorting mechanism for these two different pathways must await further investigation.

Histochemically, we have found that the pattern of alpha-amylase secretion in germinating barley seeds appears analogous to that in rice seeds (Okamoto, Kitano and Akazawa 1980; Akazawa and Miyata, 1982). However, in contrast to the remarkable advancement of our knowledge on the molecular biology of the barley alpha-amylase, such as the sequencing of DNA coding the enzyme molecule (Akazawa and Hara-Nishimura, 1985; Chandler et al., 1986), its cell biology remains relatively unexplored. Our recent experiments indicate that the nature of the carbohydrate moiety in the barley seed alpha-amylase molecule differs considerably from the rice enzyme described above. Currently we have been attempting to establish its structure in relation to the mechanism of the intracellular transport of the enzyme in the scutellar tissues of barley seeds.

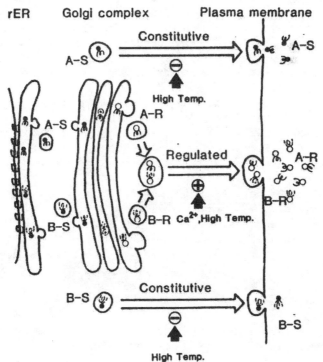

FIGURE 1. Hypothetical scheme of biosynthesis and intracellular transport and alpha-amylase isoforms in the rice seed scutellar epithelium.

ACKNOWLEDGEMENT

The authors record their thanks for financial support from Mombusho (Japanese Ministry of Education, Science and Culture) and the Mitsubishi Foundation (Tokyo).

REFERENCES

Akazawa, T. and Miyata, S. 1982. Essays in Biochem. 18, pp. 40-78.

Akazawa, T. and Hara-Nishimura, I. 1985. Annu. Rev. Plant Physiol. 36, pp. 441-472.

Bewley, J. D. and Black, M. 1978. Physiology and Biochemistry of seeds. I. Development, Germination and Growth. Springer-Verlag, Berlin/Heidelberg/ New York.

Blobel, G. 1977. In: International Cell Biology. 1976-1977. Brinkley, B. R. and Porter, K. R. (eds.). pp. 318-325. The Rockefeller Univ. Press, New York.

Chandler, P. et al. 1986. (this volume)

Gibbons, G. 1979. Carlsberg Res. Commun. 44, pp. 353-356.

Lennarz, W. J. 1981. The biochemistry of glycoproteins and protoglycans. Plenum Press, New York.

MacGregor, A. W., MacDougall, F. H., Mayer, C. and Daussant J. 1984. Plant Physiol. 75, pp. 203-206.

Mitsui, T., Christeller, J. T., Hara-Nishimura, I. and Akazawa, T. 1984. Plant Physiol. 75, pp. 21-25.

Mitsui, T., Akazawa, T., Tartakoff, A. M. and Christeller, J. T. 1985. Arch. Biochem. Biophys. 241, pp. 315-328.

Mitsui, T. and Akazawa, T. 1986. (submitted for publication)

Mundy, J., Brandt, A. and Fincher, G. B. 1985. Plant Physiol. 79, pp. 867-871.

Okamoto, K. and Akazawa, T. 1979. Plant Physiol. 63, pp. 336-340.

Okamoto, K., Kitano, S. and Akazawa, T. 1980. Plant Cell Physiol. 21, pp. 201-204.

Palade, G. E. 1975. Science. 189, pp. 347-358.

Ranki, H. and Sopanen, T. 1984. Plant Physiol. 57, pp. 710-715.

Salisbury, F. B. and Ross, C. W. 1985. In: Plant Physiology. (3rd Ed.) pp. 324-326, Wadsworth Publ. Co., California.

Tartakoff, A. M. 1983. Cell. 32, pp. 1026-1028.

Thimann, K. V. 1977. Hormone action in the whole life of plants. Univ. of Mass. Press, Amherst, USA.

Varner, J. E. and Ho, D. T-H. 1976. In: Plant Biochemistry
    Bonner, J. and Varner, J. E. (eds.). pp. 713-770.
    Academic Press, New York.
Yomo, H. 1960. Hakko Kyokaishi. 18, pp. 600-602.

(1-3, 1-4) -Beta-Glucanases in Germinating Barley

G. B. Fincher, *Department of Biochemistry,*
La Trobe University, Bundoora, Victoria,
3083, Australia.

INTRODUCTION

The major constituents of barley (<u>Hordeum</u> <u>vulgare</u>) endosperm cell walls are the $(1\to3,1\to4)$-β-glucans (approx. 70%) and arabinoxylans (approx. 20%) (Fincher, 1975). The cell walls are degraded during germination and it is generally assumed that dissolution of the endosperm walls in germinating barley greatly facilitates rapid mobilization of storage proteins and starch by providing degradative enzymes secreted from the surrounding aleurone layer and/or scutellum free access to their substrates within the cells of the starchy endosperm. While this assumption is likely to have a sound basis in fact, it has not been experimentally demonstrated. Primary cell walls of other plants are at least partly permeable to globular proteins of M_r 20,000 to 60,000 (Carpita <u>et</u> <u>al</u>., 1979; Tepfer & Taylor, 1981) and in some cases molecules of much higher molecular weight pass through the wall (Fincher & Stone, 1986). Although the endosperm cell walls of germinating wheat and barley appear to be completely degraded (Fincher & Stone, 1986), the framework of sorghum endosperm cell walls appear to remain intact (Glennie <u>et</u> <u>al</u>., 1983).

A number of enzymes capable of depolymerising cell wall $(1\to3,1\to4)$-β-glucan are detected in germinating barley, including $(1\to3,1\to4)$-β-glucan glucanohydrolase (EC 3.2.1.73), referred to here simply as $(1\to3,1\to4)$-β-glucanase), $(1\to4)$-β-glucan 4-glucanohydrolase (cellulase, EC 3.2.1.4) and $(1\to3)$-β-glucan 3-glucanohydrolase $((1\to3)$-β-glucanase, EC 3.2.1.39) (Bathgate <u>et</u> <u>al</u>., 1974; Manners & Marshall, 1969; Luchsinger <u>et</u> <u>al</u>., 1960). Levels of cellulase, much of which originates from microbial contaminants on the grain, are relatively low in germinating barley (Hoy et al., 1981). Degradation of (1→3,1→4)-β-glucan by (1→3)-β-glucanases probably depends on the presence of contiguous (1→3)-β-glucosyl residues in the polysaccharide. While blocks of contiguous (1→3)-linkages have been reported in some (1→3,1→4)-β-glucans (Igarashi & Samurai, 1966; Fleming & Kawakami, 1977; Bathgate et al., 1974), they are not always present (Woodward et al, 1983; Woodward & Fincher, 1983) and it might be expected that (1→3)-β-glucanases would effect only limited hydrolysis of cell wall (1→3, 1→4)-β-glucans. The role of the (1→3)-β-glucanase, which is present in large amounts in barley grain (Ballance et al., 1976), remains unknown. In any case the (1→3,1→4)-β-glucanases, which are abundant in the endosperm of germinating barley (Manners & Marshall, 1969) and rapidly depolymerize (1→3,1→4)-β-glucans (Luchsinger et al., 1960), are probably the most important enzymes for endosperm cell wall degradation.

Investigations of barley (1→3,1→4)-β-glucanases have been stimulated to a large extent by their commercial importance in the malting and brewing industries (Luchsinger, 1962; Bamforth, 1982). Cell wall (1→3,1→4)-β-glucans extracted from malt not only cause filtration difficulties in the brewing process, but they can also contribute to haze and gel formation in finished beer (Bamforth, 1982). These problems, which are due largely to the asymmetrical conformation of the polysaccharides and thus to their tendancy to form aqueous solutions of high viscosity (Woodward et al., 1983; Woodward & Fincher, 1983), will not arise if levels of (1→3,1→4)-β-glucanases in germinating grain are high enough to rapidly depoly-merize the cell wall (1→3,1→4)-β-glucans. Rapid rates of (1→3,1→4)-β-glucanase development and high absolute levels of the enzymes are therefore essential for the commercial utilization of malt and, as a result, represent important selection criteria in breeding programs aimed at improving malting barleys.

Earlier attempts to purify the (1→3,1→4)-β-glucanases of germinating barley led to some confusion as to the number of enzymes present; two were identified in extracts of "green malt" (Luchsinger et al., 1960; Ballance et al., 1976) whereas only one enzyme could be detected in extracts of kiln-dried malt (Manners & Marshall, 1969; Manners & Wilson, 1976). One of the original objectives of the work described here was simply to establish the number of (1→3,1→4)-β-glucanases in germinating barley

and, if more than one enzyme was indeed present, to purify
them for comparison of their chemical and enzymic
properties.

PURIFICATION AND PROPERTIES

Two $(1\rightarrow3,1\rightarrow4)$-β-glucanases were ultimately purified
approx. 470-fold (enzyme I) and 290-fold (enzyme II) from
soluble extracts of 4 day germinated barley by
conventional ammonium sulphate fractional precipitation,
ion-exchange and gel filtration chromatography (Woodward &
Fincher, 1982a). The most striking difference between the
two enzymes lay in their different isoelectric points
(Table I), which enabled a clear separation of the two by
ion-exchange chromotography. Both are basic, monomeric
proteins. Furthermore, they can be separated on the basis
of molecular size and of particular practical importance
here is their resolution by sodium dodecyl sulphate
polyacrylamide gel electrophoresis (Table I). Enzyme II
is a glycoprotein containing 3.6% carbohydrate, of which
three residues are probably N-acetylglucosamine, while
enzyme I contains traces of associated carbohydrate (Table
I; Woodward & Fincher, 1982a). Purified enzyme II is
appreciably more stable than enzyme I at temperatures
above 40°C (Woodward & Fincher, 1982 b). The effect
becomes very pronounced in the germinated grain itself;
finished malt dried at temperatures up to 80°C in a
commercial kilning program retained approx. 40% of the
$(1\rightarrow3,1\rightarrow4)$-β-glucanase activity detected in the green malt
and this activity was attributable entirely to enzyme II,
since enzyme I disappeared completely during initial
drying at 60°C (L. Loi, P.A. Barton and G.B. Fincher,
unpublished). The preferential loss of enzyme I during
kilning almost certainly explains why only one
$(1\rightarrow3,1\rightarrow4)$-β-glucanase could be detected in extracts of
kiln-dried malts (Manners & Marshall, 1969; Manners &
Wilson, 1976).
    In most other respects, the two $(1\rightarrow3,1\rightarrow4)$-
β-glucanases are remarkably similar. Both show an
absolute specificity for $(1\rightarrow3,1\rightarrow4)$-β-glucans; they
hydrolyse $(1\rightarrow4)$-β-glucosyl linkages only where the
glucosyl residue is substituted at (O)3, as follows:

$$G\ 3\ G\ 4\ G\ 4\ G\ 3\ G\ 4\ G\ 4\ G\ 4\ G\ 3\ G\ 4\ G\ \cdots (RED)$$

where G denotes a β-glucosyl residue, 3 or 4 the position
of the glycosidic linkage and (red) the reducing end of
the polysaccharide. The two enzymes exhibit characteristic endo-hydrolase action patterns and are therefore
true (1→3,1→4)-β-glucan 4-glucanohydrolases which fall
into the enzyme classification EC 3.2.1.73. Their pH
optima and kinetic properties are also very similar (Table
I; Woodward & Fincher, 1982a). It should be emphasised
that the viscometric assay used in the purification and
characterization of the enzymes is not absolutely
specific, since cellulases and possibly also (1→3)-
β-glucanases can hydrolyse (1→3,1→4)-β-glucans.
Viscometric assays using CM-cellulose and CM-pachyman must
be routinely performed to ensure that viscosity changes in
the (1→3,1→4)-β-glucan assays can be attributed solely to
(1→3,1→4)-β-glucanase activity.

Table I
Comparison of the two (1→3,1→4)-β-glucanase isoenzymes in
germinating barley[a]

Property	Isoenzyme I	Isoenzyme II
Mr	28,000	30,000
pI	8.5	10
Carbohydrate	0.7% (w/w)	3.6% (w/w)
Km (mg/ml)	3.0	3.4
V_{max} (μmol glc released/ min/μmol)	$7.1\% \times 10^3$	11.6×10^3
pH optimum	4.7	4.7

[a]Woodward and Fincher (1982a,b)

The amino acid compositions of the two purified
(1→3),(1→4)-β-glucanases are also very similar and
polyclonal antibodies raised against the enzymes are
mutually cross-reactive (Woodward & Fincher, 1982a).
Subsequent comparison of sequences of the 40 NH_2-terminal
amino acids revealed that these observations resulted from
extensive amino acid sequence homology (more than 90%)
between the two proteins (Woodward et al., 1982). Since
barley is predominantly self-fertilizing and plants of
established cultivars are essentially homozygous, it is
unlikely that the enzymes originate from multiple alleles
at a single locus (Briggs, 1978; Woodward et al., 1982).
It has been suggested that the enzymes are encoded by

separate genes and thus represent true genetic isoenzymes; this is consistent with the detection of two DNA fragments carrying $(1\rightarrow3,1\rightarrow4)$-β-glucanase sequences in digests of barley DNA (Fincher et al., 1986). The extensive sequence homology may result from the duplication of a common ancestral gene (Woodward et al., 1982).

The complete amino acid sequence of $(1\rightarrow3,1\rightarrow4)$-β-glucanase isoenzyme II has now been defined from amino acid sequences of tryptic peptides and from the nucleotide sequence of a cDNA encoding the enzyme (Fincher et al., 1986). The sequence confirms the basic nature of the protein, but reveals no sequence homology at either the amino acid or nucleotide level with the $(1\rightarrow3,1\rightarrow4)$-β-glucanase from Bacillus subtilis (Murphy et al., 1984; Fincher et al., 1986). This suggests that the barley and Bacillus $(1\rightarrow3,1\rightarrow4)$-β-glucanases have evolved via completely separate routes. However, the amino acid sequence of the barley $(1\rightarrow3,1\rightarrow4)$-β-glucanase shows extensive homology with the sequence of a $(1\rightarrow3)$-β-glucanase from tobacco, with homology values of 50% or more in large regions of the NH_2- and COOH-termini of the two enzymes (Mohnen et al., 1985, Fincher et al., 1986). In view of the phylogenetic disparity of these two plants and the different specificities of the enzymes, the extent of this homology is somewhat surprising, but is presumably related to similarities in the catalytic and substrate-binding domains of the two enzymes, which catalyse the hydrolysis of structurally similar polysaccharides. The availability of the complete amino acid sequence of the barley $(1\rightarrow3,1\rightarrow4)$-β-glucanase, together with the detailed information on the chemistry and conformation of its substrate (Woodward & Fincher, 1983), opens up exciting possibilities for investigating the overall conformation of the enzyme, the stereospecificity of substrate binding, and the molecular mechanism for active site catalysis.

DEVELOPMENT DURING GERMINATION

In a number of studies it has been demonstrated that, after a lag of 1 to 2 days, $(1\rightarrow3,1\rightarrow4)$-β-glucanase activity increases from very low levels in ungerminated grains to a maximum after 4 to 6 days (MacLeod et al., 1964; Ballance et al., 1976; Bamforth & Martin, 1983). The difference in molecular size of the two $(1\rightarrow3,1\rightarrow4)$-β-glucanases now permit the development of each isoenzyme to be quantitatively assessed using the highly sensitive Western

transfer procedure (Stuart & Fincher, 1983). In this
procedure, proteins in crude homogenates of germinating
barley are separated by gel electrophoresis and
transferred electrophoretically from the gel to
nitrocellulose paper. Both $(1\rightarrow3,1\rightarrow4)$-β-glucanases can be
detected by probing the paper with cross-reactive
polyclonal antibodies raised against either isoenzyme,
followed by ^{125}I-protein A (Stuart & Fincher, 1983).
Comparison of radioactivity bound to the enzymes in the
separated grain extracts with that bound to known
quantities of purified enzyme standards enables the two
$(1\rightarrow3,1\rightarrow4)$-β-glucanase isoenzymes to be quantitated in the
0 to 10 pmole (0 to 0.3 μg) range (Stuart & Fincher, 1983).
The protein transfer method will measure both active and
inactive enzyme, provided the proteins have not been
hydrolysed, but it can be applied in the measurement of
total, potential $(1\rightarrow3,1\rightarrow4)$-β-glucanase isoenzymes in
samples of germinating barley containing as few as 5
grains.

In germinating barley (cv. Clipper), $(1\rightarrow3,1\rightarrow4)$-
β-glucanase isoenzymes I and II develop in approx.
equimolar proportions up to 40 to 60 pmole/grain 5 days
after the initiation of germination and then decrease
(Stuart & Fincher, 1983). Similar results have been
obtained with other barley varieties, although some
variations in absolute levels and the ratio of isoenzymes
I and II are observed. In view of the very low levels of
the $(1\rightarrow3,1\rightarrow4)$-β-glucanases in sound, ungerminated grain
and the sensitivity of the immunological assay, it seemed
likely that the procedure could be used as an indicator of
precocious germination. In preliminary experiments barley
samples ranging from very sound to visibly sprouted were
examined. Levels of $(1\rightarrow3,1\rightarrow4)$-β-glucanase protein and
activity are compared with falling number values in Table
II. High levels of $(1\rightarrow3,1\rightarrow4)$-β-glucanase were measured in
badly sprouted samples (samples 1 to 6) and very low
levels in sound barley (samples 16 to 18). However, a
number of discrepancies are apparent (e.g. sample 13 and
14). When the samples listed in Table II were germinated
for 4 days, no correlation between sprouting and
subsequent $(1\rightarrow3,1\rightarrow4)$-β-glucanase production was apparent.
Indeed, pre-germinated samples were able to synthesize
very high levels of the enzymes (Table II).

Table II

(1→3,1→4)-β-Glucanase Levels in Pre-germinated Barley Samples

Sample Number	Cultivars	Grain Falling Number (sec)[a]	Enzyme in ungerminated grain activity[b]	pmole[c]	Enzyme in 4 day germinated grain activity[b]	pmole[c]
1	Lara	60	0.36	10.5	1.46	21.9
2	Clipper	61	0.27	4.6	1.48	8.2
3	Clipper & Erex	61	0.20	2.6	0.97	9.7
4	Con/Clip 61	70	0.16	2.0	1.93	19.2
5	Con/Clip 102	81	0.18	3.3	0.69	11.6
6	Grimmett	83	0.15	-	-	11.0
7	Bandulla	120	0.19	2.0	1.26	15.6
8	Corvette	173	0.13	1.3	1.36	10.6
9	Clipper	188	0.20	2.1	1.24	6.2
10	Galleon	200	0.13	1.3	1.46	14.2
11	Grimmett	231	0.10	1.7	1.32	11.2
12	Clipper	242	0.14	2.7	0.62	7.4
13	Grimmett	270	0.03	0.5	1.51	23.6
14	Clipper	271	0.13	3.7	1.03	12.0
15	Grimmett	310	0.13	1.1	1.17	8.9
16	Clipper	408	0.01	0.8	1.10	12.3
17	Grimmett	546	0.01	0.8	0.85	8.7
18	Clipper	-	0.01	0.2	1.38	17.5

[a]Performed by Mr. D.P. Law, Queensland Wheat Research Institute, Toowomba, Queensland, Australia, [b]Units/grain, [c]Total of isoenzymes I and II.

SITE OF (1→3,1→4)-β-GLUCANASE SYNTHESIS

The synthesis of many of the hydrolytic enzymes responsible for endosperm mobilization in germinating barley occurs in the aleurone layer and is enhanced by the phytohormone gibberellic acid (GA) (Chrispeels & Varner, 1967; MacLeod et al., 1964; Taiz & Honigman, 1976). There is an increasing body of evidence to suggest that cells of the scutellar ephithelial layer also secrete hydrolytic enzymes which partipate in endosperm mobilization (Gibbons, 1980; MacGregor et al., 1984; Okamoto et al., 1980; Ranki & Sopanen, 1984). To define the site of (1→3,1→4)-β-glucanase synthesis in germinating barley, aleurone layers and scutella isolated from ungerminated grain were incubated in vitro in 10 mM $CaCl_2$ with and without 1 μM GA_3 (Stuart et al., 1986). The medium surrounding the isolated tissue fragments was examined for (1→3,1→4)-β-glucanase activity using the viscometric assay, and for the presence of individual (1→3,1→4)-β-glucanase isoenzymes using the immunological assay (Stuart et al, 1986).

It was clearly demonstrated that both aleurone layers and isolated scutella are capable of synthesizing and secreting (1→3,1→4)-β-glucanase; activity secreted per isolated scutellum represents up to 40% of that secreted by an aleurone layer. The activity secreted from the aleurone layer is almost entirely attributable to isoenzyme II, which is enhanced approx. 5-fold by GA_3 in the presence of Ca^{2+}. Isoenzyme I is detected at relatively low levels both in the medium surrounding aleurone layers and in tissue homogenates. Although variations in the relative abundance of isoenzymes I and II are observed (Loi, L. and Fincher, G.B., unpublished) the isoenzyme pattern in Clipper aleurone layer secretions is in marked contrast to whole germinating Clipper barley, where isoenzymes I and II are secreted in approx. equimolar proportions (Stuart & Fincher, 1983). This discrepancy might be explained in a number of ways. Firstly, some isoenzyme I might pre-exist in cells of the starchy endosperm and be activated during germination in a manner analogous to cereal β-amylases (Adams et al., 1975). However, in experiments with embryoless half-grains, which contain both the aleurone and starchy endosperm, levels of isoenzyme I remained low compared with isoenzyme II and it is therefore unlikely that cells of the starchy endosperm represent a significant source of either isoenzyme.

Secondly, it is possible that isoenzyme I is indeed secreted from isolated aleurone layers at levels approx. equal to those of isoenzyme II, but is preferentially degraded by proteases released by the layers; any temporal and hence spatial separation of $(1{\to}3,1{\to}4)$-β-glucanases and proteases in germinating whole grains (MacLeod et al., 1964) would be lost in experiments with isolated tissue. A third possibility is that removal of the embryo results in the loss of gibberellins or growth factors other than GA_3 necessary for the aleurone responses which occur in vivo (MacGregor, 1983; Atzorn & Weiler, 1983). These possibilities serve to emphasise the inherent dangers in extrapolating of results obtained with isolated tissues to the events which occur in vivo in whole, germinating barley grains. It should be noted that most of the $(1{\to}3,1{\to}4)$-β-glucanase secreted from both the aleurone and scutellum will accumulate in the starchy endosperm of germinating barley (cf. Ballance et al., 1976; Palmer et al., 1985).

Finally, the reduced levels of isoenzyme I relative to isoenzyme II in isolated aleurone layers may reflect a lower rate of expression of the gene encoding $(1{\to}3,1{\to}4)$-β-glucanase isoenzyme I in the aleurone; the approx. equal amounts of each in whole grain might result from the balancing effect of increased isoenzyme I gene expression in the scutellum. Although isoenzyme I is relatively abundant in scutella secretions, levels remain much lower than those detected in whole grain and than levels of isoenzyme II secreted from the aleurone. (Table III; Stuart et al., 1986).

In isolated scutella, isoenzyme I secretion is enhanced to some extent by GA_3. No isoenzyme II can be detected. Instead, the immunological assay reveals the presence of a protein of Mr approx. 32,000 in the medium surrounding isolated scutella (Stuart et al., 1986). This protein is recognised by the antibodies, but it corresponds to neither isoenzyme I (Mr 28,000) or isoenzyme II (Mr 30,000). On the basis of its cross-reactivity with the antibody (Stuart et al., 1986) and its activity on $(1{\to}3,1{\to}4)$-β-glucan when separated from isoenzyme I, this protein has been tentatively designated $(1{\to}3,1{\to}4)$-β-glucanase isoenzyme III. Re-examination of aleurone secretions suggested that low levels of isoenzyme III may also be present. The origin and nature of isoenzyme III remains uncertain, since it is not detected in homogenates of germinating grain. One possibility is that it represents an incompletely or incorrectly processed form of isoenzyme I or II, or that its appearance is somehow related to the wounding of the tissues which occurs during their excision.

Table III
Levels of $(1\to3,1\to4)$-β-glucanase isoenzymes in whole germinating barley and isolated tissue pieces

	Whole grain[a] (4 days)	Aleurone layers[b] (2 days)		Isolated scutella[b] (2 days)	
		$-GA_3$	$+GA_3$	$-GA_3$	$+GA_3$
Activity (U/tissue piece)	(Not determined)	0.7	3.8	0.9	1.1
EI (pmol/ grain or fragment)	32	0	0	3	4
EII (pmol/ grain or fragment)	30	7	42	0	0

[a]Stuart & Fincher (1983)
[b]incubated with 10 mM $CaCl_2$ (Stuart et al., 1986)

Effect of Ca^{2+}

When Ca^{2+} is withdrawn from the medium surrounding isolated aleurone layers or scutella, $(1\to3,1\to4)$-β-glucanase secretion decreases to very low levels and the preferential disappearance of isoenzyme II and III (Stuart et al., 1986) may be compared with the disappearance of α-amylase group B isoenzymes following Ca^{2+} withdrawal from isolated aleurone layers (Jones & Jacobsen, 1983). No significant GA_3-enhancement of $(1\to3,1\to4)$-β-glucanase secretion is obtained in the absence of Ca^{2+}. Whether Ca^{2+} participates directly and specifically in the expression of $(1\to3,1\to4)$-β-glucanase genes, in GA_3 action or in $(1\to3,1\to4)$-β-glucanase secretion is not known. It is possible that Ca^{2+} may be required for more general cellular functions, such as the fusion of cellular membranes or as a cofactor in unrelated enzymic reactions, and thus exerts its effect on $(1\to3,1\to4)$-β-glucanase secretion indirectly.

SUMMARY AND CONCLUSIONS

The two $(1\rightarrow3,1\rightarrow4)$-$\beta$-glucan endohydrolase isoenzymes from germinating barley have identical action patterns and substrate specificities, and exhibit very similar kinetic properties. However, they differ in amino acid sequence, carbohydrate content, heat stability, isoelectric point and apparent molecular size. Amino acid sequence information suggests that the isoenzymes are derived from two genes which have evolved separately after duplication of a common ancestral gene.

The difference in molecular weight of the isoenzymes, together with the cross-reactivity of antibodies, enables the isoenzymes to be individually quantitated in grain or tissue extracts by a sensitive immunological assay based on Western transfer techniques. This immunological assay has been used not only to monitor the development of the two isoenzymes in germinating barley, but also to measure their secretion from isolated aleurone layers and scutella. Results of the isolated tissue experiments can be summarised as follows;

(i)   both the scutellum and the aleurone have the potential for $(1\rightarrow3,1\rightarrow4)$-$\beta$-glucanase synthesis and secretion.

(ii) in the presence of Ca^{2+}, GA_3 significantly enhances secretion of isoenzyme II from isolated aleurone layers.

(iii) patterns of isoenzyme secretion from isolated aleurone differ from those from isolated scutella.

(iv) a third, hitherto undetected isoenzyme III is secreted from isolated scutella.

(v)   the appearance of $(1\rightarrow3,1\rightarrow4)$-$\beta$-glucanases is likely to result largely from de novo synthesis rather than from the activation of precursor forms or from the release of pre-formed, active enzyme.

(vi) very little $(1\rightarrow3,1\rightarrow4)$-$\beta$-glucanase is secreted in the absence of Ca^{2+}.

These observations provide prima facie evidence for the presence of multiple $(1\rightarrow3,1\rightarrow4)$-$\beta$-glucanase genes that are subject to developmental and possibly tissue-specific control. Structural analysis of the $(1\rightarrow3,1\rightarrow4)$-$\beta$-glucanase genes could provide important information on the molecular mechanisms of plant gene expression and its regulation by GA in specific tissues at specific stages of development.

ACKNOWLEDGEMENTS

This work was supported by grants from the Australian Research Grants Scheme and the Barley Industry Research Council. The invaluable contributions of Dr. J.R. Woodward, Mr. I. M. Stuart and Ms. Lin Loi are gratefully acknowledged.

LITERATURE CITED

Adams, C.A, Watson, T.C. and Novellie, L. (1975) Phytochemistry 14, 953-956.
Atzorn, R. and Weiler, E.W. (1983) Planta 159, 289-299.
Ballance, G.M., Meredith, W.O.S. and LaBerge, D.E. (1976) Can. J. Plant Sci. 56, 459-466.
Bamforth, C.W. (1982) Brew. Dig. 57, 22-27, 35.
Bamforth, C.W. and Martin, H.L. (1983) J. Inst. Brew. 89, 303-307.
Bathgate, G.N., Palmer, G.H. and Wilson, G. (1974) J. Inst. Brew. 80, 278-285.
Briggs, D.E. (1978) in: Barley pp 419-480, Chapman and Hall, London.
Carpita, N., Sabularse, D., Montezinos, D. and Delmer, D.P. (1979) Science 205, 1144-1147.
Chrispeels, M.J. and Varner, J.E. (1967) Plant Physiol. 42, 1008-1016.
Fincher, G.B. (1975) J. Inst. Brew. 81, 116-122.
Fincher, G.B. and Stone, B.A. (1986) Adv. Cereal Sci. Technol. (Y. Pomeranz, ed.) AACC, in the press.
Fincher, G.B., Lock, P.A., Morgan, M.M., Lingelbach, K., Wettenhall, R.E.H., Mercer, J.F.B., Brandt, A. and Thomsen, K-K. (1986) Proc. Natl. Acad. Sci. (USA), in the press.
Fleming, M. and Kawakami, K. (1977) Carbohyd. Res. 57, 15-23.
Gibbons, G.C. (1980) Carls. Res. Commun. 45, 177-184.
Glennie, C.W., Harris, J. and Liebenberg, N.V.D. (1983) Cereal Chem. 60, 27-31.
Hoy, J.L., Macauley, B.J. and Fincher, G.B. (1981) J. Inst. Brew. 87, 77-80.
Igarashi, O. and Sakurai, Y. (1966) Agr. Biol. Chem. 30, 642-645.
Jones, R.L. and Jacobsen, J.V. (1983) Planta 158, 1-9.
Luchsinger, W.W. (1962) Cereal Chem. 39, 225-235.

462

Luchsinger, W.W., Cochrane, D.G. and Kneen, E. (1960) Cereal Chem. 37, 525-534.

MacGregor, A.W. (1983) Ann. Proc. Phytochem. Soc. Eur. 20, 1-34.

MacGregor, A.W., MacDougall, F.H., Mayer, C. and Daussant, J. (1984) Plant Physiol. 75, 203-206.

MacLeod, A.M., Duffus, J.H. and Johnston, C.S. (1964) J. Inst. Brew. 70, 521-528.

Manners, D.J. and Marshall, J.J. (1969) J. Inst. Brew. 75, 550-561.

Manners, D.J. and Wilson, G. (1976) Carbohydr. Res. 48, 255-264.

Mohnen, D., Shinshi, H., Felix, G. and Meins, F. Jr. (1985) EMBO J. 4, 1631-1635.

Murphy, N., McConnell, D.J. and Cantwell, B.A. (1984) Nucl. Acids Res. 12, 5355-5367.

Okamoto, K., Kitano, H. and Akayawa, T. (1980) Plant Cell Physiol. 21, 201-204.

Palmer, G.H., Gernah, D.I., McKernan, G., Nimmo, D.H. and Laycock, G. (1985) Am. Soc. Brew. Chem. 43, 17-28.

Ranki, H. and Sopanen, T. (1984) Plant Physiol. 75, 710-715.

Stuart, I.M., and Fincher, G.B. (1983) FEBS Lett. 155, 201-204.

Stuart, I.M., Loi, L. and Fincher, G.B. (1986) Plant Physiol., 80, 310-314.

Taiz, L. and Honigman, W.A. (1976) Plant Physiol. 58, 380-386.

Tepfer, M. and Taylor, I.E.P. (1981) Science 213, 761-763.

Woodward, J.R. and Fincher, G.B. (1982a) Eur. J. Biochem. 121, 663-669.

Woodward, J.R. and Fincher, G.B. (1982b) Carbohydr. Res. 106, 111-122.

Woodward, J.R. and Fincher, G.B. (1983) Brew. Dig. 58, 28-32.

Woodward, J.R., Morgan, F.J. and Fincher, G.B. (1982) FEBS Lett. 138, 198-200.

Woodward, J.R., Phillips, D.R. and Fincher, G.B. (1983) Carbohyd. Polymers 3, 143-156.

Proteinases and Peptidases in Germinating Cereal Grains

J. Mikola, University of Jyväskylä,
Department of Biology, SF-40100,
Jyväskylä, Finland.

Six types of peptide bond hydrolyzing enzymes have been detected in germinating cereal grains. First, there are two groups of proteinases or endopeptidases, one of which is already present in resting grains, the other appearing during germination. Secondly, there are four types of peptidases or exopeptidases. Acid carboxypeptidases sequentially liberate C-terminal amino acids mainly from the soluble peptides produced by the proteinases. Neutral aminopeptidases hydrolyze small peptides, including dipeptides at pH 6 to 8: these enzymes are characterized by their ability to hydrolyze beta-naphthylamides and p-nitroanilides of various amino acids around pH 7. Alkaline aminopeptidases, in contrast, show minimal or no activity against the last-mentioned substrates, but they rapidly hydrolyze peptides at pH 8 to 10. Finally, dipeptidases hydrolyze dipeptides optimally at pH 7 to 9. All of these enzyme groups have been extensively studied in only a few cereals. However, because the same enzyme groups occur in other monocots, in several dicots and even in gymnosperms (Mikola, J. 1983; Mikola and Mikola 1986), it is probably safe to assume that any enzyme detected in and characterized from one or a few cereals is probably present in other cereals as well.

In germinating barley, the degradation of the grain reserve proteins seems to occur in three stages (Mikola, 1983). First some special storage proteins, mainly globulins, are hydrolyzed in the embryo and aleurone layer. This primary hydrolysis provides amino acids for the synthesis of new hydrolytic enzymes in the scutellum and aleurone layer and their secretion into the starchy endosperm. The pH optima of the enzymes present in the starchy endosperm are 4 to 6, and the internal pH is maintained close to pH 5 (Mikola and Mikola 1980). In these conditions the proteins in the starchy endosperm, mainly prolamins and glutelins are rapidly hydrolyzed to a mixture of amino acids and small peptides. Both the amino acids (Sopanen et al., 1980) and the peptides (Sopanen 1979) are taken up by the scutellum. In the third stage of protein hydrolysis the peptides are hydrolyzed in the scutellum apparently by the neutral and/or alkaline amino- and dipeptidases to amino acids before possible metabolic conversions and the "long-distance" transport to the growing tissues of the seedling. A number of observations suggest that the general course of the mobilization of grain reserve protein in other cereals resembles that in barley. Consequently, to understand the functions of the different proteinases information is required not only of their properties but also of their occurrence in the different parts of the grains at the different stages of germination and early seedling growth.

PROTEINASES PRESENT IN RESTING GRAINS

Resting grains of all the cereals studied contain low proteinase activity with a pH optimum around pH 4. The enzymes responsible for these activities seem to be different from the main proteinases present in germinating grains.

In rice the activity seems be attributable to a single enzyme (Doi et al., 1980d). It hydrolyzes hemoglobin and casein optimally at pH 3 to 5, has a molecular weight of about 65,000, and is completely inhibited by 1 µM pepstatin. This inhibition together with the acid pH optimum suggest a catalytic mechanism similar to that of mammalian pepsin. The enzyme activity is localized partly in the embryo and partly in the "9% bran" (roughly corresponding to the aleurone layer). As the storage globulins of rice have a similar distribution and they are rapidly hydrolyzed in the beginning of germination, it is tempting to speculate that the acid pepstatin-sensitive proteinase acts in the first stage of reserve protein hydrolysis.

In wheat the situation seems more complicated although it has been less extensively studied (Morris et al., 1985). Extracts prepared from the embryos or endosperms hydrolyze insulin B-chain optimally at pH 4 but only the embryonic activity is inhibited by pepstatin. The result suggests the presence of two acid proteinases, one in the embryo and another in the endosperm (aleurone layer?). The pepstatin-sensitive enzyme shows some activity towards gluten proteins dissolved in 0.2 M acetic acid (Kawamura and Yonezawa 1982), but whether it is able to act on them in the natural insoluble state has not been established.

Extracts of embryos and endosperms of ungerminated barley grains also hydrolyze insulin B-chain optimally at pH 4, and the major part of the activity is inhibited by pepstatin (Morris et al., 1985). However, the embryo extracts in particular also show relatively high activity at pH 7-8, suggesting the presence of a third enzyme.

Resting grains of all cereals contain low hydrolytic activity against the chromogenic substrate of trypsin, benzoyl-L-arginine p-nitroanilide or BAPA, at pH 8 to 9. Only the enzyme from rice grains has been adequately purified and characterized (Shibata and Doi 1984a,b). It shows no activity against proteins but it acts on peptides conttaining arginine or lysine residues by hydrolyzing the peptide bonds involving the carboxyl group of those amino acids. The combination of low activity and high specificity suggest some regulatory role for this enzyme.

PROTEINASES APPEARING DURING GERMINATION

The proteinase activities of extracts of germinating cereal grains, when assayed using hemoglobin, casein, or gelatin substrates, increase about 20-fold during germination for 3 to 5 days. The increases in activity seem to be attributable to proteinases synthesized in the scutellum and aleurone layer and secreted into the starchy endosperm (Okamota et al., 1980). The newly synthesized enzymes are -SH proteinases which are able to rapidly hydrolyse prolamins and glutelins at pH 4 to 5, and the fraction is characterized by inhibition by leupeptin. The enzymes are labile in aqueous solutions. Therefore, only a few attempts at purification have been reported.

The best-known enzyme of this group is proteinase A purified from germinating wheat (Shutov et al., 1985). The enzyme is an -SH proteinase with a molecular weight of about 22,000. The enzyme preparation, which was apparently free of other proteinases and all exopeptidases, was able to hydrolyze the proteins of gluten to a mixture of di-, tri-, and tetrapeptides. This result suggests that in wheat endosperm a single enzyme is able to completely hydrolyse both prolamins and glutelins to small peptides which can either be further hydrolyzed by acid carboxypeptidases or be taken up by the scutellum.

-SH proteinases with molecular weight of 20,000 - 25,000 have also been shown to be the main proteinases in germinating grains of maize (Abe et al., 1977, 1978) and

rice (Doi et al., 1980c). Both of these enzymes are inhibited by leupeptin. The maize enzyme can hydrolyze peptide
bonds involving the carboxyl groups of glutamine and glut-
amic acid. The abundance of these amino acids in prolamins
and glutelins is a partial explanation for their rapid and
extensive hydrolysis by the respective proteinases.

ACID CARBOXYPEPTIDASES

Acid serine carboxypeptidases, characterized by pH
optima at 4 to 6 and complete inactivation by di-isopropyl-
fluorophosphate are the best-known peptide bond hydrolyzing
enzymes present in cereal grains. In germinating barley
(Mikola, L. 1983) and wheat (Mikola 1986) there are five
enzymes with different and complementary substrate specificities against various benzyloxycarbonyldipeptides (Z-dipep-
tides). Carboxypeptidases I, II, and III liberate various
C-terminal amino acid residues including proline provided
that there is not a proline residue in the penultimate position. Carboxypeptidases IV and V, in contrast, resemble
mammalian prolylcarboxypeptidases which act only on substrates with a penultimate proline residue.
Carboxypeptidase I has been isolated in pure form from
germinating barley (Visuri et al., 1969; Breddam et al.,
1983). The enzyme is a dimeric glycoprotein with a mol. wt.
of about 100,000. It is most active against Z-dipeptide
substrates with a large hydrophobic side chain in the penultimate position; typically it shows very high activity
against Z-Phe-Ala, the substrate most rapidly hydrolyzed by
most plant extracts.
Carboxypeptidase II has been isolated from both wheat
bran (Umetsu et al., 1981) and from germinating barley
(Breddam et al., 1985). It is also a dimeric glycoprotein
with a mol. wt. of about 120,000. Carboxypeptidase II has
a wide specificity range and is able to liberate basic,
acidic and uncharged amino acid residues including proline
from Z-peptide substrates (Umetsu et al., 1981; Umetsu and
Ichishima 1983; Mikola 1986). Z-Ala-Arg is a specific
substrate suitable for the assay of carboxypeptidase II both
in barley (Mikola L. 1983) and in wheat (Mikola 1986). For
the hydrolysis of longer peptides the wheat enzyme at least
is able to "go through" a proline residue (Umetsu et al.,
1981; Yabuuchi et al., 1973). The wide substrate specificity of carboxypeptidase II partially explains its ability
to relatively rapidly hydrolyse proteins (acid-denatured
hemoglobin and gluten dissolved in dilute acetic acid.

(Preston and Kruger 1976; Mikola 1986)

Carboxypeptidase III has not been isolated. However, its presence in germinating barley (Mikola, L. 1983) and wheat (Mikola 1986) can easily be demonstrated by gel chromatography because its mol. wt. is only about 40,000. It prefers Z-dipeptide substrates with large uncharged side-chains in the C-terminal position and the most specific sub-strate known up to now is Z-Ala-Phe (Mikola 1986).

Carboxypeptidases IV and V can be assayed with Z-Pro-Ala, Z-Gly-Pro-Ala, and Z-Met-Ala. However, the activities, at least when assayed with these synthetic substrates are much lower than those of the other carboxypeptidases · (Mikola, L. 1983, 1986).

Rice contains a pair of carboxypeptidases resembling carboxypeptidases I and II in several respects. The enzyme resembling carboxypeptidase II has been purified from rice bran (Doi et al., 1980b). Its molecular weight is about 110,000. During the hydrolysis of a heptapeptide the pure enzyme was able to "go through" a proline residue. The other enzyme which hydrolyzed A-Phe-Ala and, like barley carboxypeptidase I (Mikola and Pietilä 1972) showed high esterase activity against Ac-Phe-O-Et, was detected in germ-inating seeds (Doi et al., 1980a).

The behaviour of the acid carboxypeptidases during germination is remarkable. In wheat the protein-hydrolyzing carboxypeptidase II is already abundant in resting grains and its activity slowly decreases during germination whereas carboxypeptidases I and III are absent from resting grains but high activites appear in the beginning of germination (Mikola 1986). The main site of carboxypeptidase synthesis seems to be the scutellum (Mundy et al., 1985). In barley the course of events is probably similar and in germinating grains the bulk of all carboxypeptidase activities is local-ized in the starchy endosperm (Mikola and Kolehmainen 1972; Mikola and Mikola 1980).

In rice also, the enzyme resembling carboxypeptidase II (called CPAase-4) is present in resting grains, both in the bran and in the "starchy endosperm" (Doi et al., 1980b). During germination the activity of CPAase-4 decreases while the activity of CPAase-5 appears (Doi et al., 1980a).

Germinating grains of rye and oats contain high carb-oxypeptidase activities similar to barley and wheat (Winspear et al., 1984; Mikola and Saarinen 1986). In contrast, corresponding activities in germinating rice, maize, and sorghum are only 1-5% of those present in the temperate zone cereals. This difference probably reflects a major difference in the degradation of grain storage proteins in the two groups of cereals.

NEUTRAL AMINOPEPTIDASES

Neutral aminopeptidases are apparently present in the grains of all cereals (Mikola and Mikola 1986). These enzymes are characterized by their ability to liberate N-terminal amino acids from both small peptides, including dipeptides, and the beta-naphthylamides and p-nitroanilides of various amino acids at roughly similar rates at pH 6 to 8. As in the case of the acid carboxypeptidases, there are 4-5 enzymes with different and complementary substrate specificities. The best researched cereal in this respect is maize from which all the enzymes have been separated by ion-exchange chromatography (Vodkin and Scandalios 1979). Suitable marker substrates for the different enzymes include Phe-beta-NA, Leu-beta, Arg-beta-NA and Ala-beta-NA. The enzyme with high activity against Phe-beta-NA has also been purified from barley grains (Kolehmainen and Mikola 1972) and from wheat leaves (Waters and Dalling 1984). The three enzymes closely resemble each other. A separate enzyme acts on Pro-beta-NA. It has been purified only from wheat leaves (Waters and Dalling 1983).

In barley grains the neutral aminopeptidases are localized in the embryo and aleurone layer (Mikola and Kolehmainen 1972). The activities, however, are much lower than those of the other peptidase groups and during germination they increase only in the seedling. The role of these enzymes in the mobilization of grain reserve proteins is probably a minor one.

ALKALINE AMINOPEPTIDASES

An aminopeptidase from germinating barley which hydrolyzes Leu-Gly-Gly, Leu-Tyr, and a number of other uncharged peptides at pH 8-10 has been purified and characterized (Sopanen and Mikola 1975). A corresponding enzyme activity is present in different parts of senescing wheat shoots (Waters et al., 1980). As the enzyme is present in the seeds of some dicots and gymnosperms (Mikola and Mikola 1986), it is probably also present in other cereals. In resting barley grains the activity is localized in the living tissues like those of the neutral aminopeptidases, but it is higher and it increases in the scutellum during germination. The high pH optimum raises the question as to

the cellular compartment where it functions, but the high
activity suggests a role in the hydrolysis of the peptides
taken up by the scutellum.

DIPEPTIDASES

A dipeptidase which can be specifically assayed using
Ala-Gly as substrate at pH 8 has been purified from germinating barley (Sopanen 1976). The enzyme, like the alkaline
aminopeptidase, is relatively abundant in resting grains,
is localized in the embryo and aleurone layer, and during
germination shows a further activity increase in the scutellum (Mikola and Kolehmainen 1972). Corresponding enzyme
activity is present in the leaves and glumes of senescing
wheat (Waters et al., 1980). The purified dipeptidase does
not hydrolyze dipeptides of the type X-Pro. However,
extracts prepared from scutella of germinating barley
rapidly hydrolyze Ala-Pro (Mikola and Mikola 1980). This
observation suggests that the hydrolysis of peptides taken
up by the scutellum is catalyzed by the alkaline aminopeptidase, two dipeptidases, and possibly by the neutral aminopeptidases.

CONCLUDING REMARKS

As described above, sound ungerminated grains of different cereals contain relatively high activities of several
peptidases and small but definite acid proteinase activity.
Most of these enzymes are localized in the embryo and aleurone layer. Consequently, the bulk of them are removed by
milling. Secondly, there is no direct proof that any of
these enzymes is able to attack prolamins and/or glutelins
in their native physical state.
Two similar enzymes , rice CPAase-4 and wheat carboxypeptidase II deserve special attention. The rice enzyme is
the only proteolytic enzyme which has been shown to be present in the starchy endosperm of resting grains (Doi et al.,
1980b); the other enzyme is abundant in ungerminated wheat
and it is possibly partly localized in the starchy endosperm.
We can probably assume that either the various enzymes
present in sound, resting grains are harmless or the processes utilizing different grains have been adapted to their
presence.
With respect to damage due to hydrolysis of proteins

both during pre-harvest germination and in the various uses of partly germinated grains, the key enzymes are without doubt the -SH proteinases synthesized during germination. These enzymes occur in the starchy endosperm and are able to extensively hydrolyse both prolamins and glutelins. Their decisive role in protein hydrolysis was demonstrated with barley malts containing high but variable activities of both acid proteinases and acid carboxypeptidases (Sopanen et al., 1980a); in laboratory scale mashings proteinase activity was the rate-limiting factor both in solubilization of total nitrogen and in production of free amino acids.

REFERENCES

Abe, M., Arai, S. and Fujimaki, M. 1977. Purification and characterization of a protease occurring in endosperm of germinating corn. Agric. Biol. Chem. 41, pp. 893-899.

Abe, M., Arai, S. and Fujimaki, M. 1978. Substrate specificity of a sulfhydryl protease purified from germinating corn. Agric. Biol. Chem. 42, pp. 1813-1817.

Breddam, K., Bech Sørensen, S. and Ottesen, M. 1983. Isolation of a carboxypeptidase from malted barley by affinity chromatography. Carlsberg Res. Commun. 48, pp. 217-230.

Breddam, K., Bech Sørensen, S. and Ottesen, M. 1985. Isolation of carboxypeptidase II from malted barley by affinity chromatography. Carlsberg Res. Commun. 50, pp. 199-209.

Doi, E., Komori, N., Matoba, T. and Morita, Y. 1980a. Some properties of carboxypeptidases in germinating rice seeds and rice leaves. Agric. Biol. Chem. 44, pp. 77-83.

Doi, E., Komori, N., Matoba, T. and Morita, Y. 1980b. Purification and some properties of a carboxypeptidase in rice bran. Agric. Biol. Chem. 44, pp. 85-92.

Doi, E., Shibata, D., Matoba, T. and Yonezawa, D. 1980c. Evidence for the presence of two types of acid proteinases in germinating seeds of rice. Agric. Biol. Chem. 44, pp. 435-436.

Doi, E., Shibata, D., Matoba, T. and Yonezawa, D. 1980d. Characterization of pepstatin-sensitive acid protease in resting rice seeds. Agric. Biol. Chem. 44, pp. 741-747.

Kawamura, Y. and Yonezawa, D. 1982. Wheat flour proteases and their action on gluten proteins in dilute acetic acid.   Agric. Biol. Chem.   46, pp. 767-773.

Kolehmainen, L. and Mikola, J. 1971. Partial purification
and enzymatic properties of an aminopeptidase from
barley. Arch. Biochem. Biophys. 145, pp. 633-642.

Mikola, J. 1983. Proteinases, peptidases, and inhibitors of
endogenous proteinases in germinating seeds. In:
Daussant, J., Mosse, J. and Voughan, J. (eds.) Seed
Proteins, Academic Press, London. pp. 35-52.

Mikola, J. and Kolehmainen, L. 1972. Localization and activity of various peptidases in germinating barley.
Planta 104, pp. 167-177.

Mikola, J. and Pietilä, K. 1972. Hydrolysis of ester substrates of trypsin and chymotrypsin by barley carboxypeptidase. Phytochemistry. 11, pp. 2977-2980.

Mikola, L. 1983. Germinating barley grains contain five
acid carboxypeptidases with complementary substrate
specificities. Biochem. Biophys. Acta 747, pp. 241-
252.

Mikola, L. 1986. Acid carboxypeptidases in grains and
leaves of wheat, Triticum aestivum L. Plant Physiol.,
in press.

Mikola, L. and Mikola, J. 1980. Mobilization of proline in
the starchy endosperm of germinating barley grain.
Planta. 149, pp. 149-154.

Mikola, L. and Mikola, J. 1986. Occurrence and properties
of different types of peptidases in higher plants. In:
Plant Proteolytic Enzymes Vol. I. Dalling, M. J. (ed.)
(in press). CRC Press, Boca Raton.

Mikola, L. and Saarinen, S. 1986. Occurrence of acid and
neutral carboxypeptidases in germinating cereals.
Physiol. Plant., in press.

Morris, P. C., Miller, R. C. and Bowles, D. J. 1985. Endopeptidase activity in dry harvest-ripe wheat and barley
grains. Plant Science. 39, pp. 121-124.

Mundy, J., Brandt, A. and Fincher, G. B. 1985. Messenger
RNAs from the scutellum and aleurone of germinating
barley encode (1-3, 1-4)-beta-D-glucanase, alpha-
amylase and carboxypeptidase. Plant Physiol. 79, pp.
867-871.

Okamota, K., Kitano, H. and Akazawa, T. 1980. Biosynthesis
and excretion of hydrolases in germinating cereal
seeds. Plant and Cell Physiol. 21, pp. 201-204.

Preston, K. R. and Kruger, J. E. 1976. Purification and
properties of two proteolytic enzymes with carboxypeptidase activity in germinated wheat. Plant Physiol.
58, pp. 516-520.

Shibata, D. and Doi, E. 1984a. Purification and characteri-

472

zation of rice embryo BAPase (benzoyl-L-arginine p-nitroanilide hydrolase). Plant and Cell Physiol. 25, pp. 1411-1419.

Shibata, D. and Doi, E. 1984b. A novel type of substrate specificity of rice BAPAase (benzoyl-L-arginine p-nitroanilide hydrolase) with mixed endopeptidase and carboxypeptidase. Plant and Cell Physiol. 25, pp. 1421-1429.

Shutov, A. D., Beltei, N. K. and Vaintraub, I. A. 1985. A cysteine proteinase from germinating wheat seeds: partial purification and hydrolysis of gluten. Biochemistry USSR. 49, pp. 1004-1010.

Sopanen, T. 1976. Purification and partial characterization of a dipeptidase from barley. Plant Physiol. 57, pp. 867-871.

Sopanen, T. 1979. Development of peptide transport activity in barley scutellum during germination. Plant Physiol. 64, pp. 570-574.

Sopanen, T. and Mikola, J. 1975. Purification and partial characterization of barley leucine aminopeptidase. Plant Physiol. 55, pp. 809-814.

Sopanen, T., Takkinen, P., Mikola, J. and Enari, T.-M. 1980a. Rate-limiting enzymes in the liberation of amino acids in mashing. J. Inst. Brew. 86, pp. 211-215.

Sopanen, T., Uuskallio, M., Nyman, S. and Mikola, J. 1980b. Characteristics and development of leucine transport activity in the scutellum of germinating barley grain. Plant Physiol. 65, pp. 249-253.

Umetsu, H., Abe, M., Sugawara, Y., Nakai, T., Watanabe, S. and Ichishima, E. 1981. Purification, crystallisation and characterization of carboxypeptidase from wheat bran. Food Chem. 7, pp. 125-138.

Umetsu, H. and Ichishima, E. 1983. Action on peptides by wheat carboxypeptidase. Phytochemistry. 22, pp. 591-592.

Umetsu, H., Mori, K. and Ichishima, E. 1984. Subunit structure and immunological properties of wheat carboxypeptidase. Phytochemistry. 11, pp. 2435-2438.

Visuri, K., Mikola, J. and Enari, T.-M. 1969. Isolation and partial characterization of a carboxypeptidase from barley. Eur. J. Biochem. 7, pp. 193-199.

Vodkin, L. O. and Scandalios, J. G. 1980. Comparative properties of genetically defined peptidases in maize. Biochemistry. 19, pp. 4660-4667.

Waters, S. P. and Dalling, M. J. 1983. Purification and characterization of an aminopeptidase from the primary

leaf of wheat (Triticum aestivum L.). Plant Physiol.
73, pp. 1048-1054.

Waters, S. P. and Dalling, M. J. 1984. Isolation and some
properties of an aminopeptidase from the primary leaf
of wheat (Triticum aestivum L.). Plant Physiol. 75,
pp. 118-124.

Waters, S. P., Peoples, M. B., Simpson, R. J. and Dalling, M.
J. 1980. Nitrogen redistribution during grain growth
in wheat (Triticum aestivum L.). I. Peptide hydrolase
activity and protein breakdown in the flag leaf, glumes
and stem. Planta. 148, pp. 575-579.

Winspear, M. J., Preston, K. R., Rastogi, V. and Oaks, A.
1984. Comparisons of peptide hydrolase activities in
cereals. Plant Physiol. 75, pp. 480-482.

Yabuuchi, S., Doi, E. and Hata, T. 1973. The mode of
action of a carboxypeptidase from malt. Agric. Biol.
Chem. 37, pp. 687-688.

A Review of some Properties of an Endogenous Inhibitor of Cereal Alpha-Amylase

R. D. Hill, A. W. MacGregor[1], R. J. Weselake[2]
and J. Daussant[3]. Department of Plant Science,
University of Manitoba, Winnipeg, Manitoba,
Canada.

INTRODUCTION

The breakdown of cereal endosperm starch occurs through
the action of alpha-amylase released by aleurone and scutellar tissue (MacGregor et al., 1984). Gibberellins produced
in the scutellum (Radley, 1968) regulate the synthesis of
the enzyme. Other than product inhibition, there is no
evidence of metabolic regulation of cereal alpha-amylase as
exists for mammalian glycogen phosphorylase.

Early work by Kneen and Sandstedt (1943) demonstrated
the existence of soluble proteins in wheat and rye flours
that inhibited alpha-amylases. These and similar inhibitors
from other species are active against mammalian and some
bacterial alpha-amylases but have no action against higher
plant alpha-amylases (Silano, 1978). They, therefore, have
no function in the regulation of hydrolysis of endosperm
starch reserves and are not a factor in the regulation of
the germination process. Heat stable inhibitors of cereal
alpha-amylases have been reported in durum and winter wheat
(Warchalewsik, 1977) and maize kernels (Blanco-Labra and
Iturbe-Chinas, 1981).

Two laboratories (Weselake et al., 1983a, 1983b and
Mundy et al., 1983) have recently isolated a heat labile
protein from barley that specifically inhibits a group of
alpha-amylase isozymes synthesized in the aleurone during
germination. In this presentation, the properties of the
protein and its interaction with alpha-amylase will be
reviewed. Its distribution amongst the cereals and location
in tissues will be presented along with a discussion of its
possible physiological function.

PROPERTIES

Malted barley extracts heated to 70°C show different
alpha-amylase isozyme patterns than their non-heated counterparts (Frydenburg and Nielsen, 1965; MacGregor and
Ballance, 1980). Heating causes a disappearence of some
bands and an enrichment of others. The additional bands
that are present in unheated extracts are due to the association of the high pI isozymic forms (alpha-II) of alpha-
amylase with a heat labile protein producing a third group
(alpha-III) of bands on isoelectric focusing (FIGURE 1).
This protein, purified from mature barley by salt fractionation and column chromatography (Weselake et al., 1983b),
specifically inhibits the high pI forms of the enzyme while
having no effect on the low pI forms (alpha-I).

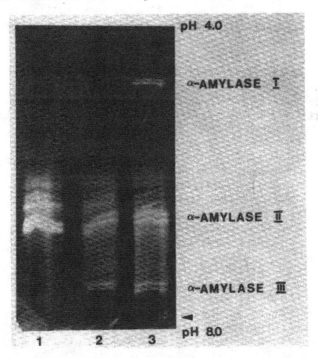

FIGURE 1. Isoelectric focusing zymograms demonstrating the
effect of inhibitor on alpha-amylase II. Lane 1. alpha-
amylase II; 2. alpha-amylase II plus inhibitor; 3. green
malt extract. Arrow indicates sample application point.
Approximately equal enzyme activities were applied in each
lane.

TABLE 1. Properties of barley alpha-amylase inhibitor and its complex with alpha-amylase II.

Molecular weight	20,000 daltons
Isoelectric point	7.3
pH optimum for inhibition	$\doteq$ 7.0
Inhibitor-enzyme dissociation	
constant pH 7.0	1×10^{-6}M
pH 5.5	7×10^{-6}M
Inhibitor : Enzyme binding ratio	2:1

TABLE 1 summarizes the properties of the inhibitor protein. The molecular size of the protein and its heat lability clearly distinguish it from the heat stable albumin proteins that inhibit non-cereal alpha-amylases. It has an isoelectric point near neutrality and optimally inhibits near its isoelectric point. The inhibitor binds strongly to alpha-amylase and is approximately seven times more effective at pH 7.0 than 5.5, the pH optimum of alpha-amylase. The interaction between inhibitor protein and enzyme is largely ionic as high salt concentration reduces inhibition. A 15-fold increase in salt concentration reduces inhibitor binding approximately 50-fold. Gel filtration and U.V. difference spectra studies (Halaydo and Hill, unpublished observations) indicate that two moles of inhibitor bind per mole of enzyme. This suggests that there are either two sites for binding on alpha-amylase or that the inhibitor functions as a dimer. U.V. difference spectra indicate that only one tryptophan per mole of alpha-amylase is affected during binding suggesting that the inhibitor binds as a dimer.

Starch granule hydrolysis by alpha-amylase II is more sensitive to inhibitor than is amylose hydrolysis (Weselake et al., 1985b). The inhibition is independent of starch granule concentration (Weselake et al., 1985b) suggesting that the inhibitor could be effective in blocking alpha-amylase II action in the endosperm, where starch granule concentration is high. Examination of the effects of inhibitor on alpha-amylase II hydrolysis indicates that there is a slight alteration of product composition when starch granules are hydrolyzed in the presence of inhibitor (MacGregor et al., 1986). This is in contrast to digestion of amylose by alpha-amylase II where the presence of inhibitor reduces the rate of hydrolysis but has no effect on product composition.

Mundy et al., (1983) have pointed out that the alpha-
amylase inhibitor is identical to a previously described
subtilisin inhibitor. Subtilisin is a serine protease and
it might be expected that the alpha-amylase inhibitor would
act against carboxypeptidase, also a serine protease, produced by the germinating seed. This, however, does not
occur as carboxypeptidase is not a metal requiring serine
protease.

DISTRIBUTION AND LOCATION

The inhibitor protein is distributed throughout the
mature barley endosperm (Weselake et al., 1985a). It is
also present in the aleurone and embryo. FIGURE 2 shows
an immuno-blot analysis of extracts of Conquest barley endosperm, embryo and aleurone tissue. Aleurone protoplasts
have been used to reduce the possibility of contamination of
aleurone tissue by the endosperm. On a total protein basis,
embryo and aleurone contain approximately half the inhibitor
protein relative to endosperm tissue in Conquest barley.
Preliminary observations indicate that there are species
variation in the proportion of inhibitor in aleurone protoplasts.
Concentrations of the inhibitor vary widely amongst barley cultivars (Weselake et al., 1985a; Münck et al., 1985)
and comprise from 0.3 to 1.3 per cent of the seed protein.
Antibodies prepared against the barley inhibitor show partial immunochemical identity with components present in
kernel extracts of tetraploid and hexaploid wheat, rye and
hexaploid triticale (Weselake et al., 1985a). These
extracts also have the capacity to inhibit alpha-amylase II
activity. No anti-amylase activity nor evidence of a pre-
cipitin reaction with antibody could be found in extracts
from sorghum, oats, millet, rice and maize. It is interesting to note that, although oats lack such an inhibitor protein, oat amylase has been reported to be inhibited by both
the barley and wheat inhibitor (Mundy et al., 1984).
There appears to be only one major species of inhibitor
protein in barley, although there is some indication that
small amounts of other forms may exist (Sadowski et al.,
1986). Structural genes for the inhibitor were identified
in the A genome and identical genes may exist on the B and D
genomes of wheat. The wheat and rye forms of the inhibitor
are electrophoretically distinct (FIGURE 3). In triticale
lines containing the full complement of the R genome, both
forms can be distinguished by immunoblotting. In triticale

478

FIGURE 2. Immuno-blot assay of alpha-amylase inhibitor in
extracts of embryo (a), endosperm (b) and aleurone protoplasts (c) of Conquest barley. Rows 1-6 of lane d contain
200, 100, 50, 20, 10, 5 μg respectively of purified barley
alpha-amylase inhibitor. Row 1 of lanes a, b and c contain
1.8 μg of extract protein. Rows 2-6 contain 5, 10, 20, 40
and 80 fold dilutions of row 1.

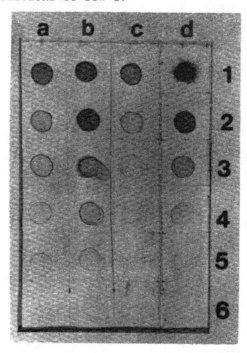

lines lacking chromosome 2R only the wheat-type inhibitor is
present suggesting that the rye inhibitor gene is located on
that chromosome.

In developing barley kernels, the inhibitor can first
be detected at about 12 days post anthesis (Laurière et al.,
1985). Its synthesis terminates about 30 days after pollination (Münck et al., 1985). Cultivar variations in the
final content of inhibitor appear to be due to differences
in the rate of synthesis of the protein rather than its
duration of synthesis. No systematic studies have been
performed to assess the effect of environmental conditions
on the proportions of inhibitor in the mature seed.

FIGURE 3.   Alpha-amylase inhibitor in various triticale
lines.   Immunoblotting after isoelectricfocusing.

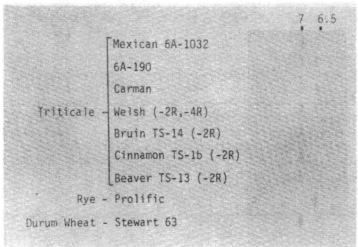

PHYSIOLOGY AND FUNCTION

The physiological function of the inhibitor protein
remains an enigma.   It obviously can not be a primary agent
in dormancy as its inhibitory action occurs well along in
the germination process.   There are, however, several factors which suggest its possible function as a germination
attenuator.

The high pI forms of alpha-amylase are specifically
activated during germination.   It is these forms of the enzyme that are inhibited and, since the inhibitor is already
present in the mature grain, it is reasonable to assume that
it does have some action in vivo.   The low dissociation constant of the enzyme: inhibitor complex and the evidence
that it may be particularly effective at inhibiting starch
granule hydrolysis suggest that it could restrict the
release of soluble dextrins during the early stages of germination.   This obviously could only be effective for a
limited time period as continued production of alpha-amylase
would eventually bind all of the available inhibitor.

Mundy (1984) has demonstrated that protein synthesis of
alpha-amylase inhibitor in barley half grains is reduced by
gibberellins while abscisic acid must have been exceedingly
low, however, as fluorography required 60 days to obtain
visible bands from immunoprecipitation of the inhibitor
protein.   It is therefore unlikely that de novo synthesis of
inhibitor could have a significant effect in reducing the

copious amounts of alpha-amylase activity that can arise in circumstances where the seed is returning to a quiescent state after germination has been initiated. Nevertheless, this response to growth regulators is consistent with possible involvement of the inhibitor in germination attenuation.

If the protein does attenuate germination, it is more likely to play a role either during kernel maturation or during early seed germination, as a pre-existing component in the seed. It is conceivable that the inhibitor may retard the release of dextrins from starch granules during germination until such time as events in the embryo are sufficiently advanced for sugar utilization. This would prevent large decreases in osmotic potential and subsequent rapid swelling of the grain. It would also reduce the availability of sugars to microorganisms in the event that the seed returned to a dormant state.

The protein is synthesized during the period of storage protein synthesis and its level may be influenced by environmental factors. According to Münck et al., (1985), who looked at variation in alpha-amylase inhibitor levels amongst barley cultivars of differing degrees of dormancy, there is no correlation between pregermination incidence and level of alpha-amylase inhibitor. This argues against a heritable relationship between sprouting resistance and inhibitor levels but does not preclude the possibility that, within a variety, environmental influences during maturation could vary the inhibitor protein level in concert with other events which affect sprouting resistance.

REFERENCES

Blanco-Labra, A. and Iturbe-Chinas, F. A. 1981. Purification and characterization of an alpha-amylase inhibitor from maize (Zea mays). J. Food Biochem. 5, pp. 1-17.

Frydenburg, O. and Nielsen, G. 1965. Amylase isozymes in germinating barley seeds. Hereditas. 54, pp. 123-139.

Kneen, E. and Sandstedt, R. M. 1943. An amylase inhibitor from certain cereals. J. Amer. Chem. Soc. 65, p. 1247.

Lauriére, C., Mayer, C., Renard, H. A., MacGregor, A. W. and Daussant, J. 1985. Maturation du caryopse d'orge: évolution des isoformes des alpha-et beta-amylases, de l'enzyme débranchante, de l'inhibiteur d'alpha-amylases chez plusiers variétes. Proc. European Brew. Congress, pp. 675-681.

MacGregor, A. W. and Ballance, D. 1980. Quantitative determination of alpha-amylase enzymes in germinated barley after separation by isoelectricfocusing. J. Inst. Brew. 86, pp. 131-133.

MacGregor, A. W., MacDougall, F. H., Mayer, C. and Daussant, J. 1984. Changes in levels of alpha-amylase components in barley tissues during germination and early seedling growth. Plant Physiol. 75, pp. 203-206.

MacGregor, A. W., Weselake, R. J., Hill, R. D. and Morgan, J. E. 1986. Effect of an alpha-amylase inhibitor from barley kernels on the formation of products during the hydrolysis of amylase and starch granules by alpha-amylase II from malted barley. J. Cereal Sci. (in press).

Münck, L., Mundy, J. and Vaag, P. 1985. Characterization of enzyme inhibitors in barley and their tentative role in malting and brewing. Am. Soc. Brew. Chem. J. 43, pp. 35-38.

Mundy, J. 1984. Hormonal regulation of alpha-amylase inhibitor synthesis in germinating barley. Carlsberg Res. Comm. 49, pp. 439-444.

Mundy, J., Svendsen, I. B. and Hejgaard, J. 1983. Barley alpha-amylase/subtilisin inhibitor. I. Isolation and characterization. Carlsberg Res. Comm. 48, pp. 81-90.

Mundy, J., Hejgaard, J. and Svendsen, I. 1984. Characterization of a bifunctional wheat inhibitor of endogenous alpha-amylase and subtilisin. FEBS Letters. 167, pp. 210-214.

Radley, M. 1968. Production of gibberellin-like substances in barley seed and seedlings. Soc. Chem. Ind. Monograph No. 31, pp. 53-69.

Sadowski, J., MacGregor, A. W. and Daussant, J. 1986. Alpha-amylase inhibitor in cereals. Comparison of the protein in different dry, wheat and triticale seeds by using immunoblotting. Electrophoresis (accepted).

Silano, V. 1978. Biochemical and nutritional significance of wheat albumin inhibitors of alpha-amylase. Cereal Chem. 55, pp. 722-731.

Warchelewski, J. R. 1977. Isolation and purification of native alpha-amylase inhibitors from winter wheat. Bull. Acad. Pol. Sci. Sev. Sci. Biol. 25, pp. 725-729.

Weselake, R. J., MacGregor, A. W. and Hill, R. D. 1983a. An endogenous alpha-amylase inhibitor in barley kernels. Plant Physiol. 72, pp. 809-812.

Weselake, R. J., MacGregor, A. W., Hill, R. D. and Duckworth, H. W. 1983b. Purification and characteristics of an endogenous alpha-amylase inhibitor from barley kernels. Plant Physiol. 73, pp. 1008-1012.

482

Weselake, R. J., MacGregor, A. and Hill, R. D. 1985a.
Endogenous alpha-amylase inhibitor in various cereals.
Cereal Chem. 62, pp. 120-123.
Weselake, R. J., MacGregor, A. and Hill, R. D. 1985b.
Effect of endogenous barley alpha-amylase inhibitor on
hydrolysis of starch under various conditions. J.
Cereal Sci. 3, pp. 249-259.

[1]Grain Research Laboratry, Canadian Grain Commission,
Winnipeg, Manitoba,Canada.
[2]Dept. of Plant Science, University of Manitoba, Winnipeg,
Manitoba, Canada.
[3]Laboratoire de Physiologie des Organes Végétaux, CNRS,
Meudon, France.

Degradation of Starch Granules in Maturing Wheat and its Relationship to Alpha-Amylase Production by the Embryo

B. A. Marchylo and J. E. Kruger, Grain Research Laboratory, Canadian Grain Commission, 1404-303 Main Street, Winnipeg, Manitoba R3C 3G8, Canada.

SUMMARY

Four spring wheat varieties exhibiting different degrees of dormancy were sampled during the latter stages of maturation and studied with a scanning electron microscope (SEM). Evidence of starch granule as well as protein matrix degradation was observed in the starchy endosperm adjacent to the embryo but not the aleurone. The extent of degradation differed among the four varieties. Excised wheat embryos from each variety were germinated and the alpha-amylase produced was analyzed by polyacrylamide gel isoelectric focusing.

INTRODUCTION

Previous studies (Gale et al., 1983; Marchylo et al., 1980a; Sargeant, 1980) have shown that alpha-amylases characteristic of germination can be detected in the wheat kernel in the final stages of maturation. Gale et al., (1983) also have suggested that alpha-amylase production prior to full maturity, in the absence of visible sprouting, may be a general phenomenon. Studies on tissues dissected from immature wheat kernels (Marchylo et al.. 1980a) indicated that in the latter stages of maturation, alpha-amylase is present primarily in the endosperm and embryonic tissues. It also was shown that the embryo (scutellum) and to a lesser extent the endosperm (aleurone) could synthesize both low and high pI alpha-amylases normally associated with

[1] Paper No. 572 of the Canadian Grain Commission, Grain Research Laboratory, 1404-303 Main St. Winnipeg, Manitoba, R3C 3G8.

germination (Marchylo et al., 1980b). More recent studies with germinating cereals have suggested that in the early stages of germination, the embryo rather than the aleurone is the major site of alpha-amylase production (Mares, 1983; MacGregor and Matsuo, 1982; Okamoto, et al., 1980; Gibbons 1979, 1980).

The purpose of this study was to carry out a microscopic examination of the endosperms of four wheat varieties during development to pinpoint any areas of initial starch degradation. Presumably, the location of this degradation should reveal the source of alpha-amylases, characteristic of germination, found in the wheat kernel in the latter stages of maturation.

MATERIALS AND METHODS

Grain Samples

Four varieties of wheat including the hard red spring wheats Neepawa, Cypress, Rl 4137 and the soft white spring wheat Idaed, were planted May 12, 1980 at the Agriculture Canada experimental plots, Glenlea, Manitoba. Wheat heads were excised shortly after flowering at 10 intervals beginning July 17 and finishing on August 27. The flowering date for each variety was approximated by examining the crop as a whole. Idaed flowered earliest (July 8) followed by Neepawa (July 10), RL 4137 (July 14) and Cypress (July 15). A portion of the heads were frozen and, where necessary, freeze dried in preparation for analysis with the scanning electron microscope (SEM). Kernels from freshly harvested heads were used for moisture determination and germination studies.

Moisture

The air oven method (A.A.C.C., 1962) was used to determine the moisture content of 25 kernel portions randomly selected from two freshly harvested heads of each wheat variety.

Germination Studies

Twenty-five kernel portions were randomly selected from six freshly harvested heads. The kernels were evenly distributed, crease side down, on moist filter paper in a

petri dish and were incubated at 18°C and 96% relative
humidity for five days after which the number of sprouted
kernels was noted.

Scanning Electron Microscopy

Wheat kernels were examined using a JEOL 35C scanning
electron microscope (SEM) at an accelerating voltage of 10
kV. For each sample period and each variety (freeze dried
where necessary), a single kernel was selected from the
central portion of each of twelve heads. The kernels were
cracked open longitudinally through the crease and both
halves were fixed to microscope slides with a mixture of
clear nail polish and colloidal graphite. The stubs then
were coated with gold prior to SEM analysis. Six sample
periods in the latter stages of maturation (August 7, 12,
15, 19, 22 and 27) were analyzed and representative photo-
micrographs were taken on Plus X Pan Kodak film.

Alpha-amylase Activity Analysis

Twenty kernels were ground using a Udy Cyclone sample
mill. The ground grain was extracted for 1 hour with stirring into 5mls of 0.02 M sodium acetate buffer pH 5.5 containing 10^{-3} M calcium chloride. Extracts were centrifuged
at 25,000 x g. Alpha-amylase activities then were determined using a Perken-Elmer model 191 Grain Amylase Analyzer
as described by Kruger and Tipples (1981).

Preparation and Incubation of Wheat Embryos

Kernels from mature samples of the four wheat varieties
were steeped in excess distilled water for 4h at 20°C.
Embryos including scutella then were excised and the edges
of the scutella were removed to prevent contamination by
aleurone. Groups of 25 embryos were placed on sterilized
sand (2g in small weighing boats) saturated with 10^{-3}M
$CaCl_2$. The weighing boats were placed in petri dishes and
were incubated for 3 days at 18°C and 96% relative humidity.

Analysis of alpha-amylase Isoenzymes

The 25 embryos and sand were extracted in a mortar with

10ml of imidazole-HCl buffer (0.025 M pH 7.4). The extracts
then were centrifuged (23,000 xg) at 4°C for 15 min.

The alpha-amylase isoenzymes were separated by polyacrylamide gel isoelectric focussing (PAG-IEF) and were
detected as described previously (Marchylo and Kruger,
1983). Incubation time for the beta-limit dextrin-polyacrylamide sandwich was 30 min.

Densitometry

Zymograms were scanned using a Shimadzu CS 910 TLC dual
wavelength scanning densitometer in conjuction with a Hew-
lett Packard 3390A integrator. Reference and sample wave-
lengths of 460nm and 720nm, respectively, were used with a
slit width of 0.02mm, a slit height of 2.0mm and a scan
speed of 2.5mm/min. The chart speed of the 3390A was set at
1cm/min.

RESULTS AND DISCUSSION

As illustrated in FIGURE 1, the moisture contents of
the four wheat varieties decreased in a similar fashion
during maturation. In order to obtain some indication of
dormancy for each variety in the latter stages of maturation, wheat kernels harvested at four sample periods (August
7, 15, 22, 27) were incubated under germination conditions.
As illustrated in TABLE 1, the red spring wheat cultivar
RL 4137 exhibited the highest degree of dormancy whereas the
soft white spring wheat variety Idaed showed virtually no

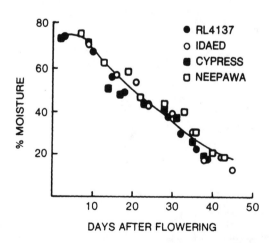

FIGURE 1. Change in
moisture content dur-
ing maturation for
three red spring wheat
varieties (Neepawa,
Cypress, RL 4137) and
one white spring wheat
variety (Idaed).

Table I

Percent Germination of Wheat Samples
Harvested During Maturation

Variety	Harvest Date	Stage of Maturity (days after flowering)	% Germination (after 5 days)
RL4137	August 7	24	0
	15	32	4
	22	39	0
	27	44	20
Neepawa	August 7	28	0
	15	36	0
	22	43	12
	27	48	64
Cypress	August 7	23	0
	15	31	52
	22	38	32
	27	43	40
Idaed	August 7	30	100
	15	38	100
	22	45	96
	27	50	92

dormancy. The red spring wheat varieties Neepawa and
Cypress were intermediate in dormancy. Dormancy of the
three red spring wheats decreased with grain drying which
is consistent with previous work (King 1971). SEM analysis
of wheat endosperms, at various stages of maturation, then
were undertaken to localize any areas of starch degradation.
None of the kernels selected showed any indication of
sprouting. In addition, alpha-amylase activity analysis
revealed that low levels (around 0.4mg Maltose/min. x 10^{-3}/
kernel) were associated with the spring wheat varieties.
Idaed, however, exhibited higher levels of activity, particularly in the final three stages (about 20mg Maltose/min x
10^{-3}/kernel), which probably is indicative of incipient
sprouting. All kernels studied were split lengthwise,
through the crease, using a scalpel. The kernels were
cracked rather than sliced to prevent smearing of surfaces.
The kernel areas of interest are illustrated in FIGURE 2.
This study focused on the regions of the starchy endosperm
adjacent to the embryo and the aleurone layer. The starchy
endosperm is separated from the embryo by a crushed layer of
cells (MacGregor and Matsuo, 1982) which is adjoining a
single layer of embryonic tissue called the scutellar epithelium. SEM analysis revealed that harvest ripe kernels
of Idaed, Neepawa and Cypress (August 27) exhibited exten-

488

sive areas of starch granule and protein matrix degradation
adjacent to the crushed layer of cells. These areas of degradation were evident at the crease end and middle portions
of the endosperm-embryo junction. In addition, they did not
extend very far into the endosperm. A typical example of
the type of degradation observed is shown in FIGURE 3A for
the variety Idaed. Degradation of this nature has been
shown to be characteristic of 24h germinated wheat and barley (MacGregor and Matsuo, 1982). Idaed also displayed
extensive degradation, as shown in FIGURE 3B, in kernels
harvested at earlier stages (August 22, 19 and 15). Neepawa
and Cypress exhibited some evidence of starch granule and
protein matrix degradation in samples harvested at these
earlier stages but the extent of degradation typically was
substantially less severe. Some evidence of starch degradation next to the embryo also was evident in Idaed for samples harvested at earlier stages of maturation (August 7,
12). However, as illustrated in FIGURE 3C, the extent of
starch and protein matrix degradation was low. Virtually no
evidence of degradation was apparent in Neepawa and Cypress
at these earlier stages.

By comparison, RL 4137 showed virtually no evidence of
starch or protein matrix degradation at any harvest stage
analyzed. Only intact starch granules surrounded by protein
matrix, as shown in FIGURE 3D, were observed in this
variety.

For the four varieties studied, SEM analysis of the
endosperm adjacent to the aleurone layer indicated no evidence of protein matrix or starch degradation at any of the
harvested stages analyzed.

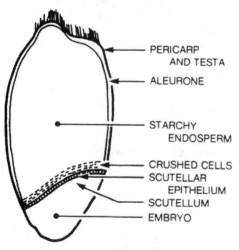

PERICARP
AND TESTA

ALEURONE

STARCHY
ENDOSPERM

CRUSHED CELLS
SCUTELLAR
EPITHELIUM
SCUTELLUM
EMBRYO

These results then
suggest that in the
latter stages of matura-
tion, the site of alpha-
amylase synthesis con-
comitant with starch
degradation is the
embryo and not the aleu-
rone. Previous studies

FIGURE 2. Longitudinal
section of a wheat ker-
nel cracked open through
the crease edge.

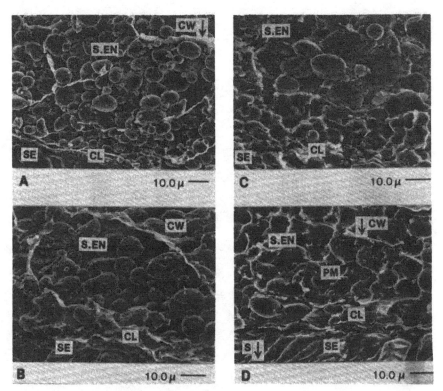

FIGURE 3.   Scanning electron micrograph of wheat endosperms.
A, B, and C Idaed harvested August 27, 15, 7 respectively;
D RL 4137 harvested August   27.   S = scutellum;   SE = scutellar epithelium;   CL = crushed layer of cells;   S.EN =
starchy endosperm;   PM = protein matrix;   CW = cell wall.

with germinating cereals (Mares 1983;   MacGregor and Matsuo,
1982;   Okamoto et al., 1980;   Gibbons, 1979, 1980) have provided substantial evidence to show that the embryo (scutellar epithelium) also is the major site of alpha-amylase synthesis in the early stages of germination.   It also is of
interest that the wheat variety with the least dormancy,
Idaed, also exhibited the largest degree of starch degradation whereas RL 4137, the most dormant variety, exhibited
virtually no degradation.   Possibly evidence of starch degradation adjacent to the embryo may be related to dormancy
characteristics of a variety and therefore could be used as
a screening procedure.
      Although this evidence suggests that the embryo is the

490

most likely source of the alpha-amylase detected in the final stage of maturation, little is known about the alpha-amylase synthesized by this tissue. Therefore, embryos excised from mature grains of the four wheat varieties were germinated and the alpha-amylase produced was analyzed by PAG-IEF. As shown in FIGURE 4, a visual comparison of the alpha-amylase isoenzymes produced in germinated embryos and germinated wheat indicates that different relative proportions of high and low pI components are present.

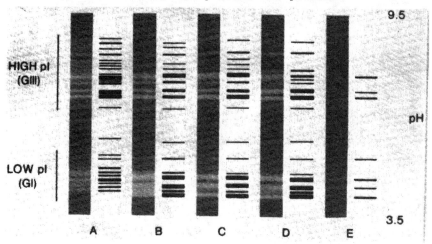

FIGURE 4. Zymograms of alpha-amylase components from: A, 3-day germinated wheat (cv. Neepawa); 3-day germinated excised embryos from B, Neepawa; C, RL 4137; D, Cypress; E, Idaed, resolved by PAG-IEF using a pH 3.5-9.5 gradient.

Densitometric analysis of the zymograms was undertaken to obtain an estimate of the relative proportions of the two main groups of alpha-amylase components. As shown in TABLE II, the low pI alpha-amylases in the three red spring wheat germinated excised embryos were similar and made up greater than 50% of the total activity. By comparison, a previous study has shown that the low pI components contribute less than about 13% of the total activity in germinated wheat. The low pI proportion was lower for Idaed, but excised embryos from this variety did not germinate well and consequently lower levels of alpha-amylase were obtained in the zymogram (FIGURE 4E) which hampered densitometric analysis. Recent analysis of alpha-amylase synthesized by excised barley embryos has shown similar results in that greater than 75% of the activity is in the form of the low pI enzyme

TABLE II. Relative proportions of alpha-amylase isoenzymes
in 3-day germinated wheat embryos.

Variety	Isoenzyme group	
	GI[a]	GIII[a]
Neepawa	54.3	45.7
Cypress	50.5	49.5
RL 4137	55.0	45.0
Idaed	36.6	63.4

[a] As percent of total activity determined by densitometric
analysis of zymograms.

forms as compared to less than 6% in the germinated whole
barley kernel (Marchylo et al., 1985). These findings suggest that alpha-amylase present in the final stages of
maturation would consist of higher proportions of low pI
enzyme components. To test this theory, PAG-IEF and densitometric analysis of alpha-amylase present in the latter
stages of maturation was carried out for these varieties.
The analysis of the three red spring wheat varieties was
hampered somewhat by the very low levels of activity.
Because of this, long incubation times (about 50 min) of the
beta-limit dextrin-polyacrylamide gel sandwich were required. This resulted in diffusion and smearing of enzyme
bands and an uneven background which in turn hampered
quantitation. However, the mean proportion of low pI, GI
components was about 40%, which would appear to agree with
this theory. Visual re-examination of previously published
zymograms of alpha-amylase components present in ten wheat
varieties in the final stages of maturation (Marchylo et al.
1980; Marchylo, 1978) also lends further support. The
results from analysis of Idaed alpha-amylase did not agree
with this theory. The estimated proportion of the low pI,
GI components was about 8%. This is comparable to the
proportion present in the germinated whole seed. As noted
previously, the activity levels in this variety were substantially higher. This might suggest that incipient
sprouting, concomitant with activation of the aleurone, had
begun. SEM analysis of the starchy endosperm adjacent to
the aleurone did not reveal evidence of hydrolytic breakdown but this may simply indicate that enzyme has not yet
been excreted from the aleurone. Further quantitative
analysis of the alpha-amylase present in the latter stages
of maturation should be carried out to unequivocally sub-
stantiate this theory.

REFERENCES

American Association of Cereal Chemists. 1962. Approved
methods (7th ed.), AACC, St. Paul, Minn.

Gale, M. D., Flintham, J. E. and Arthur, E. D. 1983. Alpha-
amylase production in the late stages of grain develop-
ment - An early sprouting damage risk period? In:
Proc. Third Int. Symp. on Pre-harvest Sprouting in
Cereals. Kruger, J. E. and LaBerge, D. E. (eds.).
Westview Press, Boulder, Co. USA. pp. 29-35.

Gibbons, G. C. 1979. On the localization and transport of
alpha-amylase during germination and early seedling
growth of Hordeum vulgare. Carlsberg Res. Comm. 44,
p. 353.

Gibbons, G. C. 1980. On the sequential determination of
alpha-amylase transport and cell wall breakdown in
germinating seeds of Hordeum vulgare. Carlsberg Res.
Comm. 45, p. 177.

Kruger, J. E. and Tipples, K. H. 1981. Modified procedure
for the use of the Perkin-Elmer model 191 Grain Amylase
Analyzer in determining low levels of alpha-amylase in
wheats and flours. Cereal Chem. 58, p. 271.

MacGregor, A. W. and Marchylo, B. A. 1986. Alpha-amylase
components in excised, incubated barley embryos. J.
Inst. Brew. (in press).

MacGregor, A. W. and Matsuo, R. R. 1982. Starch degradation
in endosperms of barley and wheat kernels during initial stages of germination. Cereal Chem. 59, p. 210.

Marchylo, B. A. 1978. The alpha-amylase isoenzyme system of
immature Canadian-grown wheat. Ph. D. dissertation,
University of Manitoba, Winnipeg.

Marchylo, B. A. and Kruger, J. E. 1983. Separation of wheat
alpha-amylase isoenzymes by chromatofocusing. In:
Proc. Third Int. Symp. on Pre-harvest Sprouting in
Cereals. Kruger, J. E. and LaBerge, D. E. (eds.)
Westview Press, Boulder, Co. USA. pp. 96-104.

Marchylo, B. A., MacGregor, A. W. and Kruger, J. E. 1985.
Production of alpha-amylases in germinating whole and
incubating de-embryonated barley kernels. J. Inst.
Brew. 91, p. 161.

Marchylo, B. A., LaCroix, L. J. and Kruger, J. E. 1980a.
Alpha-amylase isoenzymes in Canadian wheat cultivars
during kernel growth and maturation. Can. J. Plant
Sci. 60, p. 433.

Marchylo, B. A., LaCroix, L. J. and Kruger, J. E. 1980b.
The synthesis of alpha-amylase in specific tissues of
the wheat kernel. Cereal Res. Comm. 8, p. 61.

Mares, D. J. 1983. Alpha-amylase production in wheat grains in relation to pre-harvest sprouting damage. In: Proc. 33rd Annual Conference R.A.C.I. Cereal Chemistry. p. 55.

Okamoto, K, Kitano, H. and Akazawa, T. 1980. Biosynthesis and excretion of hydrolases in germinating cereal seeds. Plant Cell Physiol. 21, p. 201.

Sargeant, J. G. 1980. Alpha-amylase isoenzymes and starch degradation. Cereal Res. Comm. 8, pp. 77-86.

Rate and Location of Production of Alpha-Amylase in relation to Pre-Harvest Sprouting in Wheat

D. J. Mares, The University of Sydney,
Plant Breeding Institute, Narrabri,
NSW., Australia.

INTRODUCTION

The production of alpha-amylase and other hydrolytic enzymes, during the early stages of germination in cereal grains is essential for the provision of substrate and building blocks for the developing plantlet and the successful establishment of a cereal crop. Unfortunatley this same process can spell disaster for the cereal industry when it occurs prematurely in the ear in the field prior to harvest. Even relatively low levels of these enzymes are incompatible with the production of most high quality, human food products. Recent work (Okamoto, Kitano and Akazawa, 1980; Gibbons, 1981) has provided strong evidence that the initial site of alpha-amylase synthesis in germinating cereal grains is the scutellum and that the aleurone does not start to synthesize and secrete this enzymes until a much later stage of germination. Some controversy still exists, however, and opponents of the scheme have pointed out that definitive evidence of alpha-amylase synthesis in the scutellum of some cereals is lacking and that the presence of enzyme in the endosperm adjacent to the scutellum could also be explained by diffusion from the aleurone tissue along the plane of the scutellum/endosperm interface. Other aspects which are still unclear include the relative amount of enzyme produced by the scutellum and its relation to the amount required to affect processing quality of wheat, the stages in germination at which the different tissues are active, the extent of genetic variation for rate of alpha-amylase production and its practical significance and the possible application of genetic mutants, such as Rht 3 dwarfs, in which enzyme production by the aleurone is affected.

A number of experimental approaches were designed to provide further information on these questions.

MATERIALS AND METHODS

Wheat varieties, Songlen, Suneca, Kite, Shortim and the GA insensitive dwarf wheat Tordo were sown in field plots and harvested at maturity. Grain was hand-threshed from the wheat ears and stored in a seed room at 12°C for 6 months prior to use.

Germination of Grain. Wheat grains were surface sterilized for 10min. with 1% Biogram (Gibson Chemicals Ltd, Australia), washed with three changes of deionized water and placed in petri dishes on filter paper moistened with deionized water containing 100 g/ml streptomycin and 100 g/ml mycostatin. Grain was incubated at 4°C for 24h in an attempt to abolish any remaining vestiges of dormancy and possibly prime the grains to commence germination as synchronously as possible when transferred to 20°C. Samples, 3 x 3 grains, were harvested at intervals commencing 8h after the grains were transferred to 20°C. Following incubation grains were ground in a solution containing 0.085M NaCl and 1.4×10^{-3}M $CaCl_2$ and alpha-amylase activity determined by a method similar to that described by Barnes and Blakeney (1974).

Isolation and culture of embryo/scutellum explants. Surface sterilized grains were incubated as described above. Following 24h at 4°C the embryo and scutellum tissues were gently disected from the imbibed grains and placed on plugs (1cm diam. x 2mm) of agar and incubated at 20°C. For the determination of alpha-amylase both the germ tissue and the agar plug were homogenized.

RESULTS AND DISCUSSION

Production of alpha-amylase by wheat grains. Whole wheat grains started to produce measurable quantities of alpha-amylase beginning just after 12h of incubation at 20°C. Thereafter activity increased as shown in FIGURE 1. There was no significant difference in lag phase but there were significant differences in the initial rates of enzyme production (Shortim < Kite < Tordo, Songlen and Suneca). Following 30-36h of incubation, alpha-amylase production in the

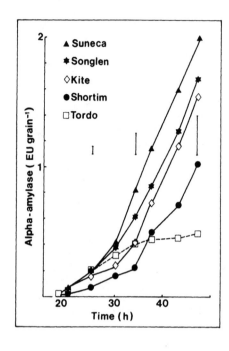

FIGURE 1. Production of alpha-amylase by germinating grains of wheat varities Suneca, Songlen, Kite, Shortim and Tordo.

GA-insensitive dwarf wheat, Tordo, reached a plateau whereas the other varieties were characterized by continuously increasing levels of activity (FIGURE 1).

Alpha-amylase production by germ tissue explants was compared with that of intact grains incubated under identical conditions. In all varieties examined the rate of increase in alpha-amylase activity in germ tissue explants was equal to, or greater than, the corresponding intact caryopses for the first 30-36h of incubation (FIGURE 2). Activity in germ tissue explants reached a maximum at approximately 36h after the start of incubation i.e. it mirrored the production of alpha-amylase by the variety Tordo which has been previously shown to have a GA-insensitive aleurone. Further incubation of germ tissue explants led to a decline in total alpha-amylase activity.

Microscopic examination of excised germ tissue revealed that there was still some aleurone tissue adhering to the surface of the scutellum. This aleurone tissue consisted of cells which were significantly smaller than the bulk of the aleurone which covered the surface of the endosperm. Unfortunately investigations of structural changes which occur in the aleurone tissue during germination have not included reference to the aleurone tissue at the surface of the scutellum. It has been suggested (Palmer, 1982) that alpha-amylase apparently released from the surface of the scutellum in barley may in fact be derived from the contaminate aleurone. Two observations appear to contradict this possibility, at least in wheat. Firstly, the production of alpha-amylase by germ tissue explants, unlike intact grains, de-embryonated grains and isolated aleurone tissue, did not appear to be stimulated by exogenous gibberellic acid. Secondly, the aleurone of GA-insensitive wheats such as the

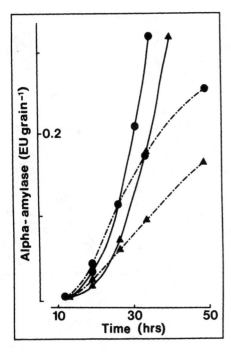

FIGURE 2. Production of alpha-amylase by germinating grains (——) and cultured germ tissue explants (–·–) of wheat varieties Songlen (●) and Shortim (▲).

Rht 3 dwarf Tordo was not stimulated by exogenous gibberellic acid yet both production of alpha-amylase by germ tissue explants and the initial production by intact grains was similar to that of GA-sensitive varieties.

In a further series of experiments, intact grains were incubated at 20°C and at intervals the germ tissue was removed, the germless grains incubated for a further 24h and the alpha-amylase activity compared with intact grains which had been incubated for the same period. Where germ tissue was removed in the first 36h there was not production of alpha-amylase by the de-embryonated grain in the subsequent 24h. At later stages, however, alpha-amylase activity continued to increase after the removal of the germ (FIGURE 3) except in the Rht 3 dwarf Tordo. The results suggest-

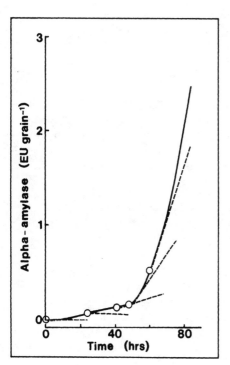

FIGURE 3. Production of alpha-amylase by germinating grains (——) and grains which were de-embryonated at various stages of germination (–––) (Cuv. Songlen).

ed that a message from the germ tissue was required to acti-
vate the aleurone alpha-amylase production system. By 36-
40h sufficient message had passed from the germ to the aleurone to allow some enzyme synthesis to continue in the
absence of the germ tissue. At even later stages the rate
of enzyme production following germ removal became progressively closer to that observed in the intact grains.

Taken together the results obtained in these investigations add further credence to the currently popular hypothesis that production of alpha-amylase in cereal grains is
biphasic i.e. during the initial stages of germination enzyme is synthesized in the germ tissue and that at a later
stage synthesis commences in the aleurone. In the present
study alpha-amylase synthesis commenced in the germ tissue
of wheat grains, probably in the scutellum, at about 12h
after the start of incubation whereas the aleurone only
became active after 36-40h. The synthetic capacity of the
germ tissue was relatively limited in comparison with the
much larger aleurone tissue. Consequently at progressively
later stages of germination the aleurone would be expected
to contribute more and more of the total grain alpha-amylase
activity.

Alpha-amylase production in relation to pre-harvest sprout-ing damage and pre-harvest sprouting tolerance.

Alpha-amylase activities of whole grain flour prepared
from a large number of weather-damaged samples (Falling
number: 560 - 60sec) were determined and the data used to
establish a relationship between falling number and alpha-
amylase (FIGURE 4). The line of best fit was used to calculate the approximate levels of enzyme activity which corresponded to the minimum falling numbers specified by the
Australian Wheat Board for wheat grades in NSW. These
values were approximations only since there was considerable
variation both with variety and with site. Alpha-amylase
activities corresponding to sound grain (FN > 350sec) and
the minimum Falling Number specifications for prime hard
(FN = 350sec), Northern hard No. 1 (FN = 300secs), Northern
hard No. 2 (FN = 250sec) and general purpose (FN = 200sec)
grades were 25-35, 40, 55, 80 and 140 m EUg^{-1} respectively.

The data illustrated in FIGURE 1 indicates the huge
capacity for production of alpha-amylase by germinating
wheat grains. By contrast the levels of activity corresponding to the minimum Falling number specifications
represent only a very small fraction of this potential and

FIGURE 4. Relationship between falling number and alpha-amylase. Horizontal bars mark the falling number limits of 350, 300, 250 and 200sec used in the Australian Wheat grading system.

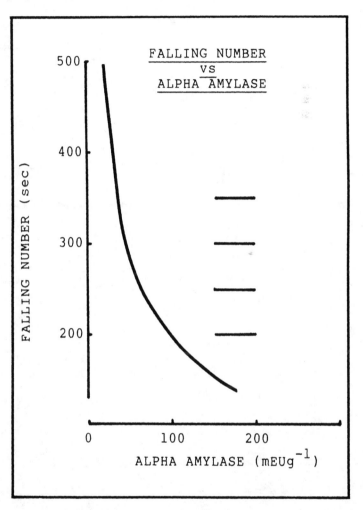

in situations where all kernels are germinating these levels would be attained within a very short time after the onset of enzyme synthesis. In practice, however, germination is not normally synchronous. Micro-environmental and physio-logical effects together with maturity variation with plants and crops usually ensures that there is a wide variation in germinability of individaual caryopses. Using the initial rates of alpha-amylase production shown in FIGURE 1, the time taken for samples of Suneca (high initial rate) and

500

TABLE 1. Relative germination times required for samples of
Shortim and Suneca to produce levels of alpha-amylase activity equivalent to falling numbers of 350, 300, and 200sec.
Rates of enzyme production taken from FIGURE 1.
[t = germination time (Suneca) - germination time
                                                    (Shortim)]

| Percent sprouted | | t (hours) | |
Kernels	FN = 350	FN - 300	FN = 200
100	0.4	0.6	1.0
50	0.6	1.0	1.9
10	2.3	3.4	5.0
1	8	10	12

Shortim (low initial rate) to reach alpha-amylase levels
equivalent to Falling Numbers of 350, 300 and 200sec were
calculated (TABLE 1).

Where only a small proportion of the grains had
germinated, the time differential approached the magnitude
required for variation in rate of alpha-amylase synthesis
production to be of practical significance in pre-harvest
sprouting tolerance. It must be remembered, however, that
after-ripening and weather conditions which result in a
loss of dormancy would lead to higher proportions of sprouted grains. Varieties with much lower rates of enzyme production than Shortim or a longer lag phase would be of considerable interest although none have been identified as
yet. The possible use of low rates of alpha-amylase production to provide tolerance to pre-harvest sprouting is still
rather academic since (a) it presupposes that production of
other enzymes e.g. proteases, would be similarly affected,
(b) it assumes that reduced enzyme production will not
adversely affect seedling establishment of the following
crop, and (c) in Australia at least, current receival
standards specify an upper limit for visibly sprouted grains
in addition to falling number. For ASW, the major Australian Wheat grade, there is nil tolerance.

Further calculations were made to determine the relative roles of germ tissue and aleurone in pre-harvest
sprouting in wheat grains. Based on a maximum production of
alpha-amylase by the germ tissues of between 200 and 400 mEU
grain[-1] (FIGURES 1 and 2) a grain weight of 35mg and an
alpha-amylase activity in sound grains of around 30mEU[-1]
(1mEU grain[-1]) then a sample containing 1% sprouted kernels
could have an alpha-amylase activity between 90 and 150

$mEUg^{-1}$ and a falling number in the range 250 and 190sec. Thus only at levels of sprouting less than 1% would the aleurone enzymes be of practical significance in pre-harvest sprouting in wheat. Genetic mutants, such as the Rht 3 dwarf wheats with a GA-insensitice aleurone, would consequently appear to be of little use in a sprouting tolerance breeding program.

SUMMARY

During the early stages of germination of wheat grains production of alpha-amylase by the germ tissue, presumably the scutellum, appeared to account for all the enzyme produced in the grain and it was only at a later stage of germination that the aleurone tissue also began to synthesize enzyme. Whilst the capacity of the germ tissue for alpha-amylase synthesis was limited, by comparison with the aleurone system, it would nevertheless be sufficient to result in sound grain being downgraded to feed grade wheat in most instances. Thus in the control of pre-harvest sprouting damage and breeding for pre-harvest sprouting tolerance only the scutellar enzyme production would appear to be relevant.

REFERENCES

Barnes, W. C. and Blakeney, A. B. 1974. Determination of cereal alpha-amylase using a commercially available dye labelled substrate. Staerke 26, pp. 193-197.

Barnes, W. C. and Blakeney, A. B. 1974. Determination of cereal alpha-amylase using a commercially available dye labelled substrate. Staerke 26, 193-197.
Gibbons, G. C. 1981. On the relative roles of the scutellum and aleurone layer in the production of hydrolases during germination of barley. Carlsberg Res. Comm. 46, pp. 215-225.
Okamoto, K., Kitano, K. and Akazawa, T. 1980. Biosynthesis and excretion of hydrolases in germinating cereal seeds. Plant Cell physiol. 21, pp. 210-204.
Palmer, G. J. 192. A reassessment of the pattern of endosperm hydrolysis (modification) in germinated barley. J. Inst. Brew. 88, pp. 145-153.

Some Aspects of the Synthesis of Enzymes and Enzyme Inhibitors in Barley

L. Munck and J. Mundy, Department of
Biotechnology, Carlsberg Research Laboratory,
10 Gamle Carlsberg Vej, DK-2500 Valby
Copenhagen, Denmark.

INTRODUCTION

The new powerful protein separation and characterization techniques now available have expanded our concept of
seed proteins originally based on Osborne's (Osborne 1924)
classification - albumins, globulins, prolamins and glutelins (TABLE 1). Classically, the easily soluble proteins,
the albumins and globulins, were considered to be more of a
functional enzymatic character while the structural protein
classes, the prolamins and glutelins, were considered as
storage proteins. Another functional class of proteins,
enzyme inhibitors (Boisen et al., 1981; Shewry and Miflin
1985), seems now to be quite prevalent in cereals seeds and
could be roughly estimated to contribute up to 5-10% of the
total seed protein content.

The inhibitory action of these inhibitors may be exerted against endogenous enzymes such as alpha-amylase-2 in
barley as well as against enzymes from bacteria, fungi and
mammals tentatively protecting the seeds from infestation
and digestion. The inhibitors in barley are both low molecular weight peptides (M = 9-12,000) such as chymotrypsin
(subtilisin) inhibitors 1 and 2 (CI-1 and CI-2, Hejgaard and
Boisen 1980; Jonassen 1980), trypsin inhibitor 1 and 2
(Shewry and Miflin 1985) and PAPI (Probable Amylase/Protease
Inhibitor, Mundy and Rogers 1986) while ASI [alpha-amylase-2
/Subtilisin Inhibitor in barley (Mundy et al., 1983) and in
wheat (Mundy et al., 1984)], PSI Putative Inhibitor for synthesis of protein in cell free animal systems (Asano et al.,
1984) and protein Z (Hejgaard and Boisen 1980) associated
with beta-amylase range in molecular weight form 21,000 to
40,000.

The original idea of a clear-cut conceptual separation

TABLE 1. Barley Proteins.

Osborne Protein Classes	Structural	Enzymes	Inhibitors
Albumins			Chymotrypsin 1 and 2 Trypsin 1 and 2 PAPI
Globulins		beta-amylase alpha-amylase 1 and 2 beta-glucanase Carboxypeptidase 1 and 2 etc.	ASI PSI Protein Z
Prolamins	A B Hordeins C D		
Glutelins	Non extractable Hordeins ?		

between storage proteins and enzymes seems difficult to up-
hold. Enzymes such as beta-amylase and enzyme inhibitors
such as CI-1 and CI-2 (Jonassen et al., 1981) are synthesized on the same template - the rough endoplasmic reticulum -
as are the hordeins. They are also stimulated in their
synthesis by a late application of nitrogen as are the hordeins (Giese and Hejgaard 1984). The trypsin inhibitors
have an amino acid sequence resembling part of the hordein C
chain (Shewry and Miflin 1985). The enzymatic, the inhibitory and the storage function seem thus to be different
parts of the same package of these proteins synthesized together on the rough ER in the endosperm cells. Since the
last pre-harvest sprouting symposium (Gibbons 1983), we have
expanded our basic research on the synthesis and tissue
specificity of the seed proteins (Mundy 1985) in collabora-
with other laboratories.
    The basis for such a work are specific polyclonal anti-
bodies obtained from pure preparations of the polypeptides
to be studied. The specific antibodies can be used to study
the synthesis in vivo as well as by quantifying the specific
protein translated from extracted mRNA in vitro in a rabbit
reticulocyte system. The separation of the tissues of

interest, germ, aleurone and endosperm is relatively easy to perform in the growing seed from 15 to 30 dpa. In the ripe seed this separation is more difficult and has to be performed by dry milling (decortication). There are several possible ways of estimating the purity of the separated botanical fractions, one is based on their autofluorescence (Jensen et al., 1981) in the fluorescence microscope combined with further examination at high magnification in the SEM. Another technique which can be used to identify the different botanical components is to analyse, HPLC with a fluorimetric detector, the tocopherol pattern of the fat which can be extracted from the dissected components (Barnes and Taylor 1981). Alpha- and beta-tocopherol are specific for the germ whereas alpha-tocotrienol predominates in the aleurone (85% of the seed content) and beta-tocotrienol is more equally distributed in the aleurone and endosperm. Proteins which are found exclusively in one of the botanical fractions also serve as an internal control for the precision of the separation.

THE ASI PROTEIN - A POSSIBLE MODEL FOR A CLASS OF PROTEINS THE SYNTHESIS OF WHICH IS STIMULATED BY ABSCISIC ACID

This double headed inhibitor with two separate sites against alpha-amylase-2 and subtilisin was identified simultaneously by our group (Mundy et al., 1983) and a research team in Winnipeg, Canada (Weselake et al., 1983). We sequenced the inhibitor (Hejgaard et al., 1983) and found it to partially resemble the soybean trypsin inhibitor of the Kunitz type as well as a subtilisin inhibitor from rice (Kato et al., 1972). ASI complexes with alpha-amylase-2 to give a heat labile electro-phoretic band previously called alpha-amylase-3. ASI inhibitors also prevail in wheat and rye (not in sorghum) but their complexes with alpha-amylase -2 are less stable than that of barley.
The alpha-amylase inhibiting action of ASI is rapidly inactivated above 50°C (Munck et al., 1984) which is of importance for the mashing process in the brewing of beer. We have checked, in a preliminary experiment, whether ASI is associated with dormancy and pre-harvest sprouting in barley varieties which vary widely in this respect (TABLE 2, Munck et al., 1984). ASI measured with a sensitive ELISA-method varied from 450 to 180 mg/kg and was not directly related either to dormancy or to pregermination as checked by germination experiments as well as with the FDB pregermination method (Jensen et al., 1984). There was no significant dif-

TABLE 2. Pregermination, alpha-amylase activity and BASI
(Barley alpha-amylase/subtilisin inhibitor) content of
barley varieties at late harvest.

Variety	Pregermination % (FDB-test)	Alpha-amylase activity Blue starch test	BASI mg/kg (ELISA)
Varde	16	+ + +	285
Kristina	14	+ +	205
Domen	8	+	190
Lise	4	+	250
Bode	4	+	240
M-268	4	(+)	450
Nordlys	2	+	225
Klages	2	+	195
Minerva	2	+	210
Stange	0	(+)	240
Agneta	0	−	325
Moyan	0	−	220
Triumph	0	−	180
Nordal	0	−	180

TABLE 3. Mean content of BASI in 50 individual kernels and
in the subset exhibiting alpha-amylase activity.

Variety	BASI (mg/kg) Mean of 50 kernels	No. of kernels exhibiting alpha-amylase activity	BASI (mg/kg) mean of kernels with alpha-amylase activity
Varde	286.5	14	284.0
Kristina	203.6	13	210.6
Nordal	190.7	0	−

ference in the ASI content of sprouted and unsprouted kernels from the same variety (TABLE 3). ASI thus seems to be
just one of the many factors which control alpha-amylase
activity and seed degradation during germination. It is
synthesized in the endosperm (and apparently not in the
aleurone) during seed development (Mundy 1985). The inhibitor is thus available in situ around the starch granules
protecting them against degradation from limited amounts of
alpha-amylase-2. It is interesting that although ASI is not
produced in the aleurone layer during seed growth, synthesis
of the inhibitor may be induced in vitro with abscisic acid

(ABA) in organ cultured aleurone layers (Mundy 1983) or in
vitro from translation of extracted mRNA (Mundy 1985) while
gibberellic acid is inhibitory. This is directly opposed to
the situation for alpha-amylase-2 and 1→3,1→4-beta-D glucanase where GA₃ is a promotor and ABA is an inhibitor (Mundy
et al., 1985).

The synthesis of carboxypeptidase in the aleurone and
in the embryo seems less dependent on plant hormones (Mundy
and Munck 1984). In an in vitro translation trial with mRNA
extracted from embryo (scutellum) and aleurone from germinated barley, alpha-amylase and beta-glucanase were predominantly synthesized in the aleurone while carboxypeptidase
showed up in approximately equal amounts in both the aleurone and embryo tissue with a slight difference in mobility
(Mundy et al., 1985). When alpha-amylase and beta-glucanase, immunoprecipitated from aleurone translation products
in this investigation, were compared with that immunoprecipitated from approximately 20 times as much scutellar mRNA
product, the mobilities and the intensities of the immunoprecipitated products were similar.

The synthesis of different enzymes in the aleurone thus
seems to be stimulated/inhibited by either GA or ABA or is
not affected by these hormones. Several polypeptides have
been shown to increase as a result of ABA administration.
So far only one of them ASI, has been identified (Mundy et
al., 1983). ASI and homologous inhibitors in other cereals
are thus interesting as markers in physiological studies of
the synthesis of those proteins which are stimulated by ABA.

FURTHER STUDIES ON THE EXPRESSION OF SYNTHESIS OF 8 SPECIFIC
PROTEINS IN THE BARLEY ENDOSPERM AND ALEURONE DURING SEED
SYNTHESIS AND GERMINATION

Both aleurone and the starchy endosperm are derived
from the same triploid tissue. While the endosperm is
essentially dead in the ripe grain, the aleurone is able to
produce a range of hydrolytic enzymes and inhibitors during
the subsequent germination. Dormancy and pre-harvest
sprouting are essential phenomena dependent on the balance
of enzymes and inhibitors finally assembled in the ripe
desiccated seed as well as on the potential induction of de
novo synthesized polypeptides. In TABLE 4 a mRNA translation experiment in vitro is outlined (Mundy 1985; Mundy et
al., 1986). mRNAs of several proteins are synthesized both
in the endosperm and aleurone at 20 to 30 dpa. This
includes protein Z, a protein closely associated to beta-

TABLE 4. Relative levels of in vitro synthesized 35S methionine labelled polypeptides translated from mRNA extracted from endosperm and aleurone tissues and precipitated with specific polyclonal antibodies and electrophoresed on SDS slab gels followed by fluorography and evaluation in a densitometric scanner (Mundy et al., 1986).

| Protein | Mr 10³ | No. of Met. | Rel. Moles/Lane[1] | | | | | | | | |
| | | | 20 dpa[2] | | 30 dpa[2] | | mature[3] | | Organ-cultured aleurone[4] | | |
			endo.	al.	endo.	al.	endo.	al.	CON	ABA	GA
Protein Z	40	4	1.12	0.51	0.70	0.30	-*	-	-	-	-
Beta-amylase	57	15	0.21	0.06	0.10	-	-	-	-	-	-
CI-1	9	2	0.66	0.40	0.75	0.43	0.03	0.14	0.04	0.04	0.04
CI-2	9	1	0.18	0.08	0.16	0.04	-	-	-	-	-
PSI	31	4	-	-	0.04	-	-	-	-	-	-
ASI	21	2	0.07	-	0.11	-	-	-	0.06	0.28	0.02
PAPI	12	1	-	0.04	-	0.07	0.06	0.19	0.05	0.06	0.05
Alpha-amylase	41	6	-	-	-	-	-	-	0.16	0.01	0.76

1. Relative Moles/Lane = $\dfrac{\text{Peak area}}{(\text{No. of met.}) \cdot (\text{35s methionine decay factor}) \cdot (\text{days exposure})}$

*   No detectable radioactivity

2. Endosperm/aleurone separation obtained by squeezing

3. Endosperm/aleurone spearation obtained by dry milling

4. Embryoless half grains

amylase, CI-1 and CI-2. In contrast, ASI and PSI mRNA are
only synthesized in the endosperm while PAPI mRNA is made in
the aleurone except for a small amount of PAPI mRNA found in
the ripe endosperm which had been separated by the less
exact dry milling techniques, the result of which should be
verified. The only mRNAs of the proteins studied which
survived desiccation are from CI-1 and PAPI. It is tempting
to ask why. mRNAs extracted from organ cultured aleurones
do show activity from CI-1, ASI and PAPI. Interestingly,
ASI is produced in organ cultured aleurone in spite of the
fact that it is not available at 20 or 30 dpa as well in the
ripe seed in this tissue. CI-1 and PAPI mRNA synthesis is
stimulated by ABA and inhibited by GA$_3$ in contrast to that
of alpha-amylase which is included for comparison. The in
vitro studies on mRNA level thus show both differences in
timing and in tissue specificity. The results are partially, but not fully, confirmed by in vivo studies. Thus ASI
and alpha-amylase synthesis patterns in vivo or in organ
cultured aleurone give the same response to GA$_3$ and to ABA
as in the mRNA rabbit reticulocyte translation system. However, in the case of PAPI, Mundy and Rogers,(1986), found in
both a hybridization experiment, with mRNA and DNA probe,
and in in vivo studies that the amount of PAPI mRNA should
be as abundant in the germination aleurone as alpha-amylase-
-2 mRNA in the active aleurone tissue. There was no
improvement in the in vitro translation of the PAPI mRNA in
a wheat germ in vitro system, and Mundy and Rogers concluded
that PAPI mRNA may be translated with less efficiency in
vitro in the presence of other mRNAs present in abundance in
organ cultured aleurone tissue. It is additionally tempting
to speculate that the apparent longevity of PAPI mRNA which
allows it to survive desiccation in the mature aleurone
might have some relationship with its poor translation in
the presently available in vitro systems.

Both PAPI and also PSI are examples of brainchildren
produced by current applied molecular techniques for which
there is as yet no verified biological function in the seed.
PSI was found to inhibit protein synthesis in animal cell
free systems (Asano et al., 1984) pointing out the possible
importance of this molecule in stopping translation if its
mRNA is available in the in vitro translation experiments.
The PAPI sequence was retrieved from a cDNA library as a
650 bp sequence which was shown in the computer to be
closely related to a known animal alpha-amylase inhibitor
from finger millet. Other parts of the sequence were also
homologous to certain plant trypsin inhibitors, but more
imprtantly, none of these inhibitory functions could be

verified in trials with the purified protein. Thus PAPI exists and is produced in significant amounts in the barley aleurone but we still do not know its function - only traces of its possible phylogenetical past. We have thus still another protein which is searching its function. It is obvious that when we have pinpointed the 20-200 key inhibitors and enzymes of importance for seed dormancy and pregermination and obtained specific antibodies against these, systematic studies of the basic character, discussed here, could be rewarding from both a fundamental and from a practical point of view. Such experiments will, however, never be easy to complete and interpret but it is the only way forward. The living cell and its physiology is our dynamic battle field. Isolated DNA and RNA are quite interesting as a source of information but they are isolated essentially dead and we therefore have to study them as part of the whole system, the living cell or a complete organism.

REFERENCES

Asano, K., Svensson, B. and F. M. Poulsen 1984. Isolation and characterization of inhibitors of animal cell-free protein synthesis from barley seeds. Carlsberg Res. Commun. 49, pp. 619-626.

Barnes, P. J. and Taylor, P. W. 1981. Tocopherol in barley germ. Phytochemistry 20, (7), pp. 1753-1754.

Boisen, S., Andersen, C. Y. and Hejgaard J. 1981. Inhibitors of chymotrypsin and microbial serine proteases in barley grains. Physiol. Plant. 52, pp. 167-176.

Gibbons, G. C. 1982. The action of plant hormones on endosperm breakdown and embryo growth during germination of barley. In: Proc. Third Inter. Symp. on Pre-harvest Sprouting in Cereals. Kruger, J. E. and LaBerge, D. E. (eds.). Westview Press, Boulder, Co. pp. 169-180.

Giese, H. and Hejgaard, J. 1984. Synthesis of salt-soluble proteins in barley. Pulse-labeling study of grain filling in liquid-cultured detached spikes. Planta 161, pp. 172-177.

Hejgaard, J. and Boisen, S. 1980. High-lysine proteins in Hiproly barley breeding: Identification, nutritional significance and new screening methods. Hereditas. 93, pp. 311-320.

Hejgaard, J., Svendsen, I. and Mundy, J. 1983. Barley alpha-amylase/subtilisin inhibitors of the soybean trypsin inhibitor (Kunitz) family. Carlsberg Res. Commun. Vol. 48, pp. 91-94.

Jensen, S. Aa., Munck, L. and Martens, H. 1981. The botanical constituents of wheat and wheat milling fractions. I. Quantification by autofluorescence. Cereal Chem. 59, pp. 477-484.

Jensen, S. Aa., Munck, L. and Kruger, J. E. 1984. A rapid fluorescence method for assessment of pre-harvest sprouting in cereal grains. J. Cereal Sci. 2, pp. 187-201.

Jonassen, I. 1980. Characteristics of Hiproly barley II. Quantification of two proteins contributing to its high-lysine content. Carlsberg Res. Comm. 45, pp. 59-68.

Jonassen, I., Ingversen, J. and Brandt, A. 1981. Synthesis of SP II albumin, beta-amylase and chymotrypsin inhibitor CI-1 on polysomes from the endoplasmic reticulum of barley endosperm. Carlsberg Res. Commun. 46, pp. 175-181.

Kato, I., Tominaga, N. and Kihara, F. 1972. In: Proc. of the 23rd Conference on Protein Structure. Iwai, K. (ed.). pp. 53-56. (in Japanese)

Munck, L., Mundy, J. and Vaag, P. 1984. Characterization of enzyme inhibitors in barley and their tentative role in malting and brewing. Proc. Brewing Congress of the Americas, St. Louis, Miss. (ASBC/AACC) Sept. 1984.

Mundy, J., Svendsen, I. and Hejgaard J. 1983. Barley alpha-amylase/subtilisin inhibitor. I. Isolation and characterization. Carlsberg Res. Commun. Vol. 48. pp. 81-90.

Mundy, J., Hejgaard, J. and Svendsen, I. 1984. Characterization of a bifunctional wheat inhibitor of endogenous alpha-amylase and subtilisin. FEBS Lett. 167, pp. 210-214.

Mundy, J. and Munck, L. 1984. Synthesis and regulation of hydrolytic enzymes in germinating barley. In: New Approaches to Research on Cereal Carbohydrates. Hill, R. D. and Munck, L. (eds.). pp. 139-148.

Mundy, J. 1985. Biochemistry and physiology of cereal grain and malt proteins - alpha-amylase and an alpha-amylase/subtilisin inhibitor. Thesis. Institute of Biochemical Genetics, University of Copenhagen, Øster Farimagsgade 2 A, DK-1353 Copenhagen K, Denmark.

Mundy, J., Fincher, G. B. and Brandt, A. 1985. Messenger mRNAs from the scutellum and aleurone of germinating barley encode (1→3, 1→4) beta-D-glucanase, alpha-amylase and carboxypeptidase. Submitted to Plant Physiol.

Mundy, J., Hejgaard, J., Hansen, A., Hallgren, L., Jørgensen

K. G. and Munck, L. 1986. Differential synthesis <u>in vitro</u> of barley aleurone and starchy endosperm proteins Submitted to Plant Physiol.

Osborne, T. B. 1924. The vegetable proteins. 2. ed. Longmans, Green and Co., London.

Shewry, P. R. and Miflin, B. J. 1985. Seed storage proteins of economically important cereals. In: Advances in Cereal Science and Technology. Pomeranz, Y. (ed.). American Ass. Cereal Chem., Vol. VII: pp. 1-83.

Weselake, R. J., MacGregor, A. W., Hill, R. D. and Duckworth, H. W. 1983. Purification and characteristics of an endogenous alpha-amylase inhibitor from barley kernels. Plant Physiol. 72, pp. 809-812.

The Pathway of Secretion of Enzymes from Isolated Barley Aleurone Layers

A. E. Ashford and F. Gubler, Botany School, University of NSW., Kensington, NSW., 2033, Australia.

INTRODUCTION

Barley aleurone offers a model system to study protein secretion in plants. Aleurone layers produce and secrete a range of enzymes in response to gibberellic acid (GA_3). Most are hydrolases with an obvious role in endosperm mobilization and for some both synthesis and secretion are totally under GA_3 control. The aleurone response is generally well characterised biochemically, especially for alpha-amylase, but our understanding of how the proteins are secreted from these or indeed any other plant cells is not clear. In aleurone newly synthesised proteins are associated with the microsomal fraction and alpha-amylase has been shown to have a signal sequence. However, direct evidence for intracellular transport via the lumen of the endoplasmic reticulum or any other organelle is lacking (see Ashford and Gubler, 1984). By localising GA_3-induced enzymes in situ during the secretory phase using histochemical and immunological techniques we have attempted to demonstrate the pathway of intracellular transport of enzyme proteins destined for secretion. The localisation of alpha-amylase and peroxidase will be examined here.

MATERIALS AND METHODS

Tissue and antibody preparation

Aleurone layers were isolated from Hordeum vulgare cv. Himalaya (1979 and 1981 harvests, Washington State University, Pullman, Washington USA) and incubated for 16h at 25°C according to the modified method used by Jacobsen and

Higgins (1982), with 10^{-6} M GA_3 plus 0.01 M $CaCl_2$. Barley alpha-amylase antibodies were prepared and purified by Dr. J. V. Jacobsen, CSIRO Division of Plant Industry (see Higgins et al., 1982; Jones and Jacobsen, 1982). Their specificity was checked against crude aleurone extracts using the Western blot technique (see Gubler et al., Planta, in press), when they were shown to be reactive against alpha-amylase but no other proteins in the extract.

Localisation of alpha-amylase

The aleurone layers were fixed in 2% glutaraldehyde in 0.03 M PIPES buffer (pH 7.5) and 0.01 M $CaCl_2$ for 3h at 0°C, dehydrated through an ethanol series and embedded in Lowicryl K4M (Carlemalm et al., 1982) at -45°C (see Ashford et al., 1986 for details of the embedding procedure). Sections were reacted with alpha-amylase antibodies and labelled with protein A-gold according to Gubler et al. (1986). The specificity of staining was evaluated by the following controls: sections treated with antibody solution pre-absorbed with 50 fold excess antigen and then reacted with protein A-gold; sections treated with rabbit anti-goat IgG antibodies (not raised against alpha-amylase) followed by protein A-gold; sections treated with buffer alone, followed by protein A-gold.

Peroxidase localisation

This was carried out on tissue slices of aleurone after fixation by prior to embedding. Fixation was in 3% glutaraldehyde in 0.025 M potassium phosphate buffer (pH 7.0) for 3h, followed by three buffer rinses, all at 0°C. The incubation solution consisted of 1mg ml^{-1} diaminobenzidine and 0.02% hydrogen peroxide in 50mM acetate buffer at pH 5.0. After staining the tissue slices were rinsed three times in acetate buffer at 0°C, post-osmicated (1% OsO_4 in 0.025 M potassium phosphate buffer) at 25°C for 2h, rinsed again in buffer and then dehydrated through an ethanol series into Spurr's resin. All sections illustrated were stained in alcoholic uranyl acetate (5min) and lead citrate (4min).

Controls were as follows: H_2O_2 omitted from the incubation medium; enzyme inactivated (blocks placed in water at boiling point for 10min prior to incubation); sodium azide (10^{-3} M) included in the incubation medium.

RESULTS

Alpha-amylase localisation

Immunocytochemical localisation of alpha-amylase with protein A-gold is shown in Plate 1 (a-d). The protein A which coats the gold particles specifically recognises immunoglobulins, particularly the IgG fraction, and there- fore binds to those areas of the section where rabbit anti- bodies have bound (Roth, 1984). Gold particles should therefore occur only where alpha-amylase protein with reactive antigenic sites is present. Since labelling is carried out post embedding on these sections, problems of penetration of antibodies across membranes are mostly avoided, but labelling only occurs at the surface of the section and the numbers of reactive antigen sites will have been considerably reduced by tissue preparation.

In the aleurone tissue embedded at low temperature in Lowicryl K4M, ultrastructural preservation was generally good. As with other non-osmicated material embedded in the acrylate-methacrylate resins, membranes are shown in nega- tive contrast but the rough endosplasmic reticulum (rough ER) is clearly recognisable by the rows of ribosomes. The ultrastructure is characteristic for GA_3-treated aleurone cells, with extensive, stacked sheets of rough ER and enlarged protein bodies throughout the cytoplasm. Large patches, with reduced electron opacity are present in the walls indicating that cell digestion is occuring (cf. Jones, 1972; Ashford and Jacobsen, 1974).

Gold particles were present over both cytoplasmic and wall areas. The highest levels of cytoplasmic labelling occurred over the rough ER and Golgi bodies (PLATE 1, a and b). The gold particles were either over the lumen or mem- brane regions of the rough ER (PLATE 1, a), but rarely over the cytoplasm between the cisternae. The labelling of the Golgi occured over both the central and peripheral regions of all the cisternae (PLATE 1, b). Some labelling was found in vesicles. Labelling over other organelles was very low and counts of gold particles (presented in Gubler et al., 1986) showed that labelling over the rough ER was nine times and over the Golgi twenty-five times higher than any other organelle. In the cell walls gold particles were very pre- cisely localised over the digested regions (PLATE 1, d), indicating that alpha-amylase, like other hydrolases is specifically released via digested wall channels.

Controls showed a similar level of background labelling to that in active sections, but did not show concentration of labelling over any organelle (PLATE 1, c).

Peroxidase localisation

Localisation of peroxidase extends and confirms the observations of the role of the Golgi. In GA$_3$-treated aleurone a large amount of reaction product is deposited in the cell walls, specifically in digested regions where wall carbohydrates have been removed. As for other enzymes not all digested wall areas are reactive. Specific regions around individual cells are affected. There is no reaction product over most areas of the wall that have not been digested.

There is a general background staining over the cytoplasm of most cells: this is variable from cell to cell and its status is not clear. (A similar background is seen in minus H$_2$O$_2$ controls.) Against this background staining sites of high peroxidase activity stand out clearly, these being much more heavily stained. The very high amount of deposit in some regions is indicative of a turnover reaction. Reaction product is present in the lumen of the rough ER, recognisable by its ribosomes (PLATE 2, a). Not all the rough ER is stained: dense reaction product only occurs in highly localised ER regions. The majority has little or no labelling in the lumen. Where there is labelling in the rough ER almost invariably the Golgi apparatus is also labelled (e.g. PLATE 2, a). There are some classic profiles with the Golgi in transverse section (PLATE 2, b). The entire stack is labelled and peripheral as well as central regions of the cisternae are reactive. Vesicles around the Golgi periphery are also full of reaction product. This reactivity in the rough ER and Golgi is generally found in regions where there is also reactivity in digested wall areas. It is highly variable from cell to cell and restricted to particular areas of individual cells.

As well as the vesicles specifically associated with the Golgi, there are also circular profiles filled with reaction product (PLATE 1, d). These are not cross sectional profiles of rough ER as they are not surrounded by ribosomes. Since long profiles of smooth ER were not found these circular profiles are interpreted to be vesicles rather than cross-sections of elongated structures. They are not very common and are generally found in cytoplasmic areas where reactive Golgi bodies also occur and are interpreted to be Golgi-derived vesicles. Some are very close to regions of the cell membrane where reaction product has accumulated in the periplasmic space.

Controls (PLATE 1, c) do not show concentrations of reaction product in any of the sites described above. The ultrastructure of the heat-inactivated control was so damaged that this was difficult to interpret.

DISCUSSION

Since the synthesis of alpha-amylase and its release from the aleurone are totally under GA3 control and all enzyme produced is destined for secretion, localisation of alpha-amylase in any organelle is direct evidence for involvement of that organelle in the secretory pathway. Localisation of the label almost exclusively in the lumen of the rough ER and Golgi bodies in sections reacted in full medium but not the controls strongly implicates both these organelles in secretion. The discovery of alpha-amylase in the rough ER is consistent with biochemical evidence which indicates that alpha-amylase carries a signal sequence and is sequestered in microsomes (Jones and Jacobsen, 1982; Rogers and Milliman, 1983; Chandler et al., 1984). On the other hand, the label in the Golgi is the first direct evidence for involvement of this organelle in protein secretion from aleurone, or indeed any other plant cell. Furthermore, concentration of the label in the Golgi compared with the rough ER, indicates that more antigen is present in this location and therefore that the alpha-amylase is being concentrated in the Golgi relative to the ER. This is consistent with the view that the Golgi comes later in the pathway. We were not able with the protein A-gold labelling to determine how material is transferred between the rough ER and Golgi or from the Golgi to the plasmamembrane. However, some indication of how this may occur is given by the peroxidase data.

Interpretation of the peroxidase localisation is much more difficult since only a proportion of the total peroxidase is secreted and synthesis is not completely under GA3 control (Gubler and Ashford, 1983). We cannot therefore assume, as we could with alpha-amylase, that sites of peroxidase localisation necessarily represent stages along the secretory pathway. However, the close association spatially between reactive rough ER, Golgi bodies, vesicles and reactive digested wall channels indicates a functional relationship. This contrasts with the situation in most areas of the aleurone tissue, where either digested wall channels were not present or were empty and the rough ER and Golgi

bodies were not stained for peroxidase. This correlation
indicates an involvement of rough ER and Golgi in the secretion of peroxidase. Reaction product in vesicles around the
periphery of the Golgi (some of them still attached) and in
similar free vesicles some distance away and often close to
deposits of reaction product outside the plasmalemma indicates that transfer of peroxidase from the Golgi to the
plasmalemma is most likely to be via Golgi-derived vesicles.
Similarly the close association of reactive ER with the cis-
face of the Golgi is indicative of a fairly direct transfer
between the two organelles. The localisation obtained for
both enzymes in digested wall areas, but not elsewhere in
the walls, confirms the earlier observations by Ashford, and
Jacobsen (1974) that the digested wall areas act as channels
for release of aleurone enzymes and that the undigested
walls are relatively impermeable. Alpha-amylase is now
added to the growing list of released enzymes preferentially
found in these regions.

   Taken together these data allow us to construct a general pathway for the secretion of enzymes from aleurone
cells. This involves insertion into the lumen of the rough
ER, transport to the Golgi complex and, via Golgi-derived
vesicles, to the plasmalemma. Release is then effected by
digestion of the cell walls. This pathway involving the
Golgi is essentially similar to that described for all
animal cells so far studied (Farquhar and Palade, 1981;
Kelly, 1985). This points to a universal mechanism in
eukaryotes.

FIGURE LEGENDS

PLATE 1. Localisation of alpha-amylase in isolated aleurone, treated with GA3 for 16h, with protein A-gold.
(a), (b) and (d) are full reacted sections, (c) control
reacted with rabbit anti-goat IgG antibodies followed by
protein A-gold. (a) Gold particles over all regions of
Golgi bodies and in vesicles. They are also present over
tangentially sectioned areas of rough ER. (c) Control
showing level of background staining. (d) Labelling of
alpha-amylase in the cell walls. Gold particles occur
specifically over areas where wall material has been lost,
(D). Magnification bar = 0.4um.

518

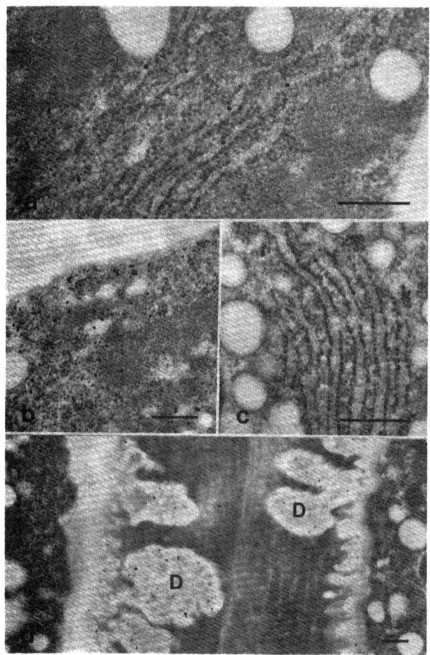

PLATE 1.

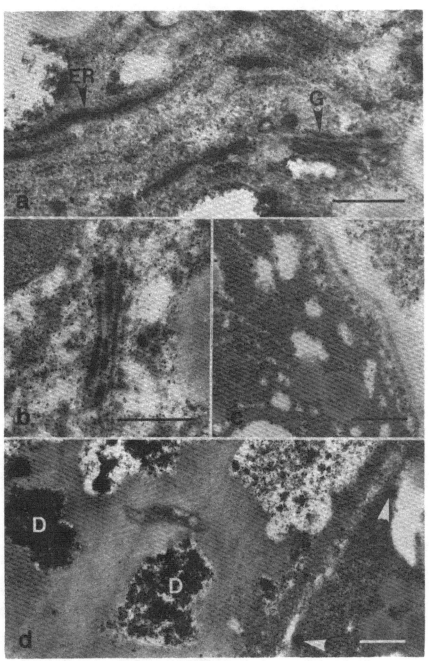

PLATE 2.

520

PLATE 2. Localisation of peroxidase in 24h, GA₃-treated isolated aleurone layers. (a), (b) and (d) reacted for 35 min in full medium. (c) Control incubated 35 min, minus H_2O_2. (a) Reaction product is deposited in the rough ER, lumen (ER), and in adjacent Golgi bodies (G). (b) The reaction in the Golgi is in the lumen of all the cisternae and in associated vesicles. (c) Control shows a non-specific background deposit. (d) Reaction product in digested wall regions, (D). Note the vesicle profiles containing reaction product (arrows), adjacent to the plasmalemma. Magnification bar = 0.4um.

ACKNOWLEDGMENTS

We are grateful to the electron microscope units of New South Wales and Sydney for providing facilities for this work and we thank in particular Lorraine Van der Lubbe and Suzanne Bullock for assistance with sectioning. The research was supported by funds provided by ARGS to A.E.A.

REFERENCES

Ashford, A. E., Allaway, W. G., Gubler, R., Lennon, A. and Sleegers, J. 1986. Temperature control in Lowicryl K4M and glycol methacrylate during polmerisation - is there a low temperature embedding method? J. Microsc. (submitted).

Ashford, A. E. and Gubler, F. 1984. Mobilization of polysaccharide reserves from endosperm. In: Seed Physiology Vol. 2. Germination and Reserve Mobilization. Murray, D. R. (ed.). Academic Press, Sydney.

Ashford, A. E. and Jacobsen, J. V. 1974. Cytochemical localization of phosphatase in barley aleurone cells: the pathway of gibberellic-acid-induced enzyme release. Planta 120, pp. 81-105.

Carlemalm, E., Garavito, R. M. and Villiger, W. 1982. Resin development for electron microscopty and an analysis of embedding at low temperature. J. Micros. 126, pp. 123-143.

Chandler, P. M., Zwar, J. A., Jacobsen, J. V., Higgins, T. J. and Inglis, A. S. 1984. The effects of gibberellic acid and abscisic acid on alpha-amylase mRNA levels in barley aleurone layer studies using an alpha-amylase cDNA clone. Plant Molec. Biol. 3, pp. 407-418.

Farquhar, M. B. and Palade, G. E. 1981. The Golgi apparatus
(complex) - (1954-1981) - from artifact to center
stage. J. Cell Biol. 91, pp. 77s-103s.

Gubler, F. and Ashford, A. E. 1983. Changes in peroxidase
in isolated aleurone layers in response to gibberellic
acid. Aust. J. Plant Physiol. 10, pp. 87-97.

Gubler, F., Jacobsen, J. V. and Ashford, A. E. 1986.
Involvement of the Golgi apparatus in the secretion of
alpha-amylase from GA₃-treated barley aleurone cells.
Planta (submitted).

Higgins, T. J., Jacobsen, J. V. and Zwar, J. A. 1982. Gibberellic acid and abscisic acid modulate protein synthesis and mRNA levels in barley aleurone layers.
Plant Molec. Biol. 1, pp. 191-215.

Jacobsen, J. V. and Higgins, T. J. 1982. Characterization of
the alpha-amylases synthesized by aleurone layers of
Himalaya barley in response to gibberellic acid. Plant
Physiol. 70, pp. 1647-1653.

Jones, R. L. 1972. Fractionation of the enzymes of the barley aleurone layer: evidence for a soluble mode of
enzyme release. Planta 103, pp. 95-109.

Jones, R. L. and Jacobsen, J. V. 1982. The role of the
endoplasmic reticulum in the synthesis and transport
of alpha-amylase in barley aleurone layers. Planta
156, pp. 421-432.

Kelly, R. B. 1985. Pathways of protein secretion in
eukaryotes. Science 230, pp. 25-32.

Rogers, J. C. and Milliman, C. 1983. Isolation and sequence
analysis of a barley alpha-amylase cDNA clone. J.
Biol. Chem. 258, pp. 8169-8174.

Roth, J. 1984. The protein A-gold technique for antigen
localisation in tissue sections by light and electron
microscopy. In: Immunolabelling for Electron Microscopy. Polack, J. M. and Varndell, I. M. (eds.).
Elsevier Science Publishers,N.Y.

Quantitative Analysis of Multiple Forms of Alpha-Amylase produced in Germinating Cereals

B. A. Marchylo, A. W. MacGregor and J. E. Kruger,
Grain Research Laboratory, Canadian Grain
Commission, 1404-303 Main Street, Winnipeg,
Manitoba R3C 3G8, Canada.

INTRODUCTION

Information on the heterogeneous nature of germinated wheat, barley and other cereals has been obtained primarily by electrophoretic methods such as agar gel electrophoresis (Smith and Bennett, 1974; Olered and Jonsson, 1970; Frydenberg and Nielsen, 1965); polyacrylamide gel electrophoresis (Goldstein and Jennings, 1975; Kruger, 1972) and by polyacrylamide gel isoelectric focusing (Marchylo et al., 1980; Sargeant, 1980; MacGregor, 1977). Polyacrylamide gel isoelectric focusing in particular, has provided excellent resolution of the many alpha-amylase components produced during germination. While these separatory techniques can provide useful information of a qualitative nature, they unfortunately are not easily amenable to quantitative analyses due to limitations inherent in the zymogram enzyme visualization procedure. Chromatofocusing, which was introduced at the previous symposium as an alternative method for separating germinated wheat alpha-amylases (Marchylo and Kruger, 1983), has proved to be a suitable technique for the quantitative analysis of individual or groups of alpha-amylase components.

The purpose of this paper is to discuss the results of studies which have used chromatofocusing to quantify changes in alpha-amylase components in germinating whole and incubating de-embryonated wheat and barley kernels as well as germinating barley embryos. In addition, preliminary information on a comparison of alpha-amylases produced in a number of germinating cereals will be discussed.

[1] Paper No. 571 of the Grain Research Laboratory, Canadian Grain Commission, 1404-303 Main St., Winnipeg, Manitoba, Canada, R3C 3G8.

RESULTS AND DISCUSSION

Germinated wheat alpha-amylase is resolved by PAG-IEF
into three heterogeneous groups of components (FIGURE 1)
which have been called the GI, GII and GIII alpha-amylases
(Marchylo and Kruger, 1983; Marchylo et al., 1980).   The
high pI GIII group of components, which focuses between
about pH 6.1-6.9, contributes the major proportion of the
total activity as determined by visual analysis of zymograms.   The low pI GI group of components, which focuses
between about pH 4.7-5.1, visually contributes significantly
less activity, while the GII group consists of only a few
components with pI's intermediate to the other groups.   Gale
(1983) has indicated that the GIII components are encoded
by the alpha-Amy-1 genes whereas the GI and, it also would
appear, the GII components are encoded by the alpha-Amy-1
genes whereas the GI and, it also would appear, the GII
components are encoded by the alpha-Amy-2 genes.

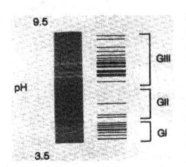

FIGURE 1.   Zymogram of five-day
germinated wheat alpha-amylase
separated by PAG-IEF using a pH
3.5-9.5 gradient.

The germinated wheat alpha-
amylases are resolved by chromatofocusing into a number of peaks as
illustrated in FIGURE 2.   As described previously (Marchylo and
Kruger, 1983), no overlap of the
GI + GIII components
was evident.   Elution
profiles were highly
reproducible and en-
zyme recoveries were
in the region of 75-
80%.

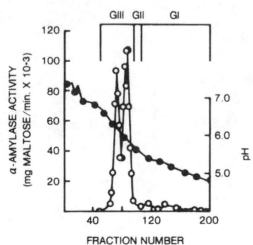

FIGURE 2.   Separation
of germinated wheat
alpha-amylase by
chromatofocusing (o).
pH gradient obtained
during chromatofocus-
ing (●).

PAG-IEF also separates barley alpha-amylase into three heterogeneous groups (FIGURE 3) designated alpha-amylase I, II and III (MacGregor, 1977) with isoelectric points of 4.9-5.2, 6.1 and 6.6, respectively (MacGregor, 1983). Recent studies have shown that alpha-amylase III in reality represents an alpha-amylase II - low molecular weight protein complex (Mundy et al. 1983; Weselake et al., 1983a, b) and therefore, barley alpha-amylase essentially is composed of one low and one high pI group of components. These also have been shown to be encoded by two alpha-amylase genes or groups of genes (Jacobsen and Higgins, 1982).

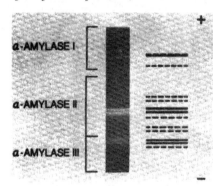

FIGURE 3. Zymogram of five-day germinated Bonanza barley alpha-amylase resolved by PAG-IEF using a pH 3.5-9.5 gradient.

Chromatofocusing also provides excellent resolution of these components as shown in FIGURE 4 (Marchylo and Mac-Gregor, 1983). It should be noted that the alpha-amylase II and III components are grouped together for quantitative analysis purposes. Reproducibility of chromatograms and enzyme recoveries are comparable to those obtained for wheat alpha-amylase.

FIGURE 4. Separation of Bonanza green malt alpha-amylase by chromatofocusing (o).

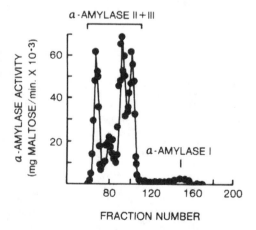

Quantitative analysis of alpha-amylase components produced
in germinating/incubating, wheat/de-embryonated wheat

Previous qualitative analyses, by PAG-IEF, of alpha-
amylase produced in germinating wheat and incubating de-
embryonated wheat (cv. Cypress) suggested that activity
attributable to the low pI GI alpha-amylases relative to
high pI GIII alpha-amylases was significantly higher in the
incubated de-embryonated wheat grain (Marchylo et al.,
1981). Chromatofocusing now has been used to obtain a
quantitative understanding of the relationship between these
components (Marchylo et al., 1984).

The results of this study confirmed that during germination, the major proportion of alpha-amylase activity in
the wheat kernel (>84%) was contributed by GIII components
(FIGURE 5A). This proportion decreased somewhat with time
concomitant with an increase in GI + GII activity to about
16% (FIGURE 5B). Although application of exogenous GA₃
induced higher levels of activity, the proportions of these
components were comparable to those obtained in the absence
of GA₃ throughout germination (FIGURE 5C, D).

Embryo excision both altered the proportion of components synthesized and dramatically decreased the potential
of the aleurone to produce alpha-amylase such that only
about 7% of the activity produced in the whole grain was
synthesized in the de-embryonated grain (FIGURE 6A).
Quantitative analysis of this alpha-amylase showed that up
to 63% was attributable to the GI + GII components (FIGURE
6B) as compared to at most 16% in the whole seed (FIGURE
5B). Addition of GA₃ induced higher levels of alpha-amylase
in the de-embryonated grain (FIGURE 6C) but the maximum
level reached was only 40% of that attained in the whole
seed in the presence of GA₃. The contribution of the GI +
GII components decreased to about 24% concomitant with an
increase to 76% for the GIII components. These proportions
remained constant during incubation (FIGURE 6D).

A comparative analysis of these results indicated that
production of GIII alpha-amylase activity was reduced by
about 96% in the incubated grain as compared to the germinated whole seed. In contrast, synthesis of GI + GII activity
was reduced by only 75%. This gave rise to the observed
increase in the relative intensity of the GI + GII bands as
visualized in zymograms. This result in combination with
the massive increase of the GIII components (20 fold) relative to the GI + GII components (4 fold) in the presence of
exogenous GA₃ suggests that synthesis of the high pI GIII
as compared to the low pI GI + GII components is controlled

FIGURE 5. A, C, Formation of alpha-amylase activity in germinated wheat. Activities of GIII and GI + GII groups of
alpha-amylases were determined following chromatofocusing
and were normalized to 100% recovery. A = in the absence of
GA₃, C = in the presence of GA₃. B, D, Proportions of GIII
and GI + GII groups of alpha-amylases relative to total GIII
+ GII + GI activity. GI, GII, and GIII alpha-amylases are
as denoted in FIGURE 2. B = in the absence of GA₃, D = in
the presence of GA₃ (reproduced from Marchylo et al., 1984).

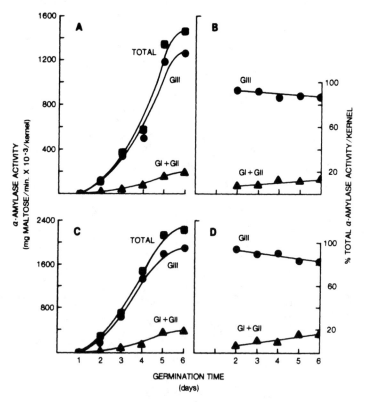

differentially by GA₃. Similar findings have been published
for Himalaya barley (Jacobsen and Higgins, 1982).

Quantitative analysis of alpha-amylase components produced
in germinating/incubating, barley/de-embryonated barley

A comparable quantitative analysis also was undertaken
with germinated barley (cv. Bonanza) alpha-amylase (Marchylo
et al., 1985). The surprising result of this study was the
lack of response to exogenously applied GA₃ for both ger-

FIGURE 6. A, C, Formation of alpha-amylase activity in
incubated de-embryonated wheat. Activities of GIII and GI +
GII groups of alpha-amylases were determined following
chromatofocusing and they were normalized to 100% recovery.
A = in the absence of GA3, C = in the presence of GA3. B,
D, Proportions of GIII and GI + GII groups relative to total
GIII + GII + GI activity. GI, GII, and GIII alpha-amylases
are as shown in FIGURE 2. B = in the absence of GA3, D = in
the presence of GA3 (reproduced from Marchylo et al., 1984).

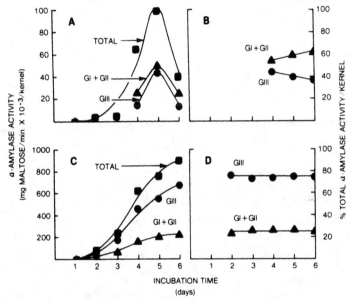

minating barley and incubating de-embryonated barley (FIGURE
7). As seen in FIGURE 7, embryo excision decreased significantly alpha-amylase production which is consistent with the
generally-accepted theory that the embryo serves as a source
of gibberellins required for induction of alpha-amylase synthesis by the aleurone. Application of exogenous GA3 then
should induce higher levels of alpha-amylase in the de-
embryonated seed; however this was not observed. The most
obvious explanation for this result is that the embryo supplies factor(s) other than GA3 which are required for optimal synthesis of alpha-amylase by the aleurone. It also is
of note that although application of GA3 to de-embryonated
wheat (Marchylo et al., 1984) did induce production of
higher levels of alpha-amylase, the response was limited to
40% of that obtained with germinated wheat. This result
also would concur with the concept that other factor(s) are
involved.

Quantitative analysis of alpha-amylase produced in both

528

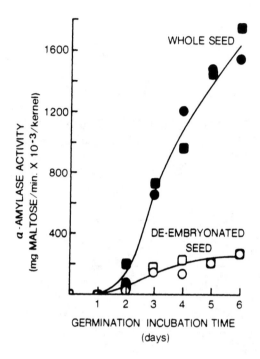

FIGURE 7. Formation of alpha-amylase activity in germinated barley in the absence (●) and Presence (■) of GA₃ and in incubated de-embryonated barley in the absence (o) and presence (□) of GA₃ (reproduced from Marchylo et al., 1985).

germinating barley and incubating de-embryonated barley showed that the major proportion of alpha-amylase activity synthesized was in the form of the high pI components, alpha-amylase II + III, i.e. alpha-amylase II. This proportion decreased in the whole seed during germination concomitant with a linear increase in the low pI, alpha-amylase I components both in the presence and absence of GA₃. This behaviour was comparable to that obtained for wheat. In barley, however, a smaller proportion of the total activity was attributable to the low pI components (<6%) as compared to wheat (<16%). As with wheat, excision of the barley embryo also resulted in an increase in the relative proportion of the low pI components to about 13% (FIGURE 8).

Comparative analysis of the results in this study revealed that alpha-amylase II + III production was reduced by 87% in the incubated de-embryonated as compared to the germinated whole grain. In contrast, alpha-amylase I activity was reduced by only 61%. The differential effect of embryo excision is similar to the results obtained for wheat; however unlike wheat, GA₃ did not induce synthesis of alpha-amylase in Bonanza barley. Therefore, for this variety, it may be that some factor(s) other than GA₃ differentially controls the expression of the two alpha-amylase genes or groups of genes that give rise to the high and low pI alpha-amylase components.

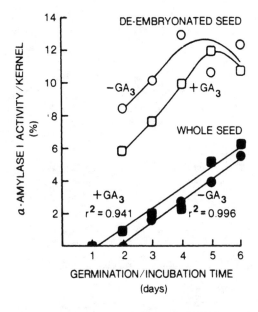

FIGURE 8. Changes in alpha-amylase I activity relative to total alpha-amylase activity during germination/incubation of barley/de-embryonated barley (abstracted from Marchylo et al., 1985).

Quantitative analysis of barley embryo alpha-amylase

Recent studies have presented evidence which indicates that the embryo can synthesize alpha-amylase during germination (MacGregor et al., 1984; Gibbons, 1980). Therefore, study was carried out to quantitatively determine the alpha-amylase composition of germinating excised barley embryos (MacGregor and Marchylo, 1986). It should be noted that the embryos were prepared in a manner designed to minimize aleurone contamination. As shown in FIGURE 9, barley alpha-amylase I comprised the major proportion (>75%) of the total alpha-amylase produced by three-day germinated excised embryos from the varieties Bonanza and Betzes. This result is in marked contrast to the small proportion (<6%) found in ger-

FIGURE 9. Separation of embryo alpha-amylase by chromato-focusing. A. Bonanza embryos; B.Betzes embryos (abstracted from MacGregor and Marchylo., 1986).

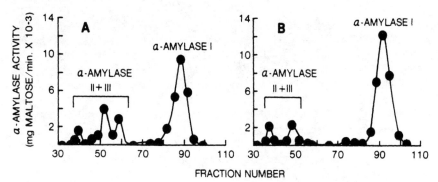

minated (MacGregor and Ballance, 1980) or incubated de-embryonated barley (Marchylo et al., 1985). It would appear then that the composition of alpha-amylase produced by barley embryos is significantly different from that produced by the aleurone. Qualitative analysis also suggests that the low pI components form a major proportion of the alpha-amylase produced in germinating wheat embryos (Marchylo and Kruger [these proceedings]; Marchylo et al., 1980b). This then lends further support to the contention that the embryo as well as the aleurone can synthesize alpha-amylase. It should be noted, however, that alpha-amylase activity produced by the excised embryos would constitute about 10% of the total alpha-amylase activity found in germinated kernels.

In conclusion, beyond the obvious results obtained, these studies should serve to point out that care must be taken when extrapolating results obtained by the study of alpha-amylase synthesis and control in isolated tissues, to the situation in the whole seed.

Preliminary comparison of alpha-amylase synthesized in a number of germinating cereals.

Synthesis of alpha-amylase has been studied in many cereals using a variety of separatory techniques. This at times has led to some confusion when comparing results. Therefore, a comparative qualitative and quantitative analysis, by PAG-IEF and chromatofocusing, of alpha-amylases produced in a number of germinating cereals has been undertaken. The cereals studied include red spring wheat, amber durum wheat, barley (two-rowed), triticale, oats, rye, sorghum, rice, millet and maize. Preliminary results using PAG-IEF have been obtained which indicate that these cereals can be divided into two groups based upon their alpha-amylase compositon. As shown in FIGURE 10, one group consisting of red spring and amber durum wheat, triticale, barley and rye exhibit both low and high pI groups of alpha-amylase components. In contrast, alpha-amylase from germinated oats, sorghum, rice, millet and maize is comprised of only the low pI alpha-amylase components. It is of interest to note that sorghum (Aisen et al, 1983) and rice (Akazawa and Mitsui, 1985), which reportedly synthesize alpha-amylase during germination primarily in the embryo, fall into this group; particularly in view of the high proportion of low pI alpha-amylase which appears to be produced by wheat and barley embryos.

FIGURE 10.   Zymogram of germinated cereal alpha-amylases
separated by PAG-IEF using a pH 3.5-9.5 gradient.   1. red
spring wheat   2. amber durum wheat   3. triticale   4. rye
5. barley (two-rowed)   6. oats   7. millet   8. sorghum
9. maize   10. rice

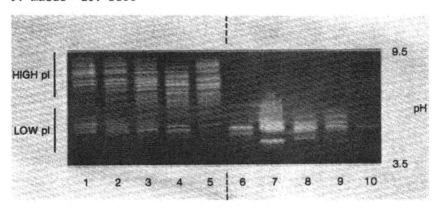

Preliminary chromatofocusing analysis of those cereals
with both high and low pI alpha-amylases has revealed some
variation in the relative proportions of these two groups of
components.   Rye appears to exhibit the highest proportion
of low pI components at about 20% (FIGURE 4).

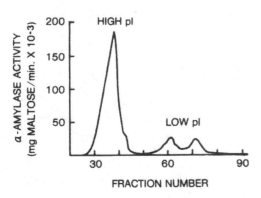

FIGURE 11.   Separation
of germinated rye alpha-
amylase by chromato-
focusing using a pH 7.4-
4.2 gradient.

REFERENCES

Aisen, A. O., Palmer, G. H. and Stark, J. R. 1983.   The
    development of enzymes during germination and seedling
    growth in Nigerian sorghum.   Starch 35, p. 316.
Akazawa, T. and Mitsui, T. 1985.   Biosynthesis, intracel-
    lular transport and secretion of alpha-amylase in:

rice seedlings. In: New Approaches to Research on Cereal Carbohydrates. Hill, R. D. and Munck, L. (eds.) Elsevier Science Publishers B. O., Amsterdam. p. 129.

Frydenberg, O. and Nielsen, G. 1965. Amylase isozymes in germinating barley seeds. Hereditas, 54, p. 123.

Gale, M. D. 1983. Alpha-amylase genes in wheat. In: Proc Third Int. Symp. on Pre-harvest Sprouting in Cereals. Kruger, J. E. and LaBerge, D. E. (eds.). Westview Press, Boulder, Bo. USA. pp. 105-110.

Gibbons, G. C. 1980. Immunohistochemical determination of the transport pathways of alpha-amylase in germinating barley seeds. Cereal Res. Comm. 8, p. 87.

Goldstein, L. D. and Jennings, P. H. 1975. The occurrence and development of amylase enzymes in incubated, de-embryonated maize kernels. Plant Physiol. 55, p. 893.

Jacobsen, J. V. and Higgins, T. J. V. 1982. Characterization of the alpha-amylases synthesized by aleurone layers in Himalaya barley in response to gibberellic acid. Plant Physiol. 70, p. 1647.

Kruger, J. E. 1972. Changes in the amylase of hard red spring wheat during germination. Cereal Chem. 49, p. 391.

MacGregor, A. W. 1977. Isolation, purification, and electrophoretic properties of an alpha-amylase from malted barley. J. Inst. Brew. 83, p. 100.

MacGregor, A. W. 1983. Cereal alpha-amylases: Synthesis and action pattern. In: Seed Proteins. Daussant, J., Mossé, J. and Vaughan, J. (eds.). Academic Press Inc. (London) Ltd., London. p. 1.

MacGregor, A. W. and Ballance, D. 1980. Quantitative determination of alpha-amylase enzymes in germinated barley after separation by isoelectricfocusing. J. Inst. Brew. 86, p. 131.

MacGregor, A. W., MacDougall, F. H., Mayer, C. and Daussant, J. 1984. Changes in levels of alpha-amylase components in barley tissues during germination and early seedling growth. Plant Physiol. 75, p. 203.

MacGregor, A. W. and Marchylo, B. A. 1986. Alpha-amylase components in excised, incubated barley embryos. J. Inst. Brew. 91, (in press).

Marchylo, B. A. and Kruger, J. E. 1983. Separation of wheat alpha-amylase isoenzymes by chromatofocusing. In: Proc. Third Int. Symp. on Pre-harvest Sprouting in Cereals. Kruger, J. E. and LaBerge, D. E. (eds.). Westview Press, Boulder, Co. USA. pp. 96-104.

Marchylo, B. A. and Kruger, J. E. 1986. Degradation of starch granules in maturing wheat and its relationship

to alpha-amylase production by the embryo. In:
These proceedings.

Marchylo, B. A. and MacGregor, A. W. 1983. Separation of
barley malt alpha-amylase by chromatofocusing. Cereal
Chem. 60, p. 311.

Marchylo, B. A., Kruger, J. E. and MacGregor, A. W. 1984.
Production of multiple forms of alpha-amylase in germinated, incubated, whole, de-embryonated wheat kernels. Cereal Chem. 61, p. 305.

Marchylo, B. A., LaCroix, L. J. and Kruger, J. E. 1980a.
The synthesis of alpha-amylase in specific tissues of
the immature wheat kernel. Cereal Res. Comm. 8, p. 61.

Marchylo, B. A., LaCroix, L. J. and Kruger, J. E. 1980b.
Alpha-amylase isoenzymes in Canadian wheat cultivars
during kernel growth and maturation. Can. J. Plant
Sci. 60, p. 433.

Marchylo, B. A., LaCroix, L. J. and Kruger, J. E. 1981.
Alpha-amylase synthesis in wheat kernels as influenced
by the seed coat. Plant Physiol. 67, p. 89.

Marchylo, B. A. MacGregor, A. W. and Kruger, J. E. 1985.
Production of alpha-amylase in germinating whole and
incubating de-embryonated barley kernels. J. Inst.
Brew. 91, p. 161.

Mundy, J., Svendsen, I. and Hejgaard, J. 1983. Barley
alpha-amylase/subtilisin inhibitor isolation and characterization. Carlsberg. Res. Comm. 48, p. 81.

Olered, R. and Jönsson, G. 1970. Electrophoretic studies of
alpha-amylase in wheat. II. J. Sci. Food Agric. 21,
p. 385.

Sargeant, J. G. 1980. Alpha-amylase isoenzymes and starch
degradation. Cereal Res. Comm. 8, p. 77.

Smith, J. B. and Bennett, M. D. 1974. Amylase isozymes of
Oats (Avena sativa L.). J. Sci. Fd. Agric. 25, p. 67.

Weselake, R. J., MacGregor, A. W. and Hill, R. D. 1983a.
An endogenous alpha-amylase inhibitor in barley kernels. Plant Physiol. 72, p. 809.

Weselake, R. J., MacGregor, A. W., Hill, R. D. and Duckworth, H. W. 1983b. Purification and characteristics
of an endogenous alpha-amylase inhibitor from barley
kernels. Plant Physiol. 73, p. 1008.

Cultivar Effects on Falling Number

*J. H. Moss, Bread Research Institute of
Australia, PO Box 7, North Ryde, NSW,
2113, Australia.
and C. P. Stiles*

SUMMARY

Wheat cultivars differ in their inherent Falling number
in the absence of pre-harvest rain damage, and also in their
sensitivity to changes in alpha-amylase activity. Environmental conditions during the development of the grain may
also affect the Falling Number.

INTRODUCTION

Quality standards for wheat received from farmers or
sold either to local mills or for export include, inter
alia, a minimum Falling Number. In Australia for Prime Hard
wheat this is 350, for No. 1 Hard and A.S.W. it is 300, with
No. 2 and No. 3 Hard grades set at 250 and 200 respectively.
These standards are well understood, and equitable. Many
rapid methods for determining suitability for processing
rely on determining alpha-amylase activity, rather than
paste viscosity. When discrimination based on these methods
diverges from the discrimination of the Falling Number test,
the benefits of speed prove to be illusory. Substantial
differences in wheat cultivars in respect of the relationship between alpha-amylase activity and Falling Number were
described in Australian wheats by Moss, Derera and Balaam
(1972). Systematic cultivar effects were described by Ringlund (1983), while protein content also seems to affect
Falling Number and Amylograph peak (Moss and Miskelly, 1984;
Meredith and Pomeranz 1982). These effects are of some
importance where there may be disagreement as to whether
sprouting damage is characterised by loss of paste viscosity
or by increase in alpha-amylase activity.

534

Paste viscosity is governed by the amylopectin content of the total alpha glucan (Loney, Jenkins and Meredith, 1975). This, in turn, bears a spurious relationship to protein because environmental factors such as high ambient temperature or limited time for grain filling lead to lower starch synthesis relative to protein (Bhullar and Jenner 1985) and to a higher proportion of amylopectin (Duffus and Murdoch, 1979). These same conditions may lead to a higher fibre content of the grain, due to less starch formation, and hence an even higher falling number because of the water absorbed by this material.

While inherent pasting properties govern much of the variation of Falling Number among sound Australian wheats, the amylase variation is not observed except in some seasons at some places. Different relationships are noted in the northern hemisphere. Morris and Poulsen (1985) found that nitrogenous fertiliser increased the degree of alpha-amylase generation after rain in susceptible cultivars, a situation which would give a negative relationship between protein content and Falling Number.

Asoka and co-workers (1984) found that a lower environment temperature increased the amylose content of rice starches, the period 5 to 15 days after anthesis being the most critical. There are also cultivar differences in amylose-amylopectin ratio in wheat. (Moss and Miskelly, 1984).

Starch granules vary in size, and the distribution according to size varies according to variety and to environment, with an interaction between the two effects. The implications in respect of pasting properties are not clear. Gelatinisation temperature depends on granule size, being higher with smaller granules. In maize (Knutson et al., 1982) rate of hydrolysis of such granules by alpha-amylase was greater, as the surface area per unit of mass was greater. They also pointed out that the smallest granules had the highest amylose content, a factor complicating other observations.

Lipids and gums also affect pasting properties, but the climatic, edaphic and inherent quality effects through these sources of variation are not established.

Grain texture also affects pasting properties. Although mechanical damage, associated with hard wheats, makes starch more susceptible to enzyme attack, in fact hard wheats as a class appear to have higher Falling Numbers than soft wheats. For instance, over the last five years Prime Hard wheat shipped to Japan has had a mean Falling Number of 553 versus 409 for Australian Standard White, while Hard Winter Ordinaries (388) and Red Spring (406) have higher

Falling Numbers than Western White at 352 for the corresponding period.

Differences in the ability of wheats to generate alpha-
amylase are not considered here. However, we did investigate the Falling Number of some wheat cultivars free of any
rain damage, and the regression of Falling Number against
alpha-amylase activity where some pre-harvest rain damage
had occured.

EXPERIMENTAL

(1) Hard versus soft wheat.

Three pairs of near isogenic cultivars, one hard and
one soft-grained in each pair, grown under similar conditions at eight sites, were tested for Falling Number. In
each case the hard grained cultivar gave a higher Falling
Number than the soft grained one, as shown in TABLE 1. The
Insignia derivatives were grown at different sites to the
others.

(2) Method of determining alpha-amylase activity.

Three methods of determining alpha-amylase activity
were compared by testing each of forty wheat samples and the
corresponding flours, in duplicate. The methods were the
dye liberation or Phadebas method (Ceska et al., 1969), the
nephelometric method (Campbell 1980) and the loss of iodine
blue or S.K.B. method (Perten, 1966).

Variance ratios were determined in respect of wheat and
flour for each method, and are shown in TABLE 2. A correlation matrix of all figures, including Falling Number on
each sample, and the logarithm of each figure was constructed. This indicated a difference between the relationship
with Falling Number in respect of each method of determining
the enzyme. In almost all cases the logarithm of the Falling Number gave the best fit, but the Phadebas figures had
also to be transformed to logarithms to achieve linearity.
TABLE 2 also shows the linear correlation between alpha-
amylase and log Falling Number in each case. All further
work was carried out by the Nephelometric method.

(3) Liquefaction numbers.

Samples of pure cultivars taken from agronomists' plots
in the rain-free 1982 harvest, and the severely affected

TABLE 1.   Comparison of near isogenic Hard and Soft lines

| | Falling number | | Significance |
	Hard	Soft	
Insignia back crosses			
	EMBLEM 534	HERON 467	P<0.05
Anza back crosses			
	OXLEY 397	EGRET 335	P<0.01
Condor back crosses			
	OSPREY 437	QUARRION 388	N.S.

TABLE 2.   Comparison of three methods for alpha-amylase determination

		Variance ratio	Correlation with Log F.N.
Phadebas	wheat	469	-0.86
	flour	209	-0.77
Nephelos	wheat	1201	-0.88
	flour	273	-0.79
S.K.B.	wheat	332	-0.93
	flour	322	-0.80

1983 harvest were ground in a Falling Number 3100 mill, and examined for Falling Number and alpha-amylase activity with a Perkin-Elmer Nephelometer (Campbell 1980). Linear regressions of Liquefaction Number (Perten, 1964) against alpha-amylase were constructed for each cultivar series and for all (90) samples. It was found that greater linearity was achieved assuming a gelatinisation time of 40 seconds than the 50 seconds recommended by Perten. The respective figures are shown in TABLE 3. A gelatinisation time of 40 seconds is consistent with the obvious increase in the viscosity of the system after 30 to 35 seconds stirring in the Falling Number test.

(4) Regressions for individual cultivars

Linear regressions in the form $\dfrac{6000}{F.N. - 40} = a + b$ (nephelos units)

were constructed for seven cultivar series, based on samples
from the 1982/83, 1983/84 and 1985/86 harvests. Falling
Number at zero alpha-amylase units, and alpha-amylase units
at a Falling Number of 300 were estimated and are shown in
TABLE 4. It is obvious that the paste viscosity varies considerably according to cultivar, and that cultivars vary in
their sensitivity to the enzyme.

TABLE 3. Linear correlation co-efficient linking alpha-
amylase units with liquefaction number for six cultivars
assuming gelatinisation times of 40, 50 and 60 seconds.

	CORRELATION CO-EFFICIENT		
Gelatinisation Time (sec)	40	50	60
Cultivar			
Sunelg	0.9947	0.9945	0.9934
Sundor	0.9963	0.9915	0.9830
Songlen	0.9561	0.9502	0.9414
Oxley	0.9915	0.9686	0.8061
Egret	0.9882	0.9916	0.9555
Cook	0.9902	0.9797	0.9485
All samples	0.9598	0.9398	0.7388

TABLE 4. Some Australian cultivars classified according to
Falling Number at zero alpha-amylase and the alpha-amylase
units required to give a Falling Number of 300. (From
regression of liquefaction number versus alpha-amylase).

Cultivar	Falling number at zero nephelos	Nephelos-1 units at falling number = 300
Songlen	529	488
Sundor	511	282
Oxley	475	289
Sunelg	461	284
Sunstar	458	281
Banks	430	250
Egret	381	168

539

DISCUSSION

Falling Number is governed by many factors, the chief of which is alpha-amylase activity. However, the relationship between alpha-amylase and Falling Number is also governed by many factors which combine to introduce considerable errors in predicting one from the other. Even the method of determining alpha-amylase can influence the apparent relationship.

Standards of soundness based on a Falling Number test are becoming commonplace in receival of wheat from farmers into public storage, or to flour mills. Certainly specifications based on Falling Number apply to contracts for the supply of flour. It may well be that this provides an adequate indication of the prospective performance of the flour in question in manufacturing process. Where it does not there are some points in the foregoing that are relevant.

(1) Cultivars vary in their Falling Numbers, even at the same alpha-amylase level.
(2) Soft wheats give lower Falling Numbers than equivalent hard wheats.
(3) Where blending is required, and liquefaction numbers are calculated, the gelatinisation time appropriate to the wheats in question should be determined.
(4) Notwithstanding (1) and (2) cultivars cannot be described as inherently high or low in Falling Number unless their ability to generate or not generate the enzyme is known.

CONCLUSION

Cultivars vary sufficiently in their inherent wheatmeal paste viscosity, and inherent susceptibility to alpha-amylase to make it impractical to classify wheats for receival according to alpha-amylase activity if they are to be sold according to a paste viscosity specification. Grain hardness and other grain characteristics seem to affect Falling Number in addition to the alpha-amylase effects.

REFERENCES

Bhullar, S. S. and Jenner, C. F. 1985. Aust. J. Plant Physiol. 12, p. 363.
Campbell, J. D. 1980. Cereal Res. Comn. 8, p. 107.

Ceska, M., Hultman, E. and Inglemar B. 1969. Experientia. 25, p. 555.

Duffus, C. M. and Murdoch S. M. 1979. Cereal Chem. 56, p. 427.

Knutson, C. A., Khoo, U., Cluskey, J. E. and Inglett G. E. 1982. Cereal Chem. 59, p. 512.

Loney, D. P., Jenkins, L. D. and Meredith, P. 1975. Stärke. 27, p. 145.

Meredith, P. and Pomeranz, Y. 1982. Cereal Chem. 59, p. 355.

Morris, C. F. and Poulsen, G. M. 1985. Crop Science. 25, p. 1028.

Moss, H. J., Derera, N. F. and Balaam, L. N. 1972. Aust. J. Agric. Res. 23, p. 769.

Moss, H. J. and Miskelly D. M. 1984. Food Tech. in Aust. 36, p. 90.

Perten, H. 1966. Cereal Chem. 43, p. 336.

Perten, H. 1964. Cereal Chem. 41, p. 127.

Ringlund, K. 1983. In: Proc. Third Int. Symp. on Pre-harvest Sprouting in Cereals. Kruger, J. E. and LaBerge, D. E. (eds.). Westview Press, Boulder Co. U.S.A. pp. 111-118.

Section VIII
Assay Methods and Objective Testing

The Canadian Grading System for Wheat and its Implications in the Monitoring of Sprout-Damage

J. E. Kruger, Grain Research Laboratory,
Canadian Grain Commission, 1404-303 Main
Street, Winnipeg, Manitoba, R3C 3G8,
Canada.

Quality is maintained in Canadian bread wheats by the Canadian grading system. All wheat initially is examined visually by the elevator agent upon delivery to a country elevator and subsequently by Canadian Grain Commission inspectors as it arrives at terminal elevators and upon discharge into cargoes. The visual examination includes the degree of soundness of the wheat which includes many factors such as sprouting, immaturity, frost and mildew. Limits are placed on the type and degree of damage that is allowable in a particular grade of wheat. The grade in turn determines the economic value of the wheat and the financial return to the farmer.

There are three grades of hard red spring wheat: No. 1, 2 and 3 Canada western red spring (CWRS). There are five grades of durum wheat: No. 1, 2, 3, 4 and 5 Canada western amber durum (CWAD). There are three grades of red winter wheat: No. 1, 2 and 3 Canada western red winter (CWRW). Soft white spring has three grades: No. 1, 2 and 3 Canada western soft white spring (CWSWS). There are two grades of utility wheat: No. 1 and 2 Canada utility (CU). Finally, there are two grades of a newly licensed class of wheat: No. 1 and 2 Canada prairie spring (CPS). Wheat of any class with the exception of amber durum that does not meet the grade specifications is designated as Canada feed wheat.

Although this paper will focus on sprouted kernels as a grading factor, I would just like to mention there are many types and degrees of damage other than sprout damage as well as quality requirements that must be met for a particular

1 Paper No. 575 of the Canadian Grain Commission, Grain Research Laboratory, 1404-303 Main Street, Winnipeg, Manitoba R3C 3G8.

grade. For example, No. 1 and 2 CWRS wheats must be of a variety having bread-making quality better than or equal to our standard variety Marquis. The No. 1 grade would have to be reasonably well matured and free from damaged kernels having a minimum test weight of 75kg per hectolitre and 65% hard vitreous kernels. The No. 2 grade must be fairly well matured but can be moderately bleached or frost damaged although it must be reasonably free from severely damaged kernels. The maximum tolerances of other primary grade determinants for No. 1 and 2 CWRS would be as shown in TABLE 1.

TABLE 1. Maximum tolerances of primary grade determinants for No. 1 and 2 CWRS wheat.

Grade Name	Foreign Material Other Than Wheat		Wheats of Other Classes or Non-prescribed Varieties	
	Matter Other Than Cereal Grains	Total Including Cereal Grains	Contrasting Classes	Total Including Contrasting Classes
No. 1 CW Red Spring	About 0.2% including max. 20K inseparable seeds	About 0.75%	About 1.0%	3.0% including not more than 1.0% non-prescribed varieties
No. 2 CW Red Spring	About 0.5% including max. 50K inseparable seeds	About 1.5%	About 3.0%	6.0% including not more than 2.0% non-prescribed varieties

As you can see, although many of the tolerances are quite objective, i.e. % degermed, a number are also quite subjective, for example, reasonably well matured. To assist inspectors each year in grade evaluations, primary standard samples are prepared to be used as visual guides in order to reflect the minimum quality which might be expected for the grade. Because of the blending which occurs throughout the transportation system, the overall quality will be considerably better than the minimum requirements for the grade. Thus, the effect of, say, frost damage in one part of the country will be considerably diluted by mixing with wheat of the same grade which has not suffered such damage. This

Table 1 (cont'd)

Maximum tolerances of primary grade determinants
for No. 1 and 2 CWRS wheat

Grade Name	Sprouted	Smudge	Total Smudge and Blackpoint	Degermed	Grass Green	Pink Kernels	Insect Damage		Dark Immature	Natural Stain
							Sawfly Midge	Grasshopper Army Worm		
No. 1 CW Red Spring	0.5%	30K	10.0%	4.0%	0.75%	1.5%	2.0%	1.0%	1.0%	0.5%
No. 2 CW Red Spring	1.5%	1.0%	20.0%	7.0%	2.0%	5.0%	8.0%	3.0%	2.5%	2.0%

Grade Name	Artificial Stain No residue	Binburnt Severe Mildew Rotted Mouldy	Total Heated Including Binburnt	Fireburnt	Stones	Ergot	Sclerotinia	Shrunken and Broken		
								Shrunken	Broken	Total
No. 1 CW Red Spring	Nil	2K	0.1%	Nil	3K	3K	3K	6.0%	6.0%	7.0%
No. 2 CW Red Spring	5K	5K	0.75%	Nil	3K	6K	6K	10.0%	10.0%	11.0%

* K refers to kernels or kernel size pieces in 500 grams

is reflected in export standard samples which are prepared
each year to reflect average grade quality within a grade.
  Now returning specifically to the tolerances for
sprout-damage in the various grades, they are as shown in
TABLE II. Canada feed has no limits set on sprout-damage.
  Such limits for the different classes are intended to
reflect the effect of sprout-damage on end-product quality.
Thus, the CWRS wheats intended for breadmaking have lower
limits than CWAD wheats intended for pasta products.

TABLE II. Sprout tolerances (%) for wheat in different
grades and classes of Canadian wheat

Grade	Class					
	CWRS	CWAD	CWRW	CWSWS	CU	CPS
1	0.5	1.0	0.5	1.0	1.5	1.5
2	1.5	5.0	2.5	5.0	5.0	5.0
3	5.0	8.0	8.0	8.0		
4		12.0				
5		no limit				

  Let us now examine how this system works in practice,
particularly with regard to sprout-damage as a degrading
factor.
  In the discussion to follow, we will be examining
wheats grown primarily in western Canada. Wheats grown in
eastern Canada are generally softer and used primarily for
domestic consumption. The overall transportation system for
movement of grain in weatern Canada is shown in FIGURE 1.
Farmers deliver their grain to approximately 2,900 primary
elevators. The grain then moves to terminal elevators at
Vancouver, Prince Rupert, Churchill or Thunder Bay by two
main rail lines. From here, the wheat is loaded onto ships
for export (Thunder Bay shipments must first move through
the Great Lakes system to the Atlantic coast or must move by
rail during the winter months to transfer elevators).
  When a farmer brings a truckload of wheat to a primary
elevator, he is given a grade for the wheat by the country
elevator agent. Quite often wheat from a particular area
will be of the same grade. Competition from neighbouring
elevators usually ensures that a fair grade is given to the
farmer. If the farmer is unsatisfied with the grade, however, he may submit a representative sample to the Canadian

FIGURE 1. Movement of grain in weatern Canada showing terminal and transfer elevators.

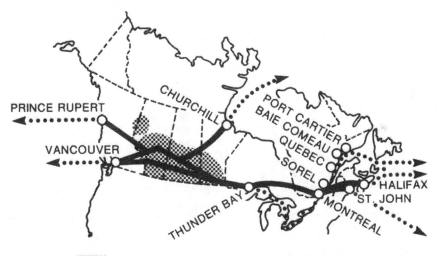

MAJOR GROWING AREAS OF WESTERN CANADA

Grain Commission for grading. Payment to the producer is
then based on the Commission's grade.

How can a busy primary elevator agent hope to analyze
a sample according to all the degrading factors at harvest
time? Well, in practice, he doesn't. First of all, he will
go through a stepwise examination which would include determination of test weight, varietal purity, vitreousness,
limits of foreign material and degree of soundness. In
examining soundness, the bottom line would be the factors
which are limiting for the grade. If it did not meet a
particular criteria, i.e. less than a certain percentage of
sprout-damage, it would be downgraded without vigorously
assessing for frost damage, etc. Usually, it is the entire
sample which will be the tip-off as to what should be examined closely. For example, sprout-damage also would have
associated with it factors such as bleaching, mildew, etc.
An elevator agent, seeing such a sample, would examine it
more closely for these factors. In many cases, the mildew
or severely bleached kernels may place the wheat in a lower
grade before the overall tolerances for sprout-damage are
reached.

Wheats of the same grade and class are then binned
together in the elevator, loaded onto boxcars and moved to
the terminal elevator. At this point, the individual box-

cars of wheat from across the prairies are graded by CGC inspectors located at terminal elevators. Elevator personnel
then clean and bin it according to the appropriate grade.
1 and 2 CWRS wheat is further segregated by protein content
on site at the terminal elevators by NIRS. The segregations
are 12.5, 13.5 and 14.5% protein and are guaranteed by the
CGC at export. Once binned, the grades of 1 and 2 CWRS
wheat cannot be mixed. Finally, the wheat is sampled continuously by the CGC as it is being loaded onto a ship. At
regular increments throughout the cargo is given a "Certifi-
cate final", a buyer's guarantee of quality.

   But that is not the end of the story. Samples are retained from cargos of each class and grade and composited at
three month intervals. These are then examined for a wide
range of laboratory quality tests in order to have an on-
going picture of how well the quality of shipments are being
maintained and to provide a guide to our export customers.
In relation to sprout-damage, the tests that are determined
are the Falling Number and nephelometric determination of
alpha-amylase on wheat. Amylograph, gassing power and
alpha-amylase (nephelometric) are determined on the flour.
The quality information gathered has provided some insights
into our monitoring and has shown that the overall quality
has been protected in our top grades of wheat. For example,
the Falling Number values of No. 1, 2 and 3 CWRS wheats in
cargo composites over the last 12 years emanating from the
Pacific coast are shown in FIGURE 2. It can be seen that
the falling number values have remained fairly constant over
the last 10 years for the No. 1 CWRS. Only in 1978, a very
wet year, was there a drop. No. 2 CWRS is, in general,
lower in falling number than 1 CWRS but again fairly acceptable. Of necessity, the sprouted wheat must end up some-
where. As a consequence, the 3 CWRS and ultimately Canada
feed must take the burden in wet years.

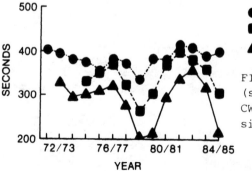

FIGURE 2. Falling number
(seconds) of 1, 2 and 3
CWRS pacific cargo compo-
sites from 1972 to 1985.

In the 1985-86 crop year, a large scale monitoring operation was initiated in which all lots of wheat moving to terminal elevators were analyzed by the Falling Number method after compositing by grade, class and primary elevator location. The study is not complete yet. What we have established, however, is that the grading system does protect the integrity of the top grades of wheat. For example, in the months of November and December, 1,669 composited samples of No. 1 CWRS wheat were analyzed, representing 7,423 cars. The weighted estimated falling number was 360 seconds. Of these, only 7.5% were less than 320 seconds.

In the immediate future, it is unlikely that Canada will change from the present system which has been so successful. In the case of sprout-damage, some refinements to the system might be made if it appears that lower tolerances for certain grades or a change in the assessment procedure can better reflect end-use quality. Present objective laboratory methods for measuring sprout-damage are too slow and expensive to be adopted in the large number of primary elevators across western Canada. If an inexpensive and rapid "black Box" for measuring sprout-damage does appear in the future, however, it will certainly be evaluated. At present, plant breeders are working on the breeding of pre-harvest sprout resistance into our wheat cultivars such that the overall amount of sprouted wheat in a wet year can be expected to decrease in the future. This, hopefully, will mean that improvement in segregation procedures will not be necessary.

Falling Number Prognoses on Rye and Wheat in Sweden

*S. Larsson, Research and Developmental
Department, Svalöf AB, S-268 00 Svalöv,
Sweden.*

When the falling number method was introduced in 1964
as a standard analysis for price determination in wheat and
rye there was an immediate need for information about optimum harvest date from the FN point of view. For this purpose an investigation was started in order to find out if it
would be possible to produce FN prognoses. This investigation was organized by the Swedish Seed Association in col-
laboration with the Scanian Farmers Cooperative Society and
covered the provice of Scania (FIGURE 1). The aim was to
find a suitable technique for obtaining representative data
for field sampling.

The prognostic FN service

The investigation was made during the years 1964-66.
As a result a prognostic service was started in 1967. This
service has been available every year since then. Other
provinces outside Scania have now adopted this prognostic
service, but Scania has today the most developed service
system. Scania has been divided into eight districts
(FIGURE 1). This division has been made according to climatic and soil conditions, in order that these factors are
as uniform as possible within each district. All together,
nearly 200 farmers serve as hosts for sampling. The fields
are not randomly selected. They are chosen as typical for
the district in question. Thus, the varieties grown and the
farming procedures must be typical for the district. For
instance this includes sowing date and amount and mode of
fertilizing. Approximately 30% of the samples are rye, 50%
winter wheat, and 20% spring wheat.

Sampling and determination of FN

In the afternoon, shortly before 6 p.m., the farmer
collects a sample of about 100 ears from a small square
(10 x 10m) in the field. The sampling is made from the same
square every time and is done according to special instruc-
tions. The ears are put into a paper bag which is transported to the laboratory the same night. The samples are
stored there until the next day. The following morning all
samples are threshed, the water contents determined and the
degree of ripeness determined. This determination of ripeness is very important since it is necessary for making a
correct judgement of the FN trend. Especially in the beginning of the period of ripeness it is necessary to dry the
samples before grinding and FN determination. The drying
temperature is maintained at about 50°C.

The FN information

The FN results are ready at about 12 o'clock noon and
the figures from the eight districts are immediately summarized and prognoses are worked out. The FN data and the
prognoses are recorded in an automatic telephone answering
service together with the local weather forecast from the
meterological institute. After 2 p.m. anybody interested
can dial the number of the recorder and receive the information. The information can also be given on the farmer's
home computer.

We call this "prognosis" though it is in fact a very
limited kind of forecast. It is a prognosis in the sense
that it gives information about the FN trend: if it is
increasing, stable, or decreasing. The rest is information
of the actual situation. In a single so called "prognosis"
the grain species, variety, locality, degree of ripeness,
water content, FN, and the tendency of the FN are metioned.
In some cases extra advice is given, for instance, a recom-
mendation to harvest as soon as possible. Since samples are
received every day during the most critical period, a good
control of the FN level and development is achieved.

How useful has the prognostic service been for the farmers?

It is a difficult question to answer as the aim of the
prognosis is an advisory service. One thing, however, is
obvious: the bread grain quality has been remarkably better

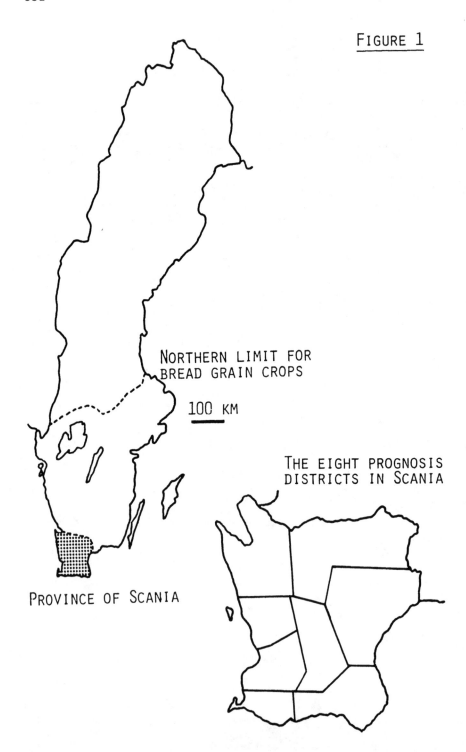

FIGURE 1

NORTHERN LIMIT FOR
BREAD GRAIN CROPS

100 KM

THE EIGHT PROGNOSIS
DISTRICTS IN SCANIA

PROVINCE OF SCANIA

in recent years. The 1936-85 annual FN variation of Scanian
rye is shown in FIGURE 2. Until 1975 the presented data
was collected from the annual quality inventories, which
were performed by the Chemical Division at the Swedish Seed
Association by commission of the National Swedish Agricultural Marketing Board. Data after 1975 were taken from the
Statistical Year book of Agriculture in Sweden. The diagram
shows which part of the total harvest will receive an extra
bonus or at least not suffer from price reductions. The
payscale is based on the present standard.

FIGURE 2. The percentage of rye in Scania with falling number $\geq$ 100, 1936-1985.

LOTS WITH FN $\geq$ 100

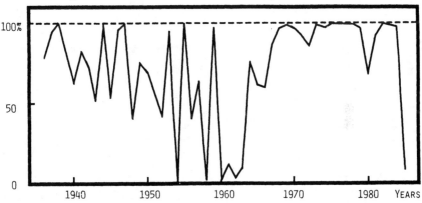

In the later part of the 30's rye had a rather uniform
and good quality. During the 40's and 50's the sprouting
damage gradually increased. During the same years there was
a steady widening of the variation between years. This
variation between very good and extremely bad years is
particularly striking during the 50's. The bottom level
with regards to sprouting damage was reached during the four
year period 1960-1963. After this period the situation
improved again.
The continuity of the diagram is remarkable. Since
1967, FN has remained at a very high and uniform level,
with some exceptions. This is unique for the 50-year
period. Is it a coincidence that the last 20 years corresponds with the period of FN prognoses in Scania? The
answer to this is probably that there are a number of different factors responsible for the improvement. First of
all the harvest weather has been unusually favourable during

these years. Secondly, varieties with better resistance to sprouting damages have partly replaced older, more suscept-able varieties. The greatly increased harvesting capacity has also made it possible to harvest more grain at the right time. We believe, however, that the FN prognoses have facilitated the correct choice of harvest time and thereby contributed to the pronounced quality improvement.

ACKNOWLEDGEMENTS

The author is very grateful to Dr. T. Hummel-Gumaelius and Dr. L. Lehmann for critical reading of the manuscript and very helpful discussions.

Assessment of Weather Damage under Field Conditions in Northern NSW

T. B. Keene, Prime Wheat Association Ltd.,
Narrabri, 2390, NSW., Australia.

INTRODUCTION

The Prime Wheat Association is an organisation of 4,000 wheatgrower members in Northern and Western New South Wales. The principal activity of the Association is the classification of wheat according to quality at the time of delivery by farmers to selected receival sites in this area. The environmental conditions are conducive to the production of high quality wheats and together with varietal control, objective assessment of wheat quality enables segregation of grades at the point of receival ex farm.

The Association is an independent grower organisation operating under an agreement with the Australian Wheat Board with the objective of maximising returns to farmers. Operating at some eighty receival sites and employing over one hundred staff during the harvest period a significant capital investment is required for both equipment and facilities to ensure that the correct segregation of grades is achieved.

The policy of the Association is determined by a Board of grower Directors who are elected to represent and implement the recommendations of the members in relation to the objective classification of wheat quality and the other activities of the Association. Formed in 1958 the Prime Wheat Association has always been in the forefront of developing systems and procedures for wheat classification with the emphasis being on the Australian Prime Hard grade of wheat which must meet stringent standards prior to acceptance. The average receivals of this grade is 650,000 tonnes per year with the record being established in the 1984/85 season when over one million tonnes of Prime Hard were segregated.

A further important area of activity is the Association's involvement in wheat research. This entails financial support for many projects related to our objectives including support for the breeding of varieties resistant to pre-harvest sprouting damage.

FIELD CONDITIONS

The area of operation of the Association incorporates half of the wheat growing belt of New South Wales, from Dubbo and Nyngan in the Central West north to the Queensland border. This is the area which has the environment suitable for the production of the Prime Hard grade. It is the higher returns achieved through the sale of this grade that ensures the cost effectiveness of objective testing at receival points in all seasons irrespective of whether weather damage has occurred.

Each grade above Australian Standard White has a minimum protein specification and this is measured using near-infra red analysis equipment. The effect of weather damage is measured using Falling Number equipment on site by casual staff trained in the operations under harvest conditions. The conditions under which the equipment operates are extreme with temperatures ranging from 25°C to 40°C and consistently being over 35°C. Thd facilities used are not air-conditioned and at times are extremely dusty which together with long and consistent working hours make conditions difficult for staff.

There are also logistical problems caused by the large geographical area both in training staff and setting up equipment. Therefore pre-harvest staff training is conducted for all casual operators but our experience demonstrates that this needs to be reinforced in the field at the time of actual employment. The Falling Number equipment is only placed in the field when our laboratory tests and visual assessment show that weather damage has occurred.

Our area of operations, in addition to producing some of the best quality white wheat in the world, also historically suffers regularly from pre-harvest sprouting damage to varying degrees due to the prevalent summer rainfall pattern.

RECEIVAL AND CLASSIFICATION

We have been using Hagberg Falling Number equipment for the field assessment of weather damage for over ten years, however it was only in 1983/84 that a concentrated trial programme or evaluation was undertaken by our Association using twenty units in conjuction with the Australian Wheat Board and the Grain Handling Authority of New South Wales.

The initiative for this pilot scheme came directly from the farmers who were dissatisfied with the previous method of visual assessment for sprouting and the nil tolerance for sprouted grains in the higher grades irrespective of any analytical test. A review of the receival standards established by the Australian Wheat Board allowed a 2% visual tolerance for Prime Hard provided that the Falling Number result for an individual farmers load exceeded 350 seconds. This trial preceeded changes to the standards for other grades such as Australian Hard and General Purpose.

The successful implementation of this pilot scheme in 1983/84, in operational terms, has created a greater awareness throughout the industry of the need for the objective assessment of weather damage. It has provided the impetus for funding and research into analysis methods and varietal resistance to provide longer term solutions to the problem.

The use of Hagberg Falling Number equipment has only gone part of the way towards solving the problem of field assessment of weather damage. The experience gained in 1983/84 and the demands of the industry resulted in an increased investment in this type of equipment for field operations by the Australian Wheat Board. Fortunately for growers the 1984/85 harvest was extremely dry and it was not necessary to place units at receival sites.

However this harvest just completed (1985/86) saw once again a return to unfavourable conditions over a critical period of maturity for a proportion of the crop in northern and western New South Wales. Operating over forty units in this area the Association conducted in excess of 35,000 Falling Number tests during an eight week period in November and December. This method of assessing weather damage is far from ideal under field conditions because of both the lengthy duration of the test and its operation.

Receival intakes at any individual site vary considerably but during the peak of harvest range from 2,000 to 10,000 tonnes per day of all grades. Therefore with hundreds of trucks being received it is impossible to provide load by load testing of weather damage using a system which takes 5 to 10 minutes per test. Given these field conditions testing is concentrated on loads which are either disputed by the farmer or visually borderline between grades. Each grade with the exception of Australian Standard White has as part of the standards a visual percentage tolerance for sprouted grain combined with a minimum falling number reading. An integral part of the operation is the regular monitoring of grades on-site to ensure that they meet these minimum standards established by the Australian Wheat Board.

However the most important aspect is the system of sampling both to ensure correct testing procedure and to confidently develop a record of the quality of each grade in each receival bin prior to transportation and shipping to the various markets.

LABORATORY CONTROL

The cornerstone to the successful field application of objective testing of wheat quality is a well established central laboratory combined with a sampling system which enables quick and accurate checks of the field operations. This is essential both to monitor the performance of the equipment used and also the staff responsible. The daily checking of field results for all grades at all sites ensures the correct application of standards, equipment, staff performance and sampling procedures.

Daily samples for each grade from each site are compiled proportionately from a probed sample of each individual truck. These are returned to the central laboratory for analysis of all relevant quality parameters. This information is supplied to the Australian Wheat Board and is critical for both marketing purposes and their confirmation of the quality of each grade. In seasons such as 1985/86 when there are significant quantities of weather damaged wheat the rapid accessibility to this quality information becomes even more important.

PROBLEMS OF WEATHER DAMAGE ASSESSMENT

Although the impetus for the objective assessment of weather damage emanated from the growers, problems associated with consistency in testing combined with the resultant financial significance has reduced the creditability of this type of operation amongst many farmers. This is not to say there is a preference for visual assessment, but rather a strong feeling towards the rapid development of technology to enable faster and more accurate testing. There is no arguement that can be put which will support a return to or a continuation of visual appraisal of sprouted grain under field conditions without the use in conjunction of objective testing.

Our experience has clearly demonstrated the difficulty of relating a percentage of sprouted grain to a Falling Number result. This relationship is obviously both seasonal and varietal and can provide no more than a guide as to the extent of weather damage. Obviously market acceptability becomes a factor in determining the percentage of sprouted grain in each grade and thus should be determined for this reason and not in an effort to correlate with some perceived falling number result.

When pre-harvest sprouting damage occurs, initially the field falling number test will confirm the acceptance of the wheat in the grades existing prior to the rain. However as further damage occurs this damage may not be obvious visually but the objective test then has a tendency to downgrade wheat which otherwise may have been accepted in a higher grade. This demonstrates the difficulty of visual assessment and the problems growers have in accepting a system which has in fact two methods of measuring weather damage.

Despite these problems it needs to be understood by all in the industry that objective assessment of weather damage is not necessarily designed to upgrade or downgrade wheat but to ensure that each load is correctly graded, so that it can be marketed effectively and in competition with other exporting countries. This maximises the returns to farmers delivering into each grade without cross-subsidisation in payments between grades.

IMPORTANCE OF OBJECTIVE TESTING

Under the present Wheat Marketing Plan in Australia the payment system is designed so as to reflect the quality of the wheat delivered into each grade by growers. Therefore correct grading initially enables benefits to be achieved in the sale of that grade and growers to maximise their returns. Thus those growers fortunate to be able to produce the higher quality wheats receive higher payments through this system. Conversely as wheat is downgraded the price received by growers is reduced significantly and the cost to the industry can be enormous.

In 1983/84 there was over two million tonnes of wheat downgraded by varying degrees in northern and western New South Wales through weather damage at a cost to growers of over $30 million. Based on receivals for the 1985/86 season over the same area of operations the following quantities of weather damage grades were received. (TABLE 1).

TABLE 1. Receival of weather damaged wheat in northern and western NSW in 1985/86.

Grade	Tonnes	Cost per tonne (approx) A$	A$ Total cost
Australian Hard No. 2	111,813	15	1,677,195
General Purpose 1	184,609	33	6,092,097
Feed Wheat	218,011	51	11,118,561
			18,887,853

These cost per tonne figures are estimates only as actual costs to the industry will not be known until the crop has been sold and priced. These costs also include an estimate of the proportion of the wheat in each grade which would have been received as Australian Standard White, Australian Hard No. 1 or Prime Hard if weather damage had not occurred.

These significant losses to the industry demonstrate the importance of ensuring the correct grading through objective testing.

SUMMARY AND FUTURE DEVELOPMENTS

The production by default of large quantities of low milling and feed quality wheat through weather damage in some years has significance beyond even the financial returns to those growers. There are international effects in the market place on the prices of both feed grains and lower quality wheat from other exporting countries. The assessment and correct grading of weather damage is essential but does not provide the solution to the problem. The breeding of varieties which are resistant to pre-harvest sprouting is and should be the ultimate objective of all associated with the industry.

Financial support for varietal research at a level which reflects the cost to the industry is essential. The establishment in Australia of a National Weather Damage Centre to co-ordinate both research and material should be a high priority.

Whilst this research towards the solution is being undertaken, resources should also be provided for the development of technology to improve the field classification of weather damage wheat both in terms of speed and accuracy.

This conference provides an International forum for the exchange of information but its success ultimately will be measured by the industry on the results that are forthcoming. The wheat industry must understand the cost of weather damage and react accordingly by supplying the resources essential to providing the solutions.

The Logic of Testing for Sprout Damage

P. Meredith, Chemistry Division, D.S.I.R.,
Ilam Research Centre, Christchurch,
New Zealand.

SUMMARY

Confusion concerning sprout damage test results from failure of industry to clearly distinguish purchase test and quality prediction test as conflicting goals where no one test can completely satisfy both situations. Only by thinking of the interactions between kind of test and kind of material being tested can we resolve the quandary of the industrial laboratory that, for example, Falling Number does not always agree with Mathewson Units or with Amylograph indications. Can we translate a complex situation into a business man's desire for a simple numerical assessment?

INTRODUCTION

We still await Betty Sullivan's "Pinch of magic powder" that will annul the effects of sprout damage in the bakery and similar situations. It seems that one or two workers have found naturally occurring inhibitors within the cereal grains, but we have had no dramatic announcements of these preparations at an economic price yet, and there is no doubt that such a discovery could be worth big money.

We have shown that the pasting properties of wheat starches vary with cultivar, season, and location, and that their susceptibilities to amylase attack also vary. We know there are a variety of amylases and other enzymes produced in sprouting, at varying locations; life is not simple. So the changes due to sprouting may express themselves in a variable manner.

During the pasting and gelling processes of starch, the granules swell irreversibly as the temperature of the suspension in water is raised. Eventually, as swelling becomes great, another process occurs; exudation of a high molecular weight but soluble form of starch from the granules into the surrounding water. As heating and stirring continues, some fragmentation of the swollen granules takes place. Amidst all this, the surfaces of the swollen granules behave as though sticky, so that the granules interact actively with one another within the viscous exudate-rich medium, as well as by merely rubbing against one another in the crowded conditions. Such is the complex nature of the hot starch paste we find in baking bread or in hot sauces.

When this hot paste cools, it sets to a gel, and the properties of the gel are primarily due to the exudate solution setting to a tangled web of long strand-like molecules. So the properties of the gel depend mainly on concentration of exudate and on the lengths of its molecules.

The results of attack by alpha amylases are not so dramatic in the hot paste as in the cooled gel, because the major effect of amylase attack is a breakdown of the long molecules in the exudate. In the hot paste, this only affects a part of the several mechanisms we have just described causing paste thickness. On the other hand, in the cold gel, where the exudate molecules are responsible for most of the strength, attack by alpha amylase has severely reduced the size of the molecular chains and so their ability to entangle with one another; hence a dramatic diminution of gel strength.

THE AIMS OF TESTING

I will be absolutely dogmatic in distinguishing two situations, PURCHASE TEST and QUALITY PREDICTION TEST, which have conflicting requirements; no test can satisfy both situations. We must make up our minds what we are trying to do.

A purchase test starts with a sample of the material being purchased, generally grain. A quality prediction test starts with material appropriate to the end-use process for which we are trying to predict quality; in other words, for processes that use flour, we make the test on flour; for processes that use wholemeal or ground grain, we make the test on a wholemeal. So we have immediately distinguished two subcategories of quality prediction test, and we now have three tests:
PURCHASE SAMPLE TEST GRAIN QUALITY TEST FLOUR QUALITY TEST
For tests on flour, where we first have to mill grain in the

laboratory, then the reliability of our test depends very much on how well our laboratory milling parallels (imitates) the commercial milling in fractionation of the anatomic parts of the grain and in the particulate nature of the flour, always bearing in mind that commercial mills vary greatly one from another in the kind of flour they will produce from a similar wheat. There is no more a standardised process called commercial milling than there is a standardised laboratory milling process.

A purchase test is designed to enquire into the past history of the grain; how badly has it been affected by the conditions of its growing, harvesting, storage, and transport? By contrast, a quality prediction test is an enquiry into the future, a scientific crystal ball, to try to tell us how our grain or flour will perform in the end-use process we have chosen.

THE VARIABLES OF THE GRAIN

Changes have taken place within a cereal grain that lead us to call it sprouted. It is how these changes affect the ways in which we use the grain that lead us to speak of sprout DAMAGE. I make that distinction because the changes of sprouting are not necessarily harmful unless processor or consumer finds them so.

The changes are not uniform; they are a function of the variation between and within individual plants interacting with a complex, and variable microclimate. Thus one ear is not like the next, nor is one grain like the one next to it in the ear. So we have to think about the following variables of the grain which may predominate in our considerations:

(1) uneven amylase distribution between grains,
(2) uneven amylase distribution within the grains,
(3) degree to which enzymes are bound, physically or chemically, in the particles,
(4) differing amylases, such as alpha, beta, and debranching kinds. There are many variants (isoenzymes), rather than single enzymes, varying in their interactions with substrates,
(5) the amylases present during grain development decay to varying extents in ripening,
(6) activators and inhibitors are present,
(7) varying substrate properties in autolytic tests,
(8) varying substrate susceptibility in autolytic tests.

THE VARIABLES OF THE TESTS

The kinds of tests for sprout damage have been recently
reviewed (Meredith and Pomeranz, 1985), so we will concentrate on the few most likely ones, of two kinds. A
FUNCTIONAL TEST imitates the end-use process under controlled conditions, examples being the baking tests and
micromalting. In general, the results by this will be as
good as the degree to which we have, in the laboratory,
paralleled the industrial process variables, and as good as
the final measurements that we make to assess the variability of results in numbers. The main sprout problem in
bread baking is an unusual degree of stickiness in the crumb
of the loaf. This is readily appreciated by the consumer
and less readily appreciated in slight degree by the baker
until the severity of stickiness creates problems at the
slicer. Stickiness is not easily measured instrumentally.

AN EMPIRICAL TEST depends on assaying some effect of
the enzyme alpha-amylase, for this is believed to be the
main offender. The effect may be measured as a change in
the substrate, starch or a derivative, or as products of the
digestion.

The empirical tests have three important variables when
it comes to selecting which to use:

(1) Diffusion of the enzyme from the grain particles will
    be a function of particle size, compactness of struc-
    ture, and time and temperature of extraction; it does
    not happen instantaneously. But an important char-
    acteristic of autolytic methods is that the enzyme is
    more closely in proximity to its substrate.
(2) Specificity of the substrate in a non-autolytic test
    may have a quite different reactivity toward some or
    all of the enzymes occurring in the sprouted grain.
    If it is an autolytic test, we are immediately in a
    quandary. When we use the grain's own starch we are
    using a variable substrate; it varies in its original
    properties of pasting and gelling, and it varies in its
    susceptibility to the actions of the enzymes. Both
    variations may occur giving a complex interaction for
    both reactions and susceptibility may vary from one
    form of enzyme to another.
(3) Sample size decides how representative the sample is of
    the parcel being tested, and how reproducible will be
    the result. Principles of good sampling MUST be fol-
    lowed. The grain to grain variation of enzyme may be
    very large. The final grain sample must properly

represent the total parcel. A sufficiently large sample of the grain must be ground before further subsampling. It is generally believed that a minimum of 300g grain must be ground and well subsampled if representative and reproducible results are to be obtained.

CRITIQUE OF THE COMMON TESTS

Looking first at the FALLING NUMBER TEST, it is clearly an empirical test designed to give a rapid but only approximate idea of the past history of the sample. It clearly cannot predict an end use with any precision. This is because in such an AUTOLYTIC situation we have a varying substrate that may indeed have been itself modified by the premature germination process. We paste this substrate and allow the enzyme to act under conditions that are far removed from most industrial processes. We get a number that has a numerical reproducibility problem to which many workers have addressed themselves over the years. However reproducible we make the test itself, we still have major problems associated with sampling for only a relatively small sample is used, 7g of ground material. And the nature of the grinding affects the interlaboratory reproducibility.

THE DYE LIBERATION METHODS were also covered in the review paper and the principal three versions attributed to Fuller, to Barnes, and to Mathewson. Each uses an arbitrary substrate formed by covalently linking a soluble dyestuff to a starch derivative. The details of the substrate and the associated buffers and co-factors clearly influence the specificity of the test for the target enzyme, alpha-amylase (Kennedy 1977). Instrumentation plays a large part in deciding which version of the test is most suitable for a particular situation, but the reproducibility is largely a function of the sample size and sampling technique. Enzyme does not instantaneously diffuse out of a granule of ground grain, so the extraction time is important.

THE GRAIN AMYLASE ANALYSER, depending on the decrease of turbidity of a starch solution, in its described form has similar sampling and extracting problems though the instrumentation of the method is elegant and the time for test good.

THE PENETROMETER TEST described by Hagberg overcomes most of the criticisms of sample size and extraction time. It is a functional test in its ability to predict the sprout damage aspects of baking performance because it is the cold

gel strength that is finally measured, in some ways par-
alleling the situation in the breadcrumb, though not a
measure of the stickiness.

The available tests are compared in TABLE 1.

	Grain purchase test	Grain quality prediction	Flour quality prediction
Functional tests			
various		+	
bake			bread
penetrometer			bread
amylograph			soups
Empirical tests			
falling number	?		soups
dye liberation	+	(malting)	
nephelos	+	(malting)	

For purchase tests, which are an arbitrary assessment,
the empirical non-autolytic tests of dye liberation or
nephelos are clearly most suitable, whilst Falling Number
includes some variables resulting from autolysis that
introduce doubts into the conclusions.

In choosing which, we need to look at costs of over-
heads and capital, manpower, and time to perform tests.
Reproducibility of assessing a parcel of grain clearly is
dependent on good sampling with a sufficiently large sample
being ground, and on the sample size and extraction time in
the test itself (TABLE 2).

TABLE 2. Comparison of test procedures.

	Ground sample weight	Extraction time
Falling number	7g	autolytic
Fuller	5	10 minutes
Barnes	10	5
Mathewson	0.2	3
Nephelos	1	5
Penetrometer	50	autolytic

It is desirable to grind a large sample and extract
that large sample for as long time as possible. If the test

is empirical, reaction may be of the dye release type or, preferably, reaction with a pregelatinised starch and measurement of the cold gel strength. A more ideal autolytic test would rapidly heat the extraction mixture to pasting temperature, hold at that for a defined period, then cool rapidly to again measure the cold gel strength.

So the answer to the business man's desire for a simple number that won't confuse him with comparisons is to decide whether he needs a test to tell the past history or a test to predict the future. Then, in the light of that decision, we must only use and quote a single test, and one that is suited to the particular circumstances. Ignorance brings peace of mind; don't let questions be asked about the relationship between one test and another, it only brings confusion.

"A little learning is a dangerous thing".

ACKNOWLEDGMENT.

I am grateful for the Fulbright Fellowship that gave me the opportunity to take this kind of overview.

REFERENCES

Kennedy, J. F. 1977. The action of alpha-amylases on dyed and natural amyloses. Staerke 29, pp. 114-117.
Meredith, P., and Pomeranz, Y. 1985. Sprouted grain. Adv. in Cereal Science and Technology, 7, pp. 239-320. American Association of Cereal Chemists, St. Paul, MN.

Falling Number Experience Reviewed

J. Perten and B. Allvin and J.Noakes*,
Falling Number, Sweden.
*Perten Instruments, Australia.

INTRODUCTION

Sprouting is a condition of grain caused by rain before and during the harvest season and may occur in wheat and rye crops on any continent. Some estimates show from 30% to 50% or more of the grains may be sprouted in a "bad" year and thus unsuitable for human consumption. Sprout damage thus causes serious problems to all sections of the grain trade: harvesting problems for the farmer, segregation problems for the handling authorities, and marketing problems for the marketing bodies. All sections suffer financial losses due to sprouting damage.

Several methods have been tried to detect sprout damage in wheat and rye, including visual inspection, colorimetric, nephelometric and enzymatic methods, but these methods have often been found to suffer from unacceptable drawbacks in the field.

FALLING NUMBER METHOD

The Falling Number (FN) method first published by Sven Hagberg in 1960, was developed technically and applied to practical use by the cereal chemist Harald Perten in 1960/ 1961.

The FN method has been extensively tested in many major international collaborative trials and has been approved by the following organisations:

1968   International Association for Cereal Science and Technology ICC Standard 107.
1972   American Association of Cereal Chemists. AACC Method 56-81 B.

1974   International Organisation for Standardization.   ISO
       3093.

Thus the FN method has become the international standard
method for the determination of sprouting and alpha-amylase
activity in grain and flour.   The method is extremely simple, in essence being the measurement of the time taken for
a plunger to fall through a rapidly gelatinized mixture of
ground grain or flour and water, which is liquifying due
to boiling temperatures.   The higher the Falling Number
Value the lower the alpha-amylase activity.   It has found
wide application in the detection of sprout damaged wheat
and rye and the segregation of sound from rain damaged
grain.

        In 1961 the Falling Number Company was established to
manufacture and market equipment for carrying out the Falling Number Test.   Recent developments of this test will be
described later.

IMPLEMENTATION

        The FN method was first introduced on a large scale in
Ireland in 1961, then in Sweden in 1964.   In Sweden it is
the only official method for measuring sprout damage.   The
magnitude of the sprouting problem can be seen from the
Swedish statistic which shows that during the period 1955-
1964, in only every second year was more than 50% of the
harvest of acceptable milling quality.   The FN method more
recently, has been introduced for wheat segregation in
Australia in 1984/85 and has been included in the New Zealand Wheat Marketing System.

USE OF THE FN METHOD

        Since the introduction of the Falling Number method in
the 1960's it has been widely adopted in many countries
round the world.   The practice varies from country to country, reflecting the agricultural policies and extent of
government regulations.

a. National standard for acceptance, segregation and
payment

        For example, in Norway, Sweden and Finland the state
guarantees farmers a minimum base price for wheat, which
varies according to quality criteria.   For a spring wheat a

FN value of 270 attracts a bonus of 4% of the base price, while for a FN value of 90 a deduction of 10.5% is made. The base price is set at a FN of 190 seconds. Recently the New Zealand Wheat Board in their Wheat Indexing and Marketing System 1986 included the FN value as part of the sprout damage index, one of seven wheat quality tests. This system imposes a 12% price variation due to sprouting, with a minimum FN value of 140.

b. Partial standard

In some countries the FN method may not be nationally prescribed on all wheat but rather adopted within certain areas of the grain industry. For example in Eastern Europe the method is used to monitor export rye in Poland and as a national standard to specify requirements for domestic flour production in East Germany.

In the 1970's the European Economic Community (EEC) now comprising the UK, West Germany, France, Belgium, Holland, Demark, Italy, Spain, Greece, Portugal, Luxembourg and Ireland included a minimum FN value of 160 for wheat of breadmaking quality in its bread wheat regulations. Although the EEC specifies limits for intervention wheat only, the FN method is in extensive use within the domestic trade, eg. in UK and W. Germany.

c. Export control and domestic milling

The FN method has been used since the 1960's in Australia, mainly by the Australian Wheat Board for monitoring the quality of export wheat and within the flour milling industry for quality control of flour production. Large scale monitoring began in 1981 with a pilot scheme in the North West of New South Wales with the Prime Wheat Association and subsequent distribution of another 65 double automatic FN instruments in 1984/85 to all wheatgrowing states in Australia.

In 1967 the USDA included minimum (200) and maximum (300) FN values in the quality requirements of straight grade flour for export shipment. Recently the International Organisation for Standardization in their minimum requirements for wheat for human consumption and international trade, ISO/DP 7970, replaced the criteria sprouted grains by alpha-amylase activity as measured by the FN method. The minimum FN value shall be above or equal to

160.

Besides national use of the FN method, major "marketing organisations" such as the Australian Wheat Board, the US Wheat Associates and the Canadian Grain Commission also publish annual quality reports including the FN value. This is partly due to market demands from importing countries where sprouting is a recurring problem. In fact the 1985 harvest was affected by sprouting in North America as well as many parts of Europe. For instance, in the UK the 1985 average FN value was 161 for wheat compared to 271 in 1984, while in North America the US export authorities have been forced to sample all export cargoes following extensive sprouting in spring wheats due to bad weather delays at harvesting.

d. Import control and domestic milling

Naturally the use of FN method has been greatest in wheat exporting countries. However wheat importing countries also use FN method to monitor imported wheat and to assure that quality specifications are fulfilled, eg. Japan, and to a lesser extent Egypt, the Peoples' Republic of China and the USSR.

APPLICATIONS

There are three main application areas in which the FN method is used namely:

1. Segregration

The main application of the FN method is for classification of grain on receipt at the silo into classes, usually 3, one that could be milled immediately, one that could be used for controlled mixing, and one that could not be used for breadmaking purposes.

The financial loss due to uncontrolled mixing can be quite considerable. Mixing 100 tons of sprouted wheat at FN 100 into 100 tones of sound wheat of FN 300 produces a mix with an FN value of 135. Using the swedish price scale instead of a 4% bonus, a 4.3% price deduction would apply. An 8.3% reduction in price on 100 tons of wheat would result in a financial loss of approximately 11.000 SEK or 2.100 AUD.

2. Milling

The second main application of the FN method is within the milling industry where it is used to control the incoming wheat loads, and also in blending calculations to ensure uniform enzyme activity in the final flour products. A FN value between 200 and 300 is generally accepted as optimum for breadmaking flour, while a value over 400 is preferred for special products eg. noodle making flour where enzyme activity influences noodle colour and thus product acceptability. The FN method can also be used in the optimisation of cereal enzyme addition to flours of low enzyme activity. Graphical charts are available to assist these blending calculations.

3. Prognosis

In Sweden and some parts of West Germany a prognosis service is provided where farmers can telephone an answering service in their district and obtain information on FN trends and a weather forecast. This service is used as an aid to determining the optimum harvest time and is based on frequent sampling from selected fields in each locality. A typical message would be winter wheat yellow ripeness - moisture content 30-35% - Falling Number 270 - tendency increasing - and a weather forecast.

RECENT DEVELOPMENTS IN THE FN METHOD

The Falling Number Company has a continuing policy of further refinement and development of their range of specialised testing equipment for the grain industry and I should like now to mention some work we have been doing recently on the FN method.

1. The FN Predictor Flag

In response to a market demand to speed up the FN test we carried out some research into the FN method reported at the 1985 Cereal Conference in Sydney and have produced a FN Predictor Flag. This simple addition to the 1600 apparatus allows prediction of the FN value after only three minutes from the position of the FN stirrers. This enables faster screening of wheat receivals at silo intake points.

2. The closed cooling system

Some remote locations have limited supplies of water and as a standard apparatus needs approximately 25 litres per hour for cooling, a closed cooling system has been designed for this application. The simple inexpensive system comprises a 50 litre tank and peristaltic pump to circulate the cooling water. Using the system the operation may continue for 12 hours during which the cooling water heats from ambient temperature to about 65°C. The system is allowed to cool at night and the operations cycle is repeated the next day. Water consumption for cooling is reduced to nil. Further research is in progress to decrease the water needed to wash the FN tubes at the end of each test.

3. Fungal Amylase Determination

Fungal amylase has become widely used in breakmaking, being an effective agent to provide sugars at fermentation temperatures while being rapidly de-activated at baking temperatures. The standard FN method does not detect fungal amylase due to its low temperature of inactivation. The FN method has been modified by measuring alpha-amylase activity at 30°C instead of at boiling point and by replacing part of the flour with pregelatinised wheat starch. As the pregelatinised starch swells so rapidly in contact with water a high force shaking machine has been developed to overcome this problem. The shaking time is set to 2 seconds.

The ICC has already started an international collaborative study to check the reliability and reproducibility of the modified FN method.

4. Communications

Greater computerisation of laboratories has led to the development of a serial data communications interface to allow the results from the FN apparatus together with sample indentification to be transferred to a main laboratory computer for convenient data handling. The printout can include LN values (Perten Liquifaction Numbers) for easy calculation of the FN of mixtures.

REFERENCES

American Association of Cereal Chemists. 1972. Approved
    Methods fo the AACC. 7th ed. Method 56-81B, approved
    November 1872. The Association, St. Paul, MN.
Commission Regulation (EEC) No. 2062/81 of 15 July 1981
    laying down the method for determining the minimum
    bread-making quality of common wheat. Official Journal
    of the European Communities L 201, Volume 24, 22 July
    1981.
DDR Fachbereichstandard Prüfung von Getreide Bestimmung der
    Fallzahl. Zuständiger Fachbereich: 88, Getreide-
    verarbeitende Industrie Ministerium für Bezirksgeleitet
    Industrie und Lebensmittelindustrie. Vertrieb:
    Institut für Getreideverarbeitung, Bergholz-Rehbrücke
    (1968)
Doty, J. W. 1981. The falling number method. A rapid tech-
    nique for malt control. Bakers Digest, April.
Hagberg, S. 1960. A rapid method for determining alpha-
    amylase activity. Cereal Chem. Vol. 37, p. 218.
International Association for Cereal Chemistry. 1968.
    Standard method of the ICC. Method 104, approved 1960,
    Method 105, approved 1980; Method 107, approved 1968;
    and Method 108, approved 1968. Verlag Moritz Schäfer,
    Detmold, West Germany.
International Organization for Standardization. 1974.
    International Standard ISO 3093, approved 1974. The
    Secretariat: Magyar Szabvanyügyi Hivatal, 1450 Buda-
    pest 9, Pf. 24, Hungary.
Matsuo, R. R., Dexter, J. E. and MacGregor, A. W. 1982.
    Effect of sprouting damage on durum wheat and spaghetti
    quality. Cereal Chem. 59,
Perten, H. 1964. Application of the falling number method
    for evaluating alpha-amylase activity. Cereal Chem.
    41, p. 127.
Perten, H. 1966. A colourimetric method for the determin-
    ation of alpha-amylase activity (ICC Method). Cereal
    Chemistry. 43, p. 336.
Perten, H. 1967. Factors influencing falling number values.
    Cereal Sci. Today. 12, p. 516.
Perten, H. 1968. Report on collaborative studies on:
    methods for determining alpha-amylase activity in
    cereal grains and flour. Cereal Sci. Today. 13, No. 5
Perten, H. 1984. A modified falling number method suitable
    for measuring both cereal and fungal alpha-amylase
    activity. Cereal Chem. 61, p. 108.

Perten, H. 1985. Shortening falling number analysis time
for measuring the sprout damage of wheat at harvest
time. Cereal Chem. 62, p. 474.

Other sources of information.

Announcement for the purchase of wheat flour blends for
export shipment. Falling Number Value Minimum 200,
maximum 300. ASCS United States Dept. of Agriculture.
1967.
A way of optimizing domestic wheat and rye supply for bread-
making purpose by using the Falling Number system for
grading of harvested grains. By Bo Allvin and Jan
Perten, Falling Number AB, Stockholm, 1981.
Falling Number prognoses on rye and wheat T. Hummel-
Gumaelius, Swedish Seed Association, Svalöf. A.
Swedish-Soviet Symposium at Svalöv. 1975.
Falling Number price scale in Norway from July 1, 1977.
Falling Number Price scale in Finland from August 15, 1978.
(Agricultural and Forestry Ministry decision for qual-
ity price for grain) Bll. 3, No. 645 (1978).
Falling Number Price scale in Sweden 1985. Price scale
according to Falling Number values used on all wheat
and rye deliveries at harvest Sweden 1985. Skånska
Lantmännens Spannmalskålender.

Rapid Screening for Weather Damage in Wheat

A. Ross, R. A. Orth* and C. W. Wrigley,
CSIRO, Wheat Research Unit.
*Bread Research Institute of Australia.
PO. Box 7, North Ryde, NSW., 2113.
Australia.

SUMMARY

A simple rapid test for objectively determining the degree of weather damage in wheat has been developed. The principle of the test is to determine the viscosity of a heat gelatinized mixture of wheatmeal and water by stirring. The power, in watts, drawn by the stirring motor is measured while the paddle rotates at a constant velocity. Different-iation between samples is observed as a high resistance to stirring in sound wheat and a low resistance in sprouted wheat. Extensive field testing showed the machines to be reliable, and robust enough for receival use. Operators found the new test easy to perform under silo conditions and they required little training before becoming adept at its use. Initial field results show excellent agreement with Falling Number assessments of wheat receivals. Faster and simpler than the Falling Number method, the new test appears to be suitable as a screening test for weather damage under grain receival conditions.

INTRODUCTION

The wide spread occurrence of weather-damaged (sprout-ed) wheat during the 1983/84 and 1985/86 Australian harvests has highlighted the need for objective testing of the degree of weather damage at the silo. The major problem with the current visual appraisal system is the inconsistency of the assessment due to its subjective nature.

Because of general concern at the extent of weather damage to the 1983/84 wheat crop, the Australian Wheat Board convened a National Forum on Weather Damage in February

1984. At the conclusion of this forum, the following criteria were agreed to as being needed in an objective field-test for weather damage.

- The test should be simple and reliable.
- It must be objective.
- It should relate closely with Falling Number analyses.
- The equipment must be robust, portable and durable.
- The test must be rapid enough not to increase the time taken to process each load.
- The test should be easily performed by non-technical operators.
- The equipment and test should be of reasonable cost.

After preliminary work, which involved evaluating a number of existing and possible tests, it became clear that the most promising approach was to devise a rapid mechanical procedure to measure the loss of starch viscosity due to amylase action. Development work concentrated on producing a system capable of quickly inducing gelatinization in a wheatmeal/water suspension, measuring viscosity changes in the resulting paste, and providing an indication of the degree of sprouting damage within two minutes. A stirred-paste system was chosen to permit continuous monitoring of of viscosity. The "Stirring Number" equipment was developed according to this principle.

A Sydney based instrument company, Newport Scientific Pty Ltd, became involved in the project, in particular in the design and development of working prototypes. Production has commenced with Newport Scientific and a provisional patent has been taken out to cover the instrument.

As a result of a promising preliminary field trial conducted during the 1984/85 Australian harvest and extensive testing of progressive prototypes in the laboratory during 1985, six prototype Stirring Number units were deployed around Eastern Australia during the 1985/86 harvest. These units were tested with the co-operation of the Australian Wheat Board, growers organizations and the University of Sydney.

MATERIALS AND METHODS

Equipment

The machine (FIGURE 1) consists of :

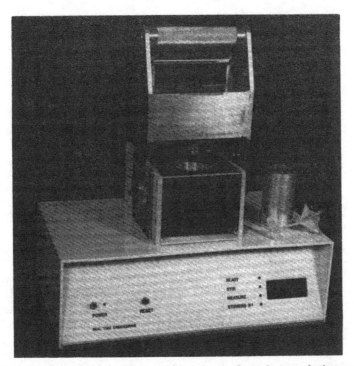

FIGURE 1. Stirring Number instrument for determining sprout damage in cereal grain. The sample container and stirrer appear to the right of the heating block and above the microprocessor control cabinet (30 cm long).

- an aluminium heating block with a well to accept the reaction vessel.
- a disposable aluminium reaction vessel.
- a disposable paddle with a single cross vane, which is automatically coupled to the motor at the beginning of the test.
- a D.C. electric motor.
- a microprocessor-based system which controls the operating parameters of the instrument. The microprocessor based system also controls the temperature of the aluminium heating block, measures the power, in watts, being consumed by the motor, and provides a digital display of the power consumption, i.e. stirring number, at the completion of the test.

Wheat samples

Grain receival trials

Wheat from the 1985/86 Australian harvest was used. Samples, taken from growers' deliveries at the point of receival, varied in Falling Number from sixty two to over six hundred seconds. Numerous varieties, over a wide range of protein contents, were tested. Wheatmeal was obtained by grinding the wheats in FN3100 mills (Falling Number A.B. Sweden).

I. A. Watson Wheat Research Centre, Narrabri, N.S.W.

Wheat samples from Dr. Mares' sprouting tolerance program were selected on the basis of Falling Numbers to provide an even distribution in the range 60-550 seconds. All samples were from the same trial site, Narrabri, 1985.

Stirring Number procedure

Determinations were made on 7.0g of wheatmeal ground in a FN3100 mill from 300g of grain. The wheatmeal was added to 25.0mL of distilled water in the can. The plastic paddle was inserted into the mixture, the can was placed into the heating block (set at 93°C) and stirring was commenced. After two minutes stirring, the final power requirement of the motor was measured as the Stirring Number.

Falling Number procedure

Determinations were made on 7.0g of wheatmeal according to ICC Specification No. 107. Where possible, the tests were performed at the same time as the Stirring Number determinations.

RESULTS AND DISCUSSION

Receival Testing

The instruments were tested under both laboratory and normal receival conditions. They proved to be reliable, easy to use and rapid enough to provide truck-by-truck assessments (22 tests per hour) if required.

TABLE 1. Relationship between Falling Number and Stirring
Number testing of wheat samples during the 1985/86 Australian Wheat Harvest.

Site	Falling Number range for samples (sec)	No. of Samples	Correlation coefficient (r)
Moree	62-571	670	0.92 ***
sub-terminal	62-300	124	0.98 ***
	200-400	213	0.73 ***
	300-571	546	0.46 ***
Temora sub-terminal	62-424	209	0.96 ***
I. A. Watson	62-636	400	0.93 ***
Wheat Research	62-300	200	0.97 ***
Centre	200-400	160	0.83 ***
	300-636	200	0.51 ***

*** indicated that the relationships are significant at the
level, p = 0.005

Moree Sub-terminal

Six hundred and seventy tests, matched by Falling Number analyses, are shown in FIGURE 2 and TABLE 1. Samples
ranged in protein content from 6.5-17.9%, and were of a number of different varieties including Suneca, Sunstar, Sun-
kota, Hartog, Takari and Kite. The overall correlation with
Falling Number values is 0.92. The Stirring Number method
appears to discriminate most effectively in the 62-300 sec
range, where the correlation coefficient is 0.98. The correlation coefficient in the range 200-400 secs of 0.73 suggests however that the effective discrimination range may
extend up to Falling Number values of 400 seconds.

Duplicate testing of samples was performed by both
methods to provide an indication of the relative consistency
of both tests. For 168 samples performed in duplicate the
mean coefficient of variation was 2.7% for Falling Number
and 2.1% for Stirring Number.

Temora Sub-terminal

TABLE 1 shows the results of two hundred and nine
tests, matched by Falling Number analyses, using samples of
Millewa, Condor, Banks, Olympic and Quarrion, among other

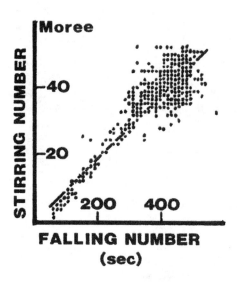

FIGURE 2. Stirring and Falling Number testing of wheat samples (listed in TABLE 1) received at Moree Sub-terminal. Points may represent more than one sample.

varieties. Samples tested at this site ranged up to a Falling Number of 424 seconds. Consistent with the data from the Moree trial, the Stirring Number results has a highly significant correlation with Falling Number (r = 0.96).

I. A. Watson Wheat Research Centre, Narrabri, N.S.W.

The results of four hundred rests, performed under laboratory conditions, are given in TABLE 1. The overall correlation with Falling Number was very high with (r = 0.93). However, as with the Moree results, the best fit was seen in the 60-300 second region.

CONCLUSIONS

In practical terms, the Stirring Number instrument satisfies the requirements set out in the introduction.

- The test is rapid enough to provide truck-by-truck assessment.
- The instrument is mechanically and electronically robust, portable and easily set up for use.

- The method is simple and the instrument easy to use.
- As sample containers and paddles are disposable, running water is not required for washing up; nor is water needed for the heating system.
- The Stirring Number instruments proved comparable to Falling Number in thir capability to assess the degree of sprouting damage in wheat.

Consequently the new machine appears to be particularly suitable as a screening test for sprouting damage at grain receival depots. Furthermore, preliminary studies indicated that the instrument is also suitable for testing other grains (barley) and for quantitative evaluation of starch properties.

ACKNOWLEDGEMENTS

Our thanks to Dr. D. Mares from the University of Sydney Plant Breeding Institute at Narrabri for his help in evaluating the instrument. Thanks also to Dr. P. Gras of the CSIRO Wheat Research Unit for his help in data analysis. Using the MSUSTAT package of Richard E. Lund, Montana State University, Montana, U.S.A. The project was partly funded by a grant from the Wheat Industry Research Council.

REFERENCE

I.C.C. (1968) Standard methods of the International Association of Cereal Chemists. I.C.C. Standard No. 107.

Nephelometric Determination of Cereal Amylases

*J. E. Kruger, Grain Research Laboratory,
Canadian Grain Commission, 1404-303 Main
Street, Winnipeg, Manitoba, Canada.
R3C 3G8.*

Approximately 7 years ago, a new method was introduced into the cereals field for the measurement of alpha-amylase. The principle of the method was nephelometry in which the rate of decrease in turbidity of substrate increased linearly with increase in amount of alpha-amylase. The Perkin-Elmer Corp., Oak Brook, Ill. introduced a machine to measure such changes called the Model 191 Grain Amylase Analyzer. The machine had already been successfully employed in the clinical field and employed amylopectin as substrate. Unfortunately, it was soon discovered that beta-amylase interfered with the determination of alpha-amylase in cereals because of its competitive hydrolysis of this substrate (Kruger <u>et al.</u>, 1979). A substrate change to beta-limit dextrin alleviated this problem and made the method specific for alpha-amylase (Campbell 1980).

A period of individual as well as collaborative testing of the instrument occurred over the next few years. One of the outcomes was that the test was adopted by the American Association of Cereal Chemists (Method 22-07). On the surface, the method seemed fairly simple. The enzyme was first extracted from ground wheat of flour. 50µL or 200µL of the extract was added to 3.00 ml of commercially available substrate at 37°C and the result read off the scale of the machine after one or two minutes. The machine was calibrated with a fungal alpha-amylase source. One of the advantages of the method, unlike some other methods such as the falling number method, was that it was possible to measure fungal alpha-amylase supplementation in a flour as well

[1] Paper No. 576 of the Canadian Grain Commission, Grain Research Laboratory, 1404-303 Main Street, Winnipeg, Manitoba, Canada, R3C 3G8.

as the natural levels.

   In spite of the widespread flurry of activity using the
Grain Amylase Analyzer, the method has not found widespread
acceptance in the flour milling and baking industries. It
is difficult to pin down one specific reason for this. Some
laboratories seemed to have difficulty in correlating the
method with other widely used methods such as the amylograph. There was difficulty in getting agreement between
the one and two minute scales of the machine. Substrate
variability due to preparation, etc., caused variation in
results. The significance of low values was in question.
The main reason for its failure was probably the attention
to detail required to ensure consistent results. This
required a fair degree of laboratory expertise in wet chem-
sitry techniques if consistent results were to be obtained.

   At the Grain Research Laboratory, we modified the
original procedure (Kruger and Tipples 1981) and have found
it reproducible and useful for many purposes. Beta-limit
dextrin substrate was prepared in our laboratory in order to
ensure a consistent and reproducible supply. The sensitivity of the machine was increased to near maximum and a
recorder was used to measure the decreases in substrate
turbidity. This also increases the sensitivity by letting
the reaction continue for a longer period for low activity
samples. In addition, any spurious changes in turbidity
due, for example, to particulate material in the enzyme
extract could be detected and the analyses repeated if necessary. The method has correlated very well with the liquefaction number and mobilities for ground wheat and flours
respectively (r = 0.93 in both cases with samples from a
number of crop years). FIGURE 1. shows the relationship
between wheat alpha-amylase and liquefaction mumber for
wheats grown in the 1983/84 crop year. The regression equation changes slightly depending on the crop year. We have
also found that wheat alpha-amylase correlates highly with

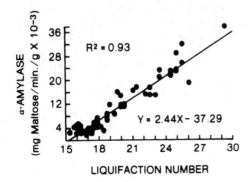

FIGURE 1. Alpha-amylase
as determined by the
modified Model 191 Grain
Amylase Analyzer proce-
dure versus liquefaction
number for Canada West-
ern hard red spring
wheats fo the 1983/84
crop year.

the amylograph (r = 0.96) if you equate the two with a poly-
nomial. The dilutions required to analyze extracts (one
part ground wheat to five parts extracting solution) of
ground wheats with different falling numbers (TABLE 1) give
some indication of the sensitivity of the method.

TABLE 1. Typical dilutions of extract necessary to analyze
wheats of varying falling number

Falling Number seconds	Dilution
> 360	undiluted
360 - 315	1:3
315 - 225	1:6
225 - 180	1:11
180 - 160	1:16

Thus, if the falling number of a wheat is less than
360, dilutions are necessary to accurately measure the
alpha-amylase activity.

The procedure for preparing the beta-limit dextrin is
rather lengthy although it does not require much actual
hands-on time. The procedure consists in (1) dissolving 40g
of waxy-maize starch in 2 L of 90% dimethylsulfoxide;   (2)
diluting to 10 l;   (3) adding beta-amylase;   (4) stirring
for 24 hr;   (5) repeat addition of beta-amylase and stirring;   (6) concentrating with an amicon ultrafiltration;
(7) dialysis and (8) freeze drying. We have tried an alternate procedure for preparing this substrate in which butanol
is used to precipitate the solubilized wazy-maize starch
prior to resolubilization and addition of beta-amylase.
This final product was less turbid on an equivalent weight
basis and, therefore, unsatisfactory for use as substrate.

The mechanism of the breakdown of beta-limit dextrin by
malted wheat alpha-amylase has been examined bu high-performance gel permeation chromatography (Kruger and Marchylo
1982). The breakdown pattern with time is shown in FIGURE
2. The most important shift in molecular weight from the
point of the nephelometric method occurs in the first 33
minutes. This shift in high molecular weight material greater than 121,000 daltons is responsible for the observed
decrease in turbidity of beta-limit dextrin. What is interesting is that there is very little formation of low
molecular weight dextrins or sugars indicating why methods
based on reducing sugar determinations would be much less

sensitive than the nephelometric method. To better understand the initial breakdown responsible for turbidity, the
reaction was decreased by adding much less enzyme. The
result is shown in FIGURE 3. Evidently, transformations
between three discrete dextrin peaks are responsible for the
turbidity loss.

Present uses of the method at the Grain Research Laboratory involve the monitoring of quarterly cargoes of
wheat, plant breeders' new lines of wheat and samples submitted by other departments of the Canadian Grain Commission
or the Wheat Board. Usually, the more widely accepted amylograph and falling number methods are also carried out with
the nephelometric method providing verification of the
results.

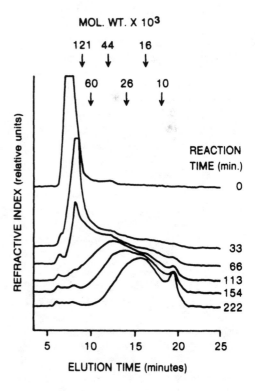

FIGURE 2. Molecular
weight distribution of
products formed at in-
creasing reaction times
due to malted hard red
spring wheat alpha-amy-
lase hydrolysis of beta-
limit dextrin (Kruger
and Marchylo 1982).

The method is especially useful for research studies.
Thus, it is routinely used to monitor alpha-amylase in
fractions resulting from a purification technique such as
ion-exchange, affinity chromatography or chromatofocusing.
In a matter of seconds, it is possible to determine whether
the alpha-amylase activity is above the linearity of the
assay facilitating the appropriate dilution.

FIGURE 3. Changes in high molecular weight products formed by malted hard red spring wheat alpha-amylase hydrolysis of beta-limit dextrin with increasing hydrolysis time (Kruger and Marchylo 1982).

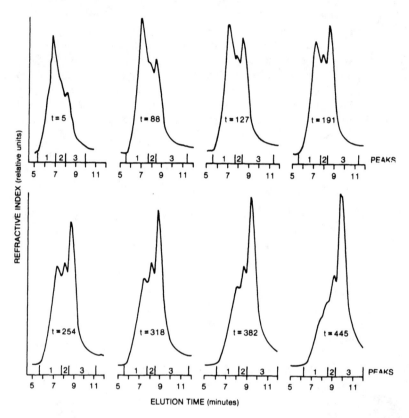

Another routine use is in the analyses of single kernels which are suspected of having sprout damage. For example, our licensed inspectors may require confirmation whether certain kernels, say with a puffy germ, should be considered as sprouted in a particular harvest year. Some typical analyses for samples of wheat with increasing degrees of visual sprout damage are shown in TABLE 2.

In general, increasing levels of sprout damage are associated with increasing levels of alpha-amylase. Furthermore, the enzymic potency of a single kernel is larger with decreasing grade. We are presently expanding this study to determine whether degrees of damage might be added to the visual grading system in order to better reflect the inherent alpha-amylase of a particular wheat sample. Not

TABLE 2. Alpha-amylase levels (mg maltose min/g) in ungerminated, degermed and various degrees of visual sprouting for different grades of CWRS wheat

| | | Visual sprouting | | | |
Ungerminated	Degermed	Degree 1	Degree 2	Degree 3	Degree 4
1 CWRS					
6.8	2.7	7.6	1.2	22	49
4.6	0	1.7	14	14	16
1.5		4.3	3.5	8.2	2508
3.6			5.4	2192	104
2 CWRS					
6.3	1.2	22	15	782	4025
30	0	17	3.0	31	1343
0	15342	2.5	4.0	69	21
0	295	1.2	8.6	12	805
3 CWRS					
5.6	4720	106	490	6505	273
14	1311	17	107	68	4387
10	19422	107	522	2313	21163
11	584	20	75	146	2116

Not surprisingly, degermed kernels can also have very high levels of enzyme. It is a particularly serious problem with rye which makes it difficult to correlate alpha-amylase by the nephelometric method with percent sprouted kernels (Kruger and Tipples 1982).

It is unlikely that the nepthelometric method could fit into a system of wheat segregation. There are a number of wet chemistry steps involved including extraction of enzyme which could not be suitably implemented at primary elevators. As the time of an analysis would be at least five minutes, this would slow grain movement. Furthermore, at most, 30 analyses could be carried out with one machine in an hour. Obtaining sufficient amounts of beta-limit dextrin would be a problem as would the cost of substrate and machines. Many other factors such as dusty conditions, different operators, etc., would also negate the use of this technique. The principle of nephelometry, however, should not be discounted in trying to devise a suitable screening technique for sprouted wheat. The change in turbidity of beta-limit dextrin by alpha-amylase is rapid and sensitive. It may be possible to develop a screening technique in which disappearnce of cloudiness could be visually examined and provide a rough segregation guide.

REFERENCES

Campbell, J. A. 1980. A new method for detection of sprout damaged wheat using a nephelometric determination of alpha-amylase activity. Cereal Res. Commun. 8, p. 167.

Kruger, J. E., Ranum, P. M. and MacGregor, A. W. 1979. Note on the determination of alpha-amylase with the Perkin-Elmer Model 191 Grain Amylase Analyzer. Cereal Chem. 56, p. 209.

Kruger, J. E. and Tipples, K. H. 1981. Modified procedure for use of the Perkin-Elmer Model 191 Grain Amylase Analyzer in detecting low levels of alpha-amylase in wheats and flours. Cereal Chem. 58, p. 271.

Kruger, J. E. and Marchylo, B. A. 1982. High-performance aqueous gel permeation chromatographic analysis of beta-limit dextrin hydrolysis by malted durum wheat, and fungal (Aspergillus oryzae) alpha-amylases. Cereal Chem. 59, p. 488.

Development of a Sensitive Immunoassay for Abscisic Acid in Wheat Grain utilizing a Monoclonal Antibody*

M. Walker-Simmons and J. Sesing, USDA-ARS, Wheat Genetics, Quality, Physiology and Disease Research, Pullman, Washington, U.S.A.

SUMMARY

An indirect enzyme-linked immunosorbent assay (ELISA) for quantitative analysis of abscisic acid (ABA) in wheat grain has been developed. With this sensitive assay +ABA amounts as low as 5 pg can be detected. The accuracy of the immunoassay was compared with high performance liquid chromatography (HPLC) analysis of ABA and similar results were achieved with each method. Good recovery of ABA added to wheat extracts was obtained. Optimum methods of ABA extraction from wheat tissue have been investigated. An extraction procedure has been determined that yields the maximum amount of ABA from wheat grain and that yields samples that do not interfere with the immunoassay. With this immunoassay method it is now possible to efficiently measure ABA levels in large numbers of wheat grain samples, even if only small amounts of tissue are available.

MATERIALS AND METHODS

All procedures involving ABA samples were performed under dim lights and incubations were done in the dark.

Extraction. Wheat (Brevor cultivar) tissue samples of dissected embryos or the remaining grain were rapidly

* Mention of a trademark of propiety product does not constitute a guarantee or warranty of the product by the USDA-ARS and does not imply its approval to the exclusion of other products that may also be suitable.

frozen and then lyophilized. For ABA extraction the freeze-dried samples were ground to a powder and immediately suspended in cold extracting methanol (80% methanol containing 100 mg/l butylated hydroxytoluene and 0.5 g/l citric acid monohydrate). [^{14}C]ABA (Amersham, Arlington Heights, IL, USA) was added to the samples as an internal standard to measure extraction efficiency and the samples were stirred overnight at 4°C. Extract samples were spun at 1000 x g, the supernatants transferred to clean test tubes and completely dried using a Savant Speed Vac Concentrator, (Farmingdale, NY, USA).

Immunoassay for abscisic acid. The assay is an indirect ELISA (enzyme-linked immunosorbent assay) utilizing a monoclonal antibody for abscisic acid. The standard buffer for the assay is TRIS-buffered saline (TBS) which contains 6.05 g TRIS [(2-amino-2(hydroxymethyl)-1,3-propandiol], 0.02 g $MgCl_2 6H_2O$, and 8.8 g NaCl per litre, pH 7.8. The washing buffer is TBS containing 0.05% Tween 20 (Sigma Chemical Co., St. Louis, MO, USA) and 0.1% bovine serum albumin, ELISA grade (Sigma Chemical Co.). The ELISA procedure is as follows :

1. ABA samples or standards are incubated overnight with the monoclonal antibody. The dried plant extracts are solubilized in 100 µl methanol and then 900 µl of TBS is added. Several serial dilutions of each sample are analyzed so that at least two and usually three of the dilutions are in the linear range of the ELISA assay. +ABA standards, +cis-trans isomer, (Sigma Chemical Co.) of 50 to 5000 pg/ml were prepared in TBS. A 400 µl aliquot of diluted plant sample or +ABA standard is pipetted into a test tube. Then 400 µl of dilute monoclonal antibody (Mertens et al., 1983) solution is added to each sample, mixed and incubated for 30 minutes at room temperature and then at 4°C overnight. Monoclonal stock solution consists of 2 mg of Idetek monoclonal antibody (Idetek, Inc., San Bruno, CA, USA) in 133.3 ml TBS containing 0.2% (wt/v) bovine serum albumin. For use in the ELISA an appropiate amount of monoclonal stock solution is thawed and diluted 1:5 in TBS.
2. Application of ABA-4'-bovine serum albumin conjugate to microtiter plates. A 200 µl aliquot of ABA-bovine serum albumin-C_4-conjugate (75 ug/ml in 0.05 M

NaHCO$_3$) is added to each well of a 96-well microtitration plate (Immulon 2 flat bottom plate, Dyntech Laboratories, Inc., Alexandria, VA, USA). The plates are incubated overnight at 4°C. The upper and lower row of wells of the microtitration plates are not used. The ABA-4'-conjugate was prepared according to Weiler (1980).

3. Addition of the ABA samples or standards to the microtiter plates. The following day the contents of the plates are discarded and plates are washed three times with TBS washing Buffer. The final washing solution is allowed to sit in the wells for 15 minutes and then discarded. Then 200 µl aliquots of the ABA samples incubated overnight with the monoclonal antibody are added to each of 3 wells on the microtitration plate and incubated for 2 hours at room temperature. +ABA standards are applied to each microtitration plate.

4. Addition of rabbit antimouse alkaline phosphatase conjugate. The contents of the plates are washed 3 times with TBS washing buffer and dumped. Then 200 µl of rabbit antimouse alkaline phosphatase conjugate (Sigma) at a 1:1000 dilution in TBS is added to each well for 2 hours.

5. Measurement of alkaline phosphatase activity. Plates are washed 3 times with TBS washing buffer and dumped. Then 200 µl of p-nitrophenyl phosphate (1mg/ml in 0.05 M NaHCO$_3$, pH 9.6) is added to each well. Plates are incubated at room temperature for about 2 hours until the absorbance at 405 nm of the sample containing no ABA is about 1.00. The incubation is stopped by adding 50 µl of 5 N KOH, and the absorbance of the samples measured. ABA levels of the samples were calculated based on the calibration curve of the +ABA standards.

High performance liquid chromatography analysis of ABA.
Four ml of water was added to the original dried plant extracts and the ABA in the samples was partitioned into ethyl acetate according to a modified method of Ciha et al., (1977). The ABA samples were then further purified by HPLC using a Bondapak C$_{18}$ reverse phase column (Waters-Millipore, Milford, MA, USA). The ABA fractions were methylated (Ciha et al., 1977) and quantified utilizing an analytical HPLC Porasil silica column (Waters-Millipore), (B. Heimbigner, M. G. Hagemann, and K. R. Gealy, USDA-ARS, Pullman, Washington, unpublished procedure). Abscisic acid methyl ester (Sigma) was used as a standard. The efficiency of the assay procedure was monitored by measuring the levels of the internal [14C]-ABA.

594

RESULTS AND DISCUSSION

Assay performance. The linear range for this +ABA assay is
between 5 and 250 pg of +ABA (FIGURE 1). There was no
cross-reactivity with abscisic acid methyl ester tested at
amounts up to 500 pg (data not shown). The ±cis-trans ABA
isomers gave a cross-reactivity of 44%.

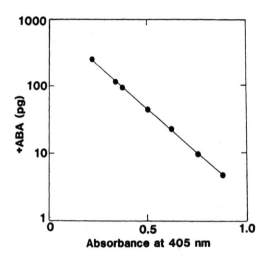

FIGURE 1. Standard
curve for +cis-trans-
ABA. Assays were com-
pleted in triplicate
and the average stand-
ard error of the mean
was less than 0.7% of
each value.

Measurement of ABA levels in wheat. Optimum extraction of
ABA from wheat grain extracts was obtained by stirring
powdered freeze-dried grain samples at a ratio of 0.01 g
sample/ 1.0 ml extracting methanol overnight. Extraction
was monitored by adding an internal standard of [14C] ABA
to the original sample preparations. After overnight stirring and drying of 7 separate extracts, an average of 100.5%
±0.04% of the original [14C] counts were recovered in the
dried samples. Shaking the plant extracts overnight or
stirring the extracts at a ratio of 0.1 g sample/1.0 ml
extracting methanol each reduced recovery of ABA by an
average of 30% as measured by both recovery of the [14C]
and by immunoassay of ABA. Addition of a constant amount of
wheat extract to increasing concentrations of +ABA standards did not interfere with linearity of the immunoassay
FIGURE 2. When ABA standard was added to wheat extracts,
and then ABA measured in the extracts alone or with added
standard, good recovery of the added ABA was found (TABLE
1).

A comparison was conducted of ELISA measurement of ABA
in the initial extract compared to ELISA of the same grain
samples after purification by ethyl acetate extraction and
by a preparative HPLC column. The purified samples were

FIGURE 2. Concentration curve for +ABA standards and for +ABA standards with 30μl plant extract (diluted 1:20 in TBS) added to each sample.

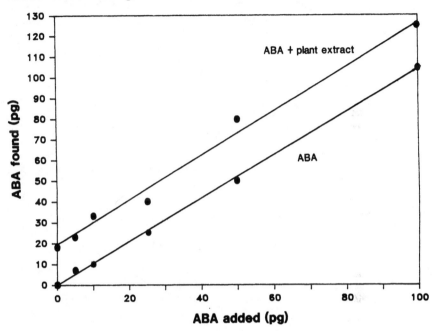

TABLE 1. Recovery of +ABA added to plant extracts.

Extract	+ABA Recovered*
	(pg)
Embryo extract + 25 pg +ABA	24 ± 1
Seed remnant extract + 25 pg +ABA	25 ± 0

* Difference between the amount of ABA measured in the plant extract + 25 pg +ABA and in the plant extract alone. Values are mean ± standard error of the mean.

measured by ELISA and also by HPLC with an analytical silica column. The similar results are presented in TABLE 2. The significant advantages of this simple ABA extraction method are that only small amounts of plant tissue are required and little sample preparation is required for the assay. An example of ABA measurement of dissected embryos and the remaining part of the grain is presented in TABLE 3. With this assay an extract prepared from as little as 5 mg of plant tissue will yield sufficient material for the ELISA assay. This enables us to examine ABA levels in individual

596

TABLE 2. Comparison of +ABA measurement of ELISA and HPLC
in initial and purified tissue extracts

| Sample preparation | ABA+ | |
	ELISA	HPLC
	(pg/mg dry wt)	
Initial extract	442 ± 23	
Extract purified by ethyl acetate partitioning and preparative HPLC	474 ± 74	444 ± 145

Values are mean ± standard error of the mean. Four replicate samples were measured.

TABLE 3. Measurement of ABA levels in the embryo and seed
remnant by ELISA at 27 and 37 days post anthesis.

Sample	DPA[1]	Moisture content of grain	+ABA
		(%)	(ng/mg-dwt)
embryo	27	53	1.49 ± .01[2]
seed remnant			1.05 ± .11
embryo	37	38	0.86 ± .002
seed remnant			0.67 ± .064

Samples were from 30 grains collected from 10 heads of the
winter wheat cultivar Brevor.
[1] DPA is days post anthesis.
[2] Values are mean ± standard error of the mean.

parts of the grain, particularly dissected embryos, and also
allows us to perform many assays during grain maturation and
in a variety of wheat cultivars.

ACKNOWLEDGEMENT

We thank Ms. Shirley Norman and Dr. Vincent Maier,
USDA-ARS Fruit and Vegetable Chemistry Laboratory, Pasadena,
CA, USA, for help in preparing the ABA-C_4'-conjugate.

REFERENCES

Ciha, A. J., Brenner, M. L. and Brun, W. A. 1977. Rapid separation and quantification of abscisic acid from plant tissues using high performance liquid chromatography. Plant Physiol. 59, pp. 821-826.

Mertens, R., Deus-Neumann, B., and Weiler, E. W. 1983. Monoclonal antibodies for the detection and quantitation of the endogenous plant growth regulator, abscisic acid. FEBS Letters 160 : pp. 269-272.

Weiler, E. W. 1979. Radioimmunoassay for the determination of free and conjugated abscisic acid. Planta 144, pp. 255-263.

Electronic Weighing and Data Collection of Plot Combine Harvesters

W. Betzwar and M. Wilson, Wintersteiger, Division Seedmech, A-4910 Reid, Austria; and International Percision Instruments, Lincoln, NE. 68507, U.S.A.

SUMMARY

Often field crop researchers find themselves short of manpower and time during the busy harvest season. Efficiency and accuracy of data collection are important parts of their operation. Automation can greatly enhance both aspects of data collection in field crop research. When applied within the routine of plot combine harvesting, the interfacing of electronic balances and moisture sensing devices with data collectors or portable microcomputers, results in significant savings of labor and time required to obtain and record harvest data. The systematic routine normally involved in such a program can substantially reduce many of the errors which often occur during weighing and manual data recording.

Experience with the NURSERYMASTER [1] one-person harvest and data acquisition system has shown it to be very efficient. A number of considerations are important in the organization of such a system. Those considerations address primarily the method of grain delivery and its freedom from excessive plant debris as well as the hardware and software of the components essential for accurate, reliable data collection and transfer. Currently the advances in hardware and software development of applicable electronic components are progressing rapidly.

Close attention to the necessary limitations and concerns of electronic data collection on plot combine harvesters while realizing the options available will result in the type of system that can be a significant advantage in a field crop research program.

[1]NURSERYMASTER is a trade name of Wintersteiger, Division Seedmech, A-4910 Reid, Austria.

INTRODUCTION

Harvest operations for small grains research has in the past typically consisted of the hand sickling of plots by various workers to be bundled, stacked and later transported to a field threshing machine operated by a crew of people. Bagged seed would then be transported to a harvest lab where samples were sorted, weighed and data hand written. These various tasks have often called for use of manpower demonstrating varied levels of skill and care. Data values are incorrect when hand cutting is less than complete, bundles are damaged during transport or spilled seed during bagging goes unreported. Under hurried conditions, conversation or transcription errors also result in incorrect seed weight data. These inaccuracies increase the coefficient of variation in spite of pains-taking efforts in the planting and growth management of a well-planned field experiment.

Since the development of plot combines that cut and thresh plots, providing clean samples with negligible seed loss, many sickles have been retired. With the more recent availability of the pure seed combine, those labor intensive operations are also unnecessary for the harvest of seed to be used for future research requiring genotypic integrity. This mechanization of harvesting conditions requires fewer people to handle samples and permits closer supervision by lead people. In spite of the extra control permitted by this type of harvest operation, frequent errors still occur in the handling of seed during the weighing and data recording operation.

An additional step to further reduce experimental error and increase efficiency of field plot research is the automation of data collection on plot combine harvesters. Such a system allows data to be collected while harvesting and requires less manpower, thereby freeing personnel for other necessary tasks during the busy harvest season. More control of the total harvest and data collection operation is realized and results may be immediately available when the harvest is complete to permit early decision making. Often this means many samples will not have to be transported from the harvest locations. The automation of harvest weight and other data collection on a plot combine harvester can be a significant benefit to a plant breeding program and the organization of such a system for each individual requires attention to a number of concerns and limitations.

PLOT COMBINE SYSTEMS

Perhaps of first concern is the availability of a combine that allows the delivery of a sufficiently clean sample to the location that will accomodate the mounting of the necessary electronic components. This is accomplished when seed is pneumatically delivered to a platform scale on a side mounted platform for a two-person system or to the right of the driver for a one-person system. These components should be mounted in such a way that an efficient flow of steps are carried out in a routine manner. This should permit the entire bagging and data collection routine to be accomplished in the time necessary for the machine to clean itself out between plots.

The level of automation and the choice of electronic components to accomplish your application is a matter for consideration. Realizing the options available and their flexability and limitations aids in these decisions. It is only an attempt to mention some of the basic concerns that have surfaced during experiences primarily with electronic data-collection on the Nurserymaster plot combine.

When selecting electronic components for use in a field environment several factors should be considered. Important to all components in use will be their readability in direct sunlight or liquid crystal display (LCD). The packaging of such equipment should be sufficient to provide environmental protection from dust and low sensitivity to temperature fluctuations. Each unit should include sufficient battery supply and back up batteries for a long day of harvest with short recharge requirements. Conversion of the 12-volt power of the plot combine is also possible. Any components planned for electronic interaction with other components must include a RS232 interfacing option and pin schematics should be provided with each instrument.

The collection of grain weight is a primary objective of plant breeding research programs. The use of an electronic platform scale permits the accurate measurements of grain weight and the direct electronic transfer of these data to a collector. When a clean sample of standarized volume can be weighed on a combine, test weight data may be collected. In choosing a scale to suit the needs of combine work the concern of weight stability must be considered against its displayed level of accuracy. The lower the level of readibility the more sensitive to vibration the scale is and the more subject to shock damage. Most scales reading to 1 gram accuracy may be too sensitive for combine use. Although modern combines do not generate a lot of vibration, some compensation must exist for these weighing conditions. An initial step is the use of shock mounts under the scale platform. In addition the internal software of a platform scale, acceptable for this type of use, should provide a weight averaging function that delivers results averaged from hundreds of readings taken over an approximately 0.25 to 0.5 second period.

Additional data collected at the time of harvest may include a percent moisture of each grain sample for standardization of weights. This may be accomplished by use of a moisture meter having its radio frequency probe mounted in the grain delivery chute of the cyclone or holding bin. Moisture data are transferred by activating a switch on the grain chute at the point of emptying the chute. Special attention must be paid to the calibration of this meter to specific levels of moisture as experience has shown this to be more difficult in low moisture regimes. Sensitivity to changes in air temperature may also be a matter of importance with some moisture sensing instruments. Typically, as air temperatures increase percent moisture decreases. Also any debris, especially green matter as would be in weed contamination, can greatly bias the moisture readings.

The selection of an instrument for capturing data transferred from measuring devices can be very important. Storage capacity is of upmost concern and the proper design of nursery codes can aid in reducing the amount necessary for a full days work. In the past, data loggers have been used to store data transferred from scales and moisture meters mounted on a combine. While data loggers have adequate data storage capacity, they are lacking in programmability. Typically, harvest sequences and nursery information must be downloaded from a computer. Data when properly matched with plot numbers is then transferred back to the computer by telephone modems or a direct RS232 transfer.

The relatively recent availability of hand held computers provides new options for data collection. The main advantage of the small microcomputer over the data logger is its programmability. Most microcomputers have built-in software that allows text and file manipulation, mathematical calculations and can run programs written in Basic language. They readily receive data in an ASCII form and the newer models have enough random access memory (RAM) to store small programs and data files collected from a complete days harvest. Examples of such units include the Radio Shack TRS80 Model 100[2] and the Epson HX-20[2]. Additional features built-in to these new portable computers include; autodial modem for telecommunications, parallel and serial port access, paper tape printer and a microcassette drive.

602

access, paper tape printer and a microcassette drive.

The hard copy printer provides safety against memory failure prior to a more permanent storage. A microcassette drive makes it possible to load programs in the field. These programs may be used to generate the sequences of plot harvest order or for powerful but space efficient statistical analysis as provided by the ANOFT[3] package. By collection and analysis of data in the field, decisions may be possible that would otherwise be made several weeks later. The use of microcomputers is very practical since they can also be used independently for recording field notes and other information during non-harvest times.

It appears that the technology necessary for electronic weighing and data collection on plot combines is becoming more economical and practical as more options become available. At the same time more researchers are realizing its potential and becoming comfortable with the idea of its use. Emphasis should be placed on the careful selection of the plot combine automation package best suited for each application.

[2] The use of trade names in this publication does not imply endorsement by Wintersteiger Seedmech of the products named or criticism of similar products not mentioned.

[3] 'Analysis of Field Trials' authored by E. Schwarzbach.

REFERENCES

Betzwar, W. and Konzak, C. F. 1982. New field plot research equipment: For automatic precision planting of seeds, and for one person harvesting and data acquisition. 6th Int. Wheat Genetics Symp. Kyoto, 1983. Proc. pp. 769-774.

Konzak, C. F., Davis, M. A. and Wilson, M. R. 1982. One person research plot harvest and data collection system. Wheat Newsletter 38, pp. 155-156.

Konzak, C. F., Davis, M. A. and Wilson, M. R. 1983. Combine harvest and data acquisition by one person. Crop Sci. 23, pp. 1205-1208.

Palgren, G. 1984. Weighing and data acquisition in field experiments with a micro-computer. IAMFE 6, July 1984, Proc. pp. 387-391.

Portman, P. and Trevenen, H. J. 1982. Conversion of small plot combine harvesters for single person operation. Crop Sci. 22, pp. 435-437.

Schwarzbach, E. 1984. A new approach in the evaluation of field trials. EUCARPIA Meeting Cereal Section 1984. Proc.

Schwarzbach, E. and Betzwar, W. 1985. Nearest Neighbour Analysis of non-replicated standard plot trials in Monte Carlo simulations. EUCARPIA Meeting of the Cereal Section on Rye, June 1985, Proc.

Schwarzbach, E. 1985. Comparison of the weighted nearest neighbour analysis with other methods of field trial evaluation in Monte Carlo simulations. 2. Symp. on Biometric Problems in Agriculture, August 1985.

Tottman, D. R. and Pollard, F. 1984. Experimental design, data collection and analysis using microcomputers. IAMFE 6, June 1984, Proc. pp. 392-394.

NURSERYMASTER plot combine with one person harvesting device for small plots, moisture meter, data collector and platform scale.

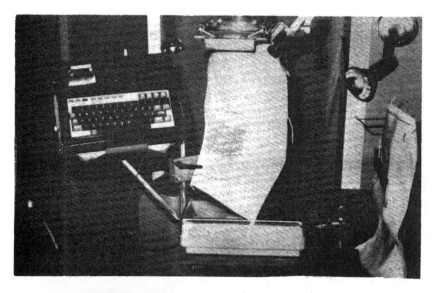

Above: Program-able hand-held microcomputer with microcassette drive and paper tape printer interfaced with an electronic platform scale for capture of weight data of small plot yields.

Left: For capture of weight data of larger plot yields, the electronic plat-form scale is at-tached on to the side bagging system using a side deliv-ery tube from the top mounted cyclone.

Participants in the Fourth International Symposium on Pre-Harvest Sprouting in Cereals

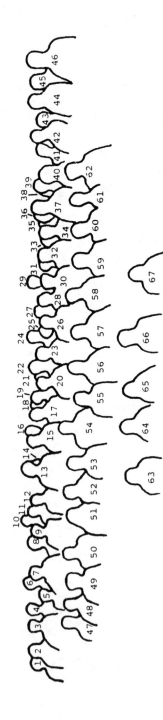

1. Ferdinard Weilenmann, 2. Zhaosu Wu, 3. Anne Ashford, 4. David Laidman, 5. Bhavna Ahluwalia,
6. Stig Larsson, 7. Takashi Akazawa, 8. Peter Chandler, 9. Kay Walker-Simmons, 10. John Noakes,
11. Bob Hill, 12. Michael Black, 13. Brian Marchylo, 14. Lynsey Welsh, 15. Andy Cairn,
16. Clive Cornford, 17. Geoff Fincher, 18. Kare Ringlund, 19. Jan Perten, 20. John Zwar,
21. Tadeusz Wolski, 22. Werner Betzwar, 23. Carol Duffus, 24. Robert Eiseman,
25. Teresa Huskowska, 26. Gitta Oettler, 27. Frank Ellison, 28. Barrie Meppem, 29. Lars Munck,
30. Ian Gordon, 31. D. Weipert, 32. Peter Nicholls, 33. Elizabeth McKenzie,
34. Heather McMaster, 35. Ewa Tymieniecka, 36. Robert Henry, 37. Graham McMaster,
38. Trevor Moor, 39. Mike Gale, 40. Wendy Hawthorn, 41. Rod King, 42. Ray Hare,
43. Ron De Pauw, 44. Edward Walsh, 45. Don Law, 46. Nick Derera, 47. John Moss,
48. Ed Czarnecki, 49. George Freeman, 50. Juhani Mikola, 51. Ellen Mosleth, 52. Lars Reiten,
53. Andy Hsiao, 54. Greg Gibbons, 55. Ben Juliano, 56. Walter Bushuk, 57. Tony Blakeney,
58. Roy Cantrell, 59. Richard Gregory, 60. Jon Mjaerum, 61. Jim Mac Key, 62. John Ronalds,
63. Peter Meredith, 64. Andrew Ross, 65. Jim Kruger, 66. Daryl Mares, 67. Paul Brennan.

Participants

AHLUWALIA, BHAVNA, Dept. of Biochemistry, La Trobe University, Bundoora, Victoria, 3083, Australia.

AKAZAWA, TAKASHI, Research Institute for Biochemical Regulation, Faculty of Agriculture, Nagoya University, Chikusa, Nagoya 464, Japan.

ASHFORD, ANNE E., Botany Dept., The University of New South Wales, PO Box 1, Kensington, New South Wales, 2033, Australia.

BETZWAR, WERNER, Wintersteiger Gesmbh and Co., Research and Development, Rechte Bahng. 30/32, POB 47, A - 1037 Wein, Austria.

BLACK, MICHAEL, Dept. of Biology, Kings College London, University of London, Campden Hill Road, Kensington, London W8 7AH, England.

BLAKENEY, A. B., Yanco Agricultural Institute NSW. Dept. of Agriculture Yanco, NSW. 2703, Australia.

BRENNAN, PAUL S., Queensland Wheat Research Institute, PO Box 5282, Toowoomba, 4350, Queensland, Australia.

BUSHUK, WALTER, Dept. of Plant Science, The University of Manitoba, Winnipeg, Manitoba, Canada R3T 2N2.

CAIRNS, A. L. P., Dept. of Agronomy and Pastures, University of Stellenbosch, Stellenbosch 7600, South Africa.

CANTRELL, ROY G., Agronomy Dept., North Dakota State University, Fargo, North Dakota, 58105-5051, U.S.A.

CHALMERS, EMMANUELLE, Plant Breeding Institute, The University of Sydney, PO Box 219, Narrabri, NSW, 2390, Australia.

CHANDLER, PETER M., C.S.I.R.O., Division of Plant Industry, GPO Box 1600, Canberra, ACT, 2601, Australia.

CORNFORD, CLIVE, Dept. of Biology, Kings College, University of London, Campden Hill Road, Kensington, London W8 7AH, England.

608

CZARNECKI, EDWARD, Agriculture Canada, 195 Dafoe Rd, Winnipeg, Manitoba, R3T 2M9, Canada.
DE PAUW, RON M., Agriculture Canada, Box 1030, Swift Current, Saskatchewan, Canada, S9H 3X2.
DERERA, NICK F., 5 Lister Street, Winston Hills, NSW, 2153, Australia.
DICK, JOEL W., Dept. of Cereal Science and Technology, North Dakota State University, Fargo, ND. 58105, U.S.A.
DUFFUS, CAROL, The Edinburgh School of Agriculture, University of Edinburgh, West Mains Road, Edinburgh, EH9 3JG, Scotland.
EISEMAN, ROBERT L., Queensland Wheat Research Institute, PO Box 5282, Toowoomba, 4350, Queensland, Australia.
ELLISON, FRANK, Plant Breeding Institute, The University of Sydney, PO Box 219, Narrabri, NSW, 2390, Australia.
FINCHER, GEOFF B., Dept. of Biochemistry, La Trobe University, Bundoora, Victoria, 3083, Australia.
FUKUNAGA, KIMIHIRA, National Agriculture Research Centre, Ministry of Agriculture, Forestry and Fisheries, Yatabe, Tsukuba, 305 Japan.
FREEMAN, W. GEORGE, "Marickfield", Edgeroi, NSW, 2391, Australia.
GALE, MICHAEL D., Plant Breeding Institute, Maris Lane, Trumpington, Cambridge, CB2 2LQ, England.
GIBBONS, GREGORY C., Australia Biotechnology Resources, 140 Pellat Street, Beaumaris, Victoria, 3193, Australia.
GORDON, IAN L., Agronomy Dept, Massey University, Palmerston North, New Zealand.
GREGORY, RICHARD S., Plant Breeding Institute, Maris Lane, Trumpington, Cambridge, CB2 2LQ, England.
HARE, RAY, Agricultural Research Centre, RMB 944, Tamworth, NSW, 2340, Australia.
HAWTHORN, WENDY, Cargill Seeds, PO Box W252, West Tamworth, NSW, 2340, Australia
HENRY, ROBERT, Queensland Wheat Research Institute, PO Box 5282, Toowoomba, Queensland, 4350, Australia.
HILL, ROBERT D., Dept. of Plant Science, The University of Manitoba, Winnipeg, Manitoba, Canada, R3T 2N2.
HSIAO, ANDREW I., Agriculture Canada, Box 440, Regina, Saskatchewan, Canada, S4P 3A2.
HUSKOWSKA, TERESA, Plant Breeding Station Laski, 05-660 Warka, Poland.
JOHNSON, RAY, Animal Research Institute, Victorian Dept. of Agriculture and Rural Affairs, Werribee, Victoria, 3030, Australia.
JULIANO, BIENVENIDO O., The International Rice Research Institute, PO Box 933, Manila, Philippines.

KEENE, TOM B., Prime Wheat Association Ltd., PO Box 146,
    Narrabri, NSW, 2390, Australia.
KING, ROD, C.S.I.R.O., Division of Plant Industry, PO Box
    1600, Canberra, ACT, 2601, Australia.
KRUGER, JIM E., Canadian Grain Commission, Grain Research
    Laboratory, 1404-303, Main Street, Winnipeg, Manitoba,
    Canada, R3C 3G8.
LAIDMAN, DAVID L., Dept. of Biochemistry and Soil Science,
    University College of North Wales, Bangor, Gwynedd, LL57
    2UW Wales.
LARSSON, STIG, Svalöf AB, S-26800, Svalöv, Sweden.
Mc EWAN, J. MARTIN, Crop Research Division, D.S.I.R.,
    Private Bag, Palmerston North, New Zealand.
MAC KEY, JAMES, Dept. of Plant Breeding, Swedish University
    of Agricultural Sciences, Box 7003, S-75007, Uppsala,
    Sweden.
Mc KENZIE, ELIZABETH, Agricultural Research Centre, RMB 744,
    Tamworth, NSW, 2340, Australia.
Mc MASTER, GRAHAM J., Australian Wheat Board, GPO Box 4562,
    Melbourne, Victoria, 3001, Australia.
MARCHYLO, BRIAN A., Canadian Grain Commission, Grain
    Research Laboratory, 1404-303 Main Street, Winnipeg,
    Manitoba, Canada R3C 3G8.
MARES, DARYL J., Plant Breeding Institute, The University
    of Sydney, PO Box 219, Narrabri, NSW, 2390, Australia.
MARSHALL, DON R., Plant Breeding Institute, The University
    of Sydney, PO Box 219, Narrabri, NSW, 2390, Australia.
MEPPEM, J. B., Prime Wheat Association Ltd., PO Box 146,
    Narrabri, NSW, 2390, Australia
MEREDITH, PETER, Ilam Research Centre, PO Box 29-181,
    Christchurch, New Zealand.
MIKOLA, JUHANI, Dept. of Biology, University of Jyvaskyla,
    Seminaarinkatu 15, SF-40100 Jyvaskyla 10, Finland.
MJAERUM, JON, Dept. of Crop Science, Agricultural University
    of Norway, Box 41, N-1432 AAS-NLH, Norway.
MOOR, TREVOR J., Barley Marketing Board (Qld), PO Box 7108,
    Toowoomba Mail Centre, Queensland, 4352, Australia.
MOSLETH, ELLEN, Dept. of Crop Science, Agricultural Univer-
    sity of Norway, Box 41, 1432 AAS-NLH, Norway.
MOSS, H. JOHN, Bread Research Institute of Australia, PO Box
    7, North Ryde, NSW, 2113, Australia.
MUNCK, LARS, Dept. of Biotechnology, Carlsberg Research
    Centre, Gamle Carlsberg Vej 10, DK-2500 Valby, Copen-
    hagen, Denmark.
NICHOLLS, PETER B., Dept. of Plant Physiology, Waite
    Agricultural Research Institute, Private Mail Bag #1,
    Glen Osmond, South Australia, 5064. Australia.

610

NOAKES, JOHN, Perten Instruments (Australia) Pty. Ltd., PO
Box 199, Strawberry Hills, NSW, 2102, Australia.
OBATA, TORU, Division of Biofunction Research, Bio-medical
Research Laboratories, The Jikei University School of
Medicine, Minato, 105 Tokyo, Japan.
O'BRIEN, TERRENCE P., Botany Dept., Monash University,
Clayton, Victoria, 3168, Australia.
OETTLER, GITTA, Landessaatzuchtanstalt, Universität Hohen-
heim, Postfach 70 05 62, 7000 Stuttgart 70, German
Federal Republic.
OSANAI, SHUN-ICHI, Kunneppu, Hokkaido, Japan, 099-14.
PERTEN, JAN, Falling Number AB, PO Box 5101, S-141 05,
Huddinge, Sweden.
REITAN, LARS, Kvithamar Agricultural Research Station
N-7500, Stjördal, Norway.
RIDD, MAX L., 109 White Street, Tamworth, NSW, 2340,
Australia.
RINGLUND, KARE, Dept. of Crop Science, Box 41, 1432 AAS-NLH,
Norway.
RONALDS, JOHN, C.S.I.R.O. Wheat Research Unit, PO Box 7,
North Ryde, NSW, 2113, Australia.
ROSS, ANDREW, C.S.I.R.O. Wheat Research Unit, PO Box 7,
North Ryde, NSW, 2113, Australia.
SCHWARTZ, C., Prime Wheat Association Ltd., PO Box 146,
Narrabri, NSW, 2390, Australia.
SING, WAYNE, Prime Wheat Association Ltd., PO Box 146,
Narrabri, NSW, 2390, Australia.
SONG, LEONARD, Saskatchewan Wheat Pool, Product Development
Office, 2625 Victoria Avenue, Regina, Saskatchewan S4T
7T9, Canada.
TYMIENIECKA, EWA, Plant Breeding Station Laski, 05-660
Warka, Poland.
WALKER-SIMMONS, M, USDA, Wheat Genetics, Quality, Physiology
and Disease Research, Johnson Hall 209, Washington State
University, Pullman, WA. 99164-6420 U.S.A.
WALSH, EDWARD J., University College Dublin, Lyons,
Newcastle, Co. Dublin, Ireland.
WELSH, LYNSEY, Yanco Agricultural Research Institute, PMB.
Yanco, NSW, 2703, Australia.
WEILENMANN, FERDINAND, Swiss Federal Research Station for
Agronomy, CH-8046, Zurich-Reckenholz, Switzerland.
WEIPERT, D., Institut für Müllereitechnologie, Bundes-
forschungsanstalt für Getreide- und Kartoffelverarbeitung
Postfach 23, D-4930 Detmold, German Federal Republic.
WOLSKI, TADEUSZ, Poznan Plant Breeders, 00-930 Warszawa,
Wspólna 30, Poland.

WU, ZHAOSU, Nanjing Agricultural University, Nanjing,
  People's Republic of China.
ZWAR, JOHN A., C.S.I.R.O. Division of Plant Industry, Box
  1600, Canberra ACT. 2601, Australia.

Notice of next meeting

The Fifth International Symposium on Pre-Harvest Sprouting
in Cereals has been tentatively scheduled for 1989 in
Norway. The committee for that meeting will be Dr. D. J.
Mares (Australia) President; Dr. K. Ringlund (Norway) Secretary; Dr. J. E. Kruger (Canada), Dr. M. Gale (England)
and Dr. F. Weilenmann (Switzerland(. Information concerning
the meeting can be obtained from any of the above organi-
zers.

Index